"...a useful and practical guide for innovators seeking to protect their ideas and bring them to the marketplace. It is an up-to-date reference source of information on the patent process and includes valuable insights into marketing new products."
>
> —*Gerald Mossinghoff*
> Assistant Secretary and Commissioner of Patents
> and Trademarks, US Department of Commerce, 1981-1985

"...written in a crisp, insightful manner, grounded in his [Levy's] substantial practical experience. He is a man of many talents who truly knows whereof he speaks and is uniquely qualified to help experienced and novice inventors alike."
>
> —*Inventors' Digest*

"Richard Levy's done it again—produced a one-of-a-kind book about the inventing process...a valuable addition to the invention literature."
>
> —*John T. Farady*
> Affiliated Inventors Foundation

"I believe no one before has pulled together so much valuable information for inventors in one book. I heartily recommend it."
>
> —*Herbert Wamsley*, Executive Director
> Intellectual Property Owners

"Richard Levy's Inventor's Desktop Companion is an invaluable tool for the amateur and professional inventor alike. It details all the important steps to follow in taking ideas to reality, a course where there can be no shortcuts to legal invention ownership. A must in every inventors library."
>
> —*Ronald O. Weingartner*
> Inventor Relations, Milton Bradley

"If information is power, Richard C. Levy's *Inventor's Desktop Companion* is turbo-charged."
>
> —*Dr. Robert Weber*
> Union Carbide Corporation

"The quintessential tool for attaining the American Dream."
>
> —*Gary Piaget,*
> Inventor of Alexander's Star

"...tips to protect your idea and your wallet."
>
> —*Pittsburgh Press*

"He's one of the country's most respected independent inventors."
>
> —*Success!*

"Simply put, he's a marketing genius. Not to mention a designer, businessman and concept engineer."
>
> —*The Learning Annex*

To my wife, Sheryl, who brings to every one of our creative adventures a great intelligence, imagination, and spirit.

To my daughter, Bettie, who sets the pace, and has taught me more about the creative process than she may ever realize.

I promise, guys, no more books for a while, just toys and games!

THE INVENTOR'S DESKTOP COMPANION:

A Guide to Successfully Marketing and Protecting Your Ideas

RICHARD C. LEVY

VISIBLE INK PRESS

NEW YORK   CHICAGO   DETROIT   LONDON

Inventor's Desktop Companion

Published by **Visible Ink Press,**
a division of Gale Research Inc.
835 Penobscot Building
Detroit, MI 48226-4094

Visible Ink Press is a trademark of Gale Research Inc.

ISBN 0-8103-7943-0

Cover Design: Cynthia Baldwin
Interior Design: Arthur Chartow,
    Bernadette M. Gornie,
    Kathleen A. Mouzakis
Front Cover Photo: Steven Brown
Back Cover Photo: Jeff Slate

First Edition

Table of Contents

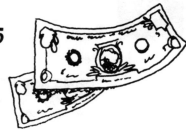

To all directory material...

FORMS AND DOCUMENTS

Dozens of forms that you may photocopy to speed the various applications and amendment processes...plus special forms that I've developed to facilitate marketing and licensing your product...

Patents

Trademarks

Copyrights

Miscellaneous

INTRODUCTION

Of the nearly five million patents granted since 1790, a few have had enormous impact on our lives while at the same time bringing fame and fortune to their inventors. The electric lamp, the transistor, the internal combustion engine, and the telephone come to mind quickly. Some have had little or no impact on our lives, such as the Pet Rock and Cabbage Patch Dolls, but have brought fame and fortune to their inventors. But the vast majority of inventions dreamed up during the last two hundred years have created no recognition or financial gain for their inventors. They simply were never **commercialized.** And they've left no trace. You can bet that when the inventors applied for their patents, the inventions probably seemed like terrific ideas. But then something happened or, more accurately, didn't happen.

What didn't happen was the sale and nothing happens until something is sold. Just as invention begins with resistance, so does the sale. When the resistance is overcome, the idea finds its successful structure or the sale is made. The *Inventor's Desktop Companion* is all about overcoming resistance.

We are all laboratories, conducting experiments with thought and extending that thought to function and utility. And what we can accomplish is bewildering. All it usually takes is the jolting of an individual out of natural laziness and the rut of habit. It is not enough to sit by and wait for things to happen. Success comes to those who are *proactive,* not reactive; to people who possess what I call the **dare to go.** The ultimate responsibility is, after all, looking for responsibility.

With this book, you'll learn how to:

- Protect your invention through patenting, trademarking, and/or copyrighting

- Market your concepts to manufacturers

- Make the best deal

- Acquire public and private funding for R&D

- Save legal expenses along the way

- Find expert advice and support—via associations, publications, and databases

In addition to learning how to license and protect your inventions, it is my hope that the *Inventor's Desktop Companion* will encourage you to:

Trust yourself more and recognize, accept, and take responsibility for the mutuality of events. Part of the adventure is recognizing and capitalizing on consequences.

Not permit yourself to be deterred by poor odds, because your mind has calculated that the opposition is too great. If it was that easy, everyone would do it.

Take your chances, not someone else's. The rewards are greater.

Ask questions. If you don't ask the question, the answer is an automatic no.

Not be afraid to make mistakes. Mistakes are the by-product of experimentation. The price of never being wrong is never being right.

Resist the herd instinct. Be yourself and be faithful to your own muse. Never give up your individuality.

Look for opportunities, not guarantees. Surely you've heard the one about death and taxes.

See rejection as a rehearsal before the big event. There can be no success without failure. Learn the value of teamwork and how much people contribute to each other's success.

And finally, learn to enjoy the hunt. For it is here, in the moment of transition, in the rushing to a goal, that the power resides.

The Better Idea and How to Market It

You don't have to be a rocket scientist to make a lot of money from inventions. And rocket scientists being who they are, it's probably their accountants and lawyers making all the money while they continue to tinker with the afterburners. This book is written for the ordinary person with an extraordinary idea and the yen to market it. After all, it's a classic American landscape: real fortunes being made by plain folks with an itch, the courage, the indefatigable, entrepreneurial spirit, and, of course, the better idea.

Apply Your Personal English

A person's judgment cannot be better than the information upon which it is based. And information by itself is worthless. Information, when combined with insight, becomes the thing that greases the wheel. I have based the *Inventor's Desktop Companion* upon my own personal experience. As such, it is based on life as it happens and not theoretical or hypothetical situations. My hope is that you'll look upon the knowledge and information imparted in this book as a chart, draw your own lines on that chart, and use it to successfully navigate the sometimes stormy waters of protecting and commercializing your invention.

Acknowledgments

A book such as this depends heavily upon the cooperation and assistance of many people. During the course of my research, I contacted an extensive number of independent inventors, Federal and state government officials, educators, patent attorneys, corporate executives, trade association directors, librarians, and members of inventor organizations.

While many of those who have helped me are mentioned at appropriate places in the text, I do want to express special thanks to certain people.

Oscar Mastin, a real pro who runs the Patent and Trademark Office Public Affairs Office, was an invaluable source of assistance then and now. He provided much of the material for the patenting section of *Inventor's Desktop Companion*, guided me through the PTO's bureaucracy, and often went above and beyond the call of duty to see that I received everything I needed to meet my tight deadlines.

At the National Institute of Standards and Technology, George Lewitt, Director, Office of Technology Evaluation and Assessment; Don Corrigan, Associate Director for Operations; and Fred Hart, Operations Research Analyst, Office of Energy-Related Inventions, provided an outstanding number of leads, background materials and suggestions. All three men are outstanding, savvy, unselfish officials who love their work and care deeply for the fate of America's inventors.

I owe much thanks to many people at the Patent and Trademark Office, including: Donald J. Quigg, former Assistant Secretary and Commissioner of Patent and Trademarks; Norma Rose, Administrative Secretary to Commissioner Manbeck; Ed Kazenske, Executive Assistant to the Commissioner; John D. Hassett, Director, Office of General Services; Lou Massel, Editor, *Manual of Patent Examining Procedure;* Carole A. Shores, Director of Patent and Trademark Depository Libraries; Martha Crocket, Technical Information Specialist, Patent and Trademark Depository Libraries; Don Kelly, Group Director, Group 320; Jeff Nase, Supervisory Petition Examiner; J. Michael Thesz, Special Program Examiner; Ruth Ann Nybold, Director of Project XL; Elizabeth Weimar, Patent Examiner, Group 180; and Anne Faris, Patent Examiner, Group 260.

At the Copyright Office, my thanks goes to Dorothy Schrader, General Counsel. Also of great assistance were: Ray Barns, Acting Director of Inventions and Innovation Programs, Department of Energy; Brenda Mack, Press Officer, Federal Trade Commission; Maureen Wood, Office of Technology Evaluation and Assessment; Keith Golden, Freedom of Information Officer, Federal Trade Commission; Marianne K. Clarke, Senior Policy Analyst, Center for Policy Research and Analysis, National Governor's Association; Jeff Norris and Joyce Hamaty, Public Affairs Specialists, National Science Foundation; Richard Sparks, Program Manager, Defense Technical Information Center; Gil Young, (Acting) Director, Minnesota Department of Trade and Economic Development; G. Thomas Cator, Executive Director, Association of Small Business Development Centers; Herbert C. Wamsley, Executive Director, Intellectual Property Owners; Ginny Panholzer and Shannon S. Jamison, Intellectual Property Owners; John T. Farady, President, Affiliated Inventors Foundation, Inc.; Joanne M. Hayes, editor, *Inventor's Digest*; Jan Kosko, Public Affairs Officer, National Institute of Standards and Technology; Robert Faris, Esq., Nixon & Vanderhye; Howard Doesher, Esq., Cushman, Darby & Cushman; John C. Sandefur, Membership Services Director, ASBDC; Dr. Martin J. Bernard III, Argonne National Laboratory; and Robert MacCollum, a good friend and expert in the field of patent drafting.

Special thanks to Michael Ross, a *gran hombre* and confidant who is always there with an ear, wise counsel and good humor.

I wish to thank Martin Connors, a gifted, dedicated and creative editor who pulled this project together in record time and made it a very enjoyable experience for me.

Richard C. Levy

That Gleam of Light

The road to success is always under construction.

"Heavier-than-air flying machines are impossible," the celebrated British mathematician and physicist William Kelvin assured everyone back in 1895.

"...man can never tap the power of the atom," said Nobel Laureate and physicist Robert Andrews Millikan, credited with being first to isolate the electron and measure its charge.

"Everything that can be invented has been invented," offered another man of vision, Charles H. Duell, Director of the U.S. Patent Office in 1899.

TRW reminded us of these profound observations in a 1985 *Wall Street Journal* ad which was tagged, "There's no future in believing something can't be done. The future is in making it happen."

What the above gentlemen did not know, quite obviously, is what inventors from fully equipped labs to basement workshops across America are proving everyday: there is no future in the word impossible. Results first. Theory second.

We are privileged! We live in a land of opportunity. Nowhere in the world do people have more freedom and encouragement to innovate, to be different and individual than in America.

We are always looking to challenge the previous and reach new levels of interest and involvement by doing things in novel ways. Our history is replete with examples of Yankee ingenuity, independent and courageous individuals who succeeded by doing things differently, dreamers who believed in themselves and their ideas.

"It takes a special kind of independence to invent something. You put yourself and your ideas on the line. And maybe people will say that you're crazy or that you're impractical," said President George Bush when, as vice president, he spoke at the National Museum of American History to commemorate the 150th anniversary of the Patent Act of 1836.

"But for over more than two centuries, millions of Americans have ignored the ridicule," he continued. "They've worked on ideas. From those ideas, they've started businesses. And many of those businesses have grown and are today our great industrial companies — companies like Xerox, Ford Motor Company, American Telephone and Telegraph and Apple Computer. Think of what America would be like if

Coming up with an idea for a new device is relatively easy. Getting it financed, manufactured, and marketed is hard work. Then there's the patent process: complicated, expensive, and time-consuming. This book will serve as a dependable raft taking you through the white water of protecting and marketing your idea.

"*A person who never made a mistake never tried anything new.*"

"*Most people look at what is; they never see what can be.*"

"*Imagination is more important than knowledge, for knowledge is limited, whereas imagination embraces the entire world.*"

—Albert Einstein

"Let us rededicate ourselves to ensuring that America is always a land in which, for those who dare to create new technologies and new businesses, the air is clear and the sky is open and the energies of man are free, where the enthusiasm for invention and the spirit of enterprise is always part of the American spirit, where men and women who want to build and dream can always find a home...Today, Tomorrow, and for all time to come."

—George Bush

"One of the reasons mature people stop learning is that they become less willing to risk failure."

—John Gardner

the skeptics had silenced the inventors."

We have been on the leading edge of uncertainty, experimentation, and exploration since the first Pilgrims set out for the New World and came ashore at Plymouth, Massachusetts in 1620.

In 1850, British subject James Nasmyth, best known for inventing the steam hammer, said about inventive Americans, "There is not a working boy of average ability in the New England States...who has not an idea of some mechanical invention or improvement...by which he hopes to better his position, or rise to fortune."

Alexis de Tocqueville, the French writer who visited America in 1831, wrote, "They [Americans] all have a lively faith in the perfectibility of man, and judge that the diffusion of knowledge must necessarily be advantageous, and the consequences of ignorance fatal. They [Americans] consider society as a body in a state of improvement, humanity as a changing scene, in which nothing is, or ought to be, permanent; and what appears to them today to be good, may be superseded by something better tomorrow."

Our heritage is rich in examples of American inventors, tinkerers, daydreamers, and gadgeteers, working from basements and garages, who dared to be different and refused to trade incentive for security. Many of their names have become celebrated: George Westinghouse created the first air brake; W.H. Carrier gave us air conditioning; Benjamin Franklin put us on bicycles; Edwin H. Land developed the 60-second camera; Charles Goodyear first vulcanized rubber; George Pullman designed double-deck sleeping cars on trains; Robert Goddard started the race to space; King S. Gillette made the first safety razor; and William S. Burroughs patented the first mechanical adding machine.

Other Americans are not so well known, but their inventions are: W.H. Carothers created nylon; Whitcomb L. Judson patented the zipper; Luther Childs Crowell made the square bottom paper bag; Walter Hunt stuck us with the safety pin; James Ritty rang up tremendous sales with the first cash register; Mary Anderson improved motoring safety with the first windshield wipers; and Garrett A. Morgan changed our driving habits and city streets with the traffic signals and is also credited with a pioneering gas mask.

Not all inventions are so conventional. Two virtually unknown American inventors, Philip Leder of Chestnut Hill, Massachusetts and Timothy A. Stewart of San Francisco, California made history on April 12, 1988 when U.S. Patent No. 4,736,866, entitled "Transgenic Non-Human Mammals" was issued. The Harvard University researchers were awarded the first patent covering an animal. Their technique introduces activated cancer genes into early- stage embryos of mice. The resulting mice are born with activated cancer genes in all their cells. These mice are extremely sensitive to cancer-causing chemicals developing tumors quickly if exposed even to small amounts. The resulting value to medical and scientific research may be immense.

As diverse a group of individuals as this appears to be, these people share many things in common, characteristics that you too will require to fulfill your aspirations and see your inventions patented, licensed, manufactured, and marketed.

On an intellectual level, they knew how to detect and follow what Emerson called that "gleam of light that flashes across the mind from within." They did not dismiss their own thoughts without notice. They abided by their own spontaneous impression. Successful inventors permit nothing to affect the integrity of their minds.

At the 1990 National Inventors Expo held at the Patent and Trademark Office's (PTO) Public Search Room in Arlington, Virginia, Deputy Commissioner of Patents and Trademarks Douglas Comer looked about at the invention exhibits and remarked, "These inventors demonstrate that the creative energy of this nation is going forward in full force and that the small inventor is the future of the American economy." He added that PTO statistics demonstrate that more than 50 percent of all patents come from small inventors.

These people shared the challenges every inventor faces in how to best protect and exploit a new product. Whether you have invented a wheel mounting device, a method for manufacturing semiconductors, an adjustable hoe attachment for a rake, fluidic rotational speed sensors, a U-joint mount, or a better mouse trap, the same basic steps apply.

This is not a theoretical book. I've personally developed or co-developed and licensed more than 70 original products and concepts to companies ranging from Fortune 500 diversified manufacturers to small independent businesses. This book will describe the pragmatic "how-to" as well as serve as a guide to the enormous amount of information and services available to the independent inventor.

In addition to the formal information on how to go about protecting your ideas through U.S. government devices such as patents, copyrights, and trademarks, the *Inventor's Desktop Companion* is crammed full of suggestions, modus operandi, and "insider" tips and pointers on invention marketing, prototyping, corporate licensing, and funding research and development. This is information every independent inventor requires, but only a fortunate few possess. **The objective is to guide the inventor down the frequently winding road leading from concept to market fulfillment.**

Do-It-Yourself Marketing; or Know Thyself

This book is a guide to locating information and doing it yourself. All the information required in this regard is provided in the chapters that follow. But remember this rule: never get so wrapped up in the marketing of the invention that you forget to market yourself.

On September 3, 1969, Patent No. 3,400,371 issued with 495 sheets of drawings and 469 pages of specifications, making it the most voluminous patent ever issued. Assigned to IBM, the patent had 16 joint inventors.

Born in 1920, Charles P. Ginsburg was the lead inventor of the modern videotape recorder. In 1952 he joined the Ampex Corporation, where he led a research team to develop a new recording machine that ran the tape at a much slower rate than existing tape machines. Ginsburg's team included a new development: the recording heads now rotated at high speed, allowing the necessary high-frequency response. The new machine was first used by CBS TV in 1956.

Creative impulses don't punch a clock (time when inventors get their best ideas):

6 a.m.-noon: 30%
Noon-6 p.m.: 14%
6 p.m.-midnight: 33%
Midnight-6 a.m.: 23%

*More than nine in 10 inventors say we're not running out of of things to invent.

*83 percent say this is the most inventive nation in the world.

*Almost 90 percent think our education system doesn't do enough to further ingenuity.

*Three in five agree that the patent process deters inventors.

Source: USA Today survey of 966 inventors who obtained U.S. patents during the first six months of 1989.

Educational Background Is No Guarantee

My university degree is in television and film with a minor in English. I have never had a course in marketing. I am a totally self-taught inventor, designer, and marketeer. I know, therefore, better than most, what the independent inventor requires in terms of information. I also know the limited resources and time restraints the independent inventor faces.

I have done my best to make this book as comprehensive as possible, designing it to deliver empirical information on what to do and how to do it. It is not just another book that tells you the where but not the how.

The Golden Rules

To be successful in the exciting business of product development it takes much more than a good idea and a strong patent.

Understanding and practicing the following six points is critical to the inventor who attempts to beat what often appear to be insurmountable odds.

1) **Don't take yourself too seriously.** Don't take your idea too seriously either. The world will probably survive without your idea. Industry will probably survive without your idea. You might need it to survive, but no one else will.

2) **You can't do it all yourself.** Remember the words of John Donne, "No man is an island, entire of itself; every man is a piece of the continent, a part of the main."

The success I have experienced is the result of unselfish, highly talented, and creative associates willing to face the frustrations, rejections, and seemingly open-ended time frames that are inherent in any product development and marketing exercise. I have also been lucky to have met and worked with very creative, understanding, and courageous corporate executives willing to believe in me and gamble on our concepts.

It is the cross-pollination and subsequent synergism that results in success, success in which all parties share. For if any link in this often complex and serpentine chain breaks, an entire project could flag.

3) **Keep egos under control.** Creative and inventive people, according to profile, hate to be rejected or criticized for any reason. They are usually highly critical of others. And they are extremely defensive where their creations are concerned.

I have always found that my concepts are enhanced by the right touch. Working together or in competition, other people contribute time and time again to making an idea more useful or marketable. Share an idea and get a better one! Products are actually a series of improvements. Inventors are people who see how to organize what exists in a novel method. They are creative people with vision. They see the ordinary in a new light. Gutenberg combined the wine press and the coin punch to give the world movable type and the printing press. Wilbur and Orville Wright would be amazed by the U.S. Space Shuttle, but it is only a modern extension of their celebrated glider that took flight near Kill Devil Hills, North Carolina in 1911.

Unchecked egocentricity can be the source for major failure in the development and licensing of new concepts. Arrogance has no place in the process.

4) **Learn to take rejection.** Rejection can be positive if it is turned into constructive growth. Don't let it shake you or your confidence.

I have rarely licensed a product to the first manufacturer who sees it. And for every product I've licensed, there are many more that did not make it.

5) **Don't just do it for the financial rewards.** You should be motivated by the gamesmanship. It may sound trite, but people who do things just for the money usually come up shortchanged.

6) **Be patient. It's going to take time.** It is a misconception that anything is made overnight other than baked goods and newspapers. Even though the market is big, the competition is ferocious. Murphy's first corollary is: Nothing is as easy as it looks. Murphy's second corollary is: Everything takes longer than you think.

The Beatles worked for several years in the clubs of Liverpool before getting a break. Michael Jackson had made zillions of dollars by age twenty two, but let us not forget he began performing professionally at age five. Ideas generally don't diminish with time, but grow and find their proper environment.

Longfellow said it best in the work, "The Ladder of St. Augustine." "The heights by great men reached and kept were not attained by sudden flight. But they, while their companions slept, were toiling upward in the night."

Information is power, but only when properly understood and utilized. This volume is packed with power and the instructions for its most effective use. If you carefully apply this information to an innovative and

Percy Lavon Julian, the grandson of a former slave, is noted most for synthesizing cortisone, used in the treatment of rheumatoid arthritis and other inflammatory conditions. Born in 1899, Dr. Julian established an international reputation in 1935 at DePauw University when he synthesized physostigmine, a treatment for glaucoma, from the calabar bean. Despite scientific acclaim, DePauw University denied him a professorship because of his race. Undeterred, he became director of research at the Glidden Company, manufacturer of paint and varnish, where he developed many synthetic substances utilizing soya proteins.

novel concept, putting your own English on the lessons implied, you'll have that added edge that could mean success.

On July 31, 1790, Samuel Hopkins of Pittsford, Vermont, received the first United States Patent, for an improvement in "the making of Pot ash and Pearl ash by a new Apparatus and Process." The grant is signed by President George Washington, as well as Edmund Randolph, Attorney General, and Thomas Jefferson, Secretary of State. The original document is in the collections of the Chicago Historical Society.

Protection Is Good for You

An unprotected idea is a lost idea.

It is smart to initiate every form of appropriate protection available under the law for an invention. First, this sends a signal to those to whom you disclose a concept, whether they be potential licensees or investors, that you are serious, committed, and willing to go that extra mile. Secondly, it tends to keep honest people honest, much in the same way locks do on doors. A dishonest person or company bent on misappropriating a piece of intellectual property will do it regardless of the protective steps you may take. Locks have never stood in the way of a good second-story man. In most cases, strong protection is a required prerequisite to any licensing agreement. You'll find that companies favor licensing protected products over nonprotected products as extra insurance against competition.

The U.S. Government provides several methods for protecting ideas: **patents, copyrights, and trademarks.** As each of these can be important to the independent inventor, I will attempt to cut through the dim "bureaucratese" and explain each in the simplest and most widely understandable of terms. I will address the questions most frequently asked and then some.

This part of the book is not intended to be a comprehensive text or total legal guide on patent, copyright, and trademark law. The source material for this writing weighed just over fifteen pounds! My purpose is to enlighten you, make you comfortable with the material and processes, share some personal experiences, and, in the end, equip you better to handle matters, make decisions, save some money, and otherwise defend yourself and your innovative concepts in what can be a very rough-and-tumble marketplace. Protecting your invention can be a tortuous journey—the steps outlined here should remove some of the tougher obstacles.

People occasionally confuse patents, copyrights, and trademarks. Although a resemblance exists in the rights granted of these three kinds of intellectual property, they are different and serve different purposes. Let's begin with patents.

Patents: Value and Definition

Do Patents Matter?

Yes! Some companies in certain industries may regard patents as unimportant, but for the independent inventor experience dictates that it is wise to proceed under the assumption that acquiring a patent is important. In point of fact, few companies will license an idea unless it has been or can be protected by a patent. "If there is no patent, what are we licensing?" is a standard query when an inventor has not made application for patent protection.

According to *Business Week,* in 1988, 6,059 intellectual property suits were filed in the U.S., 60 percent more than in 1980. Whether it's Walt Disney Company protecting Snow White or Xerox guarding a proprietary copier, almost every week a property rights case is brought to court. Chief Judge Howard T. Markey of the U.S. Court of Appeals for Washington, D.C. Federal Circuit, a patent's court of last resort, is upholding patents 80 percent of the time.

Typically, the stronger the patent protection, the better the contract you will be able to negotiate. This is because a meaningful patent inhibits any competition from manufacturing, using or selling your idea for at least a limited time. If nothing else, your manufacturer will have a head start, and manufacturing is an extremely competitive business. The winners usually have an edge, and a patent provides one.

Patents may not always be appropriate in every instance. While patents are indeed a terrific way to protect a new product or process from competition, they are also the best way to let competition in on the intricate details of your innovation. While applications for patents are held confidential, as soon as a patent issues, an abstract is immediately widely publicized in the PTO's *Official Gazette.* Anyone can obtain a full copy of your patent. Many companies subscribe to the *Official Gazette* to keep track of their competition.

For example, it has been popularly believed that The Coca-Cola Company never applied for patent protection on its precious syrup because the formula would then be revealed to the scrutiny of rival soft drink bottlers. We know the components found in the world's favorite soft drink, but no one, save for reportedly one or two senior corporate officers, knows the critical proportions.

How the Cookie Crumbles

One of the most interesting patent infringement cases, settled in 1989, involved three bakers. At issue between Duncan Hines, a division of Procter & Gamble, and its competitors, Keebler, Nabisco, and Frito-Lay, was the U.S. patent awarded to P&G for its line of Duncan

A patent for an invention is a grant of property right by the U.S. Government to the inventor (or his heirs or assigns), acting through the Patent and Trademark Office (PTO).

Hines cookies that were promoted as being "crispy on the outside and chewy on the inside" and just like homemade.

Cincinnati-based Procter & Gamble yelled foul after it introduced its new cookies in Kansas City in early 1983, and soon thereafter found similar products on the market, specifically Keebler's Soft Batch, Nabisco's Almost Home, and Frito-Lay's Grandma's Rich'n Chewy cookies. In 1984, P&G filed suit against its rivals. In the course of a five-year battle, 200 witnesses gave almost 120,000 pages of pre-trial testimony and 10,000 exhibits were submitted. On September 12, 1989, a $125 million settlement was announced, the largest ever in a patent case.

Inventor Wipes Auto Companies Clean

The man who patented an improved intermittent windshield wiper in 1968 is asking the federal court to award him up to $325 million from Ford Motor Company as compensation for copying his patented creation. Ford has countered that claim, suggesting that jurors award Robert W. Kearns, a former university professor, $2 million. In early 1990, a jury ruled that Ford had infringed upon Kearns patent. The same jury deadlocked on the damage award, creating the need for a new jury, barring an out-of-court settlement. The final verdict is expected to have sweeping implications since Kearns has similar patent infringement cases against more than 20 other automobile firms, including General Motors, Mercedes Benz, Chrysler, Toyota, and Honda.

Kearns has waged his patent war while living on a modest disability pension during the last 14 years. A complication in the case is the notion of time, specifically the 17 years granted by the original patent. The patents at the heart of Kearns's creation expired in 1988. Apart from the damages he is seeking in his case against Ford, Kearns wants to extend his patent rights and prevent the company from using the invention unless they buy it from him. "This case isn't about money," he says. "If I walk out of that courtroom with nothing more than a check, then I'm nothing more than an employee of Ford."

Experts in patent law say it is virtually impossible for Kearns or anyone to prevent others from making, using or selling an invention once the patent has expired. Kearns is undeterred. "The thing that's been taken way is time," he says. "And I'm asking for it back." At last report Kearns had been awarded $5 million, with Ford contemplating appeal.

Get the Picture?

Other stories abound. In 1985, Polaroid's patents for its instant cameras pushed Eastman Kodak out of the market and pending damage trials could cost Kodak upwards of $10 billion.

Independent inventor Jerome Lemelson was awarded $24.8 million in damages against Mattel, for utilizing his patented plastic track for the Hot Wheels vehicles.

On May 5, 1802, Mary Kies of Killingly, Connecticut became the first woman to earn a U.S. Patent. Her invention related to "weaving straw with silk or thread."

Japan's Sumitomo Electric Industries was closed down by Corning Glass Works in 1987 when Corning won a patent infringement suit over its design of optical fibers.

As Donald J. Quigg, former Commissioner of Patents and Trademarks, impressed upon new graduates of the Patent Academy, "Quality patent grants ensure an opportunity for rewards for creativity and security for investment, and they present an incomparable resource for new technological information."

The right conferred by the patent grant is, in the language of the statute and of the grant itself, "the right to exclude others from making, using, or selling" the invention. What is granted is not the right to make, use, or sell, but the right to exclude others from making, using, or selling the invention.

Types of Patents

Three types of patents are available:

1) **Utility patent**—granted for a new, useful, and nonobvious process, machine, manufactured article, composition, or an improvement in any of the above;

2) **Design patent**—available for the invention of a new, original, and ornamental design for an article of manufacture. A design patent only protects the appearance of an article and not its structure or utilitarian features;

3) **Plant patent**—provided to anyone who has invented or discovered and asexually reproduced any distinct and new variety of plant, including cultivated spores, mutants, hybrids, and newly found seedlings, other than a tuber-propagated plant or a plant found in an uncultivated state.

What Can Be Patented?

Patent law specifies the general field of subject matter that can be patented.

In the language of the statute, a person who "invents or discovers any new and useful process, machine, manufacture, or composition of matter or any new and useful improvements thereof, may obtain a [utility] patent," subject to the conditions and requirements of the law. The word "process" means a process or method, and new processes, primarily industrial or technical processes, that may be patented. The term "machine" used in the statute needs no explanation. The term "manufacture" refers to articles which are made, and includes all manufactured articles. The term "composition of matter" relates to chemical compositions and may include mixtures of ingredients as well as new chemical compounds. These classes of subject matter taken together include practically everything which is made by man and the process for making them. In a 1980 Supreme Court decision the Court indicated

Where the Twain Meets

On December 19, 1871, Mark Twain received Patent No. 121,992 for "An Improvement in Adjustable and Detachable Straps for Garments." Otherwise known as suspenders. Twain, who later lost a fortune investing money on the inventions of others, actually received three U.S. Patents, the second in 1873 on his famous "Mark Twain's Self-Pasting Scrapbook," and the third in 1885 for a game to help players remember important historical dates.

In Twain's novel, Connecticut Yankee at King Arthur's Court, his character "Sir Boss" remarks that a "country without a patent office and good patent laws is just like a crab and can't travel anyway but sideways and backwards."

Each week, increasing numbers of patents on computer software are issued. If you design such programs, don't fail to consider your work for possible patentability. You can go for it simultaneously with copyright protection.

Human beings cannot be patented because, according to the PTO, the "grant of a limited, but exclusive property right in a human being is prohibited by the Constitution."

that the PTO could not refuse a patent simply because it covered living subject matter. The PTO subsequently announced on April 17, 1987, that it would consider patents on animals after an agency appellate board ruled that oysters are patentable subject matter. This patent, however, did not issue. But it wasn't long until one did.

On April 12, 1988, patent history was made when the PTO issued the first patent covering an animal, specifically genetically engineered mice for cancer research.

What Cannot Be Patented?

Many things are not open to patent protection.

The laws of nature, physical phenomena, and abstract ideas are not patentable subject matter.

A new mineral discovered in the earth or a new plant found in the wild is not patentable subject matter. Likewise, Einstein could not patent his celebrated E=mc2; nor could Newton have patented the law of gravity. Such discoveries are manifestations of nature, free to all people and reserved exclusively to none.

The Atomic Energy Act of 1954 excludes the patenting of inventions useful solely in the utilization of special nuclear material or atomic energy for atomic weapons.

The patent law specifies that the subject matter must be "useful." The term "useful" in this connection refers to the condition that the subject matter has a useful purpose and that also includes functionality, i.e., a machine which will not operate to perform the intended purpose would not be called useful, and therefore would not be granted patent protection.

Interpretations of the statute by the courts have defined the limits of the field of subject matter which can be patented. Thus it has been held that methods of doing business and printed matter cannot be patented.

In the case of mixtures of ingredients, such as medicines, a patent cannot be granted unless the mixture is more than the effect of its components. It is of interest to note that so-called "patent medicines" are generally not patented; the phrase "patent medicine" in this connection does not mean that the medicine has been awarded a patent.

Are You Eligible for a Patent?

Patent Laws

The U.S. Constitution gives Congress the power to enact laws relating to patents, in Article I, section 8, which reads, "Congress shall have power...to promote the progress of science and useful arts, by securing for limited times to authors and inventors the exclusive right to their respective writings and discoveries." Under this power Congress has from time to time enacted various laws relating to patents. The first patent law was enacted in 1790. The law now in effect is a general revision which was enacted July 19, 1952, and which went into effect January 1, 1953. It is codified in Title 35, United States Code.

The patent law specifies the subject matter for which a patent may be obtained and the conditions of patentability. The law establishes the Patent and Trademark Office (PTO) for administering the law relating to the granting of patents, and contains various other provisions relating to patents.

Who May Apply for a Patent?

According to the law, only the inventor may apply for a patent, with certain exceptions. If you are a co-inventor, you may make a joint application. Financial contributors are not considered joint inventors, and cannot be joined in the application as an inventor. If you make an innocent mistake of erroneously omitting or naming an inventor, the application can be corrected. If a partisan who is not the inventor should apply for a patent, the patent, if awarded, would be invalid. Any person who falsely applies as the inventor is subject to criminal penalties.

If the inventor is deceased, the application may be made by legal representatives, i.e., the administrator or executor of the estate. If the inventor is not mentally competent, the application for patent may be made by a guardian. If an inventor refuses to apply for a patent or cannot be found, a joint inventor or a person having a proprietary interest in the invention may apply on behalf of the missing inventor.

Can Foreigners Apply For U.S. Patents?

Yes. The patent laws of the United States do not discriminate with respect to the citizenship of the inventor. Any inventor, regardless of citizenship, may apply for a patent on the same basis as a U.S. citizen. In fact, the number of foreigners obtaining U.S. patents continues to increase.

In 1987, foreign inventors were issued 48 percent of the U.S. utility patents, compared with 37 percent in 1976. In 1987, to no one's sur-

A Bright Idea

He received Patent No. 223,898 on January 27, 1880, for "An Electric Lamp Giving Light by Incandescence." In total, Thomas A. Edison obtained 1,093 patents, four of which were issued posthumously.

Officers and employees of the PTO are prohibited by law from applying for a patent or acquiring, directly or indirectly, except by inheritance or bequest, any patent or any right or interest in any patent.

prise, Japan placed five companies among the top 12 corporations receiving U.S. utility patents.

Terms of U.S. Patents

The term of a **utility patent** is 17 years from the date of issue, subject to the payment of maintenance fees. The right conferred by a utility patent grant extends throughout the U.S.A. and its territories and possessions.

The term of a **design patent** is 14 years from the date of issue, and is not subject to the payment of maintenance fees. The right conferred by a design patent grant extends throughout the U.S. and its territories.

The term of a **plant patent** is 17 years and is not subject to maintenance fees.

Conditions for Obtaining a Patent

In order for an invention to be patentable it must be new as defined in the patent law, which provides that an invention **cannot** be patented if:

"a) The invention was known or used by others in this country, or patented or described in a printed publication in this or a foreign country, before the invention thereof by the applicant for a patent, or

"b) The invention was patented or described in printed publication in this or a foreign country or in a public use or on sale in this country more than one year prior to the application for patent in the United States..."

For these reasons care must be taken to **not** permit any public disclosure of a patentable idea unless you are prepared to make the application before one year goes by from the date it was first disclosed.

Exhibiting a new invention at an invention exposition or new product fair, writing a magazine piece about a new idea, or permitting someone else to report in print on your idea will make it ineligible for patent protection if a patent application is not filed within one year from the time of your invention's first public exposure.

Inventors should be very careful, therefore, about disclosing their inventions prior to making patent application. Many inventors who do not have the money to patent their products think that by exposing them at expos and fairs, and in the mass media, investors and/or licensees might see them, make an offer, and cover the patent with advances. Maybe so. But if you take this dangerous route before making a patent application, be aware that the clock starts ticking down when you go public. On top of this, you have exposed an unprotected idea to the world, a dangerous and questionable move under any circumstances. Knock-off artists frequent such events.

While I support certain public events as one method to promote and publicize inventions, they are simply not recommended as a place for unprotected products that you might hope one day to patent.

Even if your idea is not exactly shown by the prior art, i.e., patents on file, and involves one or more differences over the most nearly similar thing already known, a patent may still be refused if the differences would be obvious. The subject matter you seek to patent must be sufficiently different from what has been used or described before so that it may be said to be not obvious to a person having ordinary skill in the area of technology related to the invention. For example, the substitution of one material for another, or changes in size, are ordinarily not patentable.

Patent Application Time

The time to process a patent application from filing to issue or abandonment is called the "patent pendency time." The average patent pendency time is 18.4 months for utility, reissue, and plant patents. The average patent pendency time for design patents is 31.6 months.

The process will go more rapidly if you qualify for a Special Status Patent.

What are Special Status Patents?

The PTO allows special cases to qualify for accelerated processing. No special forms are required. A letter of explanation need only be attached to your application explaining why you feel acceleration handling should be allowed.

Certain pre-set categories automatically permit special status:

1. **Age (65)**. If you are 65 years of age or older, your application may be made special by sending along a copy of your birth certificate. No fee is required with such a petition.

2. **Illness (Terminal)**. If the state of health of the applicant is such that he or she might not be available to assist in the normal prosecution of the application, it can be made special by sending along a physician's letter attesting to that state of health. No fee is required with such a petition.

3. **Infringer (Actual)**. Subject to a requirement for a further showing as may be necessitated by the facts of your particular case, an application may be made special because of actual infringement (but not for prospective infringement) upon payment of a fee and the filing of a petition alleging facts under oath or declaration to show, or indicating why it is not possible to show (1) that there is an infringing device or product actually on the market or method in use; (2) when the device, product, or method

Correspondence and Information

All correspondence concerning patents should be addressed to:

The Commissioner of Patents and Trademarks, Washington, D.C. 20231, unless you have the name of a particular official.

Call (703) 557-INFO with any questions about filing requirements.

For names of officials and their direct dial phone numbers, refer to chapter 29, Patent and Trademark Office Telephone Directory.

"Personally I am always ready to learn, although I do not always like to be taught." Winston Churchill

"Special" Status Categories

1. Age (65)
   No Fee
2. Illness (terminal)
   No Fee
3. Infringer (actual)
   Fee Due
4. Manufacture
   (ready $$)
   Fee Due
5. Energy invention
   No Fee
6. Superconductor
   technology
   No Fee
7. Enhances the
   environment
   No Fee
8. DNA oriented
   No Fee
9. Orphan Drugs
   No Fee
10. Accelerated
    Examination
    Procedure
    Fee Due

alleged to infringe was first discovered to exist; supplemented by an affidavit or declaration from your attorney or agent to show; (3) that a rigid comparison of the alleged infringing device, product, or method with the claims of the application has been made; (4) that, in the applicant's opinion, some of the claims are unquestionably infringed; (5) that the applicant has made or caused to be made a careful and thorough search of the prior art or has good knowledge of the salient prior art; and, (6) that the applicant believes all of the claims in their application are allowable.

Models or specimens of the infringing product or your product should not be submitted unless requested.

4. **Manufacture (Ready $$)**. If you have a manufacturer ready to go, this may also qualify your application for special handling. I have requested and received this on numerous occasions. You pay a fee and swear under oath or declaration: (a) that the licensee has the capital available (stating the approximate amount) and the facilities (stating briefly the nature thereof) to manufacture your invention in quantity or that sufficient capital and facilities will be made available if a patent is granted; (b) that the prospective licensee will not manufacture, or will not increase present manufacture, unless certain that the patent will be granted; and, (c) that you and/or your licensee stand obligated to produce the product immediately if the patent issues which protects the investment of capital and facilities.

Furthermore, you or your attorney must file an affidavit or declaration to show: (1) that you have searched or caused to be searched the prior art; and, (2) that you believe all the claims in your application to be allowable.

5. **Energy**. No fee is required to make special an application for an invention that will materially contribute to: (1) the discovery or development of energy resources; or, (2) the more efficient use and conservation of energy resources. Examples of inventions in category 1 would be developments in fossil fuels (natural gas, coal, and petroleum), nuclear energy, and solar energy. Category 2 would include inventions relating to the reduction of energy consumption in combustion systems, industrial equipment, household appliances, etc.

6. **Superconductors**. In accordance with the President's proposal directing the PTO to accelerate the processing of patent applications and adjudication of disputes involvingsuperconductivity technologies when requested by the applicant to do so, the PTO will comply accordingly. No fee is required.

Examples of such inventions would include those directed to the superconductive materials themselves as well as to their manufacture and application.

7. **Environment**. If the invention will materially enhance the quality of the environment by contributing to the restoration or maintenance of the basic life-sustaining natural elements—air, water, and soil—you can request to make it special. No fee is required for such a petition.

8. **DNA**. In recent years revolutionary genetic research has been conducted involving recombinant deoxyribonucleic acid (recombinant DNA). Recombinant DNA appears to have extraordinary potential benefit for mankind. It has been suggested, for example, that research in this field may lead to ways of controlling or treating cancer and hereditary defects.

A petition to make special in this category must contain an affidavit or declaration in writing to the effect that your invention is directly related to safety of research in the field of recombinant DNA. A fee must be paid.

9. **Orphan Drugs**. This category is specific to pharmaceutical research and extremely complex. For further information, contact both the PTO and the U.S. Food and Drug Administration.

A patent cannot be obtained on a mere idea or suggestions. The patent is granted for the new machine, manufacture, etc., and not for the idea or suggestion for a new machine..

A Good First Step: The Disclosure Document Program

There's always a better idea.

If you are not ready, or don't care to apply for a patent yet but want to officially evidence and register the conception date of an invention, the PTO offers a disclosure document program. For a fee of $10.00, the Office will preserve your idea on file for a period of two years. This inexpensive recognition will strengthen your case if any conflict arises as to the date of the conception of your idea, but is not meant as a replacement of an inventor's notebook or actual patent.

Program Requirements

The requirements are simple. Send the PTO a paper disclosing the invention. Although there are no stipulations as to content, and claims are not required, the benefits afforded by the Disclosure Document Program will depend directly upon the adequacy of your disclosure. Therefore, it is strongly recommended that the document contain a clear and complete explanation of the manner and process of making and using the invention in sufficient detail to enable a person having ordinary knowledge in the field of the invention to make and use the invention. When the nature of the invention permits, a drawing or sketch should be included. The use or utility of the invention should be described, especially in chemical inventions.

This disclosure is limited to written matter or drawings on paper other than thin, flexible material, such as linen or plastic drafting material, having dimensions or being folded to dimensions not to exceed 8-1/2 x 13 inches (21.6 by 33 cm.). Photographs are also acceptable. Number each page, and make sure that the text and drawings are of such quality as to permit reproduction.

Disclosure Document Request and Fees

In addition to the $10.00 fee, the Disclosure Document must be accompanied by a stamped, self-addressed envelope (SASE) and a separate paper in duplicate, signed by you as the inventor. These papers will be stamped by the PTO upon receipt, and the duplicate request will be returned in the SASE together with a notice indicating

PTO Examiners get evaluated every two weeks, which no doubt has a lot to do with the turnover of personnel: 200 to 300 annually. To keep their jobs, they must produce efficiently, accurately, and honestly.

that the Disclosure Document may be relied upon only as evidence that a patent application should be diligently filed if patent protection is desired.

Your request may take the following form:

"The undersigned, being the inventor of the disclosed invention, requests that the enclosed papers be accepted under the Disclosure Document Program, and that they be preserved for a period of two years."

Warning: The two-year retention period should not be considered to be a "grace period" during which you can wait to file a patent application without possible loss of benefits. It must be recognized that in establishing priority of invention an affidavit or testimony referring to the Disclosure Document must usually also establish diligence in completing the invention or in filing the patent application since the filing of the Disclosure Document.

Also be reminded that any public use or sale in the U.S. or publication of the invention anywhere in the world more than one year prior to the filing of your patent application on the invention disclosed will prohibit the granting of a patent on it.

The Disclosure Document is not a patent application, and the date of its receipt in the Patent and Trademark Office will not become the effective filing date of any patent application subsequently filed. It will be retained for two years and then be destroyed unless referred to in a separate letter in a related application within two years.

The program does not diminish the value of the conventional witnessed and notarized records as evidence of conception of your invention, but it should provide as more credible form of evidence than that provided by the popular practice of mailing a disclosure to oneself or another person by registered mail.

It All Begins with a Search

No pressure, no diamonds.

Why Conduct a Patent Search?

Before making an application for patent protection, it is advisable to see whether or not any prior art exists, i.e. if the same concept has been patented by someone else. This is done through what is called a **patent search**.

A thorough search is recommended for numerous reasons. First, making a search involves less expense than trying to obtain a patent and having it rejected on the basis of prior art. If the search reveals that the invention cannot be protected as engineered, the cost of preparing and filing an application as well as significant time and energy will be saved.

Further, even if none of the earlier patents show all the details of the invention, they may point out important features or better ways of doing the job. If this is the case, you may not want to get patent protection on an invention that could encounter commercial difficulty.

In the event that nothing is found in the search that would prevent or delay application, the information gathered during the search will prove helpful, acquainting you with the details of patents related to the invention.

Who Conducts a Patent Search?

You can approach the search three ways. You could have a patent attorney do it for you; engage the services of a professional patent search firm or individual; or do it yourself.

Law Firm Initiated Search

Nothing is wrong with having a law firm handle the search. Just understand that lawyers do not usually personally conduct patent searches, but typically hire a patent search firm or professional patent searcher. To whatever the lawyer is charged by the patent searcher, a premium will be added. Law firms may step on search fees 40% or more depending upon the firm and city.

If you decide to retain a lawyer for this task, remember to insist upon an estimate of the costs in advance. The fee will be based upon how far back and encompassing you want the search. If you are quot-

Advantages of the Patent Process

1. Disclosure Document and Application Filing provides instant protection of concept.

2. Preliminary search helps you avoid making infringing product.

3. Patent Grant provides excellent description of product, locks in ownership of product, and increases marketability.

"I encourage my people to do a few searches themselves to get on-the-job training and understand what a search requires. This way when they do pay money for an excellent search, they'll realize it was absolutely worth every penny," says E.D. Young, founder of The Inventors Network, Columbus, Ohio.

Chemical and electrical patent searches generally cost more, averaging eight to ten hours of searching.

ed too low a figure from a lawyer, it will probably not be much of a search. Searches done fast may not be worth much. Done correctly, a search in time saves nine.

Direct Hire Professional Search

If you want to save the lawyer's mark-up, consider hiring a patent searcher. Typically, they are listed in the yellow pages under "Patent Searchers." Be careful to contact a firm or person who specializes in this work and not an invention marketing organization listed under a toll-free number. Many such firms use the search as a hook to sucker unsuspecting inventors into their rapacious grasp. Get all the facts before committing yourself. **Always ask for an estimate of charges in advance.**

Presently in Washington, D.C., searchers charge between $30.00 and $60.00 per hour. This range depends a great deal upon whether it is an independent or large firm, with a rule of thumb being the larger the firm, the greater the bill. Some firms require a minimum in the $250 range. A simple search can run between six to seven hours for a mechanical patent. Fixed price deals for $180 are available, providing three hours and six references, copies included. As with most things in life, you get what you pay for. If the searcher stops at three and there are six pieces of prior art, you may have lost your investment. **To be useful, a search must be comprehensive.**

The searcher will charge a fee for making copies as well. Some searchers mark this up while some go at cost. I've received estimates from 40 cents per page to $3.00 per patent no matter how many pages.

The better the lawyer and/or searcher understands the invention, the better the search. Highlight the novel features of your invention. This explanation can be made through drawings or sketches, models, written description, or a discussion, or a combination of these. Take the time; it'll be worth it.

Correspondence concerning the search should be kept in a safe place, you may need it to prove dates and other facts about the invention later.

Do-it-Yourself Search

You may opt to do the patent search. Several methods are available.

The PTO runs a **Patent Public Search Room** located in Arlington, Virginia. Here, every U.S. patent granted since 1836 (over four million) may be searched and examined. Many inventors like to make at least one pilgrimage to the Crystal Plaza facility. It is located less than five minutes from National Airport by taxi and Metro Rail (Blue and Yellow Lines - Crystal City Station). Many fine hotels are within walking distance.

Upon your arrival the PTO will issue you, at no cost, a non-transferable User Pass for the day. It is wise to double check the hours of operation by calling (703) 557-2276. Depending where in the facility you want to search, the Patent Public Search Room is typically open from 8:00 a.m. to 8:00 p.m.

The Patent Public Search Room is really something to behold. You can touch and feel the original documents, including everything from Abraham Lincoln's 1849 patent (No. 6,469) for a device to buoy vessels over shoals to Auguste Bartholdi's design patent on a statue entitled, Liberty Enlightening The World (a.k.a., The Statue of Liberty).

Patent Classification System

Patents are arranged according to a classification system of more than 400 classes and 115,000 subclasses. The *Index to the U.S. Patent Classifications* is an alphabetical list of the subject headings referring to specific classes and subclasses of the U.S. patent classification system. The Classifications are intended as an initial means of entry into the PTO's classification system and should be particularly useful to those who lack experience in using the classification system or who are unfamiliar with the technology under consideration.

The Classifications are to searching a patent what the card catalog is to looking for a library book. It is the only way to discover what exists in the field of prior art. The Classifications are a star to steer by, without which no meaningful patent search can be completed. Before you begin your search, use the Classifications to prepare your direction. First, look for the term which you feel best represents your invention. If a match cannot be found, look for terms of approximately the same meaning, for example, the essential function or the effect of use or application of the object of concern. By doing some homework before you begin searching, i.e., familiarizing yourself with the *Index* and locating the class and subclass numbers for terms that pertain to your invention, you'll save time.

Once you have recorded the identifying numbers of possibly pertinent classes and subclasses, refer to the *Manual of Classification*, a loose-leaf PTO volume listing the numbers and descriptive titles of more than 300 classes and 95,000 subclasses used in the subject classification of patents, with an index to classifications. This manual is also available at the Patent and Trademark Depository Libraries.

The Classifications are arranged with subheadings that can extend to four levels of indentation. A complete reading of a subheading includes the title of the most adjacent higher heading, and so on until there are no higher headings. Some headings will reference other related or preferred entries with a "(see...)" phrase.

New classes and subclasses are continuously based upon breaking developments in science and technology. Old classes and subclasses are rendered obsolete by technological advance. In fact, if you have suggestions for future revisions of the classifications, find omissions or

The PTO's research collection is immense. More than 30 million documents are on hand, including five million U.S. patents, nine million cross-references, and 16 million foreign patents. The Scientific Library maintains a collection of 120,000 volumes and provides access to commercial data bases.

If you live in a city with one of the PTO's 68 Patent and Trademark Depository Libraries (see list below), a librarian can usually assist you in finding a list of those who make a living searching patents. Librarians are not encouraged to personally recommend any patent searcher or search organization.

errors, you are encouraged to alert the PTO. Send your suggestions to Editor, U.S. Patent Classification, Office of Documentation, U.S. Patent and Trademark Office, Washington, D.C. 20231.

Nearby the Patent Public Search Room is the **Scientific Library of the Patent and Trademark Office**. The Scientific Library makes publicly available over 120,000 volumes of scientific and technical books in various languages, about 90,000 bound volumes of periodicals devoted to science and technology, the official journals of 77 foreign patent organizations, and over 12 million foreign patents. The hours are from 8:45 a.m. to 4:45 p.m.

Patent and Trademark Depository Libraries

I think that every inventor should do at least one hands-on patent search to fully understand and appreciate the process. Obviously, not everyone can visit the Patent and Trademark Office's Public Search Room in Arlington, Virginia. If you cannot make the trip to Washington, D.C., you may inspect copies of the patents at a **Patent and Trademark Depository Library (PTDL)**, a nationwide network of 68 prestigious academic, research, and public libraries. According to Carole A. Shores, director of Patent and Trademark Depository Library Programs, it is the goal of the PTO to have PDLs in each state. "Our mission is a very simple one," explains Ms. Shores. "It is to bring more information and help to all the people out there that need it and can't afford to pay big money to get it. What makes our program special is that we really listen to their needs and when they bring requests to us we try very hard to give them what they want."

Formerly known as Patent Depository Libraries (PDLs), the PTDLs continue to be one of the PTO's most effective mechanisms for publicly disseminating patent information. PTDLs receive current issues of U.S. Patents and maintain collections of earlier issued patents. The scope of these collections varies from library to library, ranging from patents of only recent years to all or most of the patents issued since 1790. Due to the variations in the scope of patent collections among the PDLs and their hours of service to the public, you should call first to find out when it is open, to avoid any possible inconvenience.

The patent collections in the PTDLs are open to the general public and I have always found the librarians most willing to take the time to help newcomers gain effective access to the information contained in patents. In addition to the patents, PTDLs usually have all the publications of the U.S. Patent Classification System, e.g., *The Manual of Classification, Index to the U.S. Patent Classification, Classifications Definitions, Official Gazette of the United States Patent and Trademark Office*, etc. Below is an inventory of the reference materials and tools available to you at most PTDLs.

PTDL Reference Material Inventory

☞ Patents

Index to the U.S. Patent Classification Manual of Classification

Classification Definitions (microfiche)

Official Gazette-Patents (OG)

Index of Patents, Part I and Part II (comes with Patent OGs)

Patentee/Assignee Index (microfiche)

Basic Facts about Patents

General Information Concerning Patents

Attorneys and Agents Registered to Practice Before the U.S. Patent and Trademark Office

Manual of Patent Examining Procedure

Concordance, United States Classification to International Patent Classification

National Inventors Hall of Fame

☞ Trademarks

Official Gazette-Trademarks (OG) *Index of Trademarks* (comes with Trademark OGs) *Trademark Manual of Examining Procedure*

Basic Facts About Trademarks

☞ Miscellaneous

Directory of Patent and Trademark Depository Libraries

Consolidated Listing of Official Gazette Notices—Re: Patent and Trademark Office Practices and Procedures

Code of Federal Regulations, Title 37; Patents, Trademarks, and Copyrights

Patents and Trademarks Style Manual

Annual Report of the Commissioner of Patents and Trademarks

Story of the Patent and Trademark Office

☞ Current subscriptions provided with statutory fee:

Utility Patents, microfilm

Design Patents, microfilm

Plant Patents

Statutory Invention Registrations, microfilm

Changes to Patents, microfilm (Certificates of Correction Disclaimers, Reissues, Reexamination Certificates)

PTDL *Program Director Carole Shores says that her librarians are in a perpetual state of training. "The most important thing is that our librarians believe in helping their patrons," she says. I can tell you from first-hand experience that she does not exaggerate. Every inventor I know who has utilized the resources at a PTDL would agree.*

☞ Other Materials provided to Patent and Trademark Depository Libraries by the U.S. Patent and Trademark Office:

PTDL Mailing List, Representatives

PTDL Mailing List, Directors

Patent and Trademark Office Fee Schedule

U.S. Trademark Law; Rules of Practice, Forms & Federal Statutes, U.S. Trademark Association

Trademark Examination Guides

Trademark Classification of Services

Trademark Classification of Goods

"A Trademark Is Not a Patent or a Copyright" (USTA reprint) "How to Use a Trademark Properly" (USTA reprint)

"A Guide to the Care of Trademarks" (USTA reprint)

PTO Organizational Telephone Directory

IEEE Patents and Patenting for Engineers and Scientists (reprint)

CASSIS brochures

CASSIS User's Guide (card)

CASSIS User's Manual

CASSIS Location Manager's Manual

CASSIS search examples

CASSIS tips

Disclosure Document Program brochures

Sample PTDL/PTO Conference Agenda

Sample List of Conference Attendees

"Glossary of Terms" from Data Base Entry Preparation Manual Foreign Patent Information in the Scientific Library Scientific Library File of Foreign Patent Document 16mm Microfilm PTOnline

ADLIBS

Handbook on the Use of the U.S. Patent Classification System for Patent and Trademark Depository Libraries

Even though I live near enough to the Alexandria, Virginia, Public Search Room, often I opt to do my work at the University of Maryland's PTDL located within its Engineering and Physical Sciences Library. Crowded with books and not much larger than a small meeting room, the library offers only a couple of chairs. However, here I can do my search work and have access to a very extensive collection of technical publications outside the PTDL that I enjoy browsing through for ideas and technologies. Such "extras" are not available at the main government facility, and I consider them a bonus.

List of Patent and Trademark Depository Libraries

The following is a list of Patent and Trademark Depository Libraries that make available their patent collections free of charge to the public. Because of variations among the PTDLs in their hours of service and the scope of their patent collections, anyone contemplating use of the patents at a particular library is well-advised to contact the library in advance about its collection and hours, so as to avoid possible inconveniences.

State	Name of Library	Telephone Contact
AL	Auburn University Libraries	(205) 844-1747
	Birmingham Public Library	(205) 226-3680
AK	Anchorage: Z.J. Loussac Public Library	(907) 261-2916
AZ	Tempe: Noble Library, Arizona St. Univ.	(602) 965-7607
AR	Little Rock: Arkansas State Library	(501) 682-2053
CA	Los Angeles Public Library	(213) 612-3273
	Sacramento: California State Library	(916) 322-4572
	San Diego Public Library	(619) 236-5813
	Sunnyvale Patent Clearinghouse	(408) 730-7290
CO	Denver Public Library	(303) 571-2347
CT	New Haven: Science Park Library	(203) 786-5447
DE	Newark: University of Delaware Library	(302) 451-2965
DC	Washington: Howard University Libraries	(202) 636-5060
FL	Fort Lauderdale: Broward County Main Library	(305) 357-7444
	Miami-Dade Public Library	(305) 375-2665
	Orlando: Univ. of Central Florida Libraries	(407) 275-2562
GA	Atlanta: Price Gilbert Memorial Library, Georgia Institute of Technology	(404) 894-4508
ID	Moscow: University of Idaho Library	(208) 885-6235
IL	Chicago Public Library	(312) 269-2865
	Springfield: Illinois State Library	(217) 782-5430
IN	Indianapolis-Marion County Public Library	(317) 269-1741
IA	Des Moines: State Library of Iowa	(515) 281-4118
KY	Louisville Free Public Library	(502) 561-8617
LA	Baton Rouge: Troy H. Middleton Library, Louisiana State University	(504) 388-2570
MD	College Park: Engineering and Physical Sciences Library, Univ. of Maryland	(301) 454-3037
MA	Amherst: Physical Sciences Library, University of Massachusetts	(413) 545-1370
	Boston Public Library	(617) 536-5400
MI	Ann Arbor: Engineering Transportation Library, Univ. of Michigan	(313) 764-7494
	Detroit Public Library	(313) 833-1450
MN	Minneapolis Public Library and Information Center	(612) 372-6570
MO	Kansas City: Linda Hall Library	(816) 363-4600
	St. Louis Public Library	(314) 241-2288
MT	Butte: Montana College of Mineral Science and Technology Library	(406) 496-4281
NE	Lincoln: Engineering Library, Univ. of Nebraska-Lincoln	(402) 472-3411
NV	Reno: University of Nevada-Reno Library	(702) 784-6579
NH	Durham: Univ. of New Hampshire Library	(603) 862-1777

The Foreign Patent Document Collection consists of 25 to 30 million documents. The paper portion of the collection occupies more than 10,000 square feet of space, while the microfilm collection has grown to 50,000 reels.

Don Coster, President of the Nevada Inventor's Association, reports that the PTDL in Reno is "run by a person who trains the staff to the point that you cannot tell who's in charge. So, therefore, anytime you walk in you get help, and the help is for the guy that doesn't know what to do or who to write to or who to talk to. I think without those Patent and Trademark Depository Libraries we'd (independent inventors) be floating in limbo."

In an effort to reduce the cost of providing CASSIS online to the public, the PTO is providing a CD-ROM (Compact Disk Read-Only Memory) to selected PTDLs on a developmental, test-program basis.
If this experimental use of CD-ROM proves successful, a CD-ROM system may ultimately replace portions of the costlier CASSIS.

NJ	Newark Public Library	(201) 733-7782
	Piscataway: Library of Science and Medicine, Rutgers University	(201) 932-2895
NM	Albuquerque: Univ. of New Mexico General Library	(505) 277-4412
NY	Albany: New York State Library	(518) 473-4636
	Buffalo and Erie County Public Library	(716) 858-7101
	New York Public Library (The Research Libraries)	(212) 714-8529
NC	Raleigh: D.H. Hill Library, North Carolina State University	(919) 737-3280
OH	Cincinnati and Hamilton County, Public Library of	(513) 369-6936
	Cleveland Public Library	(216) 623-2870
	Columbus: Ohio State Univ. Libraries	(614) 292-6175
	Toledo/Lucas County Public Library	(419) 259-5212
OK	Stillwater: Oklahoma State University Library	(405) 744-7086
OR	Salem: Oregon State University	(503) 378-4239
PA	Philadelphia, The Free Library of	(215) 686-5331
	Pittsburgh, Carnegie Library of	(412) 622-3138
	University Park: Pattee Library, Pennsylvania State University	(814) 865-4861
RI	Providence Public Library	(401) 455-8027
SC	Charleston: Medical University of South Carolina Library	(803) 792-2371
TN	Memphis and Shelby County Public Library and Information Center	(901) 725-8879
	Nashville: Stevenson Science Library, Vanderbilt University	(615) 322-2775
TX	Austin: McKinney Engineering Library, University of Texas at Austin	(512) 471-1610
	College Station: Sterling C. Evans Library, Texas A&M University	(409) 845-2551
	Dallas Public Library	(214) 670-1468
	Houston: The Fondren Library, Rice University	(713) 527-8101
UT	Salt Lake City: Marriott Library, University of Utah	(801) 581-8394
VA	Richmond: Virginia Commonwealth University Library	(804) 367-1104
WA	Seattle: Engineering Library, University of Washington	(206) 543-0740
WI	Madison: Kurt F. Wendt Library, University of Wisconsin Madison	(608) 262-6845
	Milwaukee Public Library	(414) 278-3247

Online Patent Searches

Available at all PTDLs is the **Classification and Search Support Information System (CASSIS)**, an online computer data base.

CASSIS brings you all the data available at the Patent and Trademark Search Room. And it can streamline the manual search procedure by providing electronic access to basic patent search tools. CASSIS can help you define a "field of search" and identify the patents in that field. It provides classifications of a patent; supplies patents in a classification; displays patent titles and/or company names; structures classification titles; finds key words in classification titles and patent abstracts; and much more.

Other Databases

Databases are available that savvy searchers use to augment their work. Access through host systems such as ORBIT and DIALOG, these files include, for example, Derwent's World Patent Index covering 33 countries and technology back to 1963. The information provided by Derwent is in the form of abstracts and bibliographic background. Some databases just supply bibliographic data. If you decide to use one, find out about the scope of its information.

Currently in the option period of a five-year contract with the PTO, Derwent will be will be the only database contractor permitted to operate out of the PTO's Public Search Room for the next two years. Derwent's direct dial telephone number is (703) 486-8155.

Computer-aided searches are not by any means the final word. The best they can do is generate abstracts of patents. Someone must ultimately retrieve copies of the full patents and study them.

Patent Search Steps

No matter where you decide to conduct a search, certain steps must be taken in any patent search. Here is a brief guide to manual searches for U.S. patents.

> **Step 1:** If you know the PATENT NUMBER, go to the *Official Gazette* to read a summary of the patent. This publication is available at any of the above-mentioned patent search facilities and in many public library reference rooms.

> **Step 2:** If you know the PATENTEE or ASSIGNEE, look at the *Patent/Assignee Index* to locate the PATENT NUMBER. This is available at any of the patent search facilities. In Crystal City, it is on microfiche and in card catalogues.

> **Step 3:** If you know the SUBJECT, start with: *Index to the U.S. Patent Classification.*

> **Step 4:** Once you have jotted down the class(es) and subclass(es) out of the *Index*, refer to the *Manual of Classification* and check this information vis-a-vis the hierarchy to see if they are close to what you need. The *Manual of Classification* is available at all patent search facilities.

> **Step 5:** Using the Class/Subclass numbers you have found, look at the U.S. Patent Classification Subclass and

"I *decided fifty years ago to ask no one for answers, to find them out for myself. I suggest you do the same."* R. Buckminster Fuller

Numeric Listing and copy the patent numbers of patents assigned to the selected class/subclass. If you are at the Crystal City facility, take the class/subclass numbers into the stacks of patents and begin "pulling shoes". To pull shoes is to physically remove patent groupings from the open shelves.

Step 6: Then, using the *Official Gazette* again, look at the patent summaries of those patents. At the Crystal City facility, you will not have to go back to this publication since the actual patents are there.

Step 7: Upon locating the relevant patents, examine the complete patent in person or on microfilm—depending upon where you conduct the search.

More Help—If You Have the Time

If you want help after referring to the *Index to the U.S. Patent Classification*, you may write for free assistance to the Commissioner of Patents and Trademarks.

Sample Letter

Date:

Commissioner of Patents and Trademarks
Att: Patent Search Division
Washington, D.C. 20231

Please let me know what subject area or class(es) and subclass(es) cover my idea. The enclosed sketches on the back of this sheet, with each part labeled, show my intended invention.

My idea has certain features of structure, mode of operation, and intended uses which I have defined below.

I understand that there is no charge for this information.

Thank you for your help.

Sincerely,

(1) Features of structure, or how it is constructed.

(2) Mode of operation, or how it works.

(3) Intended uses, or purpose of idea.

(4) Rough sketches of idea, viewed from all sides, with labels to identify each part. [If necessary use extra sheets of paper, either plain or lined. These sketches may be made in pencil and need not be drawn to scale.]

If you do not locate the PATENT you are seeking, try again using another subject class and subclass.

How to Order Copies of Searched Patents (Prior Art)

You may order copies of original patents or cross-referenced patents contained in subclasses comprising the field of search from the Patent and Trademark Office. Mail your requests to: Patent and Trademark Office, Box 9, Washington, D.C. 20231.

Payment may be made by check, coupons, or money. Expect a wait of up to four weeks when ordering copies of patents from the PTO. Postage is free.

For the convenience of attorneys, agents, and the general public in paying any fees due, deposit accounts may be established in the PTO with a minimum deposit of $50.00. For information on this service, call (703) 557-3227.

What to Do with Your Search Results?

Study the results of the patent search. You're out of luck if a previously patented invention is very similar to yours; your invention may even be infringing on another invention. On the other hand, one or more patents may describe inventions that are intended for the same purpose as yours, but are significantly different in various ways. Look these over and decide whether it is worthwhile to proceed.

If the features that make your invention different from the prior art provide important advantages, you should discuss the situation with your attorney to determine whether a fair chance exists of obtaining a patent covering these features.

I have found from experience that a good patent attorney can often get some claim(s) to issue, albeit not always strong ones. A patent for patent's sake is usually possible. Whether it will be worth the paper it is printed on is yet another matter. Do not take this decision lightly because the patent process is not inexpensive. The average utility patent will cost about $2,500.

"The more information that the independent inventor can provide for himself, the smoother things will go all along the way," advises Jim Pinske of Milwaukee. Pinske is inventor of a flying toy called B'zarts. Based upon his experience searching patents at a PTDL, he learned what was required to write his own abstract and patent claims. "About 85 percent of the wording in my patent is my own," he notes proudly. "It made my attorney's job a lot easier and complete."

PTO Patent References: Quick Look-Up

CASSIS (Classification and Search Support Information System)Patent Mode (PM)
An online database used to find the one original class and subclass for each patent and all official and unofficial cross-reference class(es)/subclass(es) for that patent.

U.S. Patent Classification—Numeric Listing
Microfilm publication which lists all U.S. Patents in ascending numerical sequence, giving each patent's original class and subclass and all official and unofficial cross-reference

---

class(es)/subclass(es), current as of the date of the publication. Updated every six months. 22 reels.

Patentee/Assignee Index

Microfiche publication which lists in alphabetical order five to six year's worth of patentee's and assignees at the time each patent was issued with corresponding patent number. Updated quarterly. More than 250 microfiche.

Index of Patents, Part I—List of Patentees

Annual publication which lists in alphabetical order all patentees and assignees at the time each patent was issued with corresponding patent number. Approximately 2,000 pages per year. 1926 to date. Predated by Annual Report of the Commissioner of Patents, 1790- 1925.

"List of Patentees"

In the back of each weekly issue of the Official Gazette is an alphabetical list of patentees and assignees for the week with corresponding patent number.

Index to the U.S. Patent Classification

Alphabetical list of approximately 63,000 common, informal subject headings or terms which refer to specific class(es)/subclasses in the classification system used to categorize patents. It is intended as a means for initial entry into the classification system. Approximately 240 pages.

CASSIS Index Mode (IM)

An online database used to find class(es)/subclass(es) corresponding to common, informal terms and phrases in the Index to the U.S. Patent Classification by using a given word or word stem (or various logical combinations of words and word stems). Once the words are identified, the corresponding class(es)/subclass(es) may be obtained with the List Index (LI) command, and with the List Detail (LD [entry number]) command which provides sub-entries for a specified index entry number from the (LI) listing.

CASSIS Word Mode (WM)

An online data base used to find the class(es)/subclass whose titles include a given word or word stem (or various logical combinations of words or word stems). The file comprises the words in the titles of the Manual of Classification. Once the words are identified, the corresponding class(es)/subclass(es) may be obtained with the List (L) command.

CASSIS Patent Abstract Mode (PA)

An online database used to identify patents whose abstracts include a given word or word stem (or various logical combinations of words or word stems). The file comprises the words in the abstracts from the most recent one- to two-year's worth of patents. Once the patents are identified, the system will locate all class(es)/subclass(es) under which the patents are filed using the Get Classifications (GC) command. These class(es)/subclass(es) may then be obtained by using the list classifications (LC) command, which lists the class(es)/subclass(es) in ascending number order with the number of patents in each; or they may be obtained by using the Rank Classifications (RC) command which lists the class(es)/subclass(es) in descending order by number of patents in each.

Manual of Classification

Presents listings of the approximately 390 main groupings (classes) and approximately 113,000 smaller groupings (subclasses) into which patented subject matter is classified. Each subclass has a short, descriptive title often arranged in a specific hierarchical order designed by dots for indentation levels. About 1,400 pages.

CASSIS Title Mode (TM) Page Submode (P)

An online database used to examine the titles of the various class(es)/subclass(es) exactly as they are displayed on a page of the Manual of Classification. The level of the a subclass in the hierarchy is designated by the number of dots (called the indent level) preceding the subclass title.

CASSIS Title Mode (TM) Full-title Submode (F)

An online database used to examine the full title of a single class/subclass, which includes the titles of the higher subclass(es) and class.

CASSIS Title Mode (TM) Coordinate Submode (C)

An online database used to examine the hierarchical relationship of a given class/subclass to the next higher level of classification (the "parent") and up to eight of the subordinate subclasses (the "children").

Classification Definitions

Detailed definitions for each class and official subclass included in the Manual of Classification. The definitions indicate the subject matter to be found in or excluded from a subclass; they limit or expand in precise manner the meaning intended for each subclass title; they serve as a guide to users of the Manual to refer to the same subclass for patents on a particular technology by eliminating, as much as possible, subjective and varying interpretations of the meanings of subclass titles. The definitions provide further guidance through "search notes," which illustrate the kinds of information that can be found in a subclass and direct the searcher to other related subclasses which may contain relevant infor-

Inventor-Assistance Program News is the main information-transfer mechanism of the Study of Innovative Programs for Inventors, which is conducted by staff at Argonne National Laboratory. Its purpose is to provide an information-sharing forum among organizations that assist inventors, study those organizations in order to identify successful programs, provide technical and limited financial assistance to some of the most innovative of the organizations, and share what is learned. Available free of charge, the News is published periodically as a limited edition publication. To see if you qualify to receive it, contact Dr. Martin J. Bernard III, Building 372, Argonne National Laboratory, Argonne, IL 60439, or call (708) 972-3738.

mation. About 10,000 pages. Classification Definitions are also available on microfiche. About 360 fiche.

CASSIS Classification Mode (CM)

An online database used to list all of the patent numbers which have been assigned to particular class/subclass as an original reference, or official or unofficial cross-references.

U.S. Patent Classification—Subclass Listing

A microfilm publication which lists all patents classified in each class/subclass and digest of the U.S. Patent Classification System. Within each subclass, the patents are listed in numerical sequence and categorized by whether an original reference, or an official or unofficial cross-reference. Updated every six months. 11 reels.

"Classification of Patents"

In the back of each weekly issue of the Official Gazette, it listsall the patent numbers assigned to each class/subclass for that week.

Official Gazette of the U.S. Patent and Trademark Office (Patent Section)

Weekly publication which presents an entry according to original class/subclass designation for each of the approximately 1,500 patents issued. Each entry includes certain bibliographic information and a brief summary or abstract in the form of one or more representative patent claims (depending on date), together with (where appropriate) a reduced representative drawing. About 300 pages per week.

CDR File

Annual, cumulative computer-produced book listing patent numbers with corresponding reel and frame of the CDR microfilm for Corrections, Disclaimers, Reissues, and reexamination certificates associated with the original patent. Issued since 1973; earlier CDR material was filmed in place with the original patent. About 235 pages and four reels per year.

How Do I Apply for a Patent?

Patent Steps

You've done your search and now are ready to apply for a patent. Here are the steps: Your application must be directed to the Commissioner of Patents and Trademarks and include:

   1) A written document which comprises a specification (description and claims), and an oath or declaration. Two forms are used in association with patent applications: Power of Attorney or Authorization of Agent, Not Accompanying Application, and Revocation of Power of Attorney or Authorization of Patent Agent. These forms may be found as part of Fig. 12 in the Forms and Documents section of this book. Lawyers charge fees for providing these forms; you may reproduce from the Forms and Documents section free of charge and save yourself a considerable sum of money. Here's when to use: Power of Attorney... If you file your application with an attorney or agent, no power of attorney form is required (it will be written into the oath or declaration). If however, you decide to appoint an attorney or agent after your application, you may use the form provided.

   Revocation of Power of Attorney... After your application, if you decide to change attorney or agent, use this form.

   2) A drawing in those cases in which a drawing is required; and

   3) The correct filing fee.

The specification and oath or declaration must be legibly written or printed in permanent ink on one side of the paper only. The PTO prefers typewriting on legal size paper, 8 to 8 1/2 by 10 1/2 to 13 inches, 1 1/2 or double spaced, with margins of one inch on the left-hand side and at the top. If the papers filed are not correctly, legibly, or clearly written, the Office may require typewritten or printed papers.

The application for patent will not be sent for examination until all of the required parts, complying with the rules of presentation, are received. If the papers and parts are incomplete (I've forgotten to attach a check from time to time!), or so defective they cannot be accepted as a complete application for examination, you will be notified about the found deficiencies and given a time period in which to remedy them.

The Inventive Congressman

In May of 1949, when Abraham Lincoln was a Congressman from Illinois, he received Patent No. 6,469 for "A Device for Buoying Vessels over Shoals." The invention consisted of a set of bellows attached to the hull of a ship just below the water line. On reaching a shallow place the bellows were inflated and the ship, thus buoyed, was expected to float clear. Lincoln whittled the prototype of the invention with his own hands.

Lincoln's appreciation of inventions was later to be of great service to the nation. John Ericcson's Monitor, the ironclad ship which defeated the Merrimac, would never have been built except for Lincoln's insistence, nor would the Spencer repeating rifle have been adopted by the Army.

Patent Application Checklist

1. Transmittal letter
2. Declaration (form)
3. Filing Fee ($370 or $185)
4. Drawing
5. Specification (description and claims)
6. Small Entity Declaration (form)
7. Prior art disclosure statement and copies of relevant and material patents

Applications for patents are not open to the public. No information concerning them is released except on written authority of the applicant, assignee, or designated attorney, or when necessary to the conduct of the PTO's business.

To help make sure everything is together, a copy of a Patent Application Transmittal Letter is shown as Fig. 6. This will serve as your cover document.

The four formal responses used by the Application Division to notify inventors of defects in their applications, Notice of Informal Application, Notice of Incomplete Application, Notice to File Missing Parts of an Application—No Filing Date, and Notice to File Missing Parts of an Application—Filing Date Granted, are reproduced in the Forms and Documents section. Become familiar with them, for sooner or later you are sure to receive one if you do patent work.

Upon receiving one of these notices, a surcharge may be required. If you do not respond within the prescribed time period, your patent application will be returned or otherwise discarded. The filing fee may be refunded if the application is refused acceptance as incomplete; however, a handling fee will likely be charged.

File all the application documents together; otherwise, each part must be signed and accompanied by a letter accurately and clearly connecting it with the other parts of your application. This can be a nightmare, so send everything in together from the start and save yourself headaches.

Every application received by the Office is numbered in serial order, and you are informed of the serial number and filing date of the application by a filing receipt. This document does not mean that a patent has been awarded, only that your application is at the PTO.

What Are the Parts to the Application?

Oath or Declaration

By law the inventor must file an oath or declaration. The inventor must make an oath or declare that he/she is believed to be the original and first inventor of the subject matter of the application. Additionally, the inventor must make various other allegations required by law and by PTO rules.

If you opt for an oath, it must be sworn before a notary public or other officer authorized to administer oaths.

Specification (Description and Claims - Utility Patents)

The specification is a written description of the invention and of the manner and process of making and using it, and is required to be in such full, clear, concise, and exact language as to enable any person skilled in the technological area to which the invention pertains, or with which it is most nearly connected, to make and use it. For the record, even after all these years and a familiarity with patent descriptions, I still have a hard time fully understanding them.

The specification must set forth the precise invention for which a patent is solicited, in such a manner as to distinguish it from other inventions and from what is old. It must describe completely a specific embodiment of the process, machine, manufacture, composition of matter, or improvement invented, and must explain the method of operation or principle whenever applicable. The best way you see to carry out the invention also must be described.

In the case of an improvement, your specification must particularly point out the part(s) of the process, machine, manufacture, or composition of matter to which the improvement relates, and the description should be confined to the specific improvement and to such parts as necessarily co-operate with it or as may be required to complete understanding or description of it.

Title and Abstract

The title of your invention, which should be as short and specific as possible, should be a heading on the first page of the specification, if it does not otherwise appear at the beginning of the application.

A brief abstract of the technical disclosure in the specification must be set forth in a separate page immediately following the claims in a separate paragraph under the heading "Abstract of the Disclosure." The purpose of the abstract is to enable the PTO and the public generally to determine quickly from a cursory inspection the nature and gist of the technical disclosure. It is not used for interpreting the scope of the claims. I have included the following as an example of an Abstract:

> An electronic toy doll including electronic circuitry for selectively generating a number of simulation sounds typically associated with a mystic or science-fantasy character. In respective operational modes, the sound of wind, the sound of breathing, an eerie pseudo-random sequence of musical notes and sounds representing the operation of a weapon are selectively generated. In a further operational mode, a random one of a predetermined number of responses are provided upon generation of an actuation signal. The random response may be considered to be an answer to inquiries. Preferred circuitry for generating the simulation sounds is described, accessories, adapted for removable interconnection with the circuitry are also described.

Summary of the Invention

A short summary of your invention indicating its nature and substance, which may include a statement of the object of the invention, should precede the detailed description. Your summary should, when set forth, be commensurate with the invention you claim and any object recited should be that of the invention as claimed. I have included the following as an example of a Summary:

I favor using the declaration in lieu of an oath, making it part of the original application. A declaration does not require notarization. A sample blank application, which includes the declaration, is shown as Fig. 11. The Office will not supply blank forms.

The present invention provides a relatively inexpensive, rugged electronic doll. The doll selectively simulates the sounds typically associated with a mystic or science-fantasy character. The sounds of wind, breathing, weapons operation, and an eerie tune of pseudo-random musical notes are selectively produced. Additionally, the future is "foretold" through generation of a random one of a predetermined number of gong sounds generated in response to an actuation signal. The operational mode is controlled by a central function select logic circuit in cooperation with remote switches. The remote switches may be located in the doll body, or may be disposed in removable interconnectable remote accessories.

A sample United States Patent, which includes drawings and references, is shown as Fig. 1.

Reference to Drawings

If you have drawings, the PTO requires a brief description of several views of the drawing and the detailed description of the invention shall refer to the different views by specifying the numbers of the figures and to the different parts by use of reference letters or numerals.

Claims

I would recommend that you hire a lawyer to prepare and prosecute a utility patent. The preparation of a patent application is a highly complex business and not to be taken lightly.

If you are bent, however, on writing your own description and claims, it is best to study many patents for style and content. Then locate an updated copy of the PTO's Statutory Requirements of Claims at the nearest Patent and Trademark Depository Library. You can photocopy them from the *Manual of Patent Examining Procedure* which is available at each PTDL.

For an example of Claims, see below; also Fig. 1, United States Patent.

In the meantime, here are some points to get you going:

a) The specification section of the patent must conclude with a claim particularly pointing out and distinctly claiming the subject matter which the application regards as your invention or discovery.

b) The claims are brief descriptions of the subject matter of your invention, eliminating unnecessary details and reciting all essential features necessary to distinguish the invention from what is old. The claims are the operative part of the patent. Novelty and patentability are judged by the claims, and, when a patent is granted, questions of infringement are judged by the courts on the basis of the claims.

c) When more than one claim is presented, they may be placed in dependent form in which a claim may refer back to and further restrict one or more preceding claims.

d) A claim in multiple dependent form shall contain a reference, in the alternative only, to more than one claim previously set forth and then specify a further limitation of the subject matter claimed. A multiple dependent claim shall not serve as a basis for any other multiple dependent claim. A multiple dependent claim shall be construed to incorporate by reference all the limitations of the particular claim in relation to which it is being considered.

e) The claims or claims must conform to the invention as set forth in the remainder of your specification and the terms and phrases used in the claims must find clear support or antecedent basis in the description so that the meaning of the terms in the claims may be ascertainable by reference to the description.

A sample of Claims is shown, in part, below:

1. An electronic doll comprising: a body; function select means, disposed on said body, for selectively generating respective mode control signals; electronic signal generator means, disposed in said body and responsive to said mode control signals, including: means for generating a wind simulation signal representative of the sounds of wind; means for generating a breathing simulation signal representative of the sounds of breathing; means for generating a weapons simulation signal representative of weapons fire; means for generating pseudo-random musical notes; and means for generating one of a predetermined number of responses to an actuation signal, said one response being determined in a random manner; transducer means for generating audible output signals representative of electrical input signals applied thereto; and means, responsive to said mode control signals, for selectively applying said respective electrical simulation signals as input signals to said transducer means.

2. The electronic doll of claim 1 wherein said means for generating a wind simulation signal comprises: pseudo-random signal generator means, responsive to clock signals applied thereto for producing a pseudo-random signal, said pseudo-random signal having a digital value changing in a pseudo- random manner at a rate in accordance with said clock signal; oscillator means for generating said clock signal at a frequency in accordance with a frequency control signal applied thereto; and irregular signal generator means, responsive to said pseudo- random signal, for generating an irregular signal which alternatively rises or falls

Specification Checklist

1. Title of the invention

2. Cross-references to related applications, if any

3. Brief summary of invention

4. Brief description of drawings

5. Detailed description of invention

6. Claims (I Claim:)

7. Abstract of the disclosure

in amplitude in an irregular manner, said irregular signal being applied to said oscillator means as said frequency control signal.

3. The electronic doll of claim 2 wherein said pseudo-random signal generator means comprises:...[etc.].

Arrangement of Application Elements

The elements of the application should appear in the following order:

1) Title of the invention; or an introductory portion stating the inventor(s) name(s), citizenship, and residence of the applicant, and the title of the invention may be used.

2) Cross-references to related applications, if any.

3) Brief summary of the invention.

4) Brief description of the several views of the drawing, if there are drawings.

5) Detailed description.

6) Claim(s).

7) Abstract of the disclosure.

You are also encouraged to file an Information Disclosure Citation (Fig. 17) at the time of filing your patent application or within three months after the filing date of the application or two months after a filing receipt is received. If filed separately, be sure to tag it with the Group Art Unit to which the action is assigned as indicated on the filing receipt. This form will enable you to provide the PTO with a uniform listing of citations.

While the filing of information disclosure statements is voluntary, the form must be accompanied by an explanation of relevance of each listed item, a copy of each listed patent or publication of other item of information and a translation of the pertinent portions of foreign documents (only if an existing translation is available to you).

Examiners will consider all citations submitted and place their initials adjacent to the citations in the boxes provided on the form.

Filing Fees

Caution: Fees change from time to time. People outside the PTO information loop are seldom alerted to price hikes; front-page news it is not. There are penalties for making incorrect payments, so double check the required amount before sending in a check. The PTO Fees list, printed in the Forms and Documents section of this book, is current per publication, but still should be confirmed. The latest fee information is available by calling (703) 557-5168.

Small Entity Status Reduces Fees by Half

On August 27, 1982, Public Law 97-247 provided that effective October 1, 1982, funds would be made available to the PTO to reduce by 50 percent the payment of fees by independent inventors, small business concerns, and nonprofit organizations.

The reduced fees include those for patent application, extension of time, revival, appeal, patent issues, statutory disclaimer, and maintenance on patents based on applications filed on or after August 27, 1982.

Fees that are not reduced include petition and processing, (other than revival), document supply, certificate of correction, request for re-examination, international application fees, and certain maintenance fees.

What is an Independent Inventor?

The PTO considers an inventor as independent if the inventor (1) has not assigned, granted, conveyed, or licensed, and (2) is under no obligation under contract or law to assign, grant, convey, or license any rights in the invention to any person who could not likewise be classified as an independent inventor if that person had made the invention, or to any concern which would not qualify as a small concern or a nonprofit organization.

What is a Small Business Concern?

The PTO defines a small business as one whose number of employees, including those of its affiliates, does not exceed 500 persons. The definition also requires a small business, for this purpose, to be one which has not assigned, granted, conveyed, or licensed, and is under no obligation under contract or law to assign, grant, convey, or license, any rights in the invention to any person who could not be classified as an independent inventor if that person had made the invention, or to any concern which would not qualify as a small business concern or a nonprofit organization.

What is a Nonprofit Organization?

To be recognized as a nonprofit organization it must be so accredited by a nationally recognized accrediting agency or association or of the type described in Section 501 (c) (3) of the IRS Code of 1954 (26 U.S.C. 501 (c) (3)) and which is exempt from taxation under 26 U.S.C. 501 (a).

Facsimile Transmissions to the PTO

Effective November 1, 1988, certain papers to be filed in national patent applications and re-examination proceedings for consideration by the Office of the Assistant Commissioner for Patents, the Office of

Four PTO forms are available for claiming small entity status and are reproduced in the Forms and Documents section.

What Does "Patent Pending" Signify? These words are put on a product by a manufacturer to inform the public (and competition) that an application for patent on that item is on file at the PTO. It means stay away. The law imposes a fine on those who use these words falsely to deceive the public.

For facsimile transmissions, I recommend that that the Serial Number of the application or Control Number of the re-examination be entered as part of the sender's identification, if possible. The sending facsimile machine should also generate a report confirming transmission for each transmission session. The transmitting activity report should be retained along with the paper used as the original.

Informal communications via facsimile between applicant and the examiner, such as proposed claims for interview purposes, are permissible and encouraged. Informal communications from applicants will not be made of record in the application or re-examination and must be clearly identified as informal such as by including the word "DRAFT" on each paper. To facilitate informal communications from examiners, applicants are encouraged to supply their facsimile phone numbers on communications to the Office.

the Deputy Assistant Commissioner for Patents, and the Patent Examining Groups (Patent Examining Corps) may be submitted to the Patent and Trademark Office (PTO) by facsimile transmission.

The provision of 37 CFR 1.33(a), requiring signatures on amendments and other papers filed in applications, is waived to the extent that a facsimile signature is acceptable. The paper that is used as the original for the facsimile transmission must have an original signature, and should be retained by you or your representative as evidence of the content of the facsimile transmission. No special format, addressing information or written ratification is required for facsimile transmissions. However, the paper size must be 8 1/2 inches by 14 inches or smaller to be accepted.

A facsimile center has been established in the Patent Examining Corps to receive and process submissions. The filing date accorded the submission will be the date the complete transmission is received by the PTO facsimile unit unless that date is a Saturday, Sunday or Federal holiday within the District of Columbia, in which case the official date of receipt will be the next business day.

Each transmission session must be limited to papers to be filed in a single national patent application or re-examination proceeding. The papers, including authorizations to charge deposit accounts, which may be submitted using this procedure, are limited to those which may be filed in national patent applications and re-examination proceedings and which are to be considered by the PTO organizations named above. Examples of such papers are amendments, responses to restriction requirements, requests for reconsideration before the examiner, petitions, terminal disclaimers, powers of attorney, notices of appeal, and appeal briefs.

New or continuing patent applications of any type, assignments, issue fee payments, maintenance fee payments, declarations or oaths under 37 CFR 1.63 or 1.67, and formal drawings are excluded, as are all papers relating to international patent applications. Papers to be filed in applications that are subject to a secrecy order under 37 CFR 5.1-5.8, and directly related to the secrecy order content of the application, are also excluded.

The facsimile submissions may include a certificate for each paper stating the date of transmission. A copy of the facsimile submission with a certificate attached thereto will be evidence of transmission of the paper should the original be misplaced. The person signing the certificate should have a reasonable basis to expect that the paper would be facsimile transmitted on the date indicated. An example of a preferred certificate is:

Certification of Facsimile Transmission

I hereby certify that this paper is being facsimile transmitted to the Patent and Trademark Office on the date shown below.

_____

Type or print name of person signing certification

_____

Signature   Date

When possible, the certification should appear on a portion of the paper being transmitted. If the certification is presented on a separate paper, it must identify the application to which it relates, and the type of paper being transmitted, e.g., amendment, notice of appeal, etc.

In the event that the facsimile submission is misplaced or lost in the PTO, the submission will be considered filed as of the date of the transmission, if the party who transmitted the paper:

1) Informs the PTO of the previous facsimile transmission promptly after becoming aware that the submission has been misplaced or lost;

2) Supplies another copy of the previously transmitted submission with the Certification of the Transmission; and

3) Supplies a copy of the sending unit's report confirming transmission of the submission. In the event that a copy of the report is not available, the party who transmitted the paper may file a declaration under 37 CFR 1.68 which attests on a personal knowledge basis or to the satisfaction of the Commissioner to the previous timely transmission.

If all criteria above cannot be met, the PTO will require applicant to submit a verified showing of facts. Such a showing must show to the satisfaction of the Commissioner the date the PTO received the submission.

The facsimile center will have five facsimile units and will be staffed during the business hours of 8:30 a.m. and 5:00 p.m., Monday through Friday, excluding holidays. Although the units may normally be accessed at all times, including non-business hours, there may be times when reception is not possible due to equipment failure or maintenance requirements. Accordingly, applicants are cautioned not to rely on the availability of this service at the end of response periods.

The telephone number for accessing the facsimile machines is (703) 557-9564. In the event that the transmission cannot be accepted at the telephone number above, a backup number has been established at (703) 557-9567. The facsimile center staff can be reached at telephone number (703) 557-4277 during normal business hours.

Drawings

When you apply for a patent, it is required by law in most cases that you furnish a drawing of the invention together with your application. This includes practically all inventions except compositions of matter or processes, but a drawing may also be useful in the case of many processes. In the next chapter we'll take a closer look at patent drawings.

Patent Drawings:

WHAT THEY ARE AND HOW TO GET THEM DONE

The patent drawing must show every feature of the invention specified in your claims and is required by the PTO rules to be in a particular form. The Office specifies the size of the sheet on which the drawing is made, the type of paper, the margins, and other details relating to the making of the drawing. The reason for specifying the standards in detail is that the drawings are printed and published in a uniform style when the patent issues, and the drawings must also be such that they can be readily understood by persons using the patent descriptions.

No names or other identification will be permitted within the "sight" of the drawing, and applicants are expected to use the space above and between the hole locations to identify each sheet of drawings. This identification may consist of the attorney's name and docket number or the inventor's name and case number and may include the sheet number and the total number of sheets filed (for example, "sheet 2 of 4").

Drawings in colors other than black and white are not acceptable unless the drawing requirements are waived. Only the Deputy Assistant Commissioner for Patents can make the decision.

Color drawings are permitted for plant patents where color is a distinctive feature.

See Fig. 1, United States Patent for an example of format used in patent drawings.

Applications Filed Without Drawings

Not all applications require drawings. It has been a long approved procedure, for example, to accept a process case (i.e., a case having only process or method claims) which is filed without drawings.

Other situations where drawings are usually not considered essential for a filing date are:

1) **Coated articles or products.** If the invention resides only in coating or impregnating a conventional sheet, e.g., paper or cloth, or an article of known and conventional character with particular composition.

2) **Articles made from a particular material or composition.** If the invention consists in making an article of a particular material or composition, unless significant details of structure or arrangement are involved in the claims.

3) **Laminated structures.** If the invention involves only laminations of sheets (and coatings) of specified

Models are generally not required as part of an application or patent. They have not been for many, many years. And save for cases involving perpetual motion, or a composition of matter, you will probably never have to submit one.

However, if your invention is a microbiological invention a deposit of the microorganism is required.

If you decide to handle the drawings through a law firm, receive a guarantee that the drawings will pass PTO muster. If they are rejected for any technical reason whatsoever, you do not want to be double billed, i.e. pay for the draftsman's mistakes. On more than one occasion I have had to have extra charges taken off of my bills for the reworking of unacceptable drawings.

material, unless significant details of structure or arrangement (other than the mere order of layers) are involved in the claims.

4) **Articles, apparatus, or systems where sole distinguishing feature is presence of a particular material.** If the invention resides solely in the use of particular material in an otherwise old article, apparatus, or system recited broadly in its claim.

Photographs are not normally considered to be proper drawings. Photographs are acceptable for a filing date and are generally considered to be informal drawings. Photographs are only acceptable where they come within special categories. Photolithographs are never acceptable.

The PTO is willing to accept black and white photographs or photomicrographs (not photolithographs or other reproductions of photographs made using screens) printed on sensitized paper in lieu of India ink drawings, to illustrate inventions which are incapable of being accurately or adequately depicted by India ink drawings restricted to the following categories: crystalline structures, metallurgical microstructures, textile fabrics, grain structures, and ornamental effects.

Who Does the Drawing?

You have several options. You can do it yourself. To this end, standards for drawings are included below to guide your efforts and keep you to the letter of the required standards. I have never attempted to do the drawings for a utility patent, but I have been successful at doing a few uncomplicated design patent drawings.

You could ask that your patent attorney arrange for the drawings. This is all right, but understand that just as in the search business, the lawyer or law firm will be adding a premium of typically no less than 40 percent to the fees charged by the draftsman.

The surest and least expensive way, if you are not a draftsman, is to take bids and hire your own draftsman. Find the fair market price by calling around. Look in the telephone directory or ask at a regional patent library for candidates. Get the same guarantee for PTO acceptance from the draftsman if you contract the work yourself.

Standards for Patent Drawings

a) *Paper and ink.* Drawings must be made upon paper which is flexible, strong, white, smooth, nonshiny, and durable. Two-ply or three-ply bristol board is preferred. The surface of the paper should be calendered and of a quality which will permit erasure and correction with India ink. India ink, or its equivalent in quality, is preferred for pen drawings to secure perfectly black solid lines. The use of white pigment to cover lines is not normally acceptable.

b) *Size of sheet and margins.* The size of the sheets on which drawings are made may either be exactly 8 1/2 by 14 inches (21.6 by 35.6 cm.) or exactly 21.0 by 29.7 cm. (DIN size A4). All drawing sheets in a particular application must be the same size. One of the shorter sides of the sheet is regarded as its top.

1) On 8 1/2 by 14 inch drawing sheets, the drawing must include a top margin of 2 inches (5.1 cm.) and bottom and side margins of 1/4 inch (6.4 mm.) from the edges, thereby leaving a "sight" precisely 8 by 11 3/4 inches (20.3 by 29.8 cm.). Margin border lines are not permitted. All work must be included within the "sight." The sheets may be provided with two 1/4 inch (6.4 mm.) diameter holes having their centerlines spaced 11/16 inch (17.5 mm.) below the top edge and 2 3/4 inches (7.0 cm.) apart, said holes being equally spaced from the respective side edges.

2) On 21.0 by 29.7 cm. drawing sheets, the drawing must include a top margin of at least 2.5 cm., a left side margin of 2.5 cm., a right side margin of 1.5 cm., and a bottom margin of 1.0 cm. Margin border lines are not permitted. All work must be contained within a sight size not to exceed 17 by 26.2 cm.

c) *Character of lines.* All drawings must be made with drafting instruments or by a process which will give them satisfactory reproduction characteristics. Every line and letter must be durable, black, sufficiently dense and dark, uniformly thick and well defined; the weight of all lines and letters must be heavy enough to permit adequate reproduction. This direction applies to all lines however fine, to shading, and to lines representing cut surfaces in sectional views. All lines must be clean, sharp, and solid. Fine or crowded lines should be avoided. Solid black should not be used for sectional or surface shading.

d) *Hatching and shading.* 1) Hatching should be made by oblique parallel lines spaced sufficiently apart to enable the lines to be distinguished without difficulty. 2) Heavy lines on the shade side of objects should preferably be used except where they tend to thicken the work and obscure reference characters. The light should come from the upper left-hand corner at a 45-degree angle. Surface delineations should preferably be shown by proper-shading, which should be open.

e) *Scale.* The scale to which a drawing is made ought to be large enough to show the mechanism without crowding when the drawing is reduced in size to two-thirds in reproduction, and views of portions of the mechanism on a

For drawings, freehand work should be avoided wherever it is possible to do so.

The drawing should, as far as possible, be so planned that one of the views will be suitable for publication in the Official Gazette as the illustration of the invention.

larger scale should be used when necessary to show details clearly; two or more sheets should be used if one does not give sufficient room to accomplish this end, but the number of sheets should not be more than is necessary.

f) *Reference characters.* The different views should be consecutively numbered figures. Reference numerals (and letters, but numerals are preferred) must be plain, legible, and carefully formed, and not be encircled. They should, if possible, measure at least one-eighth of an inch (3.2 mm.) in height so that they may bear reduction to one twenty-fourth of an inch (1.1 mm.); and they may be slightly larger when there is sufficient room. They should not be so placed in the close and complex parts of the drawing as to interfere with a thorough comprehension of the same, and therefore should rarely cross or mingle with the lines. When necessarily grouped around a certain part, they should be placed at a little distance, at the closest point where there is available space, and connected by lines with the parts to which they refer. They should not be placed upon hatched or shaded surfaces but when necessary, a blank space may be left in the hatching or shading where the character occurs so that it shall appear perfectly distinct and separate from the work. The same part of an invention appearing in more than one view of the drawing must always be designated by the same character, and the same character must never be used to designate different parts. Reference signs not mentioned in the description shall not appear in the drawing and vice versa.

g) *Symbols, legends.* Graphic drawing symbols and other labeled representations may be used for conventional elements when appropriate, subject to approval by the Patent and Trademark Office. The elements for which such symbols and labeled representations are used must be adequately identified in the specification. While descriptive matter on drawings is not permitted, suitable legends may be used, or may be required, in proper cases, as in diagrammatic views and flow sheets or to show materials or where labeled representations are employed to illustrate conventional elements. Arrows may be required, in proper cases, to show direction of movement. The lettering should be as large as, or larger than, the reference characters.

h) *Views.* The drawing must contain as many figures as may be necessary to show the invention; the figures should be consecutively numbered if possible in the order in which they appear. The figure may be plain, elevation, section, or perspective views, and detail views of portions of elements, on a larger scale if necessary, may also be used. Exploded views, with the separated parts of the same figure embraced by a bracket, to show the relationship or order of assembly of various parts are permissible. When necessary, a view of a large machine or device in its entirety may be broken and extended over several sheets if there is no loss in facility of understanding the view. Where figures on two or more sheets form in effect a single complete figure, the figures on the several sheets should be so arranged that the complete figure can be understood by laying the drawing sheets adjacent to one another. The arrangement should be such that no part of any of the figures appearing on the various sheets are concealed and that the complete figure can be understood even though spaces will occur in the complete figure because of the margins on the drawing sheets. The plane upon which a sectional view is taken should be indicated on the general view by a broken line, the ends of which should be designated by numerals corresponding to the figure number of the sectional view and have arrows applied to indicate the direction in which the view is taken. A moved position may be shown by a broken line superimposed upon a suitable figure if this can be done without crowding, otherwise a separate figure must be used for this purpose. Modified forms of construction can only be shown in separate figures. Views should not be connected by projection lines nor should centerlines be used.

i) *Arrangement of views.* All views on the same sheet should stand in the same direction and, if possible, stand so that they can be read with the sheet held in an upright position. If views longer than the width of the sheet are necessary for the clearest illustration of the invention, the sheet may be turned on its side so that the top of the sheet with the appropriate top margin is on the right-hand side. One figure must not be placed upon another or within the outline of another.

j) *Extraneous matter.* Identifying indicia (such as the attorney's docket number, inventor's name, number of sheets, etc.) not to exceed 2 3/4 inches (7.0 cm.) in width may be placed in a centered location between the side edges within three- fourths inch (19.1 mm.) of the top edge. Authorized security markings may be placed on the drawings provided they be outside the illustrations and are removed when the material is declassified. Other extraneous matter will not be permitted upon the face of a drawing.

k) *Transmission of drawings.* Drawings transmitted to the Office should be sent flat, protected by a sheet of heavy binder's board, or may be rolled for transmission in a suitable mailing tube; but must never be folded. If

It is recommended that original drawings be done on bristol board so that they can be more easily corrected. Do not submit the original drawing: only paper copies of your drawings should be submitted to the Patent Office.

Good quality copies made on office copiers are acceptable if the lines are uniformly thick, black, and solid. Drawings are currently accepted in two different formats: A4 size (21 cm x 29.7 cm) for use in Patent Cooperation Treaty signatory countries; and the U.S. size, which is 8 1/2 x 14. The PTO requires that all drawings in a particular application be of the same size for ease of handling and reproduction.

Unless the examiner has approved the proposed changes, no changes will be permitted to be made on the original drawings other than correction of informalities.

received creased or mutilated, new drawings will be required.

How to Effect Drawing Changes

Often you may find yourself in need of changing a drawing after it has been submitted. For example, the PTO in checking it for standards may find that it contains mistakes, e.g. improper shading or a missing view. For this reason it is wise to keep copies of every drawing you submit. You have several options here. One is to file new drawings with the changes incorporated therein. The art unit number, serial number, and number of drawing sheets should be written on the reverse side of the drawings. You may delay filing the new drawings until receipt of the "Notice of Allowability" (PTOL-37). If delayed, the new drawings MUST be filed within the three months shortened statutory period set for response in the before-mentioned form. You may request an extension. Drawings should be filed as a separate paper with a transmittal letter addressed to the Official Draftsman.

If the mistakes are the fault of your draftsman, new drawings should be done free of charge. That is, unless the fault is yours for having left something out of the instructions, etc.

Changes to Originals Require Bonded Draftsmen

If, however, you opt to make some minor adjustments to the original drawings, such changes to an original can only be done by a PTO bonded draftsman. A bonded draftsman is the only one who can physically remove your original drawing for a time to rework it. You as the inventor can see it, but only a bonded draftsman can take it back to his or her office to do the work. **And note that the PTO will not release patent drawings to anyone if the patent application carries a filing date later than January 1, 1989.**

In any case, no changes will be permitted, other than correction of informalities, unless the examiner has approved the proposed changes.

New Regulations Effect Drawing Change Process

Under new regulations, the authorization of bonded patent draftsmen will expire as of January 1, 1991; thereafter, no one will be permitted to change original patent drawings. After this date, new drawings will have to be submitted should there be a need for changes or corrections in original patent drawings.

"The Patent Office is transferring the responsibility of correcting drawings to the applicant, by not permitting retrieval of the drawings from the Patent Office," explains Robert MacCollum, a bonded draftsman from Silver Spring, Maryland.

Should the bonded system be dismantled? MacCollum feels that the inventor should have access through a bonded system. "So I don't think that it's such a hot idea." On the other hand, he feels that not having it will do away with a lot of confusion and running around for his colleagues. "Some draftsmen dropped their bond because it was such a nuisance to have to go to the PTO in person and pick up these drawings. Also, with the bonded system, it is tough to estimate your work. It might take three trips to find and retrieve the drawings. This must be added into the cost of your work. Some out-of-towners will never send you that extra money."

List of Bonded Draftsmen

The following alphabetical roster of bonded draftsmen is current and complete as of publication. The inclusion of this list does not constitute a recommendation or endorsement by me or the PTO.

John A. Ballard
2001 Jefferson Davis Hwy.
Suite 705, Crystal Plz. 1
Arlingtion, VA 22202
(703) 685-7228

Anthony L. Costantino
17300 Lafayette Dr.
Olney, MD 20832
(301) 924-3491
Fleit, Jacobson, Cohn & Price

Jennifer Bldg.
400 7th St., N.W.
Washington, D.C. 20004
(202) 638-6666

Ellsworth G. Jackson
101 Rittenhouse St., N.E.
Washington, D.C. 20011
(202) 726-0908

Litman Law Offices, Ltd.
P.O. Box 15035
Crystal City Station
Arlington, VA 22215-1728
(301) 920-6004

Robert MacCollum
Patent Drafting Services
13108 Engelwood Dr.
Silver Spring, MD 20904
(301) 622-3940

Mason, Fenwick & Larence
1730 Rhode Island Ave., N.W.
Washington, D.C. 20036
(202) 295-2010

Thomas E. Melvin
TEM Patent Drafting
P.O. Box 15809
Arlington, VA 22215
(703) 349-8518

Mil-R Productions
3110 Mt. Vernon Ave.
Suite 100
Alexandria, VA 22305
(703) 548-3879

J. A. Mortenson
  Patent Drafting
P.O. Box 518
Northeast, MD 21901
(301) 287-8669

Gerald M. Murphy
P.O. Box 2098
Eads Street Station
Arlington, VA 22202
(703) 881-1928

Suzanne Nahmias
D S N Patent Drafting
P.O. Box 2431
Gaithersburg, MD 20879
(301) 869-0756

Edward J. Oliver
Oliver Patent Drafting Service
1205 Darlington St.
Forestville, MD 20747
(301) 336-0351

Patent Reproduction Co.
26 N St., SE.
Washington, D.C. 20003
(202)488-7096

Quality Patent Printing
P.O. Box 2404
556 S. 22nd St.
Arlington, VA 22202
(703) 892-6212

Quinn Patent Drawing Service
1416 Duke St.
Alexandria, VA 22314
(703) 521-5940

Philip Sweet T.A. and for
Oblon, Fisher, Spivak, et al.
1755 S. Jefferson Davis
Crystal Square Five, Suite 400
Arlington, VA 22202
(703) 521-5940

Do You Need A Patent Attorney?

The answer is probably yes. It is perfectly legal to prepare your own patent application. You can conduct your own proceedings in the Patent and Trademark Office. But, unless you are familiar with such matters and have studied them in detail, you could experience considerable difficulty and extreme frustration.

The preparation of an application for a utility patent and conducting proceedings at the PTO requires a knowledge of patent law and PTO practice, not to mention a familiarity with the scientific or technical matters related to a particular invention.

While a patent may be obtained in many cases by persons not skilled in such esoteric work, you would have no assurance that the patent awarded would adequately protect your particular invention. I, therefore, highly recommend that a qualified patent attorney be retained for your utility patent work. Most inventors employ the services of registered patent attorneys or patent agents.

The law gives the PTO the power to make rules and regulations governing conduct and the recognition of patent attorneys and agents to practice before the PTO. Persons who are not recognized by the PTO for this practice are not permitted by law to represent inventors in their patent actions.

The PTO will receive, and, in appropriate cases, act upon complaints against attorneys and agents. The Patent and Trademark Office has the power to disbar, or suspend from practicing before it, persons guilty of gross misconduct, etc., but this can only be done after a full hearing with the presentation of clear and convincing evidence concerning the misconduct.

The fees charged by patent attorneys and agents for their professional services are not subject to regulation by the PTO. Solid evidence of overcharging may afford a basis for PTO action, but the Office rarely intervenes in disputes concerning fees.

How Much Do Legal Services Cost?

It is not inexpensive to retain a patent attorney. The amount of time a patent attorney will put into any particular matter will depend a great deal upon the complexity of the invention. Below are some guidelines for costs, arranged according to the ordered sequence of steps a lawyer will take.

A caveat: Do not rely on any patent attorney for prototyping advice, manufacturing processes, or insights into the day-to-day complexities of marketing.

Get a handle on the photocopying charges in advance. Patent work generates a great deal of paper and it can be a gold mine for law firms. I found one large firm charging $.50 per photocopied page and sending three copies of each document (one for each of my partners). I quickly put a stop to this abuse, requesting one copy of each document. As with life, the little things add up.

Patent Search. The first thing a patent lawyer will rightfully suggest is that you authorize a patent search to see what, if any, prior art exists. The cost of a patent search will depend upon the scope of your patent. Rarely does a lawyer do the search; lawyers typically engage the services of a professional search firm and then step on the fees.

Patent Drawings. Lawyers do not personally draft patent drawings, either. They employ the services of a draftsman skilled in such matters. Sometimes the draftsmen are in-house staffers. Smaller practitioners use free-lancers, whose fees they step on by as much as 40 percent. In Washington, D.C., the average price charged by a bonded draftsman is $35 per hour. On average, three hours are needed to do one sheet. Patents comprise numerous sheets of drawings, depending upon the complexity of the invention. You can expect to pay approximately $100 per sheet.

Patent Application. To prepare a patent application, attorneys charge anywhere from $2,500 to $4,500, depending upon the complexity of the invention. The more complex the patent, the more you can expect to pay. On the other hand, simple inventions can be done for as little as $1,000.

It is a buyer's market. And just like anything else, it usually pays to shop around. Patent attorneys are able to give good estimates of expected charges once they see the scope of an invention and its claims.

When you have an acceptable estimate, get your attorney to agree in writing to that price and cap it off. If you do not do this, you may find yourself caught in what I call fee creep. Make a package deal. A price cap gives the lawyer no incentive to draw out the case.

Legal Forms. Lawyers charge fees to prepare the following forms; once again I remind you that you may reproduce from the Forms and Documents section and save yourself money in the process.

1. Declarations of Patent Application (Fig. 11).

2. Power of Attorney or Authorization of Agent, Not Accompanying Application (Fig. 12).

3. Revocation of Power of Attorney or Authorization of Agent (Fig. 12).

4. Assignment of Patent (Fig. 22).

5. Assignment of Patent Application (Fig. 22).

What to Look for in a Patent Attorney

Look for an attorney who specializes in your field of invention. Just as you would not hire a dermatologist to do the job of a cardiologist, even though both are licensed physicians, you would not want to hire a patent attorney with a background in mechanical engineering to do an electronic patent.

Larger firms offer quite an array of patent specialists from which to choose. But the smaller firms and many independent practitioners will often take anything that comes along. Do not be timid about requesting the technical qualifications of any attorney you are considering.

If your attorney cannot read your schematics, engineering drawings, or similar technical specifications, chances are your patent will not be nearly as complete and strong as it could be. Furthermore, you will be paying a premium to educate the attorney.

Bigger Is Not Always Better

At very large patent firms you will be small potatoes compared to lucrative corporate retainers. In such cases, your account may be assigned to a "spear carrier," who may be very good, but will seldom get the time to concentrate on your work. The primary responsibility of junior associates is to carry the workload for senior partners, the so-called "rainmakers," who bring in the accounts that pay for lunch, the health club, the limo, and kindred perks.

Big patent law firms make their real money defending the patents of corporate clients in court, not from two and three thousand dollar application jobs on behalf of independents. Experienced patent attorneys know that the chances are slim of an entrepreneur inventor being worth much more than what is earned on the patent application. The majority of independents are financial dead ends, while big corporate clients can be "cash cows." Seen from this perspective, you might have better chances elsewhere.

Attorneys and Agents Registered to Practice before U.S. Patent and Trademark Office

I cannot over stress that the preparation and prosecution of an application that will adequately protect your invention is an undertaking that requires knowledge of patent law and PTO practices. It also requires a knowledge of technical aspects of the invention. It is for this reason that I highly recommend that you hire only a registered patent attorney or agent to do your patent work. The PTO maintains a list of some 13,000 individuals authorized to represent inventors before the PTO.

To be admitted to this select register, a person must comply with the regulations prescribed by the PTO, which require showing that the person is of good moral character and repute, and that he or she has the legal and scientific credentials necessary to render a valuable service. Certain of these qualifications must be demonstrated by the passing of an examination. Those admitted to the examination must have a college degree in engineering or physical science or the equivalent of such a degree.

I am most comfortable with the independent specialist, even if this approach means a different attorney every time I change disciplines.

If you wish to register a complaint against an attorney or agent, contact Cam Weiffenbach, Director, Office of Enrollment, P.O. Box OED, U.S. Patent and Trademark Office, Washington, D.C. 20231; (703) 557-2012.

The PTO registers both attorneys at law and persons who are not members of the bar. The former persons are now referred to as "patent attorneys," and the latter persons are called "patent agents." Insofar as the work of preparing an application for patent and conducting the prosecution in the PTO is concerned, patent agents are typically as well qualified as patent attorneys, although patent agents **cannot** conduct patent litigation in the courts or perform various services which the local jurisdiction considers to be the practice of law. For instance, a patent agent would not be allowed to draw up a contract relating to a patent, such as an assignment or a license, if the agent resides in a state that considers such contracts as practicing law.

Legal Scam

Some individuals and organizations that are not registered advertise their services in the fields of patent searching and invention marketing and development. Such individuals and organizations **cannot** represent inventors before the PTO. Caveat emptor! Since they are not subject to PTO discipline, the Office cannot assist you in dealing with them.

While calling to acquire information on organizations that I felt might be helpful to inventors, I came across a nifty scam used by some lawyers to attract patent, trademark, and copyright clients. What appeared in telephone directories and source material to be professional councils specific to patents, trademarks, and copyrights were actually law offices! In one case, the phone numbers for three different councils led to the same law firm. The person listed as president of the three respective organizations was nothing more than a senior partner in the law firm.

The secretary who answered the phone pitched me on the fees charged for various services. This is a highbrow bait-and-switch that should be avoided.

National Council of Patent Law Associations

Patent attorneys who are active in the National Council of Patent Law Associations (NCPLA) should be particularly up-to-date on PTO matters. Established more than 35 years ago, NCPLA consists of some 40 local and regional patent law associations.

In addition to promoting exchange of information and lobbying, NCPLA keeps its members current on legislative and executive actions which might affect the nation's intellectual property system.

For more information on NCPLA, contact your nearest member association, or write to:

NCPLA, Crystal Plaza 3, Room 1DO1
2021 Jefferson Davis Highway
Arlington, VA 22202
Tel: (202) 659-1302

Getting in Touch With Local NCPLA Chapters

Here are the current NCPLA officers, addresses, and telephone numbers.

CALIFORNIA

Los Angeles Patent Law
Association

Don W. Martens
Knobbe, Martens, Olson & Bear
620 Newport Center Drive,
Suite1600
Newport Beach, CA 92660-8016
(714) 760-0404

William J. Robinson
Poms, Smith, Lande & Rose
2121 Avenue of the Stars
Suite 1400
Los Angeles, CA 90067-5010
(213) 277-8141

Orange County Patent Law
Association

Howard J. Klein
Klein & Szekeres
4199 Campus Drive, Suite 700
Irvine, CA 92715
(714) 854-5502

Bruce B. Brunda
Stetina & Brunda
24221 Calle De La Louisa, Suite 401
Laguna Hills, CA 92653
(714) 855-1246

Peninsula Patent Law Association of
California

Kate Murashige
Cotti, Murashige, Irell & Manella
545 Middlefield Road, Suite 200
Menlo Park, CA 94025
(415) 327-7250

Henry K. Woodward
Flehr, Hohbach, Test, Albritton &
Herbert
Four Embarcadero Center, Suite
3400

San Francisco, CA 94111
(415) 326-0747

San Francisco Patent and Trademark
Association

Harry A. Pacini
Stauffer Chemical Co.
1200 South 47th Street, Box 4023
Richmond, CA 94804-0023
(415) 231-1202

Neil A. Smith
Limbach, Limbach and Sutton
2001 Ferry Building
San Francisco, CA 94111
(415) 433-4150

CAROLINA

Carolina Patent, Trademark and
Copyright Law Association

John B. Hardaway III
Bailey & Hardaway
125 Broadus Avenue
Greenville, SC 29601
(803) 233-1338

Howard A. MacCord, Jr.
Burlington Industries, Inc.
P.O. Box 21207
Greensboro, NC 27420
(919) 379-4517

COLORADO

Colorado Bar Association—Patent,
Trademark, and Copyright
Section

Kenneth L. Richardson
Solar Energy Research Institute
1617 Cole Blvd.
Golden, CO 80401
(303) 231-7724

The *Inventing and Patenting Sourcebook*, published by Gale Research Inc., provides a full listing of the 13,000 registered patent attorneys and agents authorized to represent inventors before the PTO. Look to your local library for a copy.

The PTO cannot recommend any particular attorney or agent, or aid in the selection of an attorney or agent, as by stating in response to inquiry that a named patent attorney, agent, or firm is "reliable" or "capable."

CONNECTICUT

Albert W. Hilburger, Perman and
Green
425 Post Road
Fairfield, CT 06430
(203) 259-1800

Edward R. Hyde
261 Danbury Road
P.O. Box 494
Wilton, CT 06897
(203) 762-5444

Connecticut Patent Law Association

Howard S. Reiter
Corporate Patent Counsel
Colt Industries, Inc.
Charter Oak Boulevard
West Hartford, CT 06110-0651
(203) 236-0651

F. Eugene Davis IV
184 Atlantic Street
Stamford, CT 06905
(203) 324-9662

DISTRICT OF COLUMBIA

Bar Association of D.C.—Patent,
Trademark, and Copyright
Section

Archie W. Umphlett
Phillips Petroleum Co.
1825 K Street, N.W.
Washington, DC 20006
(202) 785-1252

Robert G. Weilacher
Beveridge, DeGrandi & Weilacher
Federal Bar Building West
1819 H Street, N.W.
Washington, DC 20006
(202) 659-2811

D.C. Bar
Howard D. Doescher
Cushman, Darby & Cushman
1615 L Street, N.W.
Washington, DC 20036
(202) 861-3000

Herbert C. Wamsley
2725 Fort Scott Drive
Arlington, VA 22202
(202) 466-2396

FLORIDA

South Florida Patent Law Association

Jack E. Dominik
Dominik & Saccocio
6175 N.W. 153rd Street, Suite 225
Miami Lakes, FL 33014
(305) 556-9889

Henry W. Collins
Patent Counsel
Cordis Corporation
P.O. Box 025700
Miami, FL 33102-5700
(305) 551-2707

GEORGIA

State Bar of Georgia—Patent,
Trademark, and Copyright
Section
Todd Devau
Hurt, Richardson, Garner, Todd &
Cadenhead
1400 Peachtree Place Tower
999 Peachtree Street, N.E.
Atlanta, GA 30309-3999

ILLINOIS

Patent Law Association of Chicago

Ronald B. Coolley
Arnold, White & Durkee
800 Quaker Tower
321 North Clark Street
Chicago, IL 60610
(312) 744-0090

Ladas & Parry
104 S. Michigan Avenue
Chicago, IL 60603
(312) 236-9021

INDIANA

Indiana State Bar
Association—Patent, Trademark, and
Copyright Section

Steven T. Belsheim
322 Main Street, Suite 100
Clarksville, TN 37040

Gilbert E. Alberding
Ball Corporation
Legal Department
345 S. High Street
Muncie, IN 47302
(317) 747-6422

IOWA

Iowa Patent Law Association
H. Robert Henderson
Henderson & Sturm
1213 Midland Financial Building
Des Moines, IA 50309
(515) 288-9589

MARYLAND

Maryland Patent Law Association

Frank E. Robbins
Robbins & Laramie
1919 Pennsylvania Ave., N.W.
Washington, DC 20006
(202) 887-5050

Jim Haight
Mackler and Associates
P.O. Box 2187
Gaithersburg, MD 20879
(202) 842-1690

MASSACHUSETTS

Boston Patent Law Association
John M. Skenyon
Fish & Richardson
One Financial Center
Boston, MA 02111
(617) 542-5070

Jacob N. Erlich
424 Trapelo Road, Bldg. 104
Waltham, MA 02154
(617) 377-4072

MICHIGAN

Michigan Patent Law Association

Robert F. Hess
Federal-Mogul Corporation
P.O. Box 1966
Detroit, MI 48235
(313) 354-9926

George A. Grove
General Motors Corp., Patent
Section
P.O. Box 33114
Detroit, MI 48232
(313) 974-1322

Saginaw Valley Patent Law
Association

James Bittell
Dow Corning Corporation
Patent Department - C01232
P.O. Box 994
Midland, MI 48686-0994
(517) 496-5882

Robert Spector
Dow Corning Corporation
Patent Department - C01232
P.O. Box 994
Midland, MI 48686-0994
(517) 496-5523

State Bar of Michigan—Patent,
Trademark, and Copyright
Section

Stephen A. Grace
Dow Chemical
Corporate Center, Bldg. 1776
Midland, MI 48674
(517) 636-3052

Robert A. Armitage
The Upjohn Company
301 Henrietta Street
Kalamazoo, MI 49001
(616) 385-7345

MINNESOTA

Minnesota Intellectual Property Law Association

Terryl K. Qualey
Office of Patent Counsel - 3M
P.O. Box 33427
St. Paul, MN 55133-3427
(612) 733-1940

Richard E. Brink
3M Company
P.O. Box 33427
St. Paul, MN 55133
(612) 733-1517

MISSOURI

Bar Association of Metropolitan St. Louis—Patent,
Trademark, and Copyright Section

Veo Peoples, Jr.
1221 Locust Street, Suite 100
St. Louis, MO 63103
(314) 231-9775

Edward H. Renner
Cohn, Powell, & Hind, P.C.
7700 Clayton Road, Suite 103
St. Louis, MO 63117
(314) 645-2442

NEW JERSEY

New Jersey Patent Law Association
Raymond M. Speer
Merck & Co., Inc.
P.O. Box 2000
Rahway, NJ 07065-0907
(201) 574-4481

New Jersey State Bar Association—Patent, Trademark, and Copyright Section

Arthur J. Plantamura
10 Butterworth Drive
Morristown, NJ 07960
(201) 455-3781

Teresa Cheng
Merck & Co., Inc.
P.O. Box 2000
Rahway, NJ 07068-0907
(201) 574-4982

NEW YORK

Central New York Patent Law Association

John S. Gasper
Patent Operations (Dept. N50)
IBM Corporation
1701 North Street -
 Bldg. 251-2
Endicott, NY 12188
(607) 755-3342

Eastern New York Patent Law Association

John W. Harbour
General Electric Company
Silicone Production Business Division
Waterford, NY 12188
(518) 266-2471

William S. Teoli
General Electric Company
Corporate Research & Development
P.O. Box 8
Building K1, Room 3A68
Schenectady, NY 12302
(518) 387-5872

New York Patent, Trademark & Copyright Law Association

David H.T. Kane
Kane, Dalsimer, Sulivan, Kurucz, Levy, Eisele & Richard
711 Third Avenue, 20th Floor
New York, NY 10017
(212) 687-6000

Douglas W. Wyatt
Wyatt, Gerber, Shoup, Scobey & Badie
261 Madison Avenue
New York, NY 10016
(212) 687-0911

Niagara Frontier Patent Law
Association
William G. Gosz
Occidental Chemical Corporation
P.O. Box 189
Niagara Falls, NY 14302
(716) 773-8459

Rochester Patent Law Association

Stephen B. Salai
850 Crossroads Office Building
Rochester, NY 14614
(716) 325-5553

Paul F. Morgan
Xerox Corporation
Xerox Square - 020
Rochester, NY 14644
(716) 423-3015

OHIO

Cincinnati Patent Law Association

Leonard Williamson
The Procter & Gamble Company
Patent Division
11520 Reed Hartman Highway
Cincinnati, OH 45241
(513) 530-3387

Eric W. Guttag
The Procter & Gamble Company
P.O. Box 39175
Cincinnati, OH 45247
(513) 659-2736

Cleveland Patent Law Association

James V. Tura
Pearne, Gordon, McCoy & Granger
1200 Leader
Cleveland, OH 44114
(216) 579-1700

Edwin W. (Ned) Oldham
Oldham, Oldham, Webber Co., L.P.A.
Twin Oaks Estate
1225 West Market Street
Akron, OH 44313
(216) 864-5550

Columbus Patent Law Association

Robert B. Watkins
Kremblas, Foster, Millard & Watkins
2941 Kenny Road
Columbus, OH 43221
(614) 457-5700

Richard C. Stevens
1400 First National Plaza
Dayton, OH 45402
(513) 223-2050

Dayton Patent Law Association

William Weigl
Hobart Corporation
World Headquarters
Troy, OH 45374
(513) 332-2111

Joseph J. Nauman
Biebel, French & Nauman
2500 Kettering Tower
Dayton, OH 45423
(513) 461-4543

Ohio State Bar Association—Patent,
Trademark, and
Copyright Section

Frank Foster
Kremblas, Foster, Millard & Watkins
50 West Broad Street
Columbus, OH 43215
(614) 464-2700

Ralph Jocke
Parker Hannifin Corporation
17325 Euclid Avenue
Cleveland, OH 44112
(216) 531-3000

Toledo Patent Law Section

Oliver E. Todd, Jr.
Champion Spark Plug Company
P.O. Box 910
Toledo, OH 43661
(419) 535-2364

David D. Murray
Willian Brinks Olds Hofer
Gilson & Lione
930 National Bank Building
Toledo, OH 43604
(419) 244-6578

OKLAHOMA

Oklahoma Bar Association—Patent, Trademark, and Copyright Section

Gary Peterson
Dunlap, Codding & Peterson
9400 North Broadway,
 Suite 420
Oklahoma City, OK 73114
(405) 478-5344

Allen W. Richmond
Phillips Petroleum Company
208 PLB
Bartlesville, OK 74006
(918) 661-0561

OREGON

Oregon Patent Law Association

William O. Geny
Chernoff Vilhauser McCung & Stenzel
600 Benjamin Franklin Plaza
1 S.W. Columbia
Portland, OR 97258
(503) 227-5631

Francine Gray
Kolisch, Hartwell & Dickenson
200 Pacific Building
520 S.W. Yamhill St.
Portland, OR 97204
(503) 224-6655

PENNSYLVANIA

Allegheny County Bar Association—Intellectual Property Section

Michael D. Fox
Berkman, Ruslander, Pohl, Lieber & Engle
One Oxford Centre
Pittsburgh, PA 15219
(412) 261-6161

John W. Jordan IV
Grigsby, Gaca & Davies
One Gateway Center, 10th Floor
Pittsburgh, PA 15222
(412) 281-0737

Philadelphia Patent Law Association

Eugene G. Seems
FMC Corporation
2000 Market Street
Philadelphia, PA 19103
(215) 299-6971

Paul F. Prestia
500 North Gulph Road
Valley Forge, PA 19482
(215) 265-6666

Intellectual Property Law Association of Pittsburgh

U.S. Steel Corporation
600 Grant Street
Pittsburgh, PA 15230
(412) 422-2873

Patrick J. Viccaro
Allegheny Ludlum Steel Corp.
1000 Six PPG Place
Pittsburgh, PA 15272
(412) 394-2839

TEXAS

Austin Intellectual Property Law Association

Dudley R. Dobie, Jr.
Fulbright and Jaworski
600 Congress, Suite 2400
Austin, TX 78701
(512) 474-5201

Andrea P. Bryant
IBM Corporation
11400 Burnet Road
Austin, TX 78758
(512) 838-1003

Dallas-Ft. Worth Patent Law Association

Charles Gunter, Jr.
Felsman, Bradley, Gunter & Kelly

2850 Continental Plaza
777 Main Street
Ft. Worth, TX 76102
(817) 332-8143

Robert A. Felsman
Felsman, Bradley, Gunter & Kelly
2850 Continental Plaza
777 Main Street
Ft. Worth, TX 76102
(817) 332-8143

Houston Intellectual Property Law
Association

Kenneth E. Kuffner
Arnold, White & Durkee
P.O. Box 4433
Houston, TX 77210
(713) 789-7600

David A. Rose
Butler A. Binion
1600 Allied Bank Plaza
Houston, TX 77002
(713) 237-3640

State Bar of Texas—Intellectual
Property Law Section

Robert A. Felsman
Felsman, Bradley, Gunter & Kelly
2850 Continental Plaza
777 Main Street
Fort Worth, TX 76102
(817) 332-8143

William L. LaFuze
Vinson & Elkins
3300 First City Tower
Houston, TX 77002-6760
(712) 236-2595

UTAH

State Bar of Utah—Patent,
Trademark, and Copyright Section

John Christiansen
Van Cott, Bagley, Cornwall &
McCarthy
50 South Main Street, Suite 1600
Salt Lake City, UT 84144
(801) 532-3333

Allen R. Jensen
Workemen, Nydegger & Jensen
American Plaza II, Third Floor
57 West 200 South
Salt Lake City, UT 84101
(801) 522-9800

VIRGINIA

Virginia State Bar—Patent,
Trademark, and Copyright
Section

Anthony J. Zelano
Millen and White
1911 Jefferson Davis Highway
Arlington, VA 22202
(703) 892-2200

James H. Laughlin, Jr.
Benoit, Smith & Laughlin
2001 Jefferson Davis Highway
Arlington, VA 22202
(703) 521-1677

WASHINGTON

Washington State Patent Law
Association

James P. Hamley
The Boeing Company
P.O. Box 3707, Mail Stop 7E-25
Seattle, WA 98124-2207
(206) 251-0262

WISCONSIN

Wisconsin Intellectual Property Law
Association

Ramon A. Klitzke,
 Professor of Law
Marquette University Law School
1103 West Wisconsin Avenue
Milwaukee, WI 53233
(414) 224-7094

Howard W. Bremer
Wisconsin Alumni Research
Foundation
P.O. Box 7365
Madison, WI 53707
(608) 263-2831

The Patent and Trademark Office

"The Patent Office is a curious record of the fertility of the mind of man when left to its own resources..." wrote an English lady who visited American in 1828.

The Patent and Trademark Office (PTO) administers the patent laws as they relate to the granting of patents for inventions, and performs other duties vis-a-vis patents. It examines applications for patents to determine if the applicants are entitled to patents under the law, and grants patents when they are so entitled; it publishes issued patents; records assignments of patents; maintains a search room for the use of the public to examine issued patents; and supplies copies of records and other papers. The PTO has no jurisdiction over questions of infringement and the enforcement of patents, nor over matters relating to the promotion or utilization of patents or inventions.

Today the PTO has 3,200 employees, of whom approximately half are examiners and others with technical and legal training. More than 125,000 patent applications are received annually. Over five million pieces of mail arrive each year at the PTO.

It is no surprise, then, that the PTO is understaffed and the examiners overloaded with work in certain areas. The situation was mitigated somewhat in 1989 with the hiring of 283 new patent examiners, for a total of 1,446 examiners (excluding design examiners and supervisors).

General Information and Correspondence

The Patent and Trademark Office is a huge bureaucracy. Without a map and compass, so to speak, it can be frustrating or near impossible for a beginner (especially one living outside the National Capital area), to find out who handles what and how to contact any particular person directly. To help you quickly reach the most appropriate person for your needs, see the chapter, Patent and Trademark Office Telephone Directory.

If you wish to mail a letter to someone at the Patent and Trademark Office, address your letter to the addressee care of: Patent and Trademark Office, Washington, D.C. 20231.

Since the PTO does not have resources for picking up any mail, including Express Mail, Post Office-to-Post Office Express Mail will not reach the Patent and Trademark Office. Express mail sent to a specific street address and office will arrive.

An illustration of PTO understaffing: in the biotechnology field, the shortage of examiners who understand genetic engineering is severe. The PTO has been criticized for taking up to four years to decide biotechnology cases.

More than 4.5 million patents have been awarded since the PTO was established in 1802. In Fiscal Year 1989, a record 163,306 patent applications were filed. The Office issued 102,712 patents, with utility, plant and reissue patents accounting for 97,000 of the total. Nearly 6,000 design patents were issued.

Special PTO mail box numbers should be used to allow forwarding of particular types of mails to the appropriate areas as quickly as possible. Mail will be forwarded directly to the appropriate area without being opened. If a document other than the specified type identified for each box is addressed to that box, expect delays in correct delivery.

Special Post Office Boxes

Commissioner of Patents and Trademarks, Washington, D.C. 20231

Box 3 —Mail for the Office of Personnel from NFC.

Box 4 —Mail for the Assistant Commissioner for External Affairs, or the Office of Legislation and International Inquiries.

Box 5 —Non-fee mail related to trademarks.

Box 6 —Mail for the Office of Procurement.

Box 7 —Reissue applications for patents involved in litigation.

Box 8 —All papers for Office of Solicitor except for letters relating to pending litigation.

Box 9 —Coupon orders for U.S. patents and trademarks.

Box 10 —Orders for certified copies of patents and trademark applications.

Box 11 —Electronic Ordering Services (EOS).

Box AF —Expedited procedure for processing amendments and other responses after final rejection.

Box FWC —Requests for File Wrapper Continuation Applications.

Box Issue Fee —Issue Fee Transmittals and associated fees and corrected drawings.

Box M Fee —Letters related to interferences and applications and patents involved in interference.

Box Non Fee —Non-fee amendments to patent applications.

Amendment

Box Pat. Ext. —Applications for patent term extensions.

Box PCT —Mail related to applications filed under the Patent Cooperation Treaty.

Box reexam —Mail related to re-examination.

Patent

Application —New patent application and associated papers and fees.

Trademark

Application —New trademark application and associated papers and fees.

How to Correspond Efficiently

Separate letters (but not necessarily in separate envelopes) should be written in relation to each distinct subject of inquiry, such as assignments, payments, orders for printed copies of patents, orders for copies of records, and requests for other services. None of these should be included with letters responding to Office actions in applications.

If your letter concerns a patent application, include the serial number, filing date, and Group Art Unit number. When a letter concerns a patent, include the name of the patentee, the title of the invention, the patent number, and date of issue.

If ordering a copy of an assignment, provide the book and page or reel and frame record, as well as the name of the inventor; otherwise, the PTO will assess an additional charge to cover the time consumed in making the search for the assignment.

The PTO will not send or show you applications for patents. They are not open to the public, and no information concerning patent applications is released except on written authority of the applicant, assignee, or designated attorney, or when necessary to the conduct of the PTO's business. You can write for and receive, however, records of any decisions, the records of assignments other than those relating to assignment of patent applications; books; or other records and papers in the PTO that are open to the public.

The PTO will not respond to inquiries concerning the novelty and patentability of an invention in advance of the filing of an application; give advice as to possible infringement of a patent; advise of the propriety of filing an application; respond to inquiries as to whether or to whom any alleged invention has been patented; or act as an expounder of the patent law or as a counselor for individuals, except in deciding questions arising before it in regularly filed cases.

Keep copies of everything you send the PTO. Don't forget to put your name and return address on all papers and the envelope.

In Case of Emergency...

The PTO has established a contingency plan for filing any paper or paying any fee in the Office in the event of an emergency caused by any major interruption in the country's mail service.

The Foreign Patent Document Collection consists of 25 to 30 million documents. The paper portion of the collection occupies more than 10,000 square feet of space, while the microfilm collection has grown to 50,000 reels.

Officers and employees of the PTO are prohibited by law from applying for a patent or acquiring, directly or indirectly, except by inheritance or bequest, any patent or any right or interest in any patent.

The Patent Office was completely destroyed by fire on December 15, 1836. The loss was estimated at 7,000 models, 9,000 drawings, and 230 books. More serious was the loss of all records of patent applications and grants.

Upon the determination by the Commissioner of Patents and Trademarks that such an emergency exists, the Commissioner will cause to be printed a notice of the plan in the *Wall Street Journal* and make it available by telephone at (703) 557-3158. Also, certain publications, patent bar groups, and other organizations closely associated with the patent system will be notified. Termination of the emergency program will be similarly announced. Where the postal emergency is not nationwide, the Commissioner will designate the areas of the country in which the procedures outlined will be in effect.

The plan calls for the U.S. Department of Commerce District Offices to be designated as emergency receiving stations for filing papers and paying fees in the PTO.

What Happens to Your Application at the PTO?

If your application passes initial muster, it will be assigned to the appropriate examining group, and then to an examiner. Applications are handled in the order received.

The application examination inspects for compliance with the legal requirements and includes a search through U.S. patents, prior foreign patent documents which are available in the PTO, and available literature to ensure that the invention is new. A decision is reached by the examiner in light of the study and the result of the search.

First Office Action

You or your attorney will be notified of the examiner's decision by what the PTO refers to as an "action." An action is actually a letter that gives the reasons for any adverse action or any objection or requirement. Noted will be any appropriate references or information that you'll find useful in making the decision to continue the prosecution of the application or to drop it.

If the invention is not considered patentable subject matter, the claims will be rejected. If the examiner finds that the invention is not new, the claims will be rejected; but the claims may also be rejected if they differ somewhat from what is found to be obvious. It is not uncommon for some or all of the claims to be rejected on the examiner's first action; very few applications sail through as first submitted.

Your First Response

Let's say the examiner gives you the thumbs down on all or some of your claims. Your next move, if you wish to continue prosecuting the patent, is to respond, specifically pointing out the supposed errors in the examiner's action. Patent examiners have a lot on their plates and their units are typically understaffed for the amount of work they handle. For example, the PTO reported in November of 1988 that 6,500 biotechnology patents were awaiting a decision, or about 70 per government examiner. In this area alone the PTO predicts that the number of applications will grow by 12 percent a year through the early 1990s.

Examiners must process a specific number of patents to be considered productive by their superiors for periodic job performance ratings. The bottom line is that as careful as they try to be, they make mistakes that can be reversed with careful argument.

Harry F. Manbeck, Jr. was nominated to be Assistant Secretary and Commissioner of Patents and Trademarks by President Bush in late 1989 and confirmed by the Senate in March of 1990. Manbeck joined the General Electric Company in 1949 and advanced to become General Patent Counsel in 1970, the position he held until entering the U.S. Patent and Trademark Office. A native of Pennsylvania, Manbeck has served as Chairman of the Patent, Trademark and Copyright section of the American Bar Association; President of the Association of Corporate Patent Counsel; a Director the Intellectual Property Owners, Inc., and a Director of the Bar Association of the Court of Appeals for the Federal Circuit. He is also a member of the American Intellectual Property Law Association and the Council on Patents, Trademarks and Copyrights of the U.S. Chamber of Commerce.

Your response should address every ground of objection and/or rejection. Show where the examiner is wrong. The mere allegation that the examiner has erred is not enough.

Your response will cause the examiner to reconsider, and you'll be notified if the claims are rejected, or objections or requirements made, in the same manner as after the first examination. This second Office action usually will be made final.

Feel free to call your examiner up on the telephone to discuss your case. I have always found them to be most hospitable and helpful. His or her telephone number will appear at the end of the Office action or you can look it up in the Patent and Trademark Office Telephone Directory chapter.

Depending upon how serious the matter, you or your attorney might wish to make an appointment to personally visit the examiner. Don't just drop in unannounced. It is to your benefit that the examiner has the time to prepare for your visit and get up to speed on the case. Remember that personal interviews do not remove the necessity for response to Office actions within the required time, and the action of the Office is based solely on the written record.

Final Rejection

On the second or later consideration, the rejection of claims may be made final. Your response is then limited to appeal and further amendment is restricted. You may petition the Commissioner in the case of objections or requirements not involved in the rejection of any claim. Response to a final rejection must include cancellation of, or appeal from the rejection of each claim so rejected and, if any claim stands allowed, compliance with any requirement or objection as to its form.

In determining such final rejection, your examiner will repeat or state all grounds of rejection then considered applicable to your claims as stated in the application.

The odds? As in the case of the examination by the Office, patents are granted in the case of about two out of every three applications filed.

Making Amendments to Your Application

The preceding section referred to amendments to an application. Following are some details concerning amendments:

1) The applicant may amend before or after the first examination and action as specified in the rules, or when and as specifically required by the examiner.

2) After final rejection or action, amendments may be made canceling claims or complying with any requirement of form which has been made but the admission of any such amendment or its refusal, and any proceedings rela-

---

A sample Examiner's Action is shown in the Forms and Documents section.

If the number or nature of the amendments render it difficult to consider the case, or to arrange the papers for printing or copying, the examiner may require the entire specification or claims, or any part thereof, to be rewritten.

tive thereto, shall not operate to relieve the application from its condition as subject to appeal or to save it from abandonment.

3) If amendments touching the merits of the application are presented after final rejection, or after appeal has been taken, or when such amendment might not otherwise be proper, they may be admitted upon a showing of good and sufficient reasons why they are necessary and were not earlier presented.

4) No amendment can be made as a matter of right in appealed cases. After decision on appeal, amendments can only be made as provided in the rules.

5) The specifications, claims, and drawing must be amended and revised when required, to correct inaccuracies of description and definition of unnecessary words, and to secure correspondence between the claims, the description, and the drawing.

All amendments of the drawings or specifications, and all additions thereto, must conform to at least one of them as it was at the time of the filing of the application. Matter not found in either, involving a departure from or an addition to the original disclosure, cannot be added to the application even though supported by a supplemental oath or declaration, and can be shown or claimed only in a separate application.

The claims may be amended by canceling particular claims, by presenting new claims, or by amending the language of particular claims (such amended claims being in effect new claims). In presenting new or amended claims, the applicant must point out how they avoid any reference or ground rejection of record which may be pertinent.

No change in the drawing may be made except by permission of the Office. Permissible changes in the construction shown in any drawing may be made only by the bonded draftsmen. A sketch in permanent ink showing proposed changes, to become part of the record, must be filed for approval by the Office before the corrections are made. The paper requesting amendments to the drawing should be separate from other papers.

The original numbering of the claims must be preserved throughout the prosecution. When claims are canceled, the remaining claims must not be renumbered. When claims are added by amendment or substituted for canceled claims, they must be numbered by the applicant consecutively beginning with the number next following the highest numbered claim previously presented. When the

To better organize your request for amendment, a copy of the PTO's Amendment Transmittal Letter format appears as Fig. 20.

Erasures, additions, insertions, or alterations of the papers and records must not be made by the applicant. Amendments are made by filing a paper, directing or requesting that specified changes or additions be made. The exact word or words to be stricken out or inserted in the application must be specified and the precise point indicated where the deletion or insertion is to be made.

Amendments are "entered" by the Office through the making of proposed deletions by drawing a line in red ink through the word or words canceled and by making the proposed substitutions or insertions in red ink, small insertions being written in at the designated place and larger insertions being indicated by reference.

To obtain a copy of any United States Patent, just send the patent number and a check or money order in the amount of $1.50 to:

U.S. Patent and Trademark Office Washington, D.C. 20231 Att: Copy Sales

An alternative to appeal in situations where you wish consideration of different claims or further evidence is to file a new continuation application. This requires a filing fee and the claims and evidence for which consideration is desired.

Keep copies of everything you submit. If you ask the PTO for copies of what you've submitted, there will be a charge for the service.

Channels of Ex Parte Review are illustrated as Fig. 21

application is ready for allowance, the examiner, if necessary, will renumber the claims consecutively in the order in which they appear or in such order as may have been requested by applicant.

Time for Response and Abandonment

The maximum period given for response is six months, but the Commissioner has the right to shorten the period to no less than thirty days. The typical response time allowed to an Office action is three months. If you want a longer time, you usually have to pay some extra money for an extension. The amount of the fee depends upon the response time desired. If you miss any target date, your application will be abandoned by the PTO and made no longer pending. However, if you can show whereby your failure to prosecute was unavoidable or unintentional, the application can be revived by the Commissioner.

The revival requires a petition to the Commissioner and a fee for petition, which should be filed without delay. The proper response must also accompany the petition if it has not yet been filed.

How to Make Appeals

If the examiner circles his or her wagons and begins to stonewall, there is a higher court. Rejections that have been made final may be appealed to the Board of Patent Appeals and Interferences. This august body consists of the Commissioner of Patents and Trademarks, the Deputy Commissioner, the Assistant Commissioners, and the examiners-in-chief, but typically each appeal is heard by only three members. An appeal fee is required and you must file a brief in support of your position. You can even get an oral hearing if you pay enough.

If the Board goes against you, there is yet a higher court, the Court of Appeals for the Federal Circuit. Or you might file a civil action against the Commissioner in the U.S. District Court for the District of Columbia. He won't take it personally. It goes with the territory. The Court of Appeals for the Federal Circuit will review the record made in the Office and may affirm or reverse the Office's action. In a civil action, you may present testimony in the court, and the court will make a decision.

What Are Interference Proceedings?

Parallel development is a phenomenon that should not be discounted. On numerous occasions a company executive has said to me, "I've seen that concept twice in the last month," or something to this effect. At times, two or more applications may be filed by different inventors claiming substantially the same patentable invention. A patent can only be granted to one of them, and a proceeding known as an "interference" is instituted by the Office to determine who is the first inventor and entitled to the patent. About one percent of all applications filed become engaged in an interference proceeding.

Interference proceedings may also be instituted between an application and a patent already issued, if the patent has not been issued for more than one year prior to the filing of the conflicting application, and if that the conflicting application is not barred from being patentable for some other reason.

The priority question is determined by a board of three examiners-in-chief on the evidence submitted. From the decision of the Board of Patent Appeals and Interferences, the losing party may appeal to the Court of Appeals for the Federal Circuit or file a civil action against the winning party in the appropriate U.S. district court.

The terms "conception of the invention" and "reduction to practice" are encountered in connection with priority questions. "Conception of the invention" refers to the completion of the devising of the means for accomplishing the result. "Reduction to practice" refers to the actual construction of the invention in physical form. In the case of a machine it includes the actual building of the machine. In the case of an article or composition it includes the actual carrying out of the steps in the process; actual operation, demonstration, or testing for the intended use is usually required. The filing of a regular application for patent completely disclosing the invention is treated as equivalent to reduction to practice. The inventor who proves to be the first to conceive the invention and the first to reduce it to practice will be held to be the prior inventor, but more complicated situations cannot be stated this simply.

Here is a case when it is important to be able to have evidence that proves when you first had an idea and when the prototype was made. This is why you should keep careful and accurate records throughout the development of an idea. The Disclosure Document Program was established by the PTO for this purpose.

The Patent Worksheet (fig. 23 in the Forms and Documents section) was designed to help inventors keep track of when ideas originated; for information on the Disclosure Document Program, see that chapter.

Allowance and Award of Patents

If your utility patent is found to be allowable, a notice of allowance will be sent to you or your attorney. Within three months from the date of the notice you must pay an issue fee.

What Rights Does a Patent Give You?

It's a pretty exciting moment when you get your first patent. It comes bound inside a beautiful oyster-white folder that has the United States Constitution screened in blue as its background. The large official gold seal of the Patent and Trademark Office is embossed thereon with two red ribbons furcate as a tail.

Between the covers of this folder is your patent, a grant that gives you the inventor(s) "the right to exclude others from making, using or selling the invention throughout the United States" and its territories and possessions for a period of 17 years (14 for a design patent) subject to the payment of maintenance fees as provided by law. The patent does not give you the right to make, use, or sell the invention. Any person is ordinarily free to make, use, or sell anything he or she pleases, and a grant from Uncle Sam is not required.

If you receive a patent for a new soda pop and the marketing of said beverage is prohibited by law, the patent will not help you. Nor may you market said soda pop if by doing so you infringe the prior rights of others.

Since the essence of the right granted by a patent is the right to exclude others from commercial exploitation of the invention, the patentee is the only one who may make, use, or sell the invention. Others may not do so without your authorization. You may assign your rights in the invention to another person or company.

Maintenance Fees

All utility patents which issue from applications filed on or after December 12, 1980 are subject to the payment of maintenance fees which must be paid to keep the patent in force. These fees are due at 3.5, 7.5, and 11.5 years from the date the patent is granted and can be paid without a surcharge during the six-month period preceding each due date, e.g. 3 years to 3 years and six months, etc. The amounts of the maintenance fees are subject to change every three years.

Reports Don Coster of the Nevada Inventor Association after a visit to the PTO: "Our patent applications go through a process that is so thorough and so efficient that it is hard to believe unless you see it in action. The application does not go directly to an examiner. It must first be examined for content and completeness. The drawings are checked and screened for things like military sensitivity or unlawful usage. Once accepted as legal and complete, it is classified for the proper art group. This is very critical. If the wrong examiner ends up with it on his or her table, it might be months before s/he even gets a first look at it, because applications are taken in the order that they're received... Those people are so conscientious that it rarely ever happens."

At a recent inventor conference, Commissioner Harry F. Manbeck commented on the importance of inventing: "By encouraging inventing, we maintain the force that has made America a great nation. It's that incentive of rewards that encourages inventors to create and entrepreneurs to become businessmen. Our standard of living is raised when people are encouraged to invest money in new ideas and new technology."

Patents relating to some pharmaceutical inventions may be extended by the Commissioner for up to five years to compensate for marketing delays due to federal FDA pre-marketing regulatory procedures. Patents relating to all other types of inventions can only be extended by congressional legislation.

The PTO does not mail notices to patent owners advising them that a maintenance fee is due.

If you have a patent attorney tracking your business, he or she will let you know when the money is due. An attorney gets paid every time your business moves across his or her desk. But if you are doing it by yourself and you miss a payment, it may result in the expiration of the patent. A six-month grace period is provided when the maintenance fee may be paid with a surcharge.

Can Two People Own A Patent?

Yes. Two or more people may jointly own patents as either inventors, investors, or licensees. Most of my patents are joint ownerships. Anyone who shares in the ownership of a patent, no matter how small a part they might own, has the right to make, use, or sell it for his or her own profit, unless prohibited from doing so by prior agreement. It is accordingly dangerous to assign part interest in a patent of yours without having a definite agreement hammered out vis-a-vis respective rights and obligations to each other.

Can A Patent Be Sold?

Yes. The patent law provides for the transfer or sale of a patent, or of an application for patent, by a contract. When assigned the patent, the assignee becomes the owner of the patent and has rights identical to those of the original patentee.

Assignment of Patent Applications

Should you wish to assign your patent or patent application to a third party (manufacturer, investor, university, employer, etc.), this is possible by filing the appropriate form, Assignment of Patent, or Assignment of Patent Application, which occur as part of Fig. 22.

You can sell all or part interest in a patent. If you prefer, you could even sell it by geographic region.

I consider patents as valuable properties—personal assets. Never assume that because you have been unsuccessful in selling a patent it has no value. You might sell it eventually or find someone infringing it, thus turning it to positive account.

Infringement of Patents

Infringement of a patent consists in the unauthorized making, using, or selling of the patented invention within the territory of the U.S. during the term of the patent. If your patent is infringed, it is your right to seek relief in the appropriate federal court.

When I see an apparent infringement of a patent of ours, as has occurred occasionally over the years, the first thing I do is call the company and set a meeting. I am not litigious. Things can often be worked out between parties. Thus far, I have always been able to do this. Court battles over patents can be long and expensive affairs. And, if you want to continue working in your particular field, it is wise to not make too many corporate enemies.

Several years ago I saw an infringement of a patent we hold. One call to the company's president, and a quick fax of our patent, brought immediate relief in the form of a royalty on all items made to date and in the future. It was an early Christmas. Not only that, but I was invited to submit ideas for licensing consideration.

If your friendly approach is turned away, and you are sure of your position, then the next step is to get a lawyer and decide if a Temporary Restraining Order (TRO) is appropriate. A TRO is an injunction to prevent the continuation of the infringement. You may also ask the court for a award of damages because of the infringement. In such an infringement suit, the defendant may raise the question of the validity of the patent, which is then decided by the court. The defendant may also aver that what is being done does not constitute infringement.

Infringement is determined primarily by the language of the claims of the patent and, if what the defendant is making does not fall within the language of any of the claims of the patent, there is no infringement.

The PTO has no jurisdiction over questions relating to infringement of patents. In examining applications for patent, no determination is made as to whether the patent-seeking invention infringes any prior patent.

Patent Enforcement Insurance

It has been said that the only right a patent gives the inventor is the right to defend it in a court of law. And patent infringement litigation can be costly. How much? Ask Diane B. Loisel, a nurse from Bowie, Maryland. After spending a reported to obtain a patent for a cap she for use in neonatal respiratory therapy, she claimed that a company to whom she had presented the concept had begun to manufacture and market it without her permission.

Ms. Loisel was told by her patent attorney that to litigate would cost her $250,000 in legal fees. "If you're going to get a patent, you're going to have to fight," her lawyer told her. "But he never told me it would cost so much money," she said.

To help inventors shoulder the risk and responsibility for enforcing their patents against infringers, one insurance company is beginning to market policies that are designed to reimburse the litigation expenses incurred by a patent owner in enforcing his or her U.S. patents.

"In the short time that Patent Enforcement Insurance has been on the market, we have already had dozens of very favorable responses, including a situation where one policy owner forced two larger

After a patent has expired, anyone may make, use, or sell the invention without permission of the patentee, provided the matter covered by other unexpired patents is not used. The terms may not be extended save for by a special act of Congress.

Professor Mark Spikell of George Mason University offers these pearls of wisdom to inventors:

*Your invention must have something that will prevent a competitor from producing it once you have established a market.

*You must have a very sound and conservative business plan, plus a team of expert management.

*You must be sufficiently financed.

*You must be willing to trade your personal involvement in the business for money. If a reasonable offer is presented to you to sell out your interest, you must be of a mind not to fight it.

For information on patent infringement insurance, contact Mr. Robert W. Fletcher at 1-800-537-7863; or (502) 429-8007; or write HLPM Insurance Services, Inc., 10509 Timberwood Circle, Louisville, Kentucky 40223.

companies to come to the negotiating table and ultimately take a license because they knew he had the financial backing to enforce his patent," says Robert W. Fletcher, president of HLMP Insurance Services, Inc. in Louisville, Kentucky.

Who takes out such insurance? Mr. Fletcher points out that his best customers are small corporations and that he has some independent inventors, too.

Patent attorney Robert W. Faris, a partner in Nixon & Venderhye of Arlington, Virginia, says of the policy, "One of the downsides of this type of insurance is that the insurance company is reimbursed for its expenses out of the settlement or judgement. This means that if the recovery is on the order of the legal expenses incurred for the litigation, the patent owner could come away with practically no financial recovery, although his patent rights will have been vindicated."

Faris adds that the program seems pretty risky for the insurance company. "I don't know how they are able to predict with any certainty what the risks would be beforehand. They would have to be only taking on patents whose chances for infringement are very remote." HLPM's Fletcher admits that Kemper Group and The St. Paul Companies, two other U.S. insurers, rejected the concept of Patent Enforcement Insurance because of the risk.

Under the HLMP program, the policy holder is permitted to chose patent counsel. However, before the company will open the tap and start paying bills, the policy holder must provide a written opinion from his or her attorney attesting to the fact that the matter is one that can litigated and the policy holder must show proof that the alleged infringement will cause economic damage.

Would Faris recommend Patent Infringement Insurance? "I might well recommend that certain clients look into it because it is the only way some small businesses might be able to enforce their patent rights," he concludes.

He points out that Patent Infringement Insurance **does not** cover the inventor should his or her U.S. patent infringe an existing patent. In other words, it could protect you (the inventor) from someone infringing on your patent, but is of no help to you if your patent infringes the patent of another inventor.

Claiming to have more than $100 million in coverages in force, HLMP's coverage is underwritten by the Connecticut Indemnity Co., a member of the Orion Group in Farmington, Connecticut. Policies are written for a three-year term and billed annually. A minimum deductible of $10,000 applies to every suit. The HLPM policy is not available in every state. States such as Connecticut and New York have prevented its sale.

Design Patents

If you want an inexpensive patent that will give you little actual protection because they are easy to end-run, but still meet the requirement that a patent issue, maybe a design patent is for you. It appears to be popular with lots of folks. Over 6,000 new designs were patented last year.

Inventors get design patents on just about anything, e.g. baby bibs, sweatbands, tissue box holders, dishes, ammo boxes, game boards, vending machines, telephones, pencils and pens, and even internal combustion engines.

Rubbermaid has one for a cereal container. Totes protects its umbrella handle designs. The Parker Pen Company takes them out for writing instruments. And the Ford Motor Company has them on car parts such as automobile quarter panels.

The range of ornamental appearances that have been patented during the 147-year history of design patents is most impressive. Over 270,000 designs have received design patent protection since the first one was granted to George Bruce for "Printing Type" on November 9, 1842.

For little expense and effort a design patent is another way to stake out a claim. It permits you to legally post a no trespass sign in the form of **Patent Pending, Patent Applied For, or Patent and the number of the patent.**

If you've invented any new, original, and ornamental design for an article of manufacture, a design patent may be appropriate. A design patent protects only the appearance of an article, and not its structure or utilitarian features. The proceedings relating to granting of design patents are the same as those relating to other patents with few differences.

In design cases as in "mechanical" cases, novelty and nonobviousness are necessary prerequisites to the grant of a patent. In the case of designs, the inventive novelty resides in the shape or configuration or ornamentation as determining the appearance or visual aspect of the object or article of manufacture in contradistinction to the structure of a machine, article of manufacture, or the constitution of a composition of matter. Simply put, it is the appearance presented by the object that creates an impression upon the mind of the observer.

Do You Need A Patent Attorney for Design Patents?

This is, of course, a personal decision. I have not engaged a patent attorney's services for design patent applications. Unlike the complicated business of utility patents, the design patent application process is

A utility patent and a design patent may be based upon the same object matter; however, there must be a clearly patentable distinction between them.

very easy and uncomplicated. I have found that paying a lawyer to do my design patents is like tossing money out of the window because the application form is simple, esoteric language is not required to draft the claims, and searches are actually so pleasant an experience that I will often do them myself.

A design patent search doesn't require much reading. You're just-looking at a lot of line drawings, many of which are fascinating.

! spent $500 learning that lawyers were not necessary in such matters. At the time I had been using a fine law firm for utility patents and I naturally released our first design patent to them. I knew no better.

My lawyer said that he would "write up the specifications." I was asked to provide his draftsman with a prototype of the design. It was a tricycle with a mainframe shaped like a toothpaste tube. It became Proctor & Gamble's Crest Fluorider.

When the design patent arrived, I saw for the first time the specifications (see Design Patent Specification below). Design patents are obvious cash cows for patent attorneys. I have removed the product name so that you will be able to use its format for your design patents by inserting a title. This is a sample only and, unlike other Figures in this book, should not be returned to the Patent Office. The petition, specification, and claim must be typed on legal-sized paper.

Design Patent Title

The title is of great importance in a design patent application. It serves to identify the article in which the design is embodied and which is shown in the drawing, by a name generally used by the public. The title should be to a specific definite article. Thus a stove would be called a "Stove" and not "Heating Device." The same title is used in the preamble to the specification, in the description of the drawing, and in the claim.

To allow latitude of construction it is permissible to add to the title - "or similar article." The title must be in the singular.

Design Patent Specification

To create your own design patent specification document, all you have to do is use the below example and fill in the appropriate blanks.

Be it known that I (we), _____ and _____ have invented a new, original, and ornamental design for a _____, over which the following is a specification. Reference is made to the accompanying drawing, which forms a part hereof, wherein:

FIGURE 1 is a perspective view of a _____ in accordance with the present design;

FIGURE 2 is a rear elevational view of said_____;

FIGURE 3 is a side elevational view of said _____, the side of the _____ not shown being a mirror image of the side shown in this view;

FIGURE 4 is a front elevational view of said _____.

WHAT I (we) CLAIM IS:

The ornamental design for a _____ or similar article, as shown and described.

(Note: attached to this application are the draftsman drawings)

Design Patent Drawings

Unless you are capable of doing this work yourself, hire a competent, experienced patent draftsman to make your drawings. The requirements for drawings are strictly enforced. Professional draftsmen will stand behind their work and guarantee revisions if requested by the Office due to inconsistencies in the drawings.

The claim of the design patent determines its classification, i.e. the appropriate class and subclass into which the design patent will be placed. This classification is designated as the "original classification" of the patent. Copies of a patent may be placed in other subclasses for the convenience of the examiner as an aid in searching during the examination process. Additional copies are designated "cross reference."

The PTO changes the Design Patent Classification System as required to provide an appropriate area for each patented design. Areas which show great activity are expanded while other areas, used infrequently as activity fades, are compressed into other subclasses.

Plant Patents

The law provides for the granting of a patent to anyone who has invented or discovered and asexually reproduced any distinct and new variety of plant, including cultivated sports, mutants, hybrids, and newly found seedlings, other than a tuber-propagated plant or a plant found in an uncultivated state.

Asexually propagated plants are those that are reproduced by means other than from seeds, such as by the rooting of cuttings, by layering, budding, grafting, and inarching.

With reference to tuber-propagated plants, for which a plant patent cannot be obtained, the term "tuber" is used in its narrow horticultural sense as meaning a short, thickened portion of an underground branch. The only plants covered by the term "tuber-propagated" are the Irish potato and the Jerusalem artichoke.

Elements of a Plant Application

An application for a plant patent consists of the same parts as other applications and must be filed in duplicate, but only one need be signed and executed; the second copy may be a legible carbon copy of the original. Two copies of color drawings must be submitted.

The reason for submitting two copies is that one, the photocopy, must go to the Department of Agriculture for an advisory report on plant variety. The original is retained at the PTO.

Applications for a plant patent which fail to include two copies of the specification and two copies of the drawing when in color, will be accepted for filing only. The Application Division will notify you immediately if something is missing from the filing. You'll be given one month to rectify the situation. Failure to do so will result in loss of the filing date and the fee paid.

Plant Patent Specification and Claim

The specification should include a complete, detailed description of the plant and its characteristics that distinguish the plant over related known varieties and its antecedents. The specification must be expressed in botanical terms (in the general form followed in standard botanical text books or publications dealing with the varieties of the kind of plant involved), rather than the non-botanical characterizations commonly found in nursery or seed catalogs. The specification should

A plant patent is granted on the entire plant. It, therefore, follows that only one claim is necessary and only one is permitted.

Fig. 2 is an example of a plant patent application with drawing.

The Plant Variety Protection Act (Public Law 9l-557), approved December 24, 1970, provides for a system of protection for sexually reproduced varieties, for which protection was not previously provided, under the administration of a Plant Variety Protection Office within the Department of Agriculture. Requests for information regarding the protection of sexually reproduced varieties should be addressed to Commissioner, Plant Variety Protection Office, Consumer and Marketing Service, Grain Division, 6525 Bellcrest Road, Hyattsville, Maryland 20782.

also include the origin or parentage of the plant variety sought to be patented and must particularly point out where and in what manner the variety of plant has been asexually reproduced.

When color is a distinctive feature of the plant, it should be positively identified in the specification by reference to a designated color as given by a recognized color dictionary, for example, cherry red blooms. Where the plant variety originated as a newly found seedling, the specification must fully describe the conditions (cultivation, environment, etc.) under which the seedling was found growing to establish that it was not found in an uncultivated state.

Plant Patent Oath or Declaration

Your oath or declaration as inventor, in addition to the statements required for other applications, must include the statement that you asexually reproduced the new plant variety. Where the plant is a newly found plant the oath or declaration must also state that it was found in a cultivated area.

Plant Patent Drawings

Plant patent drawings are not mechanical drawings and should be artistically and competently executed. The drawing must disclose all the distinctive characteristics of the plant capable of visual representation. When color is a distinguishing characteristic of the new variety, the drawing must be in color. Two duplicate copies of color drawings must be submitted. Color drawings may be made either in permanent water color or oil, or in lieu thereof may be photographs made by color photography or properly colored on sensitized paper. The paper in any case must correspond in size, weight, and quality to the paper required for other drawings. Mounted photographs are acceptable.

All color drawings should be mounted so as to provide a two-inch margin at the top for Office markings when the patent is printed.

Specimens

Specimens of the plant variety, its flower, or fruit, should not be submitted unless specifically called for by the examiner. If the PTO wants to inspect a plant that cannot be physically submitted, it will send an examiner to the growing site.

Fees and Correspondence

All inquiries relating to plant patents and pending plant patent applications should be directed to the Patent and Trademark Office and not to the Department of Agriculture.

Treaties and Foreign Patents

The rights granted by a U.S. patent extend only throughout the territory of the U.S. and have no effect in a foreign country. Therefore, to receive patent protection in other countries you'll have to make separate application(s) in each of the other countries or in regional patent offices. Almost every country has its own patent law.

The laws in many countries differ from our own. In most foreign countries, publication of the invention before the date of the application will bar the right to a patent. Most foreign countries require maintenance fees and that the patented invention be manufactured in that country within a certain period, usually about three years. If no manufacturing occurs within that period, the patent may be subject to the grant of compulsory licenses to any person who may apply for a license.

The Paris Convention

The Paris Convention for the Protection of Industrial Property is a treaty relating to patents which is followed by 93 countries, including the United States. It provides that each country guarantee to the citizens of the other countries the same rights in patent and trademark matters that it gives to its own citizens. The Paris Convention is administered by the World Intellectual Property Organization (WIPO) in Geneva, Switzerland.

The treaty also provides for the right of priority in the case of patents, design patents, and trademarks. This right means that, on the basis of a regular first application filed in one of the member countries, the applicant may, within a certain period of time, apply for protection in all the other member countries. These later applications will then be regarded as if they had been filed on the same day as the first application. Thus, these later applications will have priority over applications for the same invention which may have been filed during the same period by other persons. Moreover, these later applications, being based on the first application, will not be invalidated by any acts accomplished in the interval, such as, for example, publication or exploitation of the invention, sale of copies of the design, or use of the trademark.

The time frame allowed for subsequent applications in other member countries is 12 months in the case of utility patents and six months in the case of design patents and trademarks.

Patent Cooperation Treaty

Negotiated at a diplomatic conference in Washington, D.C., in June

James Fergason, inventor of the Liquid Crystal Display, has this advice on foreign patenting. "If your item is hot, file here and in the country of highest market potential simultaneously. Foreign filing is not a trivial thing. The costs are so high that you really must determine the justification before attempting it. Japan, for example, can be a nightmare to search because patent applications are allowed to languish in limbo for seven years before they must be examined."

According to Commissioner Harry Manbeck, the swelling tide of foreign patenting in the U.S. is due to several conditions. He notes that because of economic necessity, some industrial countries have poured enormous amounts of money into supporting their technology. Advancing technology means inventing and inventing means patenting. Because of the Paris Convention, says Manbeck, "We have access to foreign patent systems. They, in turn, have access to ours. We hope that the trend to issuing so many foreign patents is slowing, or may even reverse."

The PCT facilitates the filing of applications for patent on the same invention in member countries by providing, among other benefits, for centralized filing procedures and a standardized application format.

of 1970, the Patent Cooperation Treaty (PCT) came into force on January 24, 1978, and is presently adhered to by the 39 countries listed alphabetically below.

LIST OF PATENT COOPERATION TREATY MEMBER STATES

1) Austria
2) Australia
3) Barbados
4) Belgium
5) Brazil
6) Bulgaria
7) Cameroon
8) Central Africa Republic
9) Chad
10) Congo
11) Democratic People's Republic of Korea
12) Denmark
13) Finland
14) France
15) Gabon
16) Germany, Federal Republic of
17) Hungary
18) Italy
19) Japan
20) Liechtenstein
21) Luxembourg
22) Madagascar
23) Malawi
24) Mali
25) Mauritania
26) Monaco
27) Netherlands
28) Norway
29) Republic of Korea (South Korea)
30) Romania
31) Senegal
32) Soviet Union
33) Sri Lanka
34) Sudan
35) Sweden
36) Switzerland
37) Togo
38) United Kingdom
39) United States

Under U.S. law it is necessary, in the case of inventions made in that country, to obtain a license from the Commissioner of Patents and Trademarks before applying for a patent in a foreign country. Such a license is required if the foreign application is to be filed before an application is filed in the United States or before the expiration of six months from the filing of an application in the U.S.

If the invention has been ordered to be kept secret, the consent to the filing abroad must be obtained from the Commissioner of Patents and Trademarks during the period the secret is in effect.

A basic fee applies for the first thirty pages, a basic supplemental fee is charged for each page over thirty, and a designation fee is assessed per member country or region, etc.

Foreign Applications for U.S. Patents

Any person of any nationality may make application for a U.S. patent so long as that person is the inventor of record. The inventor must sign the same oath and declaration (with certain exceptions).

An application for a patent filed in the U.S. by any person who has previously regularly filed an application for a patent for the same invention in a foreign country (which affords similar privileges to citizens of the U.S.) shall have the same force and effect for the purpose of overcoming intervening acts of others. The requirement is that it be filed in the U.S. on the date on which the application for a patent (for the same invention) was first filed in the foreign country, provided that the application in the U.S. is filed within 12 months (six months in the case of a design patent) from the earliest date on which any such foreign application was filed. A copy of the foreign application certified by the patent office of the country in which it was filed is required to secure this right of priority.

If any application for patent has been filed in any foreign country prior to application in the U.S., the applicant must, in the oath or declaration accompanying the application, state the country in which the earliest such application has been filed, giving the date of filing the application. All applications filed more than a year before the filing in the U.S. must also be recited in the oath or declaration.

An oath or declaration must be made with respect to each and every application. When the applicant is in a foreign country, the oath or affirmation may be before any diplomatic or consular officer of the U.S. It may also be made before any officer having an official seal and authorized to administer oaths in the foreign country, whose authority shall be proved by a certificate of a diplomatic or consular officer of the U.S. In all cases, the oath is to be attested by the proper official seal of the officer before whom the oath is made.

When the oath is taken before an officer in the country foreign to the U.S., all the application papers (except the drawings) must be attached together and a ribbon passed one or more times through all the sheets. The ends of the ribbons are to be brought together under the seal before the latter is affixed and impressed, or each sheet must be impressed with the official seal of the officer before whom the oath was taken. If the application is filed by the legal representative (executive, administrator, etc.) of a deceased inventor, the legal representative must make the oath or declaration.

When a declaration is used, the ribboning procedure is not necessary, nor is it necessary to appear before an official in connection with the making of a declaration.

No U.S. patent can be obtained if the invention was patented abroad more than one year before filing in the U.S. Six months are allowed in the case of a design patent.

Patents Issued to Residents of Foreign Countries

The following is a list of patents issued to foreign countries in FY1989:

Arab Emirates 2

Argentina 29

Australia 568

Austria 421

Bahamas 5

Barbados 1

Belgium 373

Bolivia 1

Brazil 36

British Virgin Islands 1

Bulgaria......................23

Canada 2,077

Chile 2

China,
   Peoples' Republic of 53
   Republic of............. 674

Cocos Islands 1

Colombia...................... 4

Costa Rica 3

Cuba 1

Cyprus 3

Czechoslovakia 29

Dominican Republic 2

Denmark 242

Ecuador 1

Egypt 2

El Salvador 1

Federal Republic
  of Germany............ 8,756

Finland 265

France 3,310

German Democratic
  Republic..................... 58

Greece 23

Guadelope 1

Guatemala 1

Haiti 1

Honduras 2

Hong Kong 123

Hungary 124

Iceland 3

India 16

Indonesia 2

Iran.............................. 1

Iraq............................... 2

Ireland.......................... 64

Israel.......................... 333

Italy 1,398

Japan................... 20,907

Kuwait 5

Lebanon........................ 1

Liechtenstein 16

Luxembourg................ 36

Malaysia........................ 3

Malta	1	Singapore	17
Mexico	42	South Africa	144
Monaco	9	South Korea	175
Morocco	3	Spain	146
Netherlands	1,132	Sri Lanka	1
Netherlands Antilles	1	Sweden	993
New Zealand	75	Switzerland	1,437
North Korea	3	Thailand	4
Norway	157	Trinidad/Tobago	2
Panama	1	Turkey	4
Peru	4	United Kingdom	3,378
Philippines	6	Uraguay	1
Poland	15	U.S.S.R.	153
Portugal	8	Venezuela	23
Rhodesia	2	Yugoslavia	17
Saudi Arabia	6	**Total:**	47,950

Trademarks: Definition and Benefits

Coke. Greyhound. Scrabble. Crest. NBC's peacock. Mr. Goodwrench. The Campbell Kids. Morris the Cat. Paramount's mountain and stars. The MGM "pussy cat". All are well known trademarks and very important to their owners as tools to sell products and services.

Trademarks also can be very important to inventors as a tool for helping to sell an invention or new concept. A good trademark creates fast product identification and can help tell the product's story. I always spend a great deal of time creating an appropriate trademark for a product that requires one as part of its release package. For example, my suggestion to Procter & Gamble that it license our patented tricycle for Crest to use as a premium would not have had the same sex appeal without the trademark, Crest Fluorider.

When told the name of our proposed original ride-on, the product's brand manager immediately responded, "Wow! That sounds exciting, a Crest Fluorider. When can we see a prototype?"

PTO Says: No Trump to The Donald

To illustrate just how important a federal trademark can be, take the case of Donald Trump's trademark application for "Trump Taj Mahal." The Donald wanted this mark for his $1 billion-plus commanding and lavish casino hotel in Atlantic City, N.J. His lawyers said that the trademark would block a competitor from taking the name and profiting from Trump's reputation.

When Trump's application landed on the desk of a mid-level PTO attorney by the name of Mary Schimelfenig, she initiated a routine computer search to make sure that the trademark was available. She found Trademark No. 1,158,610, awarded to a little known businessman by the name of Raj Mallick's for his cozy Taj Mahal restaurant at 1327 Connecticut Ave. in Washington, D.C. It mattered not to Schimelfenig that Washington's Taj Mahal bears no resemblance to Trump's glitzy extravaganza. Nor was she influenced by the size of the Taj near Dupont Circle, which seats no more than 100 people. Schimelfenig did understand trademark law, which allows "no likelihood of confusion" among trademarks. Therefore, two Taj Mahal restaurants could not be. The Trump clan was advised of the objection.

"The legal issue isn't that there are hundreds of miles in between these two Taj Mahals or that one is big and one is small," Ron

TRADEMARK OFFICE

Trademarks are handled by the Trademark Office of the Patent and Trademark Office, under the directorship of an Assistant Commissioner for Trademarks. It employs a total of 150 Examining Attorneys. Last year, this office handled over 83,000 trademark applications. It disposed of about 80,000 applications. Figure 24, the workflow diagram, will give you some idea of how applications are routed and how long it takes to get them cleared and on the official trademark register. At the end of FY 1989, the average time between filing an application and its registration or abandonment was 13.8 months.

Wolfington, managing attorney in Ms. Schimelfenig's office told *The Washington Post* . "We have to ask, 'What if the (local) Taj Mahal becomes a chain and opens in Atlantic City and runs into Trump?' Then there's a likelihood of confusion."

At this writing, the Trump organization is vigorously pursuing the trademark via a new application. The lack of the trademark does not prevent him from operating under the mark, "Trump Taj Mahal." But Trump lacks the federal trademark protection he wants against possible future rivals. And as for Mr. Mallick, he may one day have a chance to "hit the jackpot" should the PTO deny Trump a second time. The importance of the trademark being what it is to Donald Trump, the owner of D.C.'s Taj Mahal restaurant may invoke the art of the deal and sell his trademark to the New York multibillionaire for a Trump-sized sum—a customary method of solving such contests.

Definition of Trademarks

Trademarks. A "Trademark," as defined in section 45 of the 1946 Trademark Act (Lanham Act) "includes any word, name, symbol, or device, or any combination thereof adopted and used by a manufacturer or merchant to identify his goods and distinguish them from those manufactured or sold by others."

Service Marks. A mark used in the sale or advertising of services to identify the services of one person and distinguish them from the services of others. Titles, character names, and other distinctive features of radio or television programs may be registered as service marks notwithstanding that they, or the programs, may advertise the goods of the sponsor.

How Long is a Trademark in Force? The term of a trademark is potentially infinite.

Certification Marks. A mark used upon or in connection with the products or services of one or more persons other than the owner of the mark to certify regional or other origin, material, mode of manufacture, quality, accuracy, or other characteristics of such goods or services, or that the work or labor on the goods or services was performed by members of a union or other organization.

Collective Marks . Trademark or service mark used by the members of a cooperative, an association, or other collective group or organization. Marks used to indicate membership in a union, an association, or other organization may be registered as Collective Membership Marks.

Trade and Commercial Names . Marks differ from trade and commercial names that are used by manufacturers, industrialists, merchants, agriculturists, and others to identify their businesses, vocations, or occupations, or other names or titles lawfully adopted by persons, firms companies, unions, and other organizations. The latter are not subject to registration unless actually used as trademarks.

Function of Trademarks

The primary function of a trademark is to indicate origin. However, trademarks also serve to guarantee the quality of the goods bearing the mark and, through advertising, serve to create and maintain a demand for the product. Rights in a trademark are acquired only by use and the use must ordinarily continue if the rights you acquire are to be preserved. Registration of a trademark in the Patent and Trademark Office does not in itself create or establish any exclusive rights, but it is recognition by the government of your right to use the mark in commerce to distinguish your goods from those of others.

Benefits of Registration

While Federal registration is not necessary for trademark protection, registration on the Principal Register does provide certain advantages:

1. The filing date of the application is a constructive date of first use of the mark in commerce (this gives registrant nationwide priority as of that date, except as to certain prior users or prior applicants);

2. The right to sue in Federal court for trademark infringement;

3. Recovery of profits, damages, and costs in a Federal court infringement action and the possibility of treble damages and attorneys' fees;

4. Constructive notice of a claim of ownership (which eliminates a good faith defense for a party adopting the trademark subsequent to the registrant's date of registration);

5. The right to deposit the registration with Customs in order to stop the importation of goods bearing an infringing mark;

6. Prima facie evidence of the validity of the registration, registrant's ownership of the mark and of the registrant's exclusive right to use the mark in commerce in connection with the goods or services specified in the certificate;

7. The possibility of incontestability, in which case the registration constitutes conclusive evidence of the registrant's exclusive right, with certain limited exceptions, to use the registered mark in commerce;

8. Limited grounds for attacking a registration once it is five years old;

The federal government has issued over one million trademarks since the passage of the Trademark Act of 1905. Today, about 620,000 are in active status.

9. Availability of criminal penalties and treble damages in an action for counterfeiting a registered trademark; and

10. A basis for filing a trademark in foreign countries.

Use in Commerce

The Trademark Law Revision Act of 1988, which provides for the most significant changes in more than 40 years in the laws governing Federal registration of trademarks, went into effect on November 16, 1989. For the first time, applicants may file for a Federal trademark registration based on their bona fide intent to use the mark in interstate or foreign commerce. This is a significant change from the policy that in the past allowed owners to register their marks only if the mark has already been used to identify goods or services moving in interstate commerce.

Intent to Use: Significant New Procedure

Under the new procedures, an applicant whose trademark meets basic criteria will have six months to use it from the date the PTO gives preliminary approval. This period may be extended in six-month intervals up to a total of 30 months. Once the applicant has furnished proof that the trademark has been used, it may be registered.

The intent-to-use system is not designed to permit or foster the hoarding of trademarks that may never be used. The actual use of a trademark is a prerequisite to registration. It will not allow people to holds marks for ransom. In other words, if you get wind that a major corporation intends to use a certain mark in commerce, the new law does not permit you to grab the mark, not use it, and then license it to the corporation in return for a fee.

The intent-to-use system is designed to let trademark owners test the protection potential and registrability of their marks before beginning the costly expense of developing and promoting new trademarks, and to align the U.S. trademark registration program with the systems of most foreign countries.

Marks Not Subject to Registration

A trademark cannot be registered if it:

a) Consists of or comprises immoral, deceptive, or scandalous matter or matter that may disparage or falsely suggest a connection with persons, living or dead, institutions, beliefs, or national symbols, or bring them into contempt or disrepute.

b) Consists of or comprises the flag or coat of arms or other insignia of the United States, or of any state or municipality, or of any foreign nation, or any simulation thereof.

The revised trademark law also shortens the term of registrations from 20 to 10 years, and thereby purges the register of defunct marks.

c) Consists of or comprises a name, portrait, or signature identifying a particular living individual except by his written consent, or the name, signature, or portrait of a deceased President of the United States during the life of his widow, if any, except by the written consent of the widow.

d) Consists of or comprises a mark which so resembles a mark registered in the Patent and Trademark Office or a mark or trade name previously used in the United States by another and not abandoned, as to be likely when applied to the goods of another person, to cause confusion, or to cause mistake, or to deceive.

Registrable Marks

Principal Register : The trademark, if otherwise eligible, may be registered on the Principal Register unless it consists of a mark which, 1) when applied to the goods/service of the applicant is merely descriptive or deceptively mis-descriptive of them, except as indications of regional origin, or 2) is primarily merely a surname.

Such marks, however, may be registered on the Principal Register, provided they have become distinctive as applied to the applicant's goods in commerce. The Commissioner may accept as prima facie evidence that the mark has become distinctive as applied to applicant's goods/services in commerce, proof of substantially exclusive and continuous use thereof as a mark by the applicant in commerce for the five years next preceding the date of filing of the application for registration.

Supplemental Register : All marks capable of distinguishing your goods and not registrable on the Principal Register, which have been in lawful use in commerce for the year preceding your filing for registration, may be registered on the Supplemental Register. A mark on this register may consist of any trademark, symbol, label, package, configuration of goods, name, word, slogan, phrase, surname, geographical name, numeral, or device, or any combination of the foregoing.

How to Search a Trademark

Just as in the case of patent application, it is advisable for you to make a search of registered marks before filing an application.

Three methods are available. Have your patent attorney do it; engage the services of a professional trademark search organization; or do it yourself.

Law Firm Trademark Search

A patent attorney will gladly handle a trademark search. The attorney will not do it personally, but instead will rely on the services of a professional trademark search firm. Law firms may step on search fees 40 percent or more depending upon what the market will bear.

If you decide to hire a lawyer for this job, receive an estimate of costs up front. Lawyers most often use computer-aided search methods, which can take a few minutes. There may be a minimum charge. To avoid surprises, find out in advance how much the lawyer will charge for copies.

Professional Trademark Search

If you want to save yourself the lawyer's mark-up, consider hiring a trademark searcher yourself. They are listed in the yellow pages under "Trademark Searchers" or "Patent Searchers." Patent searchers will often handle trademark searches as well.

Get an estimate. An average price is $75.00 per hour with a one hour minimum. It should not take anywhere near one hour to do a single mark. Significant portions of the trademark operation are already fully automated. For over five years, the public has been able to access the Office's electronic files concerning registrations and applications through the Trademark Reporting and Application Monitoring system (TRAM). Via this computer-aided search, it can take a few minutes to access bibliographic, status, and location data. Access to TRAM often obviates the need to consult the actual file.

Add to this price about $3.00 per copy of each trademark copied.

Do-it-Yourself Trademark Search

You do the search utilizing two methods:

1) The Trademark Office has a **Public Search Room** , located in Crystal Plaza 2 - 2C08, Arlington, Virginia. Once

Copies of the *Trademark Official Gazette* may be obtained from the Superintendent of Documents, Government Printing Office, Washington, D.C. 20402. A pre-paid of $9.50 is required. For more information, please call (202) 783-3238.

Lawyers and professional searchers often charge more for logo searches. In my opinion, both take an equal amount of effort. Yet you'll find trademarks at $75.00 and logos $150.00 from the same firm.

the layout is learned (which takes about five minutes), you can breeze through a manual search in 20 minutes.

2) If a visit to Arlington is inconvenient, see if there is a Patent and Trademark Depository Library is nearby. It will offer numerous trademark reference books, e .g. Gale's *Brands and Their Companies* , Compu-Mark's *Directory of U.S. Trademarks* , etc., as well as an online computer search capability (See TRAM reference above).

Depending upon how extensive the search, costs can run anywhere from $10.00 and up for computer time. Often appointments must be made to access the computer services.

T-Search

T-Search is the PTO's online trademark search system to research word marks and design elements of trademarks. The PTO currently offers an experimental public use program on this system.

How to Register a Trademark

A trademark application package consists of four items:

    1) A written application;

    2) A drawing of the mark;

    3) Five specimens or facsimiles of the mark;

    4) The required filing fee.

Trademark applications may be delivered in person to the Attorney's Window located on the lobby level of Crystal Plaza 2 or mailed to the Commissioner of Patents and Trademarks, Washington, DC 20231.

Living only 20 minutes from Crystal City, Virginia, I always hand deliver my trademark applications. I conduct my own search upstairs at the Trademark Search Library and, if I find the mark in question to be clear, go to the Attorney's Window, check in hand, and thus avoid any mail delays. A PTO representative accepts the application and payment, date stamping them in immediately.

How to Complete the Trademark Application

The Trademark Application/Declaration form (newly updated due to the Trademark Law Revision Act of 1988) is reproduced in the Forms and Documents section of this book. Trademark forms are also available from the PTO (call 703-557-4636) or Department of Commerce Offices.

The application must be written in English. The form may be used-for either a trademark or service mark application. Additional forms may be photocopied. The following explanation covers each blank, beginning at the top.

> **Heading.** Identify (a) the mark (e.g. "ERGO" or "ERGO and design") and (b) the class number(s) of the goods or services for which registration is sought. In the upper right hand corner of the application form, put down the classification number that refers to your product. An application in which a single fee is submitted must be limited to the goods or to the services comprised in a single class.

Requirements for receiving a filing date:

1. Written application in the English language containing the following information:
Your name, current address, and citizenship (state or country of incorporation or location of partnership); Identification of the goods or services covered (specify actual goods and/or nature of services performed); Verification of the facts contained in the application, with your signature.

2. A drawing of the mark. (Word marks must be typewritten in upper-case letters and design marks must be in black ink on white paper and no larger than four inches by four inches. The drawing page must include the appropriate heading.)

3. A filing fee of $175 (the filing fee is $175 for each separate class of goods or services in the application).

Trademark Notice

Once a Federal registration is issued, the registrant may give notice of registration by using the r symbol, or the phrase, "Registered in the U.S. Patent and Trademark Office," or "Reg. U.S. Pat. & Tm Off." Although registration symbols may not be lawfully used prior to registration, many trademark owners use a TM or SM (if the mark identifies a service) symbol to indicate a claim of ownership, even if no Federal trademark application is pending.

Classification is part of the PTO's administrative processing. The International Schedule of Goods and Services is used (reproduced in the Forms and Documents section). The class may be left blank if the appropriate class number is not known.

Applicant. The application must be filed in the name of the owner of the mark. Specify, if an individual, applicant's name and citizenship; if a partnership, the names and citizenship of the general partners and the domicile of the partnership; if a corporation or association, the name under which it is incorporated and the State or foreign nation under the laws of which it is organized. Also indicate the applicant's post office address.

Identification of Goods or Services. State briefly the specific goods or services for which the mark is used or intended to be used and for which registration is sought.

Use clear and precise language; for example, "women's clothing, namely, blouses and skirts," or "computer programs for use by accountants," or "retail food store services." Note that the identification of goods or services should describe the goods that you sell or services that you render, not the medium in which the mark appears, which is often advertising. "Advertising" in this context identifies a service rendered by advertising agencies. For example, a restaurateur would identify his service as "restaurant services," not "menus, signs, etc.," which is the medium through which the mark is communicated. Although this may seem to be only extreme common sense, the PTO has gone to great lengths to insure that you don't become confused.

Basis for Application. You must check at least one of four boxes to specify the basis for filing the application. Usually an application is based upon either (1) prior use of the mark in commerce (the first box) or (2) a bona fide intention to use the mark in commerce (the second box), but not both. If both the first and second boxes are checked, the Patent and Trademark Office will not accept the application and will return it to you without processing.

The last two boxes pertain to applications filed in the United States pursuant to international agreements, based upon applications or registrations in foreign countries. These bases are used relatively infrequently.

If you are using the mark in commerce in relation to all the goods or services listed in the application, check the first box and state each of the following:

The date the trademark was first used anywhere in the

country on the goods, or in connection with the services, specified in the application;

The date the trademark was first used on the specified goods, or in connection with the specified services, sold or shipped or (or rendered) in a type of commerce which may be regulated by Congress;

The type of commerce in which the goods were sold or shipped or services were rendered (for example, "interstate commerce" or "commerce between the United States and Brazil"); and

How the mark is used on the goods or in connection with the services (for example, "the mark is used on labels which are affixed to the goods," or "the mark is used in advertisements for the services").

If you have a bona fide intention to use the mark in commerce in relation to the goods or services specified in the application, check the second box and supply the requested information. This would include situations where the mark has not been used at all or where the mark has been used on the specified goods or services only within a single state (intrastate commerce).

Execution. The application form must be dated and signed.

The declaration and signature block appear on the back of the form.

The Patent and Trademark Office will not accept an unsigned application and will return it to you without processing. By signing the form, you are swearing that all the information in the application is believed to be true.

If you are acting as an individual applicant, you must execute the form; if a joint applicant, all must execute; if a partnership, one general partner must execute the application; and if a corporation or association, one officer of the organization must execute the application.

Trademark Drawings

Your drawing must be a fairly exact representation of the mark as actually used in connection with the invention. You do not have to draw a service mark if it is not capable of representation by a drawing. If your application is for registration of a word, letter or numeral, or any combination thereof, not depicted in special form, the drawing may be simply the mark typed in upper case letters on paper, otherwise complying with the requirements.

In the Forms and Documents section, the latest trademark forms are made available for your convenience. They are: Statement of Use; Amendment to Allege Use; and Request for Extension of Time. These are all for use with Declaration.

If you submit a type-written drawing, the "drawing" must be in all capital letters even if your mark uses capital letters and lower case or all lower case letters. If you wish to have your mark registered in the form in which it is actually used in commerce, you must submit a special form drawing. In that case, the showing of the mark must be no larger than four inches by four inches and should be centered on the page. The special form drawing must be black and white. Colors are designated by special lining shown in the "Basic Facts About Trademarks" booklet, available from the PTO.

Here are the exact specifications required by the PTO for trademark drawings:

Paper and Ink. The drawing must be made upon pure white durable paper, the surface of which must be calendered and smooth. A good grade of bond paper is suitable. India ink alone must be used for pen drawings to secure perfectly black solid lines. The use of white pigment to cover lines is not acceptable.

Size of Sheet and Margins. The size of the sheet on which a drawing is made must be 8.5 inches wide and 11 inches long. One of the shorter sides of the sheet should be regarded as its top. When the figure is longer than the width of the sheet, the sheet should be turned on its side with the top at the right. The size of the mark must be such as to leave a margin of at least one inch on the sides and bottom of the paper and at least one inch between it and the heading.

Heading. Across the top of the drawing, beginning one inch from the top edge and not exceeding one-fourth of the sheet, there must be placed a heading, listing in separate lines, applicant's name, applicant's post office address, date of first use, date of first use in commerce, and the goods or services recited in the application (or typical items of goods or services if a number are recited in the application). This heading may be typewritten.

Character of Lines. All drawings, except as otherwise provided, must be made with the pen or by a process which will give them satisfactory reproduction characteristics. Every line and letter, names included, must be black. This direction applies to all lines, however fine, and to shading. All lines must be clean, sharp, and solid, and they must not be too fine or crowded. Surface shading, when used, should be open. A photolithographic reproduction or printer's proof copy may be used if otherwise suitable. Photocopies are not acceptable.

Extraneous Matter. Extraneous matter must not appear upon the face of the drawing.

Linings for Color. Where color is a feature of the mark, the color or colors may be designated in the drawing by means of conventional linings as shown in the following chart:

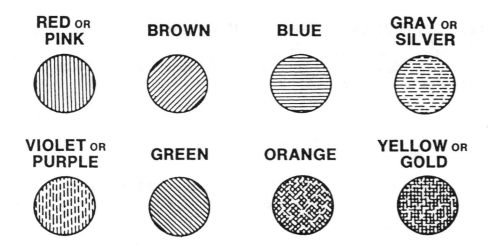

Transmission of Drawings. Other than typed drawings, you should send your artwork to the PTO flat and well protected by a sheet of heavy binder's board, rolled and posted in a suitable mailing tube.

Informal Drawings. A drawing that does not conform with these requirements may be accepted for the purpose of examination only, but it must be ultimately corrected or a new one furnished, as required, before the application will be allowed or the mark can be published.

Trademark Specimens/Facsimiles

Trademark Specimens. A trademark may be placed in any manner on the goods, or their containers, tags, labels, or displays. The five specimens submitted must be duplicates of the actual labels, tags, containers, or displays or portions thereof, when made of suitable material and capable of being arranged flat and of a size no larger than 8 1/2 x 13 inches. Three-dimensional or bulky material submitted as specimens will not be accepted, and may delay your filing date.

Trademark Facsimiles. If you must submit facsimiles of your trademark, send in five copies of a suitable photograph or other acceptable reproduction, not larger than 8 1/2 x 13 inches and clearly and legibly showing the mark and all matter used in connection with it.

Service Mark Specimens or Facsimiles. In the case of a service mark, specimens or facsimiles of the mark as used in the sale or advertising of the services must be furnished unless impossible from the nature of the mark or the manner in which it is used, in which event some other representation must be submitted.

Specimens to support use of a mark must be actual samples of how the mark is being used in commerce, i.e., actual tags, not a depiction of a tag. If it is impractical to send an actual specimen because of its size, you may submit a photograph clearly showing how the mark is used.

The average pendency period for a trademark application is approximately 14 months.

According to a recent USA *Today* survey of 1,000 inventors, fame and fortune are not the major motivating factors in their work. More than 70 percent chose personal satisfaction as a priority. The inventors, who could chose more than one category, ranked fame dead last. And only 27 percent thought profit was the major reason for their work.

If your service mark is not used in a printed or written form, only three recordings will be required.

Requirements for Receiving a Filing Date

Materials submitted as an application for registration of a mark will not be accorded a filing date as an application until all of the following elements are received:

1. Name of the applicant.

2. A name and address to which communications can be directed.

3. A drawing of the mark sought to be registered substantially meeting all PTO requirements.

4. An identification of goods or services.

5. A basis for filing:

    a) A date of first use of the mark in commerce, and at least one specimen or facsimile of the mark as used, in an application under section 1(a) of the Trademark Act or

    b) A claim of a bona fide intention to use the mark in commerce and a certification or certified copy of the foreign registration on which the application is based in an application under section 44(e) of the Act, or

    c) A claim of a bona fide intention to use the mark in commerce and a claim of the benefit of a prior foreign application filed in accordance with section 44(d) of the Act, or

    d) A claim of a bona fide intention to use the mark in commerce in an application under section 1(b) of the Act.

6. A verification or declaration in accordance with section 2.33(b) signed by the applicant.

7. The required filing fee for at least one class of goods or services. Compliance with one or more of the rules related to the elements specified above may be required before the application is further processed.

The filing date of the application is the date on which all of the above elements are received in the PTO.

If the papers and fee submitted do not satisfy all of the requirements, the papers will not be considered to constitute an application and will not be given a filing date. The Patent and Trademark Office will return the papers and any fee submitted therewith to you. The Office will notify you of the defect or defects which prevented their being considered to be an application.

Trademark Examination Procedure

Applications are docketed and examined in the order of receipt. If the examiner finds any reason that the proposed mark will not pass, you'll be advised of the objections. You are given six months to respond. If you fail to respond, the application for trademark will be abandoned.

If no problems are discovered, or if you take care of any problems in the response to the Examining Attorney's letter, the application will be approved for publication for opposition. The mark shown in the application and all pertinent information will be published in the *Trademark Official Gazette* . During the following 30 days, anyone who believes that they may be harmed by registration of the mark may file a proceeding at the Trademark Trial and Appeal Board to oppose registration of the mark.

After the mark has been published in the *Gazette* and if no opposition has been filed, the PTO will, for applications based on a bona fide intention to use the mark in commerce:

*Issue a Notice of Allowance.

*In response, you will have six months to begin use in commerce and file a Statement of Use, or,

*You may, upon written request, extend the period to begin use for another six months.

*Further six-month extensions will be granted for "good cause" not to exceed 24 months.

*Upon receipt at the PTO, the Statement of Use will be examined by an Examining Attorney, and

*If the Statement of Use complies with the Trademark Act, a Certificate of Registration will be issued.

For applications based on "use in commerce" or a foreign registration, the PTO will simply issue a Certificate of Registration.

For more information, see Fig. 24, Trademark Examination Activities. For a schedule of fees, see figure 5, PTO Fee Schedule.

Statutory Grounds for Refusal

The examining attorney will refuse registration if the mark or term applied for:

Upon receipt at the Trademark Office, your application receives a filing date, enters pre-examination processing, and has its data entered into the trademark computers (upon receipt of all the information set forth as a filing requirement). The application is then forwarded to an Examining Attorney in the Trademark Office who will review the application for compliance with the various provisions of the Trademark Act.

In FY1989, The Trademark Trial and Appeal Board had more than 4,000 cases pending.

1. Does not function as a trademark to identify the goods or services as coming from a particular source; for example, the matter applied for is merely ornamentation;

2. Is immoral, deceptive or scandalous;

3. May disparage or falsely suggest a connection with persons, institutions, beliefs or national symbols, or bring them into contempt or disrepute;

4. Consists of or simulates the flag or coat of arms or other insignia of the United States, or a State or municipality, or any foreign nation;

5. Is the name, portrait or signature of a particular living individual, unless he or she has given written consent; or is the name, signature or portrait of a deceased President of the United States during the life of his or her spouse, unless the spouse has given consent;

6. So resembles a mark already registered in the PTO as to be likely, when applied to the goods of the applicant, to cause confusion, or to cause mistake, or to deceive;

7. Is merely descriptive or deceptively misdescriptive of the goods or services;

8. Is primarily geographically descriptive or deceptively misdescriptive of the goods or services of the applicant;

9. Is primarily a surname.

A mark will not be refused registration on the grounds listed in numbers 7, 8, and 9 if you can show that, through use of the mark in commerce, the mark has become distinctive so that it now identifies to the public your goods or services. Marks which are refused registration on the grounds listed in numbers 1, 7, 8, and 9 may be registered on the *Supplemental Register* , which contains terms or designs considered capable of distinguishing the owner's goods or services, but that do not yet do so. A term or design cannot be considered for registration on the *Supplemental Register* unless it is in use in commerce in relation to all the goods or services identified in the application, and an acceptable allegation of use has been submitted. If a mark is registered on the *Supplemental Register* , the registrant may bring suit for trademark infringement in the Federal courts, or may use the registration as a basis for filing in some foreign countries. You may file an application on the Principal Register and, if appropriate, amend the application to the Supplemental Register for no additional fee.

Copyrights: the Whole Story

What do Hal David's song "Do You Know The Way To San Jose?," Cadaco's hit game "Adverteasing", *Baby Talk* magazine, "The Graduate," and this book have in common? If you said they are all protected by copyright, you are correct.

Copyrights are very different from patents and trademarks. A patent primarily prevents inventions, discoveries, or advancements of useful processes from being manufactured, used, or marketed. A trademark is a word, name, or symbol to indicate origin, and in so doing distinguish the products and services of one company from those of another. Copyrights protect the form of expression rather than the subject matter of the writing. They protect the original works of authors and other creative people against copying and unauthorized public performance.

The Copyright Office

Copyrights are not handled by the Patent and Trademark Office. For this discussion we move across the Potomac River from the PTO's Crystal City, Virginia headquarters, up Independence Avenue, and onto Capitol Hill to the **Library of Congress,** which is primarily responsible for administering copyright law.

What is a Copyright?

Copyright is a form of protection provided by the laws of the United States (Title 17, U.S. Code) to the authors of "original works of authorship" including literary, dramatic, musical, artistic, and certain other intellectual works.

What Can Be Copyrighted?

This protection is available to both published and unpublished works. I slap copyright notices on everything I create. Copyrights can be as important to an inventor as to an author, hence the reason I have included it in this book. Instructions and other written instruments such as background papers, concept papers, drawings, photographs, and the like that relate to inventions are all protected under U.S. copyright laws.

How Do You Secure a Copyright?

The way in which copyright protection is secured under the present law is frequently misunderstood. In years past it was required that you fill out special forms and send them to the Library of Congress together

Mandatory notice of copyright has been abolished for works published for the first time on or after March 1, 1989. Failure to place a notice of copyright on copies or phono-records of such works can no longer result in the loss of copyright.

The U.S. Constitution authorized Congress to establish the copyright legislation. The first federal copyright act was passed in 1790.

Beginning March 1, 1989, copyright in the works of U.S. authors will be protected automatically in all member nations of the Berne Union (there are some 80 member nations in the Berne Union). Members of the Berne Union agree to treat nationals of other member countries like their own nationals for purposes of copyright. Therefore, U.S. authors will often receive higher levels of protection than the guaranteed minimum. Works of foreign authors who are nationals of a Berne Union country are automatically protected in the U.S.

Deposit: Under the new law, you are required to deposit two copies of the published Work in the Copyright Office for use by the Library of Congress. If the work is unpublished, one copy is required. This should be done within three months of publication of any notice of copyright. Failure to do so may result in a fine. Even so, such an omission would not affect your copyright protection, just your pocketbook.

with a check and a number of copies of the original work. Today, no publication or registration or other action in the Copyright Office is required to secure copyright under the new law.

Under present law, copyright is secured "automatically" when the work is created, and the work is "created" when it is fixed in a copy or phonographically recorded for the first time. In general "copies" are material objects from which a work can be read or visually perceived either directly or with the aid of a machine or device, such as books, manuscripts, sheet music, film, videotape, or microfilm.

However, it is still prudent to make a formal application with the Library of Congress. This is to establish a "public record" of your claim; receive a certificate of registration (required if you ever have to go into a court of law over infringement); and if you receive a copyright within five years of publication, it will be considered to be prima facie evidence in a court of law.

Notice of Copyright

Before you publicly show or distribute your work, notice of copyright is required. The use of the copyright notice is your responsibility and does not need any special advance permission from, or registration with, the Copyright Office.

The notice for visually perceptible copies should contain all the following three elements:

1) **The Symbol** © or the word "Copyright," or the abbreviation "Copr."

2) **The year of first publication of said work.** In the case of complications or derivative works incorporating previously published material, the year of first publication of the compilation or derivative work is enough. The year may be omitted where a pictorial, graphic, or sculptural work, with accompanying text (if any) is reproduced in or on greeting cards, postcards, stationary, jewelry, dolls, toys, or any useful articles.

3) **The name of the owner of copyright** in the work, or an abbreviation by which the name can be recognized, or a generally known alternative of the owner.

Example: © 1991 Richard C. Levy   You should affix the notice in such a way as to give it "reasonable notice of the claim of copyright."

How Long Does Copyright Protection Endure?

A work that was created on or after January 1, 1978, is automatically protected from the moment of its creation, and is usually given a term enduring for the author's life, plus an extra fifty years after the author's death. In the case of a "joint work prepared by two or more authors who did not work for hire," the term lasts for fifty years after the last surviving author's death. For works made for hire, and for anonymous and pseudonymous works (unless the author's identity is revealed in Copyright Office records), the duration of copyright will be seventy-five years from publication or one hundred years from creation, whichever is shorter.

Works that were created before the present law came into effect, but had neither been published nor registered for copyright before January 1, 1978, have been automatically brought under the statute and are now given Federal copyright protection. The duration of copyright in these works will generally be computed in the same way as for works created on or after January 1, 1978 (the life-plus-50 or 75/100-year terms apply to them as well). However, all works in this category are guaranteed at least twenty-five years of statutory protection.

Under the law in effect before 1978, copyright was secured either on the date a work was published, or on the date of registration if the work was registered in unpublished form. In either case, the copyright endured for a first term of twenty-eight years from the date secured. During the last (28th) year of the first term, the copyright was eligible for renewal. The new copyright law has extended the renewal term from twenty-eight to forty-seven years for copyrights that were subsisting on January 1, 1978, making these works eligible for a total term of protection of seventy-five years. However, the copyright must be timely renewed to receive the forty-seven year period of added protection.

What Copyrights Do Not Protect

Ideas cannot be copyrighted. The same is true of the name or title given to a product and the method or methods for doing something.

Copyright protects only the particular manner in which you express yourself in a literary, artistic, or musical form. Copyright protection does not extend to ideas, systems, devices, or trademark material involved in the development, merchandising, or usage of a product.

Application Forms

The forms you will likely require include Form TX, Form VA, Form CA, and Form RE.

Form TX: For published and unpublished non-dramatic literary works. This comprises the broadest category, covering everything from novels to computer programs, game instructions, and invention proposals.

Form VA: For published and unpublished works of the visual arts. This would be for artwork you may have developed as an adjunct to your invention, charts, technical drawings, diagrams, models, and works of artistic craftsmanship.

Form CA: For application for supplementary Copyright Registration. Use when an earlier registration has been made in the

Renewal is Still Required

Works first federally copyrighted before 1978 must still be renewed in the 28th year in order to receive the second term of 47 years. If such a work is not timely renewed, it will fall into the public domain in the U.S. at the end of the 28th year.

If you would like to get specific information on the copyright process, call a Copyright Public Information Specialist at (202) 479-0700 weekdays between the hours of 8:30 a.m. and 5:00 p.m. These folks are a wealth of information. You can also write for information to Information and Publications Section, LM-455, Copyright Office, Library of Congress, Washington, D.C. 20559.

Other numbers of note:

Forms & Circulars
Hotline:
(202) 707-9100

Reference &
Bibliography
Section:
(202) 707-6850

Certifications &
Documents Section:
(202) 707-6787

Copyright General
Counsel's Office:
(202) 707-8380

Documents Unit:
(202) 707-1759

Licensing Division:
(202) 707-8150

Copyright Office; and some of the facts given in that registration are incorrect or incomplete; and you want to place the correct or complete facts on record.

Form RE: Renewal registration. Use when you wish to renew a copyright.

These forms are reproduced in the Forms and Documents section.

The cost is only $10.00 per registration, until January 3, 1991, at which time it increases to $20. This fee has been the same since 1978, and, unlike PTO fees, an act of Congress is needed to increase it.

How to License Your Invention

When you have something hot, burn it. When it gets cold, sell it for ice.

Has the Time Come for Your Idea?

Competition drives American industry, creating a fertile market for new inventions. No scarcity exists of progressive manufacturers willing to consider appropriate, new, and innovative products from outside, independent inventors. Many companies in every field strive to produce better and more competitive products on a continuing basis, and often rely upon the independent inventor.

What Victor Hugo wrote in 1852 has never rung truer than today: "Greater than the tread of mighty armies is an idea whose time has come." The difficulty is bridging the supply and demand gap between the creators of new ideas and those who possess the capabilities and facilities to manufacture and market them successfully.

America constantly demands new products. We are, after all, a "throw away" society. We purchase something, use it for a while, and upgrade to the next generation of the product as soon as it's affordable. Generally, we do not take good care of things, nor do we spend a lot of money fixing them when they break.

The biggest problem inventors—experienced and inexperienced—have is how to license their innovations. While many people have the ability to dream and innovate, few of them also have the ability to sell.

Two methods are available to you for manufacturing and marketing your invention. One is to license the invention to a firm that specializes in manufacturing and selling products. The other method is to raise venture capital and become the manufacturer. Both have pluses and minuses. Neither is risk free. No guarantees are offered either way, just different opportunities. Both are in the best and proudest traditions of American entrepreneurism.

I personally subscribe to the school of thought that says it is best to license inventions to companies run by executives who know more than I do about how to cost-effectively develop, manufacture, package, market, and promote products to specifically targeted users. I have no desire to go through the learning curve to acquire an expertise in all that's needed to do what it is a manufacturer does. And if you are thinking of doing this, remember Murphy. Murphy's first corollary is: Nothing

Marketing is a Science

Marketing inventions is a science; no less a discipline than electronics, mechanical engineering, or any other field of study. It can be learned. Of course, as with any talent, some natural abilities may make it easier for some, but nothing is out-of-reach if you take the time.

is as easy as it looks. Murphy's second corollary is: Everything takes longer than you think.

My aim is to create and maintain an atmosphere in which I can be most creative and productive, a situation free from supervisors, meetings, suits, power lunches, time-tested methods, office politics, and assorted "administrivia." You don't do this by raising venture capital and starting a manufacturing company. You do this by accepting the risks inherent in independent product development and licensing your creative yield.

You have three methods to address the business of invention licensing. The first is the do-it-yourself approach. I prefer this method. Over the years I have found that putting together my own team and making my own deals works best. I prefer to take my chances, not someone else's. It is more fun, I have better control, and the rewards are far greater.

Another way is to take your inventions to an inventor's exposition and run it up the flag pole for everyone to salute. I have discovered that in most such cases the only people who make money are the organizers of the exposition. They'll sell everything from admission tickets to ads in programs. What rarely sells, unfortunately, are inventions.

And finally, you can engage the services of an agent or broker. Good agents are few and far between. The best agents become so involved that they become virtual co-developers. I speak from personal experience, based on occasionally representing products other than my own. Nothing less will do. Jon W. Bayless, a partner in Sevin Rosen Bayless Venture Fund, says "A third party who is not committing his future to the business has no part to play in the process. Third party involvement is frequently a reason for a project to lose the consideration it may deserve."

Let's look at the latter first because as soon as the patent issues, you are likely to hear from one species of agent, whether you want to or not, the infamous **invention marketing service.**

PTO Can Help You Get the Word Out

The Patent and Trademark Office cannot help develop or market your invention, but it will publish, at the request of the patent owner, a notice in its *Official Gazette* that the patent is available for licensing or sale. The fee for this is a very reasonable $7.00.

Invention Marketing Services: Words to the Wise

How Do They Get Your Name?

Don't be surprised if you first learn of your patent award from an "invention marketing" company. When your patent issues, a notice of it will be automatically carried in the *Official Gazette* of the United States Patent and Trademark Office. This publication, issued weekly since 1872, comes out each Tuesday. It publicly records the following information about each new patent:

1) the name, city, and state of residence of the applicant, with the Post Office address in the case of unassigned patents;

2) the same data for the assignee, if any;

3) the filing date;

4) the serial number of the application;

5) the patent number;

6) the title of invention;

7) the number of claims;

8) the U.S. classification by class and subclass;

9) a selected figure of the drawing, if any, except in the case of a plant patent;

10) a claim or claims;

11) international classification;

12) U.S. patent application data, if any; and,

13) foreign priority application data, if any.

In the case of a reissue patent they publish the additional data of the number and date of the original patent and original application.

Invention marketing companies subscribe to the *Official Gazette* as an inexpensive way to obtain a qualified mailing list of inventors recently granted patent protection. They typically approach the independent inventors rather than their assignee corporations.

"They [invention marketeers] scare the living daylights out of me," says James Kubiatowicz, former director of product development/toys at Spearhead Industries. "They're leeches. They prey on the novices." Then pulling open his desk drawer, he adds, "I've got thirty-two post cards from these guys and I am saving the stamps. One of these days I'll steam them off and put them into a retirement fund.

"Anybody who is breathing and has a dollar can get strokes from these companies. They work on a person's ego."

The "Pros" and the "Cons"

"About 80 percent of all people claiming to help inventors build a business, market their product, or raise capital are con-men, beggars, thieves, or incompetents. Finding the other 20 percent is damn tough," says Professor Mark A. Spikell. Dr. Spikell is cofounder of the Entrepreneurship Center in the School of Business Administration at George Mason University.

Florida inventor Frederick L. Jones, a member of the Palm Beach Society of American Inventors, agrees about invention marketing companies. "They prey on people. They'll take all the money they can."

Caveat emptor! Learning to recognize the good invention marketing companies from the bad ones is not easy. The invention marketing business is rife with nonperforming, paracreative slugs who prey on and take advantage of unsuspecting, inexperienced, frustrated, and hungry independent inventors. In my experience, most invention marketing organizations that advertise via direct mail, classified ads, radio, or television spots are less than effective. Often, they carry a very appropriate appellation: "front money frauds."

Alan A. Tratner, president of Inventors Workshop International, says about such parasites, "We view the unsavory practices of many of these firms and individuals as a 'cancer' in the inventing community that needs to be eradicated immediately. Too often inventors lose large amounts of money and are derailed by the unfulfilled promises and come-ons of these companies."

Al Lawrence Smith of the Patent and Trademark Office cautions, "Beware of these people." He suggests that inventors only permit marketeers to have their patented products on a contingency, non-exclusive basis.

"Pure and simple garbage," is how Fred L. Hart, Outreach Inventor Liaison at NIST's Office of Energy Related Inventions, describes the work most invention marketing services do for inventors. In a meeting with Fred at his Gaithersburg, Maryland office, he showed me two recently submitted invention assessment reports that a Pittsburgh invention marketing firm had prepared for two different inventors and their inventions. The first was for a frost pump. The second involved a magnetic engine (the mythical perpetual motion machine). *Both reports were 34 pages in length. Both reports were identical, word-for-word, except for the name of the invention and the respective product descriptions.*

The inventors paid the invention marketing company for the same boilerplate. It was obvious that a standard text had been designed and was being used for every product. In the case of the perpetual motion machine, Fred said the company had written and submitted three separate but identical reports on behalf of three different inventors who felt they had discovered the way around gas engines. The reports were all signed by the same man who tagged his name with the initials, P.E. (Professional Engineer).

It is almost a guarantee that if you engage the services of an invention marketing company, the only ones who'll get rich will be the invention marketeers on your payroll. In an article entitled, "Dream Weavers," published in the October 1988 issue of *Venture* magazine, writer Sallie Hofmeister estimates that these scams rip inventors to the tune of $50 million annually. She labels them, "sales agents, who work mostly by phone and largely on commissions...typically quick with praise for even the most ludicrous ideas."

Hofmeister reports that a mere 48 of the 9,184 inventors that contracted with Invention Marketing, Inc. (IMI) in its first decade, before it changed its name to Invention Submission in 1984, ever made a penny from their ideas, and only 14 of them netted more than the median $1,000 they paid the company, according to documents obtained by the Minnesota attorney general.

Watch out for them. They are easy to recognize. You'll find their ads running on all night radio talk shows, at 3:00 a.m. sponsoring Million Dollar Movie reruns, and in the classified advertising sections of newspapers and magazines that appeal to do-it-yourselfers, such as *Popular Mechanics*, *Popular Science*, and *Income Opportunities*.

Direct Mail Tactics

The direct mail pieces they send might be in the form of a postal card. They usually read something like this:

IMPORTANT NOTICE
To: Owner of U.S. Patent

We have located six companies that produce, market, or sell products in a field to which your invention might apply. You may wish to contact these manufacturers if you are interested in one or more of the following:

> A. Licensing your patent.
>
> B. Finding a company to produce your invention.
>
> C. Securing marketing, distribution, and/or sales help.
>
> D. Hiring design or technical support.

Then comes the kicker. You are asked to remit a fee of $75.00 plus $1.00 for each company named.

Fat Envelopes

Other direct mail offers come in #10 envelopes overflowing with all kinds of slick fliers and official-looking confidential disclosure agreements with diploma-like borders. Rarely will you be able to make out the signature of the "authorized agent" who signed off on the standard "Dear Inventor" cover letter. You'll likely not recognize any products which they claim to have successfully licensed to industry.

And the company's toll-free number will be something like 1-800-SUCCESS.

They use all kinds of subtle techniques to compensate for the fact that they have no successful products to show off. For example, I recall seeing one promotional flier that depicted a beautiful, modern building with the caption "4th Floor, Chicago, Illinois." The company wanted the reader to assume that this was the base of its corporate headquarters.

Dave Thomas created an improved joystick for video games. His wife, Susan, saw an ad in USA Today: "Have an invention? Need help?" The advertiser they contacted in Boston was excited about promoting the device. There was just one hitch: it would take $10,000 to get it off the ground.

Then Susan's brother-in-law, who believed in the invention, was killed in an auto accident. His widow provided the $10,000 from the insurance settlement.

Dave and Susan flew to Boston. The "account representative" wined and dined them. They signed a contract and handed over a $10,000 cashier's check.

And now? The account representative has disappeared. The Massachusetts Attorney General's Office doesn't have any record of the company.

And Dave and Sue are $10,000 wiser.

Frank Smirne, president of F. Smirne Plastic Co., Allentown, Pennsylvania, suggests that inventors get advice from other inventors before dealing with invention marketeers.

Don Coster, a 67 year-old former casino worker who now heads up the Nevada Inventor's Association, speaks his mind on invention marketing companies. "I feel that the 800 numbers in the back of publications such as Mechanix Illustrated, to pick one example, reach companies that appeal to vulnerable people with ideas and no way to pursue them. In the end, these companies get the inventor's money and the inventor gets nothing."

However, no reference was made to the picture or office building anywhere in the flier's copy.

The more sophisticated invention marketing companies wrap themselves in Old Glory and apple pie. Their names incorporate words like American, Federal, and National and they have Washington, D.C. postal addresses.

How to Identify the Good Guys

One of the acid tests I use to separate the honest and serious professional invention marketing companies from those involved in unfair business practices is whether or not they want any advance payments from the inventor for anything whatsoever.

"The firm we dealt with appears to be interested in marketing our idea for us," a woman from Virginia wrote me, "but they are quoting us a price of from $3,000-$9,000 for this service. Also, they want a slice of any profits (up to 60 percent)."

The most reputable invention marketing services will not require front money from the inventor. You invested your ingenuity, time, and money in creating the product. The marketer must invest ingenuity, time, and money in causing the product to be licensed. What is more fair than that? The moment you pay for marketing services, the carrot is removed. There is no risk. With nothing to lose and a deposit already in the bank, your marketing partner has little reason to display incentive.

Charles F. Mullen, an inventor from Houston, Texas, works part time as sales manager for national accounts at an advertising agency, and says he spends the rest of his time inventing and helping others market their products. "We work straight percentage. We don't charge them anything up front," he says. "That's the trouble with the industry. Most of the people get them up front." Mr. Mullen claims to have licensed fourteen inventions in all different fields.

A reputable and active marketing company will be able to demonstrate a track record of successful products and satisfied inventor clients. References should be available. Ask for a list of clients and their telephone numbers. The products they have marketed should be available somewhere other than their office waiting rooms.

Read the small print. I saw one flier that showed numerous products, leaving the impression that they had been licensed by the invention marketing group. Then I saw the disclaimer. "These products are not intended to represent success for inventors who have worked with our firm."

They should be able to substantiate licensing agreements claimed and royalty earnings. Do they make the majority of their earnings from royalties or from inventor service fees? If it is not from royalties, you're in the wrong place.

Always check the name of the invention marketing company and/or individual marketer with the local Better Business Bureau, state agency for consumer fraud, and your Attorney General's office. See if any complaints have ever been registered against the company or its officers and representatives.

Ask fellow inventors. Word about bad apples spreads like wild fire. Many inventor organizations maintain files on the worst invention marketing organizations and frequently write them up in their newsletters.

Inventors' Digest Litmus Test

Amateur inventors, according to Joanne M. Hayes, editor of *Inventors' Digest* , a publication of Affiliated Inventors Foundation, Inc. (AIF), are full of questions, not the least of which is, "How can I find somebody to help me with my invention who won't rip me off?"

To find out if an invention is patentable and/or marketable and/or worth doing at all, an inventor has to show the idea to someone. Inventions do not develop and come to market in a vacuum. Joanne says that one of the first things many amateurs do is look in the phone book for help under "Invention" or "Patents" and call the numbers listed. Yet others find a name in a magazine or hear about an invention evaluation/marketing company on the radio.

"The invention development field is loaded with con men, rip-off outfits, and downright crooks," warns Ms. Hayes. With this in mind, her organization has designed a checklist for inventions.

AIF says that the following questions should help you to identify practices which are warning signals and typify activities operating with doubtful honesty. "If you check any of the non-bold responses, be warned that there is a strong possibility you are dealing with crooks. We recommend that you avoid doing business with any company that does not answer the following questions satisfactorily," advises Hayes.

How to Rate An Invention Marketing Company

About Recording Inventions:

Do they imply that the Disclosure Document Program (DDP) of the U.S. Patent Office offers protection or that it is a viable substitute for a U.S. patent (the Patent Office says it is not)?

yes **no**

About Evaluations:

Do they offer a meaningful evaluation of inventions at a reasonable price (less than $200)?

yes no

Empirical observation is always in order. In the name of science and boy scouts everywhere, I tested one of these invention marketing companies. After hearing a toll-free number on late-night radio, I gave the company a call. My idea was for a device that kept individual strands of spaghetti fresh. The concept was to insert spaghetti strands into clear plastic extrusions capped off at both ends. "Terrific. Lots of potential," I was told by telephone. All I had to do was pay $400 for an initial product assessment.

It is wise to check invention companies out with your State Attorney General, and even the Federal Trade Commission.

If you wish to register a complaint against an invention marketing company, write to: Stanley Ciurczak, Chief, Complaint Division, Room 692, Federal Trade Commission, Washington, D.C. 20580.

Does their evaluation report point out weaknesses as well as strengths of your invention?

yes   no

Does the evaluation report make specific recommendations concerning such important areas as patenting, model making, marketing and protection?

yes   no

About Patent Searches:

Do they recommend a patent search before recommending expensive marketing services?

yes   no

If they offer to perform a patent search, do they state specifically that the search will include an opinion of patentability by a registered patent attorney (or registered agent)?

yes   no

If you purchased a patent search, did they tell you the name of the individual responsible for the search and the name and registration number of the individual providing the opinion of patentability?

yes   no

About Marketing Services:

Do they ask for exorbitant front-end cash fees running to hundreds or thousands of dollars for their "best efforts" to sell or present your invention to industry?

yes   **no**

Do they tell you that in most cases to sell or license the rights to your invention you need a working model?

yes   no

Do they tell you anything in conversation, such as the "company is all lined up to buy your invention," which does not end up in the written contract?

yes   **no**

If they think your invention is so promising, are they willing to offer their marketing services on a no-cash, straight-commission basis?

yes   no

Do they suggest trying to market your potentially patentable invention before it reaches the "patent pending" stage?

yes   **no**

About Patent Applications:

Do they recommend filing for a patent application before beginning marketing efforts?

yes **no**

Do they quote a firm, fixed price for preparation and filing a patent application (except for patent drawings, which are priced by the sheet)?
yes no

Do they tell you about the patent application costs after filing, such as the almost-always required amendment(s), the final issue fee, maintenance fees, etc.?

yes no

About References:

Are they willing to provide you with names and addresses of inventors who have used their services?

yes no

Is the company willing to provide you with a copy of its latest financial statement indicating its net profit?

yes no

REMEMBER: If you checked even one non-bold response, you should investigate the company in greater detail before proceeding!!

Invention Consultants

Do not confuse invention marketing firms with legitimate invention consultants! I was introduced to the product development business when I was hired as a marketing consultant by an entrepreneur who had an idea for a child's computer. As you can imagine, therefore, I believe in using consultants whenever their expertise can contribute to the progress of a project.

What is consulting? I like the definition offered by England's Institute of Management Consultants. It defines consulting as:

"The service provided by an independent and qualified person or persons in identifying and investigating problems concerned with policy, organization, procedures and methods; recommending appropriate action and helping to implement these recommendations."

As many types of consultants exist as there are problems to solve. These experts can bring new techniques and approaches to bear on an inventor's work. This contribution can range from helping to bridge a technological gap to the special knowledge and talent required to successfully license or market a particular innovation.

"I actually made my consultant a partner," says Richard Tweddell III, inventor of VegiForms, a device that press molds vegetables, while still on the vine, into the likeness of human faces. A professional toy designer of Kenner Products in Cincinnati, Ohio, Tweddell credits his consultant with moving his company from ground zero. "He showed me how to license the product, he reviews all of our licensing agreements and he found companies that were interested in taking the rights to it."

Carl E. Haskett, inventor of the Haskett SunClock, an elaborate sundial that is a precise as a watch, is a former consultant at Booz. Allen. Hamilton. He advises that in dealing with consultants the inventor must "build a box so tight around the problem that there are no holes in corners." The University of Oklahoma graduate who as a B.S. degree in Engineering Physics says this means the task must be defined so carefully that the consultant cannot expand it without absolute permission.

In the case of the child's first computer project, an electronics wizard originated the basic product with help from several consultants. A former Texas Instruments engineer, for example, consulted regarding the best microprocessor to use. My services were engaged to provide information on how to best position the item within the marketplace.

Consultants Offer Objective Viewpoints

Advisors can provide impartial points-of-view by seeing challenges in a fresh light. They operate outside existing frameworks and free from existing beliefs, politics, problems, and procedures inherent in many organizations or situations. Most consultants operate on the basis of an hourly rate plus expenses. However, inventors, by the nature of their work, are often able to make equity deals whereby in return for their advice, consultants are given participation in any profits said invention might generate. Inventors should think long and hard before doing something like this because it is often less expensive to risk the cash and hold all the points possible in- house.

The Pictionary Story

One recent example of an inventor who took in consulting partners to make his project materialize is Seattle waiter, Robert Angel, inventor of the popular game *Pictionary* .

Angel knew that he had a terrific concept, but needed some graphic support. He asked artist Gary Everson to help him design the game board in return for points in the venture. Everson agreed. The then 26-year old inventor also needed assistance in making business decisions, so he gave points, as well, to an accountant named Terry Langston.

Pictionary went on to be the best-selling game of 1987, grossing more than $52 million at retail. In 1988 the product's sales soared to an astounding $120 million. To date more than 14 million units of the quick draw game have been sold.

The three partners have become wealthy, receiving royalties on every game sold.

Do not think that consultants have all the answers. They do not. Consulting is very hard work and not everything can be solved as quickly as one would like. Do not look for miracle solutions.

Shop around. Get references on any consultant or research organization you are considering. Don't be impressed by a consultant's or organization's professional association alone, (for example, if they are part of a university). Their success rate in fields related to yours is what matters. What matters is know-who as much as know-how. How much can they do with a single phone call? Results is what you want, not just paper reports.

A very good source for consultant leads is Gale Research Inc.'s *Consultants and Consulting Organizations Directory* . To find a copy, consult your local library.

Federal and State Protection

Federal Trade Commission

The Federal Trade Commission (FTC) has received so many complaints about invention marketing companies that it has a complaint category for them.

Based upon a case decided in 1978, the FTC has determined that the following practices used in the advertising and marketing of idea or invention promotion or development services are unfair or deceptive trade practices and are unlawful under Section 5 (a) (1) of the FTC Act:

1) For a seller of idea or invention promotion or development services to misrepresent, directly or indirectly, that potential purchasers will be provided with evaluations or appraisals of the patentability, merit, or marketability of ideas or inventions.

2) To represent, directly or indirectly, that the seller of idea or invention promotion or development services, or its officers, agents, representatives, or employees are registered patent attorneys or patent agents, or are qualified to practice before the U.S. Patent and Trademark Office, unless such is a fact.

3) To misrepresent, directly or indirectly, the scope, nature, or quality of the services performed to develop or refine ideas or inventions.

4) To misrepresent, directly or indirectly, the scope, nature, or quality of the services performed to introduce or promote ideas or inventions to industry.

5) To represent, directly or indirectly, that a seller of idea or invention development service has special access to manufacturers or has been retained to locate new product ideas, unless such is a fact.

6) To misrepresent, directly or indirectly, that a person, partnership, corporation, government agency, or other entity endorses or uses the services of a seller or provider of services.

7) For sellers of idea or invention promotion or development services to fail to disclose, when price information is provided to potential purchasers, all significant fees or charges that may be incurred by purchasers in connection with such services.

A general telephone number for the FTC is (202) 326-2222.

Send your Freedom of Information Requests to: Keith Golden, FOIA Officer, Federal Trade Commission, 6th Street and Pennsylvania Avenue, N.W., Washington, D.C. 20580.

8) To misrepresent, directly or indirectly, the background, qualifications, experience, or expertise of a seller or provider of services.

9) For a seller of idea or invention promotion or development services to induce through misleading or deceptive representations the purchase of services that have little or no inherent value, or to offer to provide services that grossly exceed the value of the services actually provided. It is also an unfair or deceptive act or practice to retain money from the sale of such services.

If you want to find out whether the FTC has any ongoing investigation into a particular invention marketing company, use a Freedom of Information Act (FOIA) request. The FTC will not divulge any particulars about an active investigation, legal or otherwise, or even about complaints received; however, if a consent order has been issued against the invention marketing company, you'll find out.

You may also read between the lines. If under an FOIA request you ask for copies of any and all FOIA requests for information about a specific company(s), you just may get in return copies of requests sent to the FTC by law firms representing said company(s). The law firm would not be making a request about its client unless something was amiss.

For example, I once made an FOIA request to find out whether anyone else was interested in a specific invention marketing firm. The FTC sent me copies of requests, although denied, from a television news department, and a law firm. The fact that the FTC had denied their requests told me that said invention marketing firm was being looked at for some reason.

Protective Legislation at the State Level

The invention marketing business went unregulated for years. Now, however, some states such as California, Illinois, Minnesota, North Carolina, Ohio, South Dakota, Tennessee, Texas, Virginia, and Washington have enacted protective legislation on behalf of resident inventors.

Let's take a quick look at some of the issues that the states address. This will give you a good idea of what to watch for and ask prospective invention marketing companies. If you want a full text of the legislation—which I recommend that you acquire—contact the state legislature and it will be sent to you free of charge.

Let's take a look at the laws of Minnesota and Virginia. I'll highlight some important points they make.

Minnesota (Invention Services 325A.02)

1) Notwithstanding any contractual provision to the contrary, inventors have the unconditional right to cancel a

contract for invention development services for any reason at any time before midnight of the third business day following the date the inventor gets a fully executed copy.

2) A contract for invention development services shall be set in no less than 10-point type.

3) An invention developer who is not a lawyer may not give you legal advice with respect to patents, copyrights, or trademarks.

4) The invention marketer must tell you (1) the total number of customers who have contracted with him up to the last thirty days;

(2) the number of customers who have received, by virtue of the invention marketer's performance, an amount of money in excess of the amount of money paid by such customers to the invention marketer pursuant to a contract for invention development services.

5) The contract shall state the expected date of completion of invention marketing services.

6) Every invention marketer rendering invention development services must maintain a bond issued by a surety admitted to do business in the state, and equal to either ten percent of the marketer's gross income from the invention development business during the preceding fiscal year, or $25,000, whichever is larger.

Virginia, Chapter 18, 59.1-209

1) No invention developer may acquire any interest, partial or whole, in the title to the inventor's invention or patent rights, unless the invention developer contracts to manufacture the invention and acquires such interest for this purpose at or about the time the contract for manufacture is executed.

2) The developer must tell you if they intend to spend more for their services than the cash fee you will have to pay.

3) The Attorney General has the mandate to enforce the provisions of this chapter, and recover civil penalties.

If your state does not have protective legislation for inventors yet, get the word to the appropriate elected officials. Every state should have some sort of regulation on the books that sets fair standards for invention marketing companies.

Marketing Your Inventions Via Expos, Trade Shows, and Conferences

Inventor Expos

Some inventor expositions are worthwhile and well-meaning, such as the annual National Inventors Day Exhibit sponsored by the Patent and Trademark Office or regional shows sponsored by inventor organizations. But many slick operators disguise their motives by organizing inventor expositions and fairs. Know thy promoters and their motives.

Some wolves in sheep's clothing invite inventors suffering from "sellitus" to display prototypes, even drawings or photographs of their inventions. They charge for booth exhibition space and advertising in publications that are released in conjunction with the event. The general public is charged an admission fee to see the inventions. Then to top it all off, some hit exhibitors with broker or commission fees should an invention be licensed through its exposure at the show. Another scheme is to offer the inventor a large cash buy out should a product sell, and the promoter walks off with the royalty points.

When all is said and done, my experience confirms that few if any meaningful contacts ever come out of these shows from the inventor's standpoint, while the promoters make lots of money and get publicity to boot.

Respected inventor Calvin D. MacCracken, president of Calmac Manufacturing, feels that inventor expos are generally not worth it. "My reasoning is simple," he explains. "Product should be marketed within the field for which it is intended to be used. For example, if you have something for the plumbing industry, you show it at a plumbing industry, not an inventor's, expo—no matter how good the product is."

Trade Shows

Do not confuse inventor expos with trade shows. Every industry takes part in trade fairs, including the butchers, the bakers and the candlestick makers. According to the Trade Show Bureau, more than 9,000 trade shows took place in the U.S. during 1989, 3,300 of them in excess of 10,000 square feet or 100 booths.

National, regional and local events promote the sale of almost anything you can imagine. There are trade shows for everything from hardware, consumer electronics, apparel and aircraft to nuclear medicine, dental equipment, toys, comic books and musical instru-

IMPORTANT! Should you be contemplating such a show, keep one very important thing in mind. If you intend to apply for patent protection on an invention, its display may foul your chances. Exhibiting a new invention in public for sale or otherwise will make it ineligible for patent protection if a patent application is not filed within one year from the time of the invention's first public exposure.

"I'm not too happy with inventor expos," says E. D. Young, a 25-year veteran of inventor organization management and founder of The Inventors Network. "I would rather see the inventor hire a professional to market his product. . . and I have yet to see excellent brokers—I mean good, solid, reliable brokers—at expos. . . for some reason they do not attend these things." Young adds that these shows primarily consist of inventors looking at each other's products. "And," he adds, "that's the blind leading the blind."

ments. If it is manufactured and sold, you can be sure that it is marketed at a trade show somewhere, sometime.

One-Stop Research Shopping

While I do not recommend trade shows as the best place to present or license patented inventions, they are a must for getting the beat on any particular market and its dynamics. It's all there for you to peruse at your leisure, with the convenience of one-stop shopping. Competitors line up side-by-side for the important buyers to compare products and pricing, and to look for industry innovations and trends.

If you go to the shows, remember that companies have paid many thousands of dollars to exhibit. Their primary reason for being there is to ring up sales or create leads. They are not there to license concepts, by and large.

Anyway, the sales force does not review new concepts. It is responsible for selling, not developing, product. It is both fruitless and dangerous to impose on and expose inventions to salespeople. They are, however, excellent sources of information on companies and the industry and normally delighted to chat about their products, the state-of-the-market, and so on.

Exceptions exist, of course. In some industries, R&D executives attend trade shows to get a feel for the competition as well as host "invited" outside inventors with product to show. Presentations are typically conducted in hotel suites away from the exhibition site. It is best to call the corporate headquarters in advance of the trade show and check policy.

I attend the shows to scout for new product introductions, pick-up information handouts and samples, and make personal contacts. There is no better or more cost-effective way to acquire product literature than at a trade show. Manufacturers publish flyers and information kits just for trade show distribution. And most come with a price list!

I take empty flight bags with me in which to transport home the booty that I collect. Sturdy bags. I never rely on the paper or plastic bags some of the companies supply. The material is too valuable (and heavy!). Thanks to trade shows over the past decade, I have today a most comprehensive reference library.

Trade shows are also an excellent place to meet and network with executives to whom you otherwise would not have access. They rarely take their "bodyguards" to trade shows; it is too expensive and, after all, they also go to meet new people. They even make it easier by wearing name tags!

I have made super contacts in convention hotel elevators, lobby queues, and taxi-shares to and from the exhibition centers.

The best kinds of shows at which to meet senior executives are the national or international events. The smaller regional or local trade

So many trade shows occur that a dedicated industry has evolved, complete with its own newsletters, magazines and professional organizations. One of the very best sources of information on trade shows is Gale Research's annual *Trade Shows Worldwide: An International Directory of Events, Facilities, and Suppliers*. More than 5,000 events are listed, frequently with dates and locations for the next five years or more. Check your local libraries for a copy.

Most trade shows are admission free "to the trade." All you usually need is a business card to enter the exhibition area.

shows are typically staffed by salespeople alone. Nevertheless, such shows provide a less hectic atmosphere and many of the same resource materials.

How to Locate the Right Trade Show

There are several methods for discovering where and when trade shows for any particular industry will be taking place.

1. Ask a manufacturer or distributor in your field of invention. The sales and marketing people will have such information at their fingertips.

2. Contact the trade association that covers your field of invention. More than 3,600 trade associations operate on the national level in America. A great way to start is to peruse Gale Research's annual Encyclopedia of Associations, available at most libraries. More than 22,000 active associations, organizations, and other non-profit membership groups are described, in virtually every field of human endeavor.

3. As I mentioned before, check Gale's Trade Shows Worldwide. The directory provides a full description of the shows, including number of exhibitors and significant events taking place during the show.

Conferences and Meetings

Perhaps more conferences and meetings are going on than trade shows. It doesn't take much to have either. All you technically require is so many experts sitting around a table discussing a field of interest.

The biggest difference between trade shows and conferences is that you almost always pay to attend conferences. This is because the primary reasons for a conference include hearing experts speak, picking their brains, sharing your ideas, and networking.

Conferences are excellent places to get to know the people behind the products. Socializing is encouraged and the atmosphere is calmer than that at trade shows. There is no pressure to sell. There is no pressure to buy. The object is to brainstorm and exchange ideas. Participants can increase their "know how" and "know who" at the same time.

Many trade fairs have conferences or seminars scheduled. And many conferences offer simultaneous resource fairs.

How to I Find Out About Conferences in My Area?

Several methods are available for finding out where and when conferences for a particular industry will take place.

1. Ask a manufacturer or distributor in your field of

Many exhibitors host receptions and special events for buyers. If you get lucky, you may be invited to attend.

"My success is reflected by the amount of research I have done," says Howard Jay Fleisher, New York City inventor of the Polygonzo puzzle. "I basically get everyone of the trade magazines in the gift and toy industries. I go to every trade show from the gift show to the toy show, walk around, meet the people and find out who's who and what's what.

I know what's out there and know the right company to approach. That's the biggest mistake most inventors and designers make. They don't really research who is the best company for their product...instead they just do a shotgun approach."

For information on future locations of the National Innovation Workshop, call OERI at (301) 975-5500 or write OERI, Building 202, Room 209, Gaithersburg, MD 20899.

Major companies that have recently participated in the National Inventors Expo include Amoco Corp., Coca Cola, Du Pont, General Electric, IBM, NutraSweet Co., and Proctor & Gamble.

invention. Many larger manufacturers have training departments which can provide helpful information.

2. As with trade shows, contact the trade association that covers your field of invention. Once again, check Gale's Encyclopedia of Associations.

3. Ask department heads and professors at nearby universities where your field of interest is taught. Universities, particularly those teaching engineering and kindred technical fields, will have a current schedule of conferences on hand.

National Innovation Workshops

The most reputable and best organized invention expo is an event co-sponsored by the Patent and Trademark Office. The National Inventors Expo is held annually at Crystal Plaza Building 3, 2021 Jefferson Davis Highway, Arlington, Virginia. It features some 50 exhibits by independent inventors of their patented inventions, and by an array of large and small businesses. Admission is free to the Expo, which is open to the public. For current information, contact Oscar Mastin, Office of Public Affairs, PTO, Washington, DC 20231, or call (703) 557-3341.

The National Institute of Standards and Technology's Office of Energy-Related Inventions (OERI) supports six National Innovation Workshops each year in different regions of the country to stimulate and bring inventors in contact with local innovation sources. Evolving over the years, this program provides a network for innovation and is recognized as a successful trendsetter throughout the inventor community.

The National Innovation Workshop series was initiated in the spring of 1980, with the standard format a two-day seminar. Two addresses are given each day (keynote and luncheon) by nationally known speakers. A wide variety of free how-to and technical publications are available from federal, state, and local government agencies and other organizations. Each day features three 75-minute periods of eight to 10 concurrent workshops each (48 to 60 workshops total).

Workshop topics include: Patenting and Protection; Estimating the Worth of an Invention; Licensing; Marketing; New Business Start-up; The Business Plan; R&D and Venture Financing; the DOE-NIST Energy-Related Inventions Program; and SBIR Programs.

The cost of attending is about $85 if you are nonaffiliated; $10 less if you are a member of an inventor organization.

How to Select the Right Company to Approach

Remember the old saw that says if you have a better mousetrap, the world will beat a path to your door? Maybe so back in 1889 when Ralph Waldo Emerson suggested this, but today the reverse is true. Successful inventors I know spend a lot of time beating paths to the doors of companies looking to license better mouse traps.

Selecting which companies to approach as possible licensees for your invention is a matter that requires a great deal of thought. It is not something to be taken lightly. You don't want to make the same mistake Rick Blaine (Humphrey Bogart) made in the movie *Casablanca*. "I came to Casablanca for the waters," he said. When told that he was in the middle of a desert, he adds, "I was misinformed."

You must be very well informed, indeed. Your first approach to a company requires the kind of detailed analysis, imagination, and forethought associated with championship chess. In fact, your approach resembles the game of chess insomuch as it consists of a planned attack and defense, and has as its object, the king's surrender.

Just as much effort, and often more, is required to deal with a small company as with a large one. Don't be afraid to approach the major players. Remember, eighty percent of the business in any particular industry is typically done by twenty percent of the companies. You'll want a company capable of engineering and then following through on a success. Also the larger the company, the better and faster it pays.

Put a lot of time into your decision on which manufacturer to approach. Don't make the error of insufficient options. Some rejection is to be expected, so you'll want as many targets as possible.

Study corporate product lines. Do store checks. Get new product catalogues. Companies do what they do, and bringing them something out of their discipline is usually a lost effort. You would not, for example, go to Black & Decker with a new type of record player. Black & Decker manufactures innovative tools and labor saving devices. Round pegs don't fit square holes.

Corporations exist to make money. Executives, especially those in lucrative profit-sharing plans and incentive programs, want their corporations to be successful, but at what risk? I have found that most senior executives will listen to any scheme that rings of potential profit. That profit can come in the form of a new product or a labor-saving device. But it is rare that an executive will rock the proverbial boat for

A true story: Peter Roberts at age 18 in 1964, while a clerk at a Sears store in Gardner, Massachusetts, invented a quick-release wrench in his spare time. He was awarded a U.S. patent on the item, which he licensed to Sears in 1965 for a mere $10,000 and no royalties. The retailer had told Roberts that his tool had only minor sales potential, hence the buyout.

But between 1965 and 1975, Sears moved some 19 million of Peter Robert's wrenches for a net profit of more than $44 million. Claiming that Sears purposely misled him and owed him back royalties, Roberts took the giant retailer to court. After a long battle, Roberts, then 44, settled for $8.2 million.

The Swiss, once the undisputed world leaders in the watchmaking industry, got their chimes rung when the Japanese came out with cheap, plastic timepieces utilizing microchips, capturing billions of consumer dollars in the process. Only in recent years, with the plastic Swatch, have the Swiss begun to rebound.

untested, unfamiliar products, especially those that fall outside the company's expertise or channels of distribution. Unless you are dealing with an executive in charge of new business opportunities, trying to get a company to purchase a product that is inappropriate to its line-up or retail outlet channels is not only a waste of everyone's time, but it does nothing for your reputation. And that reputation is more precious then your invention.

After a while in this business you learn that many large firms are guided by numbers more than products. Lawyers and accountants tend to become CEOs before R&D executives do. These bean counters see products in terms of SKUs (stock keeping units) only. They like to keep the pipeline filled with line extensions of already proven and profitable products. Their attitude is that a major breakthrough in medicine, for example, would be nice, but let's keep the mouthwashes and toothpastes coming. These types would rather take a popular pudding and put it on a stick than gamble on creating a new novelty food. They are so busy listening to statistics that they forget companies can create them.

One company licensed a product of ours, and was going to produce it until a consumer study showed that its popularity would upset existing business. The company opted for the status quo and dropped the item. "Why should we erode our market share in an industry we almost totally control?," reasoned the company president. "If we bring your product out, the consumers will obviously love it, but we'll just point to opportunities our competition does not realize exist."

On the other end of the spectrum, a senior research and development executive from a $1 billion plus company told me once, "It's my responsibility to waste $2 million per year on long shots. I would not be doing my job if I didn't." Unfortunately, there are too few corporate executives with this entrepreneurial attitude.

The first move is yours. It is not an easy one. Going to the wrong company with the wrong product can cost valuable time, do nothing to enhance your contacts, and even bring grief. You want your idea in the right hands because, as advertising legend Bill Burnbach said, "An idea can turn to dust or magic depending on the talent that rubs against it."

NIH Syndrome

NIH, as used here, is not the acronym for the government's National Institutes of Health. NIH is corporate jargon for "Not Invented Here," a syndrome from which many companies suffer. It means that such companies have in-house research and development staffs and do not entertain outside submissions at all, or do, but only to see what is being done independent of themselves, not to license.

It is hard to tell from the outside which companies fall into the NIH category. Those you think would, do not, and those who you would bet don't, do. This learning curve is part of every selection exercise.

The NIH issue has two sides, and corporate policy can change with different administrations. Many executives feel that no insulated group of salaried product development people, no matter how brilliant, can come up with winning products day in and day out.

Many companies cannot afford to pay engineers and designers to sit around and blue-sky ideas all day long. These kinds of companies are always worth approaching.

Other firms are against outside licensing of patented ideas because they would rather see the millions of dollars paid in royalties kept inside for its own research and development activities. They are not typically structured to interact with independent inventors.

I like to tell executives who believe they can do it all in-house that it is the spirit of the independent inventor that built America. To completely shut the independent inventor out is to severely limit one's opportunities and horizons. Then I remind them of these stories:

Kodak, America's largest manufacturer of photographic products, should have developed the instant camera. It didn't. The 60-second camera was invented and produced by Edwin Land, a maverick inventor. And when Kodak ultimately decided to imitate Land's invention, it was stopped in its tracks by the courts.

IBM, a name synonymous with computer innovation, completely missed the hand-held calculator market. It should not have. The Japanese captured the lucrative market and never gave an inch.

And the U.S. television networks, with worldwide news-gathering operations, let CNN get started because they thought no one would pay to see news 24 hours a day. CNN has become such a success that the nets are now all over the cable market.

Breaking the Code

Many companies that do work with outside developers do not encourage "unknowns" and return inquiries with a letter like this:

"Our advertising, research, marketing, and new product planning staffs are primarily responsible for creativity and development. Corporate policy precludes us from either encouraging or accepting unsolicited ideas from persons outside the Company. While an idea may seem feasible to the submitter, there are usually a number of factors that would make it impractical for us to implement. Moreover, many of the unsolicited ideas that we receive from both nonprofessional and professional sources have previously been submitted in one form or another."

Reading through and between the lines, this letter is not as negative as it initially appears to be. The first clue, that the company does not do everything internally, is that its internal staffs are not "exclusively responsible," but rather "primarily responsible" for creativity and devel-

The SEC's public reference room is located at 450 5th Street, N.W., Washington, DC 20549. It is open from 9:00 a.m. to 5:00 p.m. daily. The telephone number is: (202) 272-7450. Consumer Telecommunications for the Deaf-TTY-Voice (202) 272-7065.

opment. This means that outside people do back-up their company's research and development.

The next good news comes when the letter states that the company cannot encourage or accept "unsolicited ideas from persons outside the Company." I read this to signify that the company probably solicits ideas from a trusted base of outside creative sources. Companies that send back letters similar to this are worth a second look.

Attack of the Killer Lawyers

Another factor that is contributing more and more to the problem of approaching large manufacturers with outside submissions is the business some quick-buck lawyers are engaged in, the business of litigating patents.

What these unscrupulous attorneys do is buy from small companies and independent inventors the rights to portfolios of their existing but unlicensed patents. They tend to look for patents covering things like early bar code readers, liquid crystal display technologies and electronic musical instruments, especially synthesizers. Innovations that are heavily used in one form or another today by industry.

The sole and unfortunate purpose of these predators is to find large manufacturers of products which might be perceived to violate their newly acquired patents and quickly initiate law suits for patent infringement. Supported by paralegals, they make their money not through royalties their patents generate, but through out-of-court settlements paid by large manufacturers to get rid of them.

"I would call it a scam," says Robert W. Faris, a partner in the patent law firm of Nixon & Vanderhye (Arlington, Virginia), referring to this type of business. "They'll (attorneys) sue at the drop of a hat. Their only purpose is to exploit patents through litigation."

Faris explains that such attorneys know that the courts today are pro-patents. They also know that all anyone needs to get a patent is a well-written description. No inventor has to prove his or her idea with a prototype anymore. When a patent issues, it is assumed valid.

Public Companies vs. Private Companies

I have no preference for either one. My decision is based upon what I am able to find out about a company and whether it is best for my product. I maintain, at all costs, the mental frame of mind that I am evaluating the company rather than an inferior position of being considered by the company.

Public Companies—Where to Find Information on Them

The best place I know to obtain deep and detailed information on a publicly traded company is at the Securities and Exchange

Commission (SEC), Washington, D.C. This independent, bipartisan, quasi-judicial federal agency was created July 2, 1934 by act of Congress. It requires a public disclosure of financial and other data about companies whose securities are offered for public sale. Some 11,000 companies are registered.

All companies whose securities are registered on a national securities exchange and, in general, companies whose assets exceed $3,000,000 with a class of equity securities held by 500 or more investors must register their securities under the 1934 Act.

In the national capital area, the SEC operates a public reference room (public reference rooms are also located in Chicago and New York City). This specially staffed and equipped facility provides, for your inspection, all of the publicly available records of the Commission.

These include corporate registration statements, periodic company reports, annual reports to shareholders, tender offers and acquisition reports, and much more.

Requests by Mail

If you find it inconvenient to visit one of the Public Reference Rooms, the Commission will, upon written request, send copies of any document or information. Send a written request stating the documents or information needed, and indicate a willingness to pay the copying and shipping charges. Also include a daytime telephone number. Address all correspondence to: Securities and Exchange Commission, Public Reference Branch, Stop 1-2, 450 Fifth Street, N.W., Washington, D.C. 20549.

Bechtel Information Services also provides prompt and low-cost research and copying services. It is located at 15740 Shady Grove Road, Gaithersburg, MD 20877-1454. In Maryland, the telephone number is (301) 258-4300, Outside Maryland, phone toll free: 1-800-231-DATA.

Annual Reports: Research Pay Dirt

Corporate annual reports are on file with the SEC, or can be obtained directly from the company. There is no charge for annual reports that are ordered from the company. The SEC copies require copying. Contact the executive in charge of Investor Relations or the Senior Vice President and Chief Financial Officer at the particular company that you are researching.

Somewhere within an annual report it will say something like this: A copy of the company's annual report on Form 10-K, as filed with the SEC, will be furnished without charge upon written request to the Office of the Corporate Secretary. The 10-K is research pay dirt!

SEC reference libraries located in New York City and Chicago are open to the public. They are located at 26 Federal Plaza (212) 264-1685 and 219 South Dearborn Street (312) 353-7433, respectively.

Another way to obtain annual reports and SEC filings is to tell a stockbroker that you are interested in purchasing stock in a particular company and to acquire the company's annual report for your perusal.

On occasion, a company will attach to its 10-K exhibits such as employee stock option plans, licensing agreements, executive employment contracts, leases, letters of credit, etc. Information from any one of such documents could be important in a future negotiation. If a company denies your request for a certain contractual term, would it not be nice to be able to point out that there is corporate precedent for your receiving said stipulation?

I once had read the employment agreement between the executive with whom I was negotiating and his company. This helped me to estimate his worth to the company and what he could and could not make happen on my behalf. I even knew the type of car he was leased.

Form 10-K

I find this annual report to be the most useful of all SEC filings. In summary, it will tell you the registrant's state of business. This form is filed within 90 days after the end of the company's fiscal year. The SEC retains 10-Ks for ten years.

Part I of the 10-K reveals, among other things:

a) **When the company was organized and incorporated.** You will want to know how long the company has been in business to gauge its experience. What you would expect from an established company may vary from what you would tolerate at a start-up firm.

b) **What the company produces, percentages of sales any one item may be; seasonal/nonseasonal, etc.** It is critical to have a complete picture of company's product lines, their strengths and markets, any seasonality or other restrictions to the appropriateness of your item, and if and how your product could be positioned.

c) **How the company markets, for instance via independent sales representatives or its own regional staff offices.** It is important to know how a company gets something onto the market and where sales staff loyalty is, i.e. company employees typically have more loyalty than independent sales reps who handle more than one company's line.

d) **Whether or not it pays royalties and how much per year.** You can often see how much work the company does with outside developers and whether it licenses anything at all. An example of such wording is this from one corporate 10-K: "We review several thousand ideas from professionals outside the Company each year." I recall another 10-K that read, "The Company is actively planning to expand its business base as a licenser of its products." Statements such as these show that doors are open!

e) **What amount of money the company spends to advertise and promote its products.** If your product will require heavy promotion, and the company does not promote its lines, you may be at the wrong place. It is counterproductive to take promotional products to companies that don't advertise.

f) **Details on design and development.** You should know before approaching a company whether an internal design and development group exists and how strong it is. I found one 10-K in which a company stated, "Management believes that expansion of its R&D department will reduce expenses associated with the use of independent designers and engineers and enable the Company to exert greater control over the design and quality of its products." It could not be more obvious that outside inventors were not wanted.

g) Significant background on production capabilities. Often it is valuable to know in advance what the company's in-house production capabilities are, and what its outsourcing experiences are in your field of invention. It's no use taking a technology to a company that does not have the experience to produce it.

h) **Terms of long-term leases.** It can be important to know whether a company owns or rents its facilities as a measurement of its strength and capabilities. An inventory of real estate can also give you an excellent overview of warehouses, plants, offices, etc.

i) **If the company is involved in any legal proceedings, law suits, for instance.** You may not want to go with a manufacturer that is being sued right and left. Maybe it has just risen from a bankruptcy and is still not strong financially. All of this kind of information is an excellent indicator of corporate health.

j) **The security ownership of certain beneficial owners and management.** This is vital to understanding the pecking order and power structure. Here is where you'll see who owns how much stock (including family members), and what percentage of the company this represents. The ages and years with the company are also shown.

l) **Competition.** This section will give you a frank assessment of the company's competition and its ability to compete. One 10-K I read once admitted, "The Company competes with many larger, better capitalized companies in design and development..." It is unlawful to paint a rosy picture when it doesn't exist. The 10-K is one of the few places you can get an accurate picture. Would you want to license a product to a company that states, for example, "...most of the Company's competitors have financial resources, manufacturing capability, volume and marketing expertise which the Company does not have." This tells me to check out the competition!

Form 10-Q

The Form 10-Q is a report filed quarterly by most registered companies. It includes unaudited financial statements and provides a continuing view of the company's financial position during the year.

The 10-K and 10-Q are the two filings I find most valuable. However, the SEC has many other reports available. The best way to view a full inventory of records on file is to contact the nearest SEC regional office and pick up or have sent to you one of the Commission's booklets.

By combining the information available in 10-Ks with your personal observations, you will be much better prepared to do business.

Private Companies: Questions You Need to Ask

Detailed information on privately held companies is harder to come by. No regulations require that they fill out the kinds of revealing reports public companies must. Nevertheless, it is important to gather as much background information as possible.

Here are some questions I have answered **before** approaching a private company. The answers come to me from a combination of sources ranging from state incorporation records to interviews with competition, suppliers, retailers (as appropriate), and the owners themselves. Finding the answers requires some digging, but acquiring this background information on a company you hope to deal with is critical to your long-term success.

1) **Is the company a corporation, partnership, or sole proprietorship?** This can have legal ramifications from the standpoint of liabilities the licensee assumes. A lawyer can advise you on the pluses and minuses of each situation.

2) **When was the company organized or incorporated?** If a corporation, in which state is it registered? When a company was organized will give you some idea as to its experience. The more years in business, the more tracks in the sand are left. The state in which it is registered to do business will tell you where you may have to go to sue it.

3) **Who are the company's owners, partners, or officers?** Always know with whom you are going into business. In the end companies are people, not just faceless institutions.

4) **What are the company's bank and credit references?** How a company pays its bills is important for obvious reasons, and its capital base is worth assessing.

5) **Is the manufacturer the source for raw material?** Does it do the fabrication? Such information will help estimate a company's capabilities for bringing your invention to the marketplace.

6) **How many plants does the company own (lease), and what is the total square footage?** Does it warehouse? This kind of information will help complete the corporate picture.

7) **What products are currently being manufactured or distributed?** You don't want to waste time pitching companies that do not manufacture your type of invention. Maybe a company you thought to be a manufacturer is really only a distributor.

W hat kind of marketing and advertising support can you expect your product to receive? What kind of advertising does the company generally support? Are they active on the national, regional, or local promotional scene? Does the company generally pitch announcements on television, radio, or in print? Does the company routinely provide point of purchase promotion? If yours is a product requiring a certain type of promotion to succeed, it makes no sense to license a product to a firm that does not promote its lines in that manner.

8) **How does the company distribute?** Find out about the direct sales force. Make inquiries concerning outside sales representatives and number of jobbers. Does the company use mail order, house-to-house, mass marketing, or some other form of distribution? This information will quickly reveal how a company delivers its product and whether its system is appropriate to your product. With a mass market item, it would be foolish to approach a firm that markets door-to-door, regardless of its success.

Company Product and Corporate Profiles: Where to Find Them

The Thomas Register

One of the best sources for product and corporate profiles is the *Thomas Register* . Available in most public library reference rooms, "Thomcat", as it is known, contains information on more than 120,000 U.S. companies in alphabetical order, including addresses and phone numbers, asset ratings, company executives, the location of sales offices, distributors, plants, and service and engineering offices. If you know a brand name, you can locate it in the *Thomcat Brand Names Index* .

S&P's Register

Another excellent source is *Standard & Poor's Register of Corporations, Directors and Executives* , available at public libraries. Consisting of three volumes, it carries data on more than 45,000 corporations including their Zip Codes; telephone numbers; names, titles, and functions of approximately 400,000 officers, directors, and executives. A separate volume selects 70,000 key executives for special biographical sketches. The last volume contains a classified industrial index.

Serve Your Steak with the Sizzle

After you have read all the literature and investigated the company inside and out, you must ask yourself: can the company deliver? And will you be comfortable working with its people? The abbreviation "Inc." after a company's name is not significant. Nice offices, a few secretaries, a fax and copying machine do not a successful licensee make. Just as with your inventions, steak must be under the sizzle. And the way you are treated should be in good taste.

If you wish to purchase your own set of books, the *Thomas Register* costs $175.00 for the 18 volumes (1989). Like an encyclopedia, it requires considerable shelf space. To order call: 1-800-222-7900.

The S&P *Register of Corporations, Directors and Executives* (1988-89) costs $450.00. To order call: 1-800-221-5277.

Preparing Your Proposal

I always back-up my verbal presentations with written proposals. The inventor cannot always go out with every prototype, and as the saying goes, if it ain't on the page, it ain't on the stage.

What Are the Elements of a Successful Proposal?

All of my proposals begin with a concept summary. This is nothing elaborate, just a solid paragraph that paints a picture of the product and its overall concept and objective.

This is followed by a technical section that addresses costing and manufacture. You are well advised to have a rough estimate of what the invention will cost to manufacture. Don't expect the company to know.

I always try to provide the following data with every item (as appropriate):

1) A sheet listing all components with respective prices from various sources. Pricing from three different sources is a good bet. When possible, a mix of domestic and offshore numbers is best. Do not forget to include the volume the quotes are based upon, plus vendor contacts.

2) Note the type of material(s) desirable, e.g. polyethylene, wood, board, etc. Provide substitutions and options for consideration.

3) When you calculate the item's cost, do not forget to consider the price of assembly (if any). The quoting vendors will be helpful here.

4) If your item requires retail packaging, add an extra 15-30 percent.

5) Add an extra 20 percent for modifications and losses. At this point you have the item's hard cost.

6) To arrive at the manufacturer's selling price, add in your royalty, an amount of money for promotion (if appropriate), and a gross profit margin for the manufacturer (65 percent). You may wish to estimate the mark-up at the retail end (if appropriate). A good estimate is 30 percent.

Organizations Offer a Helping Hand

Inventor and business organizations, as well as state and university assistance programs can be of great value to the inventor preparing presentation packages for prospective licensees.

For more information, see the chapters on Associations: National and Local and University Innovation Research Centers.

Good For The Long Haul

It is important to make clear to your prospective licensee that you are not just capable of delivering the item under consideration, but that lots more can come from the same source: you.

Every written proposal should also contain (as appropriate):

1) **Detailed operating instructions:** Take nothing for granted. The worse thing that can happen is a client's inability to use your item after you have departed. Illustrate with pictures if required. No item is too easy.

2) **A marketing plan:** Highlight your item's advantages over existing product(s) and what makes it unique. Define its appeal and target audience. Suggest follow-ups, including second generations and line extensions of your product (as appropriate). Manufacturers like products that have a future, especially if they will be required to spend lots of start-up dollars in the development and launch phases.

3) **Trademarks:** Offer possible trademarks. If a trademark search has been done, include the results of your search or status of any applications. The right mark can go a long way in securing a sale.

4) **Patents:** Include an update on any patent searches, PTO actions, etc. If a patent has issued, attach a copy of it. The more detailed and comprehensive your work is here, the quicker and easier the corporate evaluation of your product should be.

5) **Ad Campaign/Copy:** Suggest advertising direction, artwork, and slogans whenever appropriate. This type of work makes presentations even more persuasive and polished.

6) **Test results:** Report the results of any formal testing or focus group sessions. Include photographs and videotapes whenever appropriate.

7) **Inventor's Background/Capabilities:** If you are unknown to the company, provide personal background and describe yourself and your capabilities to support further development and manufacture of the invention under consideration. After all, who understands an invention better than its inventor? Your experience may save the manufacturer a great deal of money.

The aim here is to show the manufacturer that your front line is strong and the bench is deep. The more confident the company feels about you the better.

Watch Your Language

You cannot be too specific in proposals. Spell out everything. Take nothing for granted. And do not assume that words and terms have the same meaning to every reader, especially when you're making presentations to potential foreign licensees.

Although much is known about how new products are developed, there is no consensus on the meaning of key terms and definitions.

"With the increased emphasis in developing new commercial products and processes to help the U.S. become more competitive internationally, the need for a common language in the innovation process is becoming even more important," says George Lewett, chief of the National Bureau of Energy-Related Inventions at National Institute of Standards and Technology (NIST).

It is of interest to note that the NIST, the Department of Energy, and the National Society of Professional Engineers are collaborating to develop a consensus language for use in describing the innovation process. The project is expected to take about one year.

So far, seven organizations have nominated members to a task group. They are: American Institute of Chemical Engineers, American Society for Engineering Education, American Society of Mechanical Engineers, Commercial Development Association, Industrial Research Institute, Institute of Industrial Engineers, and NSPE.

I hope to incorporate their glossary in a future edition of this book.

To keep track of who has seen what and when, a Patent Worksheet appears in the Forms and Documents section.

Manufacturers Don't License Ideas

It is imperative that you develop a prototype of an invention prior to disclosing it to potential licensees. A prototype is defined as an original model on which something is patterned. If you do not have the time, money, or commitment to build a prototype, the odds of licensing the item are nil.

The most effective kind of prototype is what is called a "looks like-works like" version. There are no short cuts. Manufacturers don't license "ideas". They react to physical matter. Don't count on people being able to "imagine" what your product will look like or how it will operate. Even if they could, busy executives do not usually have the time or interest to engage in such typically futile exercises.

Executives love to touch and feel prototypes. Kick the tires, so to speak. Knowing this, do your best to have prototypes that most resemble and operate like a production model. Go that extra mile to ensure the prototypes are solid and have perceived value.

Don't take any of this lightly. You must be as sophisticated and slick in your presentation to a potential licensee as it will have to be in its pitch to the trade and/or the consumer. While getting a product known is relatively easy, marketing a need is something else. And that's your ultimate goal.

Making Prototypes

If you cannot make the prototype, plenty of places will provide assistance. In some cities, prototype makers can be found in the telephone directory. Universities and engineering schools often have workshops where connections can be made to get something done. Local inventor groups are a wonderful source of information as well. See the chapters National and Regional Inventor Associations: National and Local and University Innovation Research Centers for more information.

Many invention marketing companies offer prototype-making services. Be careful. Know what you are getting into when contracting to have a product prototyped. I would not engage anyone to do any prototype work without inspecting their shop and checking references.

Multiple Submissions

If you have more than one prototype, and the situation is appropri-

Prototypes must be solid because often they take quite a beating at the hands of potential licensees. Don't be surprised to get them back broken. It happens even at the best of companies. This comes with the territory.

Shawn Tyler Brown, 47, of Bend, Oregon, is inventor of Magna Ears, a device that enhances hearing up to 50 percent in some cases. He reports from the prototype front: "The prototype I made was flat. I made it out of ABS and used a hairdryer to shape it. It was nice for an amateur job, but the professional vacuum forming gave it shape."

It is prudent to require that any prototype maker sign a Confidential Disclosure Agreement before you show or discuss your product. A sample agreement is shown as in the Forms and Documents Section.

A Fish Story

The art of product presentation is not unlike the sport of fly fishing. It takes time and patience. In casting, the lure is presented and then pulled back. The lure's movement, aided by twitches, pauses, and jerks of the rod by the angler, entices the fish to strike. In both cases the object is to hook a big one!

ate, you may wish to consider making submissions to more than one manufacturer at the same time. I have no set rule about this and take it case by case, guided by experience.

If a company asks to hold off further presentations until it has an opportunity to review the item more in depth, try to set guidelines. In all fairness to everyone, some products require a reasonable number of days to be properly considered. However, if you feel the company is asking for an unreasonable period of time, seek some earnest money to hold the product out of circulation. The amount of time and money is negotiable. Also insist that the product not be shown to anyone outside the company.

Mutual Dependency

You need the company or you would not be there. Show yourself as being independently creative, while at the same time taking the "we-approach" and not the "I-approach."

In order for your product to sustain itself through the review and development process, it will need a champion. Typically this standard-bearer will come from among those attending your first meeting. Get others involved. Turn "your idea" into "our idea."

If You Sign My Paper, I'll Sign Yours

Many companies ask that an agreement be signed before they'll accept the submission of outside ideas. This may surface when you first approach the company or happen on the day you appear to make the formal presentation. I have never had a problem with such requests. I always know with whom I am dealing and feel confident in the relationship. If I did not, I wouldn't be there in the first place.

A suspicious attitude may seriously inhibit your progress. Put your time and energies into creating concepts versus over-protecting them. Become paranoid over this and no one will ever see your ideas.

Idea Submission by Outsiders

Outside submission agreements take many forms. Figs. 40, Idea Submission Agreement I and Fig. 41, Idea Submission Agreement II are commonly used formats. Most companies use variations of these forms. Use these forms to familiarize yourself with the types of document, and as examples, in case you want to use one to protect yourself before looking at another individual's ideas.

Agreement to Hold Secret and Confidential

In some instances, it is appropriate for the inventor to have the company sign an agreement to hold a product secret and confidential. A sample agreement is shown in the Forms and Documents section.

How to Present Your Concepts

First Impressions Are Lasting Impressions

The inventor is always selling two things: the concept and the inventor. Your personal credibility is often more important than the credibility of any particular single creative concept.

It is critical that a corporate executive buy the inventor as much as the invention. You may be capable of dreaming up numerous innovative products for a company to consider, and will want to be invited back again and again. Without respect from corporate executives, your products will never be taken seriously. And you cannot put a dollar value on the ability to make an encore.

When to Make the Pitch: Avoid Cold Calls

Presentations are best when carefully choreographed and staged. Nothing is usually gained ambushing executives outside of their offices. As Agesilaus II, King of Sparta, said, "It is circumstance and proper timing that give action its character and make it either good or bad."

For every rule ever told about when to sell, another rule proves it wrong. **The best rule on when to sell ideas is whenever possible.** *Timing is, of course, everything. I operate under the principle that when you have something hot, burn it. When it gets cold, sell it for ice.*

I licensed our first electronic toy, StarBird, to Milton Bradley after Thanksgiving, and the manufacturer premiered it only weeks later at American International Toy Fair. Milton Bradley put its resources on the line to make it happen in spite of the toy's complicated structure.

I developed the game "Adverteasing," which was licensed in June and manufactured and in stores by mid-September of the same year. By early December it had racked up sales of over 250,000 units.

While these examples are exceptions to the rule, had I waited until the "best" time to make the pitch, opportunity might have been delayed or the product may never have been licensed. The marketplace is temperamental and erratic.

Curtain Up. Light The Lights.

Once you have been extended an invitation to display your concept to a manufacturer, it's major show time. And if you thought inventing was tough, you haven't yet experienced hardship. The moment you walk into a company's conference room with your invention at the

Another rule to consider is this: the faster a product is made available, the faster it begins to generate income.

Resistance, then success. Back in the 1930s, Charles B. Darrow invented a game called Monopoly. It was rejected by six or seven companies, including Parker Brothers. Parker Brothers eventually saw the light and published Monopoly, which annually sells more than one million units. Darrow became the first millionaire game inventor.

ready, you pass into an eerie twilight zone. You are the hero in a Nintendo video game and the corporate executives comprise an array of varied personalities. Some pray for you to succeed. Others are gremlins out to gobble you up, shoot you down, and otherwise obliterate your ideas faster than Octorok can toss stones at Link.

Thomas Alva Edison once observed, "Society is never prepared to receive any inventions. Every new thing is resisted, and it takes years for the inventor to get people to listen to him before it can be introduced."

"Creation is a stone thrown uphill against the downward rush of habit," another inventor said.

World history is rich in stories about people's resistance to new ideas and change. Some seem unbelievable in retrospect. I find it helpful and comforting to recall some of these stories before I begin a presentation.

Impressionist art that sells for millions of dollars today was met with onslaughts of condemnation and cavil when first introduced. The innovative Stravinsky was once labeled "cynically hell-bent to destroy music as an art form." Today he is regarded as a genius.

When railroads were established, farmers protested that the "iron horse" would scare their cattle to death and stop hens from laying eggs. The British Association for the Advancement of Science insisted the automobile would fail because a human driver "has not the advantage of the intelligence of the horse in shaping his path."

But to a great extent, **timing is everything.** Inventor Carl E. Haskett of Tyler, Texas, tells the story of how he invented a waterproof bathing cap for women about 25 years ago which came out "on the day women went to bouffant hair styles and didn't wear any more bathing caps."

Inflexibility Is Inherent in the System

Remember that in order to operate, companies must have rules and controls. Loose cannons do not last long in corporate environments. Any organization must understandably have somewhat of an established routine to survive. Therefore, executives, especially in larger companies, often fall into predictable and ordered routines. Your job is to break this routine, make people believe in you, and interest the company in buying your concept.

How to Get the Best Terms

Never fear to negotiate, never negotiate out of fear.

Negotiating contracts is a skill that can be learned. People tend to make it complex. It is not.

Do It Your Way

The most basic rule is to conduct your business according to your style. Make haste slowly. Set the pace. Do not get caught up in your prospective licensee's timetables and priorities. Things tend to get worse under pressure.

Necessity Has No Law

Lawyers tend to intimidate most people. If you let them, they'll confound you with facts, blind you with Latin, and plague you with precedents. Whether you take a lawyer with you or negotiate yourself, be sure that everything is spelled out to your satisfaction—even if it is not the way something is taught at Harvard Law. Always insist upon clarity over form.

Winning

Getting what you want does not always have to be at another person's expense. It is possible to get what you want and still let your opponent have something. After all, you are entering into what you both hope will be a long and mutually beneficial relationship. As our political process demonstrates, societies thrive best not on triumph in domestic debates, but on reconciliations. Nothing would ever be accomplished if every technical disagreement turned into a civil war.

A good deal in one in which the two sides both meet their needs. Needs can be reconciled. Compromise is okay. Unfortunately, not every person you meet at the bargaining table believes in this theory. Often you'll encounter a slick customer. They never appear to be the killers that they are. A smile on the lips. A twinkle in the eye. Holy water in the back pocket. Decent. Humble. Bible in one hand. They'll bless you and offer a prayer for your common good fortune. It is only after they've left that you feel blood trickling down your pant leg and feel the stiletto in your back.

Other people will give off signals that they are not trustworthy from the start. Terms aside, my own rule of thumb is that unless I am totally

I have a close friend who beat a large manufacturer in court to the tune of a million dollars. After splitting the award with his partner and attorney, and paying taxes and costs, he came away with about $250,000. It sounded good at first. But he soon realized that every door was closed to him in his industry. None of the major companies wanted to see his concepts. They still don't.
It was a high price to pay. The right idea with the right company can be worth millions.

How much should you ask for as an advance? I normally ask the company what it typically pays, while, at the same time, double-checking with others who may have licensed products to the same manufacturer. Inventors tend to share information with each other quite freely.

comfortable with the executives and the company, I don't even sit down to deal. **No deal is better than a bad deal.**

Agree To Agree

Prior to discussing the nuts and bolts of a specific licensing contract, I want to know two things up front:

>1) Is the company willing to pay me a standard royalty on the net sales of each unit sold and

>2) Is the company willing to pay me a to-be-negotiated advance against future royalties. No strings attached.

Once I establish a basis for negotiation, no contractual terms typically stand in the way. And they should not so long as I want to sell and the manufacturer is serious about licensing. Everything usually shakes out. But, if a question arises about wanting to pay an advance and royalty, I refuse to deal until we settle on those points.

Up-Front Monies

The advance is important because it can help you recoup a portion of your outlay for research and development. Don't look for it to cover all your expenses. Perhaps more important, it is evidence of how much value the purchaser puts on your work, and the kind of support you can expect to receive.

Generally, the size of the advance is in direct proportion to the kind of support the company will give your project. For example, a firm will do a lot more to ensure the success of the item given a $100,000 advance verse one that received $10,000. The less up-front, the less to lose if the project comes under consideration to be dropped.

Royalties

Many industries have royalty structures established. Do your homework and you can find out what to request. I have seen royalties run from 1 to 20 percent across various industries.

First, ask your prospective licensee what is fair and equitable. If you feel you deserve better than that, explain why. When you don't ask, the answer is always an automatic **NO**.

Levy's Rules of Contract Negotiation

>1. **Negotiate yourself.** In choppy seas, the captain should be on the deck. No one will do it better than you. No one has more to gain or lose.

>2. **Thou shalt not committee.** Any simple problem can be made insoluble if enough people discuss it.

3. **Don't deal with lawyers.** It is always best to negotiate with an executive who is in a decision-making position. Lawyers are not paid to make executive decisions but to set rules and follow them. They see themselves as protectors, saving the executives from themselves. Yet I have found that the most successful executives break rules all the time.

4. **Never respond to pygmies chewing at your toenails.** Don't roll over just because a lawyer says that without x, y or z the project will not be approved. The company wants to do the project with you or you would not be in negotiation. Executives, not lawyers, are responsible for profits. If your invention can boost revenues, executives will shine.

5. **Two plus two is never four.** Exceptions always outnumber rules. Established exceptions have their exceptions. By the time one learns the exceptions, no one remembers the rules to which they correspond.

6. **Written words live.** Spoken words die. As they say in the theater, if it ain't on the page, it ain't on the stage. During negotiations, confirm every conversation with a memorandum to eliminate any misunderstanding about who agreed to what.

7. **Ask when in doubt.** Asking dumb questions is far easier than correcting dumb mistakes.

8. **Keep it short and to the point.** The length of a business contract is inversely proportional to the amount of business.

9. **Do not accept standard contracts.** In any so-called standard contract, boilerplate terms should be treated as variables. Not until a contract has been in force for six months will its most harmful terms be discovered. Nothing is as temporary as that which is called permanent.

10. **HAVE FUN.** The moment I stop enjoying a negotiation, I pick up my marbles and go home. An agreement is a form of marriage and both parties must be compatible for it to succeed.

Note: Most manufacturers will not expect you to do everything for free once they officially license a product.

Something for the Next Generation: Project XL

At age 14, one schoolboy invented a rotary brush device to remove husks from wheat in the flour mill run by his friend's father. The young inventor's name? Alexander Graham Bell.

At age 16, another of our country's junior achievers saved pennies to buy materials for his chemistry experiments. While still a teenager, he set his mind on developing a commercially viable aluminum-refining process. By age 25, Charles Hall received a patent on his revolutionary electrolytic process.

Chester Greenwood, 13, had a negative experience with cold weather one December day in 1873 while ice skating. To protect his ears, he found a piece of wire and with his grandmother's help, padded the ends. At first, his friends laughed. However, before long they realized that Chester could stay out ice skating long after they had gone inside to escape the freeze. Soon they were asking him for his custom ear covers. At age 17, Chester applied for a patent and over the next 60 years became very rich due to his earmuffs.

Launched in 1985 under the leadership of then Assistant Secretary, Commissioner of Patents and Trademarks Donald J. Quigg, Project XL is an outreach program of the Patent and Trademark Office and an integral part of the U.S. Department of Commerce's Private Sector Initiative Program. It is designed to encourage the development of inventive thinking and problem-solving skills among America's youth. The Project's principal focus is the promotion of educational programs that teach critical and creative thinking and on fostering national proliferation of such programs.

Since it began, Project XL has reached more than 4,000 teachers and 100,000 students. The overall objective of Project XL is to ensure the nation's position as a world technological leader in the next century—to guarantee that Americans will have the innovative skills to meet the challenges of an increasingly competitive world.

"I believe that the schools of this great nation are filled with Edisons, Wrights, Marconis, Whitneys, and Bells, along with other potential thinkers who can change the world," said Quigg. "The very least we can do is to help them realize their potential—to nurture those young people who will inherit and build the future."

A secondary benefit from Project XL is that young people and their parents and teachers will gain an increased awareness of new technol-

M. *David Richards, the inventor of Drink-Up, a patented faucet device, likes Project XL because it strives to teach kids to believe in something they can do. "Make them invent something and the pride they'll take in it will turn around and they'll have pride in themselves."*

ogy's importance to advancing society and strengthening the domestic economy. Project XL aims to instill an increased appreciation of the contributions inventors make to our way of life and recapture the spirit of those golden years at the turn of the century when inventors were heralded as true American heroes.

Project XL Objectives

Project XL is comprised of the following components:

⇨National coordination of efforts to teach inventive thinking and problem-solving skills at every level of public and private education throughout the country.

⇨Presentation of national and regional conferences to promote the teaching of critical and creative thinking skills and the inventive process.

⇨Establishment of an Education Roundtable, an open forum, and national discussion network, drawing upon the talents and resources of public and private sector leaders to develop and promote programs in this area.

⇨Development of broad-based speakers' bureau on the topics of invention, problem-solving, creativity, thinking skills, and related topics.

⇨Dissemination of an informational guide called the Inventive Thinking Project , designed to channel students in grades K-12 into the inventive thinking process through the creation of their own unique inventions or innovations.

⇨Creation of an educator's resource guide to include programs, materials, literature, organizations, and other sources that promote thinking across all disciplines.

⇨Curriculum development for special teaching materials designed to stress problem-solving, the value of creative thinking, and the importance of American inventors.

⇨Identification of government programs and resources that focus on the development of future problem-solvers in all fields.

⇨Establishment of an Inventive Thinking Center, a collection of literature, videotapes, and other curriculum materials.

Donald J. Quigg Excellence in Education Award

In 1989, the PTO presented outgoing Commissioner Quigg with a plaque establishing Project XL's Donald J. Quigg Excellence in Education Award. It recognizes the efforts of an individual (or group) to promote the teaching of inventive thinking skills at all levels of educa-

For information on Project XL programs, please contact Ruth Ann Nybold, Administrator for Project XL, U.S. Patent and Trademark Office, Crystal Park II, Washington, D.C. 20231; telephone (703) 557-1610.

Nomination Form
Donald J. Quigg
Excellence in Education Award

Name of Nominee(s): _____

Address: _____

_____

Phone # : _____   _____

Position : _____

Business Address: (if appropriate) _____

_____

Name of Nominator: (if appropriate) _____

Address: _____

_____

Phone # : _____   _____

Position: _____

Category: (Check One)   ☐   Educator
                          ☐   Student
                          ☐   Parent
                          ☐   Business
                          ☐   Professional Society
                          ☐   Government

Selection Criteria:

1. **Accomplishments (or Professional Achievement): A description, in essay form, of specific accompishments in promoting higher order thinking skills, resulting in improvement to a program, class, or student. Include in this section a description of the nominee's outstanding abilities as a communicator and leader. (Major emphasis in judging placed on essay.)**
2. **Community Involvement: A description of any community activities of the nominee outside the professional sphere in which he or she has participated in pursuit of higher order thinking skills.**
3. **Awards and Publications: A list of any professional awards received and a list of any professional publications germane to the nominee's pursuit of higher order thinking skills.**

The Weekly Reader National Invention Contest presents awards to K-8 students nationwide for inventions or innovations that are judged on originality, usefulness in addressing real needs, workability, and clarity of presentation. The Weekly Reader awards a grand prize for elementary school students and for middle school students, and bestows awards for specific grade levels. Contact *Weekly Reader*, 245 Long Hill Road, Middletown, Connecticut 06457; (203) 638-2638.

tion. Winners will be chosen by a panel of distinguished experts in the fields relating to the PTO and Project XL.

Furthering the Creative Instinct

Irv Siegelman, editorial director for *Weekly Reader* projects, is enthusiastic about Project XL. "I think it is one of the best things the government has done in a long time. It helps kids to become better creative and inventive thinkers." He hastens to add, "It's a wonderful thing that it is not in the Department of Education, but in the Patent and Trademark Office, which gives it the added cachet of the whole inventive process." *Weekly Reader* runs an annual invention contest which Siegelman believes more than 600,000 children have participated in since its start in the 1987 school year.

For a listing of Project XL programs and publications, check Gale Research's *Inventing and Patenting Sourcebook.*

"Creativity is contagious, pass it on!"

—Albert Einstein

OFFICE OF THE ASSISTANT SECRETARY AND COMMISSIONER OF PATENTS AND TRADEMARKS

Assistant Secretary and Commissioner Harry F Manbeck Jr rm906 PK2	557-3071
Administrative Secretary Norma M Rose rm906 PK2	557-3071
Executive Assistant to the Commissioner Edward R Kazenske rm906 PK2	557-3071
Program Analyst Ann Farson rm906 PK2	557-3071
Deputy Assistant Secretary and Deputy Commissioner Douglas B. Comer rm904 PK2	557-3961
Secretary (Vacant) rm904 PK2	557-3961
Assistant Commissioner for Patents Rene D Tegtmeyer rm919 PK2	557-3811
Secretary Sherry D Brinkley rm919 PK2	557-3811
Assistant Commissioner for Trademarks Jeffrey M Samuels rm910 PK2	557-3061
Secretary Sheila G Pellman rm910 PK2	557-3061
Assistant Commissioner for Administration Theresa A Brelsford rm908 PK2	557-2290
Secretary Karon Morris rm908 PK2	557-2290
Assistant Commissioner for Finance & Planning Bradford R Huther rm904 PK2	557-1572
Secretary Vickie T Bryant rm904 PK2	557-1572
Assistant Commissioner for External Affairs Michael K Kirk rm902 PK2	557-3065
Secretary Johnell M Bersano rm902 PK2	557-3065
Assistant Commissioner for Information Systems Thomas P Giammo rm916 PK2	557-9093
Secretary Helen White rm916 PR2	557-9093
Secretary (Vacant) rm1004 PK2	557-9093

OFFICE OF THE SOLICITOR

Solicitor Fred E McKelvey rm918 PK2	557-4048
Secretary Olga M Suarez rm918 PK2	557-4048
Deputy Solicitor (Vacant) rm918 PK2	557-2317
Secretary Cheryl P Gibson rm918 PK2	557-2317
Associate and Assistant Solicitors:	
Lee E Barrett rm918 PK2	557-4035
Muriel C Crawford rm918 PK2	557-4035
John W Dewhirst rm918 PK2	557-4035
Albin F Drost rm918 PK2	557-4035
Robert D Edmonds rm918 PK2	557-4035
Harris A Pitlick rm918 PK2	557-4035
John H Raubitschek rm918 PK2	557-4035
Richard E Schafer rm918 PK2	557-4035
Linda N Skoro rm918 PK2	557-4035
Nancy C Slutter rm918 PK2	557-4035
Paralegal Specialists:	
Teresa N Byerley rm918 PK2	557-4046
Patricia D McDermott rm918 PK2	557-4031
Maryann B Volkmar rm918 PK2	557-4022
Solicitor's Library:	
Theresa Trierweiler-Cappo rm918 PK2	557-4052

OFFICE OF ENROLLMENT AND DISCIPLINE

Director Cameron Weiffenbach rm810 PK1	557-2012
Secretary Betty Kaminsky rm810 PK1	557-2012
Harry I Moatz rm810 PK1	557-2012
Marian E Ford rm810 PK1	557-2012
Patricia M Jordan rm810 PK1	557-2012
Roster Information rm810 PK1	557-1728

BOARD OF PATENT APPEALS AND INTERFERENCES

Chairman Saul I Serota rm12C12 CG2	557-4072
Secretary (Vacant) rm12C12 CG2	557-4072
Vice Chairman Ian A Calvert rm10D10 CG2	557-4000
Secretary Wanda G Banks rm10D10 CG2	557-4000

General Information
    Ex parte Appeals rm12C08 CG2 557-4101
    Interferences rm10C01 CG2 557-4007

Examiners-in-Chief
    Neal E Abrams rm12D04 CG2 557-4057
    James R Boler rm10C12 CG2 557-4009
    Raymond F Cardillo Jr rm10AD4 CG2 557-7524
    Marc L Caroff rm10C04 CG2 557-4009
    Irwin C Cohen rm12B18 CG2 557-4703
    Jerry D Craig rm10D02 CG2 ..,.. 557-4058
    Mary F Downey rm10B14 CG2 557-4065
    Stephen J Emery rm12D12 CG2 557-4023
    Donald D Forrer rm10C16 CG2 557-4061
    Charles E Frankfort rm12D10 CG2 557-4059
    Bradley R Garris rm12D06 CG2 557-7148
    Melvin Goldstein rm10D08 CG2 557-4068
    John T Goolkasian rm10A10 CG2 557-4003
    Kenneth W Hairston rm10A02 CG2 557-4112
    Paul J Henon Jr rm10C22 CG2 557-4058
    Thomas J Holko rm10B08 CG2 557-4063
    Edward C Kimlin rm10C10 CG2 557-4003
    Errol A Krass rm10B12 CG2 557-7516
    William F Lindquist rm10C18 CG2 557-4061
    Charles N Lovell rm10D06 CG2 557-4070
    William E Lyddane rm12C04 CG2 557-4073
    Thomas E Lynch rm10B04 CG2 557-7517
    Harrison E McCandlish rm12B14 CG2 557-4703
    John P McQuade rm10A20 CG2 557-4088
    James M Meister rm10B06 CG2 557-4063
    Edward J Meros rm10C20 CG2 557-4003
    Andrew H Metz rm12B10 CG2 557-4326
    Marion Parsons Jr rm12C02 CG2 557-4393
    William F Pate III rm10B10 CG2 557-7653
    Irving R Pellman rm10A16 CG2 557-4064
    Verlin R Pendegrass rm10A08 CG2 557-4067
    Alton D Rollins rm12B04 CG2 557-4023
    Gene Z Rubinson rm10A12 CG2 557-4066
    James A Seidleck rm10A22 CG2 557-4070
    William A Skinner rm12D08 CG2 557-7147
    John D Smith rm12B16 CG2 557-4326
    Ronald H Smith rm12D02 CG2 557-4057
    William F Smith rm10A06 CG2 557-4066
    Michael Sofocleous rm10C06 CG2 557-4009
    Lawrence J Stabb rm10A18 CG2 557-4087
    Robert F Stahl rm12B02 CG2 557-4393
    Arthur J Steiner rm10A14 CG2 557-4062
    Bruce H Stoner Jr rm12B12 CG2 557-4025
    Henry W Tarring II rm10B02 CG2 557-4001
    James D Thomas rm10C02 CG2 557-4061
    Norman G Torchin rm10B16 CG2 557-4009
    Stanley M Urynowicz Jr rm10C14 CG2 557-4009
    Sherman D Winters rm10D04 CG2 557-4001

Programs and Resources Administrator
    Craig R Feinberg rm12C10 CG 2 557-7169

Service Branch
    Chief Clerk of Board T Maxine Duvall rm12C06 CG2 557-4101
      Deputy Clerk Nannie B Henry rm10C01A CG2 557-4007
      Deputy Clerk Shirley A Jefferys rm12C08 CG2 557-4101
      Deputy Clerk Eunice I Price rm10C07 CG2 557-4101
      Ex parte Legal Clerk Groups 110 and 150

Eleanor R Green rm10C09 CG2	557-4107
Ex parte Legal Clerk Group 130-180	
Karen Sweeney rm12C08 CG2	557-3100
Ex parte Legal Clerk Group 120	
Donald T Harris rm12C08 CG2	557-4108
Ex parte Legal Clerk Groups 210-220-230-240-250-260-290	
Fernando Burgess rm10C09 CG2	557-4109
Ex parte Legal Clerk Groups 310-320-330-340-350	
Mabel A Neal rm12C08 CG2	557-4106
Inter partes Legal Clerk	
Olivia M Duvall rm10C01 CG2	557-4006
Inter partes Legal Clerk	
Carrie Evans rm10C01 CG2	557-4004

OFFICE OF QUALITY REVIEW

Director James D Trammell rm1100 CP6	557-3564
Secretary Carolyn D Ballard rm1100 CP6	557-3564

OFFICE OF ASSISTANT COMMISSIONER FOR PATENTS

Assistant Commissioner Rene D Tegtmeyer rm919 PK2	557-3811
Secretary Sherry D Brinkley rm919 PK2	557-3811
Special Assistant R Franklin Burnett rm919 PK2	557-3054
Secretary Donna E Ellis rm919 PK2	557-3054
Paralegal Specialist (Vacant) rm919 PK2	557-3054
Manual of Patent Examining Procedure Editor Louis O Maassel rm919 PK2	557-3070
Deputy Assistant Commissioner for Patents James E Denny rm917 PK2	557-4279
Secretary Patricia R Appelle rm917 PK2	557-4279
Patent Programs Administrator Michael J Lynch rm917 PK2	557-4279
Patent Policy and Resources Director (Vacant) rm917 PK2	557-4279
Supervisory Petitions Examiner Jeffrey V Nase rm913 PK2	557-4282
Petitions Information rm913 PK2	557-4282

Office of Patent Program and Documentation Control

Director Richard H Rouck rm925 PK2	557-4222
Secretary Carolyn Evans rm925 PK2	557-9182
Program Analyst Carolyn Arrington rm925 PK2	557-9184
Patent Academy Richard McGarr rm502 PK1	557-2086
Paper Correlating Office JoAnn Harris rm925 PK2	557-5148
Paper Correlating Office Margaret Seward rm925 PK2	557-5149
Palm Coordinator Rolf G Hille rm925 PK2	557-9175
Special Program Examination Unit	
Supervisory Special Program Examiner Manual A Antonakas rm923 PK2	557-8384

PATENT DOCUMENTATION ORGANIZATION

Administrator for Documentation William S Lawson rm300 CM2	557-0400
Secretary Jim Doyle rm300 CM2	557-0400
Data Base Administrator Philip K Olson rm300 CM2	557-0400
Deputy Administrator for Documentation Edward J Earls rm300 CM2	557-0400
Classification Support Staff Director Sally Middleton rm300 CM2	557-0400
Contract Monitoring and Reclassification Division Chiquita Clark rm967 CM2	557-8877
Projects Monitoring Unit Inez Roberts rm967 CM2	557-5164
Processing Unit I - Janice Burse rm969 CM2	557-3396
Processing Unit II - Cornell Boney rm968 CM2	557-7467
Processing Unit III - Pat Walker rm965 CM2	557-7458
Special Projects Unit Daisy Turner rm964 CM2	557-2590
Data Control Technician Division Sadie Scott rm326 CM2	557-5910
Editorial Division Vernella Crowley rm300C CM2	557-5103
New Document Processing Division Marcia Smith Lobby CP6	557-5110
Preprocessing Branch Jerry Redmond Lobby CP6	557-5116
Final Processing Branch Gail Lewis Lobby CP6	557-5114
Special Processing Natalie Jackson Lobby CP6	557-5111
Weekly Issue Mary Johnson rm1240 CP6	557-7906

Office of Documentation Planning and Support
Director George Chadwick rm300 CM2 557-0400
Office of Documentation Information
Director Jane Myers rm304 CM2 557-0400
Technology Assessment and Forecast (Vacant) rm304 CM2 557-0400
Information Services Evelyn Freeman (Acting) rm322 CM2 557-5666
Information Resources (Vacant) rm322 CM2 557-5666
Patent Index rm322 CM2 557-3951
Patent Depository Library Program Manager
Carole A Shores rm306 CM2 557-9686
Secretary Dot Jenkins rm306 CM2 557-9686
Chemical-Electrical Classification Group Director Donald J Hoffman rm901 CM2 557-2825
Secretary Sandra P Crawford rm903 CM2 557-2820
Unit I Diane B Russell rm971 CM2 557-2753
Unit II Eugene B Woodruff rm935 CM2 557-2826
Unit III Gary Solyst rm923 CM2 557-3505
Unit IV Earl Folsom rm912 CM2 557-0151
Mechanical-General Classification Group Director John W Will rm310 CM2 557-0107
Secretary Christina Boska rm310 CM2 557-0107
Unit I Donald P Rooney rm310 CM2 557-0138
Unit II Robert Craig rm310 CM2 557-0136
Unit III Harold P Smith rm310 CM2 557-2446
Special Projects Unit Director Leslie Wolf rm982 CM2 557-2781
Secretary Melvina Jarrett rm980 CM2 557-0173
Office of International Patent Documentation Director Thomas Lomont rm300 CM2 557-0667
Secretary Glenda Calhoun rm300 CM2 557-0400
Scientific Library Program Manager Henry Rosicky rm2C08 CP34 557-2955
Secretary Gail Owens rm2C08 CP34 557-2955
Administrative Librarian Irene Heisig rm2C08 CP34 557-2955
Foreign Patents Division Barry Balthrop rm2C01 CP34 557-2970
Bindery Unit Ronald Knickerbocker FERN 557-1530
Document Retrieval and Copy Branch Lendoria Roberson rm2C01 CP34 557-3545
Receipts and Records Branch Beverly Brooks Corridor Level, Suite 1821D CM2 557-0186
Reference Service Bernard Hamilton rm2C01 CP34 557-3545
Scientific Literature Division Kay Melvin rm2C06C CP3 557-2957
Technical Services Branch Jesse Gibson rm2C06A CP3 557-2961
Collection Development (Vacant) rm2C01 CP34 557-3092
User Services Branch Dora Weinstein rm2C04 CP34 557-2957
Circulation rm2C01 CP34 557-2957
Computer Searching rm2C01 CP34 557-2957
Interlibrary Loans rm2C01 CP34 557-2957
Reference Service rm2C01 CP34 557-2957
Translations Division Dean Thorne rm2C15 CP34 557-3193
Receptionist Carol Releford rm2C15 CP34 557-3193

CHEMICAL EXAMINING GROUPS

110 General, Metallurgical, Inorganic, Petroleum and Electrical Chemistry, and Engineering rm9C17 CP3 557-2517
Director Dennis E Talbert rm9D17 CP3 557-9600
Secretary Constance L Morgan rm9D19 CP3 557-9600
General Information/Receptionist rm9C17 CP3 557-2517
SAC Dorothy Dawkins rm9C17 CP3 557-3598
111 Metallurgical Methods and Apparatus, Alloys and Metal Stock
L Dewayne Rutledge rm10E02 CP3 557-6722
112 Electro-Chemistry, Process and Apparatus John F Niebling rm10B04 CP3 557-8788
113 Inorganic compounds and non-metallic elements (except radioactive); chemical gas purification processes;
beneficiating ores; hydrometullargy; magnetic and piezoelectric compositions
plaster; single crystals and crystallization John Doll rm9A15 CP3 557-2517
114 Methods for semiconductor treating and manufacturing; batteries;
photovoltaic cells and their methods of operation Brian E Hearn rm10D35 CP3 557-6728
115 Chemical compositions, dying and pigments Paul Lieberman rm10A15 CP3 557-8779

116 Mineral oil processes and products, catalytic compositions, chemistry of hydrocarbons,
fuel and igniting devices, sugar, starch and carbohydrates, cleaning and contact with solids and
carbon compounds Helen Sneed rm9D35 CP3 557-3029

118 Liquid and solid fuels, chemical and biological fertilizers, refractory
glass and cement compositions, lubricating compositions William R Dixon Jr rm10B02 CP3 557-8787

120 Organic Chemistry and Biotechnology rm8C13 CP2 557-3920
    Director Samih N Zaharna rm8A07 CP2 557-0661
        Secretary Anne Willey rm8A07 CP2 557-0661
        General Information/Receptionist rm8C13 CP2 557-3920
        SAC Helen Childs rm8C13 CP2 557-3920

121 Heterocyclic organic chemistry, nitriles and azo chemistry Mary C Lee rm9D01 CP2 557-3920

122 Nitrogen containing heterocyclic compounds, mercaptans and
phosphorus esters Donald G Daus rm8B32 CP2 557-3920

125 Medicines, poisons, cosmetics and testing compositions Albert T Meyers rm8B02 CP2 557-3920

126 Organic carboxylic acid and ester compounds, oxy, aldehyde and
keto compounds, phosphorus acid compounds Donald Moyer rm9B32 CP2 557-3920

129 Herbicides, heterocyclic and nitrogen chemistry Glennon H Hollrah rm9A01 CP2 557-3920

130 Specialized Chemical Industries and Chemical Engineering rm8C17 CP3 557-2475
    Director Robert F White rm8D19 CP3 557-3804
        Secretary Vickie Enos rm8D19 CP3 557-3804
        General Information/Receptionist rm8C17 CP3 557-2475
        SAC Ruth Lyles rm8C17 CP3 557-9854

131 Adhesive bonding and miscellaneous chemical manufacture Michael W Ball rm9E02 CP3 557-2475

132 Food or edible material, processes, compositions and products Donald E Czaja rm8D01 CP3 557-2475

133 Paper making and fiber liberation; glass manufacture; concentrating evaporators, separatory and
thermolytic distillation processes and apparatus; separating and assorting solids-froth floatation;
and compositions, method and apparatus for etching in chemical manufacture gas, heating and
illumination apparatus and process and mineral oils apparatus David L Lacey rm8B36 CP3 557-2475

135 General molding or treating apparatus; static molds; gas separation;
gas and liquid contact Jay H Woo rm8D35 CP3 557-2475

136 Liquid purification or separation Richard V Fisher rm7A09 CP3 557-2475

137 Processes of plastic and nonmetallic article shaping or treating Jan H Silbaugh rm8A01 CP3 557-2475

139 Coating processes and coating apparatus Norman Morgenstern rm9B02 CP3 557-2475

150 High Polymer Chemistry, Plastics, Coating, Photography, Stock Materials and Compositions rm11C19 CP2 557-6525
    Director James O Thomas Jr rm11A04 CP2 557-6533
        Secretary Sharon C Graham rm11A04 CP2 557-6533
        General Information/Receptionist rm11C19 CP2 557-6525
        SAC Ellen Scott rm11C19 CP2 557-6525

151 Mixed synthetic resin compositions, block and graft copolymers and
irradiation of polymers John Bleutge rm11C19 CP2 557-6525

153 Foams, condensation polymers of cellulose, phenols, isocyanates,
polyesters, polyepoxides, stabilization of polymers John Kight III rm11C19 CP2 557-6525

154 Stock materials or miscellaneous articles of manufacture George Lesmes rm11C19 CP2 557-6525

155 Polymer compositions, addition polymers, and carbohydrates Joseph Schofer rm11C19 CP2 557-6525

156 Radiation imagery chemistry - process, composition or product, and
stock materials or miscellaneous articles of manufacture Paul Michl rm11C19 CP2 557-6525

158 Coating apparatus, record receivers, stock materials or miscellaneous
articles, and selected imaging processes and products Ellis Robinson rm11C19 CP2 557-6525

180 Biotechnology rm9C13 CP2 557-0664
    Director John E Kittle rm9A09 CP2 557-3637
        Secretary Cheryl P Gibson rm9A09 CP2 557-3637
        General Information/Receptionist rm9C13 CP2 557-0664
        SAC Kathryn Perry rm9C13 CP2 557-6941

181 Chemical apparatus such as analyzers, reactors and sterilizers; processes of chemical and clinical analysis,
sterilizing and preserving; immunology and liquid purifications or separation by living
organisms Barry S Richman rm9B02 CP3 557-6629

182 Clinical chemistry, microbiology, immunology and enzymology,
purification and chemical engineering Robert J Warden rm9B32 CP2 557-7369

183 Peptide and carbohydrate chemistry, and drug, bioaffecting and body treating compositions containing peptides, carbohydrates, antibody, antigen, enzyme, or animal or plant extracts of undetermined constitution Johnnie R Brown rm8D31 CP2 557-3776

184 Multicellular organisms (animal/plant), molecular genetics, cell culture, nucleic acid assays, immunology, hybridoma, molecular biology, microbiology, fermentation and chemical engineering Charles F Warren rm10D32 CP2 557-7387

185 Molecular genetics, catalysis, enzymology, microbiology and chemical engineering Thomas G Wiseman rm9D31 CP2 557-3567

ELECTRICAL EXAMINING GROUPS

210 Industrial Electronics, Physics and Related Elements rm9C17 CP4 557-5080
    Director Gerald Goldberg rm9D19 CP4 557-2488
      Secretary Teresa E Dugan rm9D19 CP4 557-2488
      General Information/Receptionist rm9C17 CP4 557-5080
      SAC Charles B Blake rm9C17 CP4 557-7405

211 Acoustics, Electric Charge Devices and Systems, Electromagnetic Control Systems, Music, Optics, Photography, Photocopying, Motion Pictures, Electric and Magnetographic Recorders, and Mechanical Registers L Thomas Hix rm9B40 CP4 557-7674

212 Electrical Motor-generator Structure, Piezoelectric Elements and Devices, Generator Systems, Battery and Condenser Charging and Discharging, Power Supply Regulation, and Conversion Systems Patrick R Salce rm9A01 CP4 557-9695

214 Electrical Switches and Arc Suppression, Heating by Induction, Plasma and Electric Discharge Machining, Industrial Electric Heating Furnaces, Electrical Component Housing and Mounting and Protection of Electrical Systems and Devices A David Pellinen rm8B34 CP4 557-7215

215 Conductors, insulators, inductors, electric photocopying and electrical devices Arthur T Grimley rm9E16 CP4 557-7671

216 Recorders, Scales, Magnets, Magnetic and Thermal Switches, Electric Heating, Electric Welding and Resistors Elliott A Goldberg rm8B02 CP4 557-2323

217 Motor Control, Electrical Transmission Systems, Prime-mover Dynamo Plants, Electrical Elevator Controls, Code Conversion and Horology William M Shoop Jr rm9D01 CP4 557-3231

220/290 Utility and Design Applications rm10C17 CP4 557-2895
    Director Kenneth L Cage rm10D17 CP4 557-2877
      Secretary Arnette S McGill rm10D19 CP4 557-2478
      General Information/Receptionist Group 220 rm10C19 CP4 557-2895
      General Information/Receptionist Group 290 rm1106 CP6 557-2864
      SAC Joanne Hodge rm10C17 CP4 557-9151
    Licensing and Review Hilda Grimes rm10C34 CP4 557-4948
    Mildred Lawrence rm10C34 CP4 557-4949

221 Weapons (firearms, ordnance, ammunition, explosive devices), general lubrication, nuclear reactor systems, illumination, aeronautics and ships as well as all classified mechanical applications Deborah L Kyle rm10D01 CP4 557-3253

222 Radio, optic, acoustic, wave communications systems and all classified electrical applications Thomas H Tarcza rm10E16 CP4 557-4922

223 Chemical and including radioactive materials, powder metallurgy, rocket fuels, explosives and thermal and photoelectric batteries as well as all classified chemical applications John Terapane rm10E02 CP4 .. 557-4934

291 Ornamental designs in the area of industrial arts Wallace R Burke rm11E02 CP6 557-4979

292 Ornamental designs for fine arts Bernard Ansher rm11A02 CP6 557-4965

230 Information Processing, Storage and Retrieval rm11C17 CP4 557-2878
    Director Earl Levy rm11D37 CP4 557-5088
      Secretary Laura Dorsey rm11D37 CP4 557-5088
      General Information/Receptionist rm11C17 CP4 557-2878
      SAC Katherine A Nelson rm11C17 CP4 557-4174

231 General & Special Purpose Digital Data Processing, Digital Arithmetic, Speech Analysis & Synthesis & Data Presentation Systems Gary V Harkcom rm10B02 CP4 557-7128

232 General & Special Purpose Digital Data Processing Systems Archie E Williams rm11B02 CP4 557-2119

233 Static Information Storage & Retrieval & Elements of Dynamic Magnetic Information Storage & Retrieval Systems Stuart N Hecker rm11D01 CP4 557-0326

234 Ordnance or Weapon System Computers & Special Applications of Computers Including Vehicle Control, Navigation, Measuring, Testing & Monitoring Parshotam S Lall rm11D17 CP4 557-4316

235 Electrical Dynamic Information Storage & Retrieval & Record Controlled Systems (Vacant) rm11D17 CP4 557-2878

236 Computer Control Systems, Miscellaneous Applications of Computers, Computer Aided Product
   Manufacturing, Artificial Intelligence, Analog & Hybrid Computers & Error Correction &
   Detection Systems Jerry Smith rm11E10 CP4 557-8041
237 General & Special Purpose Digital Data Processing Systems Gareth D Shaw rm11B40 CP4 557-8047
240 Packages, Cleaning, Textiles and Geometrical Instruments rm6C17 CP4 557-2900
   Director Trygve M Blix rm6D37 CP4 557-2906
      Secretary Donna Purdham rm6D37 CP4 557-2906
      General Information/Receptionist rm6C17 CP4 557-2900
      SAC Doretha A Bailey rm6D15 CP4 557-2900
241 Packaging art including glass, fabric, metal, wood, paper and plastic
   receptacles plus closures Stephen Marcus rm6B02 CP4 557-4719
242 Fluid treating, presses, food apparatus, cleaning, agitating,
   centrifuges, and web feeding Harvey C Hornsby rm6E02 CP4 557-6116
243 Conduits, bathroom facilities, cleaning apparatus, filling apparatus,
   switches, and article carriers Henry Recla rm6E10 CP4 557-9891
245 Textiles, winding and reeling, pushing and pulling, bearings, and
   flexible torque transmitters Stuart S Levy rm5B24 CP4 557-6855
246 Measuring and testing, dynamic information storage or retrieval, optical
   image projectors and joint packing William Cuchlinski rm6E14 CP4 557-9894
247 Textile and leather manufacture, apparel, and textiles Werner Schroeder rm6D01 CP4 557-3302
250 Electronic and Optical Systems and Devices rm8D17 CP4 557-3311
   Director Edward Kubasiewicz rm8D19 CP4 557-2084
      Secretary Deborah P Leeper rm8D19 CP4 557-2084
      General Information/Receptionist rm8C17 CP4 557-3311
      SAC JoAnn Davis rm8C19 CP4 557-4784
251 Lasers, fiber optic devices and antennas William L Sikes rm8E02 CP4 557-2733
252 Electronic modulators, demodulators, oscillators, amplifiers, tuners and
   wave transmission lines and networks Eugene R Laroche rm7A15 CP4 557-4317
253 Semiconductor devices Andrew J James rm7B02 CP4 557-4835
254 Semiconductor and vacuum tube circuits and systems and electronic and
   electromechanical counting circuits and systems Stanley D Miller rm7E02 CP4 557-4753
255 Optical measuring and testing systems and photocell circuits Davis L Willis rm8D13 CP4 557-4339
256 Radiant energy systems Craig E Church rm7E16 CP4 557-3453
257 Optical systems and elements and vision testing and correcting
   John K Corbin rm8E16 CP4 557-2884
260 Communications, Measuring, Testing and Lamp/Discharge Group rm5D19 CP4 557-3321
   Director Stephen G Kunin rm5D19 CP4 557-7075
      Secretary Iyone L Miles rm5D19 CP4 557-7075
      General Information/Receptionist rm5C14 CP4 557-3321
      SAC Vivian C Harris rm5D21 CP4 557-2067
261 Telegraphy, telephony and audio systems Jin F Ng rm4D01 CP4 557-7739
262 Television and television facsimile James J Groody rm4B02 CP4 557-7309
263 Multiplex communications, digital communications and telecommunications
   Robert L Griffin rm5E02 CP4 557-1139
264 Electrical communications and acoustic wave systems John W Caldwell rm5E16 CP4 557-3356
265 Measuring and testing of non-electrical phenomenon Stewart J Levy rm4E02 CP4 557-7603
266 Lamp and discharge devices/systems and image analysis David K Moore rm5B02 CP4 557-6868
267 Measuring and testing of electrical phenomenon Reinhard J Eisenzopf rm5D01 CP4 557-6878
268 Condition responsive communications, measuring and testing Joseph A Orsino rm5D01 CP4 557-7956

MECHANICAL EXAMINING GROUPS
310 Handling and Transporting Media rm5D19 CP3 557-3618
   Director Bobby R Gray rm5D19 CP3 557-3677
      Secretary LaJuene Desmukes rm5D19 CP3 557-3677
      General Information/Receptionist rm5C17 CP3 557-3618
      SAC Margaret Stevens rm5D17 CP3 557-3618
311 Dispensing, article dispensing, coin handling, check-controlled apparatus, elevators, and sheet feeding
   or delivering devices Joseph J Rolla Jr rm6D01 CP3 557-6491
312 Railways and railway equipment, motor vehicle wheels and bodies and article assorting and
   handling implements Robert B Reeves rm5A01 CP3 557-6765

355 Supports, racks, fire escapes, ladders, scaffolds, flexible partitions
    Ramon S Britts rm4E02 CP3 557-6200
356 Petroleum, mining, highway and bridge engineering, well drilling, and
    endless belts Jerome W Massie rm4D35 CP3 557-6200
357 Tables, chairs, cabinets, windows, doors, buckles, buttons, clasps
    Kenneth J Dorner rm4A01 CP3 557-6200
358 Fasteners, safes, locks, closure fasteners, beds, control levers and
    linkages Gary L Smith rm4D17 CP3 557-6200

OFFICE OF THE ASSISTANT COMMISSIONER FOR EXTERNAL AFFAIRS
Assistant Commissioner Michael K Kirk rm902 PK2 557-3065
    Secretary Johnell M Bersano rm902 PK2 557-3065
  Director of Congressional Affairs Arthur E White rm902 PK2 557-1310
  Congressional Liaison Janie F Cooksey rm902 PK2 557-1310

    Office of Public Affairs Director (Vacant) rm 208B PK 1 557-3341
      Public Information Specialist Oscar G Mastin rm1B01 CP3 557-3341
    Office of Legislation and International Affairs
    Director (Vacant) rm902 PK2 557-3065
      Legislative and International Intellectual
      Property Specialists
      Judy W Goans rm902 PK2 557-3065
      H Dieter Hoinkes rm902 PK2 557-3065
      Lee J Schroeder rm902 PK2 557-3065
      G Lee Skillington rm902 PK2 557-3065
      Attorney Advisors
      Rosemarie G Bowie rm902 PK2 557-3065
      Michael S Keplinger rm902 PK2 557-3065
      Richard C Owens rm902 PK2 557-3065

OFFICE OF THE ASSISTANT COMMISSIONER FOR TRADEMARKS
Assistant Commissioner Jeffrey M Samuels rm910 PK2 557-3061
    Secretary Sheila G Pellman rm910 PK2 557-3061
  Deputy Assistant Commissioner for Trademarks Robert M Anderson rm910 PK2 557-3916
    Secretary Charlene A Rucker rm910 PK2 557-3061
  Trademark Legal Administrator Carlisle E Walters rm910 PK2 557-7464
    Secretary Carol P Smith rm910 PK2 557-3061
  Staff Assistant (Vacant) rm910 PK2 557-2222
  Trademark Program Analyst (Vacant) rm910 PK2 557-2221

Office of Trademark Program Analysis
Director Kimberly Krehely rm3C06 CP2 557-3268
Program Analyst Betty Andrews rm3C06 CP2 557-3268

TRADEMARK EXAMINING OPERATION
Director David E Bucher rm3C06 CP2 557-3268
    Secretary Tawana A Hawkins rm3C06 CP2 557-3268
  Administrator for Trademark Operations Patricia M Davis rm3C06 CP2 557-3268
    Secretary (Vacant) rm3C06 CP2 557-3268
  Administrator for Trademark Policy and Procedures Charles J Condro rm3C06 CP2 .. 557-3268
    Secretary Lisa Y Harrell rm3C06 CP2 557-3268
  Petitions and Classification Attorney
  Jessie N Marshall rm3C06 CP2 557-3268
    Petitions Assistant Annette L Pray rm3C06 CP2 557-3268
  Trademark Procedures and Special Projects Attorney James T Walsh rm3C06 CP2 557-3268
    Secretary Nina Bailey rm3C06 CP2 557-3268
  Trademark Program Assistant (Vacant) rm3C06 CP2 557-3268
  Paralegal Assistant Blake Pearl rm3C06 CP2 557-3268
  Paralegal Assistant Doris Kahn rm3C06 CP2 557-3268
  Trademark Law Offices
Managing Attorney Law Office I (Vacant) rm3C28 CP2 557-3273
  Senior Lead Attorney Deborah S Cohn

Support Staff Manager Carolyn Spriggs

Managing Attorney Law Office II (Vacant) rm2C24 CP2 557-3277
    Senior Lead Attorney Donald Fingeret
Support Staff Manager Doshie Day

Managing Attorney Law Office III Myra Kurzbard rm2C22 CP2 557-9560
    Senior Lead Attorney Robert Feeley
Support Staff Manager Gwen Stanmore

Managing Attorney Law Office IV Thomas Lamone rm3C13 CP2 557-9550
    Senior Lead Attorney David Soroka
Support Staff Manager Elsie Bradley

Managing Attorney Law Office V Paul Fahrenkoph rm2C11 CP2 557-5380
    Senior Lead Attorney Christopher Sidoti
Support Staff Manager Thurmond Streater

Managing Attorney Law Office VI Ronald E Wolfington rm3C27 CP2 557-2937
    Senior Lead Attorney Mary Sparrow
Support Manager Pearl Clements

Managing Attorney Law Office VII Lynne Beresford rm4C13 CP2 557-5237
    Senior Lead Atttorney David Shallant
Support Staff Manager Karen McCray

Managing Attorney Law Office VIII Sidney Moskowitz rm4C11 CP2 557-5242
    Senior Lead Attorney Michael Bodson
Support Staff Manager Ada Rollins

Trademark Services Division Director Doreane Poteat rm6D29 CP2 557-5249
  Supervisory Trademakr Services Assistant Seth M Cheatham rm6D29 CP2 557-5249
    Secretary Sophia Brock rm6D29 CP2 557-5249
  Quality Review Clerk Deborah Ahmed rm6D29 CP2 557-5249

Application & Classification Section Leon Jackson rm4B30 CP2 557-5255
Publication & Issue Supervisor Tony Milligan rm4C23 CP2 557-5247
Post Registration Sec Supervisor Portia Taylor rm4C24 CP2 557-1986
  Affidavit Examiners rm4C24 CP2 557-1988
  Renewal Examiners rm4C24 CP2 557-1988
Mail Reader/Messenger Lilly Mott rm4C26 CP2 557-5257
Microfilm Section Della Williams rm4D27 CP2 557-5255

TRADEMARK TRIAL AND APPEAL BOARD
Members of the Board

Chairman J David Sams rm1008 CS5 557-3551
      Ellen Seeherman rm1008 CS5 557-3551
      Robert F Cissel rm1008 CS5 557-3551
      Louise E Rooney rm1008 CS5 557-3551
      Gary D Krugman rm1008 CS5 557-3551
      Janet E Rice rm1008 CS5 557-3551
      Rany L Simms rm1008 CS5 557-3551
      Elmer W Hanak III rm1008 CS5 557-3551
    Attorney-Examiners
      Paula T Hairston rm1008 CS5 557-3551
      Beth A Chapman rm1008 CS5 557-3551
      G Douglas Hohein rm1008 CS5 557-3551
      T Jeffrey Quinn rm1008 CS5 557-3551
      Carla C Calcagno rm1008 CS5 557-7049
    Paralegal Specialist Gladys R Springer rm1008 CS5 557-3551
    Clerk of the Board Evelyn R Lopez rm1008 CS5 557-3551
    Administrator Erma S Brown rm1008 CS5 557-3551
    Legal Technician Sheila H Veney rm1008 CS5 557-3551

OFFICE OF THE ASSISTANT COMMISSIONER FOR ADMINISTRATION

Assistant Commissioner Theresa A Brelsford rm908 PK2	557-2290
Secretary Karon Morris rm908 PK2	557-2290
Deputy Assistant Commissioner for Administration Wesley H Gewehr rm908 PK2	557-3055
Secretary Diane Rich rm908 PK2	557-3055
Program Analyst Joan S Griffey rm908 PK2	557-2290

Office of General Services

Director John D Hassett rm803 PK1	557-0183
Secretary Peggy Fewell rm803 PK1	557-0183
Deputy G William Richardson rm803 PK1	557-0183
Security/Safety Officer (Vacant) rm803 PK1	557-0183
Correspondence and Mail Division Sallye Rayford rm1A03 CP2	557-1689
Deputy (Vacant) rm1A03 CP2	557-2932
Incoming-Outgoing Mail Branch Margaret LaSalle rm1A01 CP2	557-3233
Initial Review & Serializing Branch Shirley Steele rm1B03 CP2	557-3478
Correspondence Branch (Vacant) rm1A03 CP2	557-3226
Facilities Management Division Robert Randolph rm802C PK1	557-7042
Records and Property Management Branch Flo Stanmore rm802C PK1	557-0410
Space and Telecommunications Branch William Morris rm802C PK1	557-7331
Office Services Division Constant G Fearing rm803 PK1	557-0183
Travel Arrangements rm803 PK1	557-0183
Support Services Branch Joe Ragland FERN	557-3560
Transportation Unit John Holmes FERN	557-1531
File Information Unit William Satterwhite rm1D01 CP3	557-6944
Official Search Unit Daisy Johnson FERN	557-9690

Office of Patent and Trademark Services

Director (Vacant) rm7D25 CP2	557-3236
Secretary Teresa Knight rm7D25 CP2	557-3236
Deputy Mary E Turowski rm7D25 CP2	557-3236
Public Search Services Division Catherine Kern (Acting) rm1A01 CP3	557-2276
Secretary (Vacant) rm1A01 CP3	557-2276
Patent Search Branch Bernard Thomas rm1A03 CP3	557-2219
Secretary Barbara Evans rm1A03 CP3	557-2277
Patent Search Room rm1A03 CP3	557-2276
Trademark Search Branch Linda Lynch rm2C08 CP2	557-3281
Assignment Search Branch Diane Russele rm5C22 CP2	557-3826
Program Control Division Mary Brown (Acting) Lobby CP1	557-7261
Secretary Louwilda Turner Lobby CP1	557-7261
Public Service Window Lois Stevenson Lobby CP3	557-2833
Public Service Center Mary Reed rm1A02 CP4	557-5168
Secretary (Vacant) rm1A02 CP4	557-5168
Assignment/Certification Services Division (Vacant)	
Assignment Branch Annie Harrell rm7D19 CP2	557-3266
Examination Section Virginia Clark rm5C16 CP2	557-3247
Digest and Recording Lucy Stevenson rm7D19 CP2	557-3259
Quality Control and Status (Vacant)	557-8691
Certification Branch Lannie Anderson rm800 PK1	557-1552
Input and Records Control Mary Gartrell rm800 PK1	557-1587
Microfiche & Printing Frances Morris rm800 PK1	557-1603
Certification Section Dorothea Saunders rm800 PK1	557-1564
Court and Documentation (Vacant) rm800 PK1	557-1564
PCT International Services Division Jane Corrigan rm1248 CP6	557-2003
Receiving Office Branch Terry Johnson rm1248 CP6	557-2003
Searching/Preliminary Examining Authority; and Designated/	
Elected Office Branch Lauretta Zirk rm1248 CP6	557-2003
Application Processing Division Willie Bowman rm7D19 CP2	557-3256
Classification & Routing Branch Norma White rm7C20 CP2	557-3260
Administrative Examination Branch Mose Montgomery rm7C10 CP2	557-3254
Special Processing Branch Otis Quick rm7C18 CP2	557-3831

Data Entry Branch Everette Oliver rm7C24 CP2 557-5662
Quality Review and Assembly Branch Delora Dillard rm7C24 CP2 557-3763
File Maintenance & Correspondence Branch Jeanette Gatling rm7C12 CP2 557-1561
Micrographics Division Michael Johnson (Acting) rm7D25 CP2 557-3236
Application Filming Branch (Crystal) Calvin Pullen rm6C22 CP2 557-3079
Patent & Trademark Filming Branch (HCHB) Al Mundy 377-4968
Quality Review Branch (HCHB) Mary Smith 377-5501

Office of Publications
Director Richard A Bawcombe rm513 PK1 557-3794
   Secretary (Vacant) rm513 PK1 557-0698
Deputy Director (Vacant) rm513 PK1 557-3283
  Publishing Division Sylvia F Martin rm512 PK1 557-6388
  Deputy Manager (Vacant) rm513 PK1 557-6395
   Allowed Files and Assembly Branch Yvette E Simms rm513 PK1 557-6412
   Production Control Branch Willard D Ireland rm513 PK1 557-6393
   Editorial Branch Marthina Thompson rm513 PK1 ....

   Data Base Query Section (Vacant) rm513 PK1 557-6393
   Patent Copy Inspection Section Jim Alexander rm513 PK1 557-6404
  Drafting Branch Martin Baum rm510 PK1 557-1963
  Statistical Analysis Division Michael Stellabotte rm513 PK1 557-6414
  Data Base Inspection Branch Melvinia Gray rm513 PK1 557-0709
  Certificates of Corrections Branch Mary H Allen rm809 PK1 557-1992
  Technical Development Division Edwin P Hall (Acting) rm513 PK1 557-6945
  Patent Maintenance Division CH Griffen rm811 PK1 ....

Office of Equal Employment Programs
Director R Jacqueline Dees rm600A PK1 557-1692
  Secretary (Vacant) rm600A PK1 557-1692
  Secretary Assistant Shawneequa Graham rm600A PK1 557-1692
Supervisory Equal Opportunity Specialist Godfrey Beckett rm600A PK1 557-1692
Equal Opportunity Specialists
  Sharon L Carver rm600A PK1 557-1692
  Myra P Young rm600A PK1 557-1692
  Helen D Mitchell rm600A PK1 557-1692

Office of Management and Organization
Director Sara E Bjorge rm505 PK1 557-5825
  Secretary Donna Oliver rm505 PK1 557-5825
Project Managers:
  Alvin Dorsey rm505 PK1 557-5825
  Greg P Mullen rm505 PK1 557-5825
  Myra Young rm505 PK1 557-5825
  Joseph Jones rm505 PK1 557-5825

Office of Procurement
Director William J Eldridge rm806 PK1 557-0014
  Secretary Cristina M Moran rm806 PK1 557-0014
Contract Division Chief Page A Etzel rm806 PK1 557-0014
Small Purchases Division Chief Jean A McLeod rm806 PK1 557-0014

OFFICE OF THE ASSISTANT COMMISSIONER FOR FINANCE AND PLANNING
Assistant Commissioner Bradford R Huther rm904 PK2 557-1572
  Secretary Vickie T Bryant rm904 PK2 557-1572

Office of Budget
Director James R Lynch rm805 PK1 557-3875
Deputy Director Miguel B Perez rm805 PK1 557-3875
  Secretary Marna Engram rm805 PK1 557-3875

Office of Finance
Director Leonard L Nahme rm802A PK1 557-3051
Secretary Virginia R Clark rm802A PK1 557-3051
Appropriation Accounting Division John L Oliff rm802B PK1 557-2983
Fee Accounting Division Frank S Lane Sr rm1B01 CP2 557-2983
Deposit Account Branch Delores H Riley rm1B01 CP2 557-3227
Financial Management Division Robert M Kopson rm802A PK1 557-3051

Office of Long-Range Planning and Evaluation
Director Frances Michalkewicz rm507 PK1 557-1610
Program Assistant Jeannette Hawthorne rm507 PK1 557-1610
PROJECT XL rm507 PK1 557-1610

Office of Personnel
Personnel Officer Carolyn P Acree rm700 PK1 557-2662
Assistant to the Personnel Officer Nancy C Swanberg rm700 PK1 557-2662
Secretary Mildred Jeter rm700 PK1 557-2662
Classification and Employment Division Cynthia Nelson rm700 PK1 557-3631
Employee and Labor Relations Division Richard Haisch rm600 PK1 557-3643
Workforce Planning, Effectiveness Division Robert Ramig rm701 PK1 557-6327
Workforce Planning, Development and Systems Division Thomas H Neuhauser rm601 PK1 557-3431
Personnel Processing Division Beverly Boykin rm700 PK1 557-1208
Office of Labor Law Counsel Lynn Sylvester rm700 PK1 557-9684

OFFICE OF THE ASSISTANT COMMISSIONER FOR INFORMATION SYSTEMS

Assistant Commissioner for Information Systems Thomas P Giammo rm916 PK2 557-9093
Secretary Helen White rm916 PK2 557-9093
Deputy Assistant Commissioner for Information Systems Boyd Alexander rm 916 PK2 557-6000
Secretary Carla Bowman rm 916 PK2 557-6000
Technical Coordinator Linda Budney rm 916 PK2 557-9093
APS Project Manager Don LeCrone rm1002 PK2 557-6000
Secretary Paulette Whiteside rm1002 PK2 557-6000
Program Management Support Services Office L Liddle rm1002 PK2 557-6000
Secretary Michele Helms rm1002 PK2 557-6000

Office of System Test and Evaluation
Director Bruce A Reynolds rm1002 PK2 557-4114
Secretary Linda Bilbo rm1002 PK2 557-4114

Office of Automated Patent Systems
Director Jeff Cochran rm1000 PK2 557-6156
Secretary Audrey Jackson rm1000 PK2 557-6156

Office of Automated Trademark Systems
Director Raymond Rahn rm911 PK2 557-3544
Secretary Felicia Palmer rm911 PK2 557-3544

Office of Systems Engineering and Data Communications
Director Stephen Jacobson rm1004 PK2 557-7862
Secretary Eliza Davis rm1004 PK2 557-7862

Office of Electronic Data Conversion and Dissemination
Director David Grooms (Acting) rm914 PK2 557-6154
Secretary (Vacant) rm914 PK2 557-6154

Center for Automated Patent Systems
Center for Trademark and General Systems
Director William Maykrantz (Acting) rm1001 PK2 557-3646
Program Assistant Sylvia Huffman rm1001 PK2 557-3646
System Support and Operations Division John Fancovic rm1001 PK2 557-3646
Applications Systems Management Division Thomas Woomer rm1001 PK2 557-6330
Secretary Lisa O'Donnell rm1001 PK2 557-6330
Archives and Contingency Management Division J Kent Hughes rm914 PK2 557-6132
Site Manager Mel Pears Boyers PA (412) 794-3636

APS Contract Office

Director Barry Brown rm509 CP1 557-5800
    Secretary Arnett Wright rm509 CP1 557-5800

Office of Public Services

Director Patrick Rowe rm603 PK1 557-9543
    Secretary Teresa Knight rm603 PK1 557-9543
Deputy Director Mary E. Turowski rm603 PK1 557-9543
Public Search Services Division Cheryl Davis (Acting) rm1A01 CP3 557-2276
    Secretary (Vacant) rm1A01 CP3 557-2276
Patent Search Branch Robin Roark rm1A01 CP3 557-2219
Trademark Search Library Doris Kahn (Acting) rm2C08 CP2 557-3819
Assignment Search Branch Shirley Royall rm5C22 CP2 557-3826
Program Control Divison (Vacant) Lobby CP1 557-7261
    Secretary (Vacant) Lobby CP1 557-7261
Public Service Center Sharon Furbush rm208 PK1 .... 557-HELP
    Secretary Gale Randolph rm208A PK1 .... 557-HELP
Assignment/Certification Services Division Cathy Kern (Acting) rm5C11 cP2 ... 557-7770
    Secretary Margaret Bassford rm5C11 CP2 557-7770
Assignment Branch Recording Officer (Vacant) rm7D19 CP2 557-3266
    Secretary Avis Gans rm7D19 CP2 557-3266
Certification Branch Lannie C Anderson rm800 PK1 557-1552
    Secretary Wanda Meredith rm800 PK1 557-1522

Inventor Associations: National and Local

Affiliated Inventors Foundation, Inc. (AIF)
2132 Bijou St.   (303) 623-8710
Colorado Springs, CO 80909-5950   John T. Farady, Executive Director

The Foundation provides the following services to its 400 member inventors: free educational materials; free consultations; free preliminary appraisals; free invention evaluations; low-cost patent and trademark services; funding opportunities; free marketing assistance services; commission-based negotiation services (five to 15 percent range); and an invention publicity program, e.g. features member inventions in its bimonthly magazine, *Iinventor's Digest*, which is mailed to thousands of manufacturers.

Since it opened for business 15 years ago, the Foundation has evaluated a reported 18,000 plus inventions. 1989 was a busy year for AIF. It claims to have provided nearly 2,200 independent inventors with preliminary appraisals for their inventions. Of these, more than 600 were rejected in this first stage. According to executive director John Farady, nearly 30 percent of all inventions they see have fatal flaws and are rejected. "Our educational goal is to provide independent inventors with sufficient information about each phase of invention development to help them make better decisions in their own best interests," he says.

"We do not require any advance cash fees," he continues. "And our contract negotiation services are on a no-fee, straight commission basis should we be successful." The Foundation states that "unlike the front-end cash-fee companies, it is not the aim of the Foundation to offer marketing services to so many inventors that it cannot focus on anything properly. This organization keeps the membership controlled in such a way that it can be most effective."

I have always liked AIF and what co-founder Farady is trying to do for his membership. The Foundation is one national organization that has lots to offer independent inventors and its publications, especially *Inventors' Digest*, are most informative.

Write for an information package. Check out the Foundation's references with any of the following organizations: Colorado Springs Chamber of Commerce, 100 Chase Stone Center,; Colorado Springs, CO 80902; Dun & Bradstreet (any local office) Report No. DUNS-06-062-9748; and Central Bank of Colorado Springs, 2308 East Pikes Peak Avenue, Colorado Springs, CO 80909.

The toll-free number for the Foundation is 1-800-525-5885.

Akron/Youngstown Inventors Organization
1225 W. Market St.   (216) 864-5550
Akron, OH 44313   Ned Oldhan, Contact

Serves Akron/Youngstown area. Affiliated with Ohio Inventors Association. **Contact:** Alternative contact: Charles Clark, 623 Grant St., Kent, OH 44240; (216) 673-1875.

Alaska Inventorprizes
205 E. 4th Ave.   (907) 338-5484
Anchorage, AK 99501   R.J. Bret Thomas, Contact

Provides encouragement, information, and a forum for Alaska's most innovative people, thereby supporting the State's growth through the commercialization of Alaskan ideas. Members are inventors, with some invention developers and venture capitalists. **Publications:** Quarterly newsletter.

Alaska Inventors Association
P.O. Box 241801
Anchorage, AK 99524

Albuquerque Invention Club
Box 30062   (505) 266-3541
Albuquerque, NM 87190   Dr. Albert Goodman, Contact

Inventor groups appear to take root anywhere inventors practice their trade. Seeking to build relationships among themselves, independent inventors have formed organizations throughout the country that provide professional and social forums. These assemblies offer members product development support and guidance as well as resource information. A common objective is to stimulate self-fulfillment, creativity and problem solving. Inventors get together and explore professional issues, relate experiences, success stories and heartbreaks, exchange "insider information" and share personal contacts, methods, techniques and dreams. They gain strength from each other in a symbiotic, cross-circuiting of spirits and concepts.

American Association of Entrepreneurial Dentists (AAED)

420 Magazine St.   (601) 842-1036
Tupelo, MS 38801   Dr. Charles E. Moore, Exec. Director

Dentists and other dental professionals involved in research, industry, manufacturing, marketing, publication, and other entrepreneurial activities. Objectives are to: promote dental research and product development and assist entrepreneurial dentists; offer expertise in technical, professional, and scientific skills; exchange information among members on how to improve dentistry, creativity, and free enterprise; encourage high ethical standards and professional conduct in dental service, products, and marketing. Informs the public, dentists, educators, and manufacturing companies of new and beneficial techniques, products, and services; coordinates the review of specifications for dental materials and products by regulatory agencies; and encourages the dental industry to become more actively involved in public dental education. Offers consultation and placement services; evaluates and presents new ideas and products to manufacturing companies at convention trade expositions. Provides lists of foreign dental dealers and buyers, export technique information and shipping procedures, hospitality accommodations for members, and private entrepreneurial activities. Encourages creativity; sponsors competitions; bestows awards. Maintains biographical archives, and hall of fame of plaques awarded for recognition of dentist of the year, of the decade, and of the century. **Publications:** Entrepreneurial News, periodic; also publishes pamphlet. **Meetings:** Four dental meetings per year (with exhibits). **Founded:** 1983.

American Association of Inventors (AAI)

6562 E. Curtis Rd.   (517) 799-8208
Bridgeport, MI 48722   Dennis Ray Martin, President

Private inventors and other interested individuals. Provides assistance with development of ideas, patent applications, marketing, and production. Offers workshops on product development; maintains 485 volume library and speakers' bureau. Compiles statistics; bestows awards. **Members:** 6512. **Publications:** (1) The Success System, bimonthly; (2) Threw the Key Ring, bimonthly. **Meetings:** Annual conference. **Founded:** 1862.

American Copyright Council (ACC)

1600 I St., N.W.   (202) 293-1966
Washington, DC 20006   Fritz E. Attaway, Vice President

Coalition of computer, film, law, magazine, music, publishing, recording, and television organizations created to unite the copyright community. Seeks to initiate a concerted effort to preserve copyright principles and to educate the public about the value of copyrights and the harm caused by copyright infringement. Maintains speakers' bureau; compiles statistics. Publishes the Copyright Industries in the United States. **Members:** 25. **Founded:** 1984.

American Copyright Society (ACS)

345 W. 58th St.   (212) 582-5705
New York, NY 10019   Gerard Delachapelle, Managing Director

Authors, composers, and music publishers. Seeks to protect and enhance the corporate interests of members. Arranges for the collection of performance and mechanical royalties. Conducts periodic conferences, meetings, and research on copyrights. **Founded:** 1952.

American Entrepreneurs Association (AEA)

2392 Morse Ave.   (714) 261-2393
Irvine, CA 92714   Wellington Ewen, President

Persons interested in business opportunities and in starting profitable businesses. Conducts in-depth research on all types of small businesses. **Members:** 150,000. **Publications:** Entrepreneur Magazine, monthly; also publishes research reports. **Founded:** 1973.

American Inventors Council

P.O. Box 58426   (215) 546-6601
Philadelphia, PA 19102-8426   Henry H. Skillman, Treas.

Provides a means by which an experienced inventor (one who has succeeded in marketing his or her inventions and who has gained understanding from those experiences) can help other inventors with the problems, obstacles, and frustrations inventors experience in the process of turning an idea into a reality. National; members must have one product marketed or patented; associate membership is open to the public at large. **Publications:** National and local newsletters are planned.

American Society of Inventors (ASI)

P.O. Box 58426   (215) 546-6601
Philadelphia, PA 19102-8426   Henry H. Skillman, Treasurer

Engineers, scientists, businessmen, and others who are interested in a cooperative effort to serve both the short- and long-term needs of the inventor and society. Works with government and industry to improve the environment for the inventor. Goals include: encouraging invention and innovation; helping the independent inventor become self-sufficient; establishing an invention market and a communication system for inventors and businessmen to solve problems; seeking the optimization of employee disclosure agreements for use by employers; providing a voice for the independent inventor in reference to patent law changes. Acts as clearinghouse and pilot group for the formation of additional chapters; chapters develop special service projects, hold regular meetings featuring speakers, and provide a meeting place for other inventors. Sponsors periodic seminars. **Members:** 100. **Publications:** ASI Newsletter, periodic. **Meetings:** Periodic National Fall Inventor's Conference - usually November, Philadelphia, PA. **Founded:** 1953.

Arkansas Inventors Congress, Inc.

Route 3, Box 670   (501) 229-4515
Dardanelle, AR 72334   Garland Bull, President

Assists inventors and innovative persons as a support group. Members include inventors, patent attorneys, invention developers, and venture capitalists. Regular members must have patented inventions or must have patents pending; associate membership is open to anyone interest. **Publications:** Bimonthly Newsletter.

Association for Science, Technology and Innovation (ASTI)

P.O. Box 1242
Arlington, VA 22210   Ben Sands, President

Aim is to establish dialogue among different disciplines such as engineering, medicine, education, and the physical, social, and biological sciences, which share the common problem of effective management of innovation. Objectives are: to share ideas, knowledge, and experience among diverse communities; to expand and organize knowledge of the factors that effect productivity of science and technology efforts; to promote development, demonstration, and application of policies, standards, and techniques for improving management of innovation. Although the association operates primarily in the Washington, DC area, it has members throughout the U.S., Europe, Africa, and Central America. Conducts seminars; holds monthly management luncheon, roundtable, and symposium. Sponsor competitions. **Members:** 175. **Publications:** Newsletter, bimonthly. **Founded:** 1978.

Association of Collegiate Entrepreneurs (ACE)

Center for Entrepreneurship   (316) 689-3000
Box 147, Wichita State University
Wichita, KS 67208   Douglas Mellinger, Director

Student entrepreneurs, community entrepreneurs, and sponsoring firms. Seeks to enhance opportunities for student entrepreneurial pursuits by providing opportunities for students to network with faculty and entrepreneurs. Encourages venture capitalists, consultants, and other entrepreneurs to become involved as mentors in assisting students with entrepreneurial ventures. Works with the Center for Entrepreneurial Management (see separate entry). Maintains speakers' bureau; compiles statistics; bestows awards honoring the top 100 young entrepreneurs. **Members:** 2000. **Publications:** Newsletter, monthly. **Meetings:** Annual conference (with exhibits) - usually March. **Founded:** 1983.

Association of Small Business Development Centers (ASBDC)

1050 17th St., N.W., Suite 810   (202) 887-5599
Washington, DC 20036   Tom Cator, Executive Director

State centers providing advice for those planning to establish a small business. **Publications:** ASBDC News. **Founded:** 1977.

Association of Venture Founders (AVF)

805 Third Ave., 26th Floor   (212) 319-9220
New York, NY 10022   Anni J. Lipper, Executive Director

Successful entrepreneurs who seek to enhance the wealth, knowledge, and business success of members. Provides educational networking for the continuing education of members. Holds three seminars per year. **Members:** 150. **Publications:** (1) Venture Magazine, monthly; (2) Who's Who in AVF (directory), annual. **Founded:** 1979.

California Engineering Foundation

913 K St. Mall, Suite A   (916) 440-5411
Sacramento, CA 95814   Robert J. Kuntz, Contact

California Inventors Council

P.O. Box 2732   (415) 652-3138
Castro Valley, CA 94546   Lawrence Udell, Contact

Contact: Barrett Johnson, P.O. Box 2036, Sunnyvale, CA 94087 (408) 732-4314.

Center for Entrepreneurial Management (CEM)

180 Varick St., Penthouse Suite   (212) 633-0060
New York, NY 10014   Joseph R. Mancuso, President

Serves as a management resource for entrepreneurial managers and their professional advisors. Selects and makes available published materials on developing business plans, organizing an entrepreneurial team, attracting venture capital, and obtaining patents, trademarks, and copyrights. Develops, collects, and disseminates current information on business trends, new laws and regulations, and tax guidance. Conducts intensive-study courses and seminars. Has identified stages of the entrepreneurial process and, through essays and audiocassettes, addresses problems pertinent to each stage. Maintains library of small business and venture capital information. **Members:** 3000. **Publications:** The Entrepreneurial Manager (newsletter), monthly. **Founded:** 1978.

Central Florida Inventors Club

6402 Gamble Dr.   (305) 299-8598
Orlando, FL 32808   Steve Chandler, Contact

The inventor has two types of clubs from which to choose: local and national. Local clubs are real grass roots. They typically encompass only a limited geographic area, e.g. Albuquerque Invention Club or the Inventor's Council of Hawaii. The local clubs can be great fun and informative, not to mention wonderful places to strike up friendships. Ranging in size from several members to several hundred members, local clubs rarely have formal offices. Meetings are held at hotels, restaurants, or members' homes.

Local clubs do not promise to deliver anything more than you would expect. They are usually not the best place to pick up fast-breaking information on intellectual property legislation. But what I particularly like about these kinds of clubs is their lack of pretension, the sense of warmth, hospitality, community, and comradeship they engender.

Central Florida Inventors Council

4855 Big Oaks Lane — (305) 859-4855
Orlando, FL 32806 — Dr. David Flinchbaugh, President

Provides a periodic opportunity for inventors to get together, seek supportive relationships, and learn the "how to" of inventing, patenting, and starting companies. Affiliated with Central Florida Council for High Technology; Florida Engineering Society; and National Congress of Inventor Organizations. Membership is primarily inventors. **Contact:** Alternative address: P.O. Box 13416, Orlando, FL 32859; telephone (407) 857-8242.

Chicago High Tech Association

53 W. Jackson Blvd., Suite 1634 — (312) 939-5355
Chicago, IL 60604 — Ken Boyce, Executive Director

Fosters the development and application of technology, enhances the competitiveness and economic growth of the region, and serves as a membership organization for technology-related companies and institutions. In addition to Illinois-area membership, has national and international members. Affiliated with Illinois Software Association. Members include inventors, manufacturers, patent attorneys, invention developers, venture capitalists, and government officials with interest in support, development, or production of technology to promote economic development. **Publications:** Quarterly newsletter; Tech Connection, a quarterly magazine.

Columbus Inventors Association

2480 East Ave. — (614) 267-9033
Columbus, OH 43202 — Tim Nyros, Contact

Confederacy of Mississippi Inventors

4759 Nailor Rd. — (601) 636-6561
Vicksburg, MS 39180 — Rudy Paine, Contact

Serves inventors and would-be inventors in the Mississippi and eastern Louisiana area. Affiliated with National Congress of Inventor Organizations. **Publications:** Tinkers Gazette, a quarterly newsletter.

Copyright Clearance Center (CCC)

27 Congress St. — (508) 744-3350
Salem, MA 01970 — Eamon T. Fennessy, President

Photocopy users (corporations, academic and research libraries, information brokers, government agencies, and others who systematically utilize or distribute photocopy material) and publishers, authors, or other owners of copyrights. Established in response to 1978 copyright law which requires that permission of copyright owners be obtained by anyone doing systematic photocopying or photocopying not permitted under the fair use provision of the law. Provides owners with a centralized agency through which permission to use registered materials may be granted and fees may be collected. Operates Annual Authorizations Service, a photocopy license service for U.S. corporations. **Members:** 4000. **Publications:** (1) Photocopy Authorizations Report (newsletter), quarterly; (2) Publishers' Photocopy Fee Catalog (with quarterly supplements), semiannual; (3) Annual Review. **Telecommunication Access:** Fax, (617)741-2318. **Founded:** 1977.

Copyright Society of the U.S.A. (CSUSA)

New York University School of Law — (212) 998-6194
40 Washington Square, S.
New York, NY 10012 — Elizabeth Botta, Secretary

Lawyers and laymen; libraries, universities, publishers, and firms interested in the protection and study of rights in music, literature, art, motion pictures, and other forms of intellectual property. Promotes research in the field of copyright; encourages study of economic and technological aspects of copyright by those who deal with problems of communication, book publishing, motion picture production, and television and radio broadcasting. Seeks better understanding among students and scholars of copyright in foreign countries, to lay a foundation for development of international copyright. Cosponsors (with New York University School of Law) the Walter J. Derenberg Copyright and Trademark Library, which includes foreign periodicals dealing with literary and artistic property and related fields. Sponsors symposia and lectures on copyright. Encourages study of copyright in U.S. law schools. **Members:** 1000. **Publications:** Journal of the Copyright Society, quarterly. **Meetings:** Annual conference - always June. Also holds mid-winter meeting - always February. **Founded:** 1953.

Educators' Ad Hoc Committee on Copyright Law (EAHCCL)

National School Boards Association — (703) 838-6710
1680 Duke St.
Alexandria, VA 22314 — August W. Steinhilber, General Counsel

Coalition of organizations representing educators, librarians, and scholars dedicated to protecting the rights of educators and scholars to free access to copyrighted materials. Charges that legitimate photocopying of published works is being discouraged by exaggerated copyright warnings which go far beyond the restrictions in the federal copyright law. This law permits "the fair use of a copyrighted work" and defines "fair use" to include reproduction "for purposes such as criticism, comment, news reporting, teaching, scholarship or research." Opposes broad, strict copyright notices because they could "readily cause a scholar, teacher, or librarian to refrain from exercising his rights in making or obtaining a copy of material," resulting in a serious restraint on legitimate professional activity. Asserts that the field of education has a limited right to record television programs, use copyrighted material in computer-assisted instruction, and photocopy printed materials for use in the classroom. Proposes that the

Copyright Office refuse to register copyrights for works in which the published statement exceeds the terms of the law. **Founded:** 1963.

The Entrepreneurship Institute (TEI)

3592 Corporate Dr., Suite 100   (614) 895-1153
Columbus, OH 43229   Mr. Jan W. Zupnick, Executive Officer

Provides encouragement and assistance to entrepreneurs who are creating and developing new companies. Unites financial, legal, and community resources to help foster the success of new companies. Promotes sharing of information and interaction between members. Offers consulting on accounting, marketing, banking, and legal issues. Operates Community Entrepreneurial Development Projects which are designed to improve communication between businesses, develop one-to-one business relationships between entrepreneurs and local resources, provide networking, and stimulate the start-up and growth of companies. **Publications:** Brochure; also distributes list of available publications. **Meetings:** Periodic forum, with lectures and workshops. Local groups hold periodic meetings. **Telecommunication Access:** 24-hour hot line. **Founded:** 1976.

Foundation for Innovation in Medicine (FIM)

411 North Ave., E.   (201) 272-2967
Cranford, NJ 07016   Stephen L. DeFelice, M.D., Chairperson

Seeks to regenerate interest in medical discovery and innovation, which the foundation believes flourished in the U.S. in the 1940s and 1950s, but has since declined despite "vastly increased public and private expenditures in research and development." Intends to monitor the state of innovation by conducting seminars and conferences. Encourages clinical research on natural substances and substances with little commercial value. Publishes books; offers educational videocassettes. **Meetings:** Semiannual; also holds annual symposium. **Telecommunication Access:** FAX (201) 272-4583; telex, 475-4248.

High Technology Entrepreneurs Council

P.O. Box 72791   (702) 736-3794
Las Vegas, NV 89170-2791   George S. Sanders, Contact

Houston Inventors Association

600 W. Gray   (713) 520-1443
Houston, TX 77019   Greg Micek, Contact

Indiana Inventors Association

P.O. Box 2388
Indianapolis, IN 46206-2388   Randall Redelman, President

Assists the independent inventor with support, education, and networking. Members include inventors, patent attorneys, and marketing representatives. **Publications:** Indiana Inventor, a monthly newsletter.

Inno-Media

230 Tenth Ave., S.   (612) 342-4311
Minneapolis, MN 55415   Sam Koutavas, Director

Acts as a resource center for inventors, innovators, and industry.

Innovation Development Institute

45 Beach Fluff Ave., Suite 300   (617) 595-2920
Swampscott, MA 01907

Innovation Institute

Box 429   (701) 343-2237
Larimore, ND 58215   Dr. Jerry Udell, Contact

Innovation Invention Network

132 Breakneck Rd.
Southbridge, MA 01550   Samuel N. Apotstola, President

Attempts to get realistic information to inventors about what is involved in protecting intellectual property, especially inventions. Affiliated with National Congress of Inventor Organizations. **Members:** Most of the 120 members are inventors. **Publications:** Newsletter published ten times a year.

Innovators International (II)

P.O. Box 4636   (206) 842-7833
Rolling Bay, WA 98061   Paul von Minden, Acting Director

Purposes are to: formalize innovation as a profession by encouraging individuals and organizations to develop services and products through creative, innovative thinking and establishing guidelines for professional innovation; promote innovation as a means for global communication. Projects include: Business Builders, helping underemployed and unemployed individuals to create their own work; Ideacraft, providing educational programs and developing activities and tools to help people to be creative and innovative; Innovator Builder Entrepreneur, encouraging educators, counselors, and administrators to promote and provide courses and program curricula associated with innovation; Earth Plan, motivating people to be creative and innovative in their efforts to meet social responsibilities and develop alternative methods of promoting world peace and stability; Product Town, involving the development of a model environment that could eventually be established in space and utilize technologically advanced, tested, and perfected products specially created for such an environment; Seton-Language

Chuck Mullen, who bills himself as an inventor and consultant for new idea development, is chairman of the board of advisors of the Houston Inventors Association, which has 220 active members. How many members have been successful at licensing their inventions? "Not many," he says. "Maybe five or six. Most join the Association to learn."

National inventor organizations are often businesses owned and operated by business people, many strictly for profit. They run the gamut from reputable, high-powered lobbying organizations based in the nation's capital to questionable societies that make money selling magazine subscriptions, official-looking certificates, book clubs, group insurance policies, discounted car rentals, and kindred fare.

Based Services, establishing links between human and computer language and developing computer software that performs what are called intelligent tasks such as the checking of spelling and proofreading. Publishes A Time To Build. **Meetings:** Periodic. **Founded:** 1980.

Intellectual Property Owners, Inc.

1255 Twenty-Third St., N.W., Suite 850   (202) 466-2396
Washington, DC 20037   Herbert Wamsley, Executive Director

Intellectual Property Owners (IP0) is a nonprofit association representing people who own patents, trademarks, and copyrights. It was founded in 1972 by a group of individuals who were concerned about the lack of understanding of intellectual property rights in the United States. IPO's corporate members include the likes of Monsanto, Standard Oil, Westinghouse Electric, United Technologies, Procter & Gamble, IBM, Ciba-Geigy, Upjohn, AT&T, Amoco, and Union Carbide.

IPO conducts an active government and public relations program in Washington, D.C. One of its best-known programs is the Inventor of the Year Award, presented each spring to an inventor whose invention was either patented or first made commercially available during the previous year.

I believe that IPO is the flagship among such organizations, an outstanding and well-respected association. It offers its members the most current information on patent, trademark, and copyright issues, as well as a strong voice on Capitol Hill and at the PTO.

For example, during 1988, IPO mailed about 20 editions of its *Washington Briefs*, usually within a few hours after significant intellectual property events occurred. More detailed information is published in its printed newsletter, IPO News. Much of this information was available from no other source.

For those having a serious interest in patent matters and a desire to "network" within a very prestigious group of individuals representing Fortune 500 companies, they don't get better than IPO.

Telecommunication Access: Fax: (202) 833-3636. Telex: 248959.

Intermountain Society of Inventors

1395 E. Greenfield Ave.   (801) 278-5679
Salt Lake City, UT 84121   Charles Faux, Contact

Intermountain Society of Inventors and Designers

5770 Minder Dr.
Salt Lake City, UT 84121   Edward Ravigar, Contact

Informs members as to the procedures of inventing through the knowledge gained from meeting and associating with other members and guests. Affiliated with national Congress of Inventor Organizations. Open to inventors, manufacturers, patent attorneys, and invention developers.

International Association of Professional Inventors

Route 10, 4412 Greenhill Way   (317) 644-2104
Anderson, IN 46011   Jack L. Banther, Contact

   Contact: Tom Nix, 818 Westminster, Kokomo, IN 46901 (317) 459-3553.

International Copyright Information Center (INCINC)

c/o Association of American Publishers   (202) 232-3335
2005 Massachusetts Ave., N.W.
Washington, DC 20036   Carol A. Risher, Director

Formed as a result of a meeting of the Joint Study Group comprising members of United Nations Educational, Scientific and Cultural Organization (UNESCO) and Bureaux Internationaux Reunis pour la Protection de la Propriete Intellectuelle (International Bureau for the Protection of Intellectual Property) (BIRPI) in the fall of 1969, which passed a proposal calling for the establishment of national information centers on copyright clearance in major publishing countries throughout the world. INCINC serves as the U.S. center where it assists publishers in developing countries in their efforts to contact U.S. publishers regarding the licensing of translation and English-language reprint rights to U.S. books. **Founded:** 1970.

International Licensing Industry and Merchandisers' Association (ILIMA)

350 Fifth Ave., Suite 6210   (212) 244-1944
New York, NY 10118   Murray Altchuler, Executive Director

Companies and individuals engaged in the marketing of licensed properties, both as agents and as property owners; manufacturers and retailers in the licensing business. Objectives are: to establish a standard reflecting a professional and ethical management approach to the marketing of licensed properties; to become the leading source of information in the industry; to communicate this information to members and others in the industry through publishing, public speaking, seminars, and an open line; to represent the industry in trade and consumer media and in relationships with the government, retailers, manufacturers, other trade associations, and the public. Compiles statistics; bestows awards; maintains hall of fame and placement service. **Members:** 235. **Publications:** (1) Licensing Directory, annual; (2) LIMA Light (newsletter), quarterly. **Meetings:** Annual trade show, with exhibit and seminar. **Founded:** 1984.

International Patent and Trademark Association (IPTA)

33 W. Monroe   (312) 641-1500
Chicago, IL 60603   Jeremiah D. McAuliffe, President

Lawyers who have professional qualifications and interest in the international protection of patents, designs, trademarks, copyrights, and other intellectual property rights. IPTA is the American group of the International Association for the Protection of Industrial Property. Monitors international developments that may affect industrial property and related rights.

Members of the Intellectual Property Owners (IPO) include about 100 large and medium sized corporations and a growing number of small businesses, universities, patent attorneys and independent inventors. Herb Wamsley, IPO Executive Director, says that he is interested in expanding his organization's membership base to include more individual members such as independent inventors, investors, and attorneys.

Local clubs don't spend their money on slick newsletters and staff salaries. They are what you would expect from smallish, local organizations. Volunteers do their best with what they have at hand. Typically, the clubs provide access to PTO and other publications in the field, act as a liaison between the inventor and manufacturers, critique member inventions, and offer referrals.

Studies, discusses, and reports on proposed national and foreign legislation treaties and conventions that are likely to affect national and international intellectual property interests. Bestows Sydney Diamond Award. **Members:** 700. **Publications:** (1) Annuaire, 3-4/year; (2) Membership Directory, annual; also publishes minutes of committee and executive meetings. **Meetings:** Annual meeting; also holds triennial international congress - 1989 June, the Netherlands. **Founded:** 1930.

Invention Development Society

8230 S.W. 8th St.   (405) 376-2362
Oklahoma City, OK 73128   William L. Enter, Sr., President
Helps inventors develop and patent inventions, with the goal that the inventions might be manufactured in Oklahoma thereby increasing jobs for Oklahomans. Mainly active in Oklahoma, although a limited national presence is expected. Chapters in Enid, Fairview, and Oklahoma City. Members include inventors, manufacturers, patent attorneys, invention developers, venture capitalists, and others. **Publications:** Monthly newsletter.

Invention Marketing Institute (IMI)

345 W. Cypress St.   (818) 246-6540
Glendale, CA 91204   Ted DeBoer, Executive Director
Inventors (8,175) and manufacturers (700). Purpose is to help inventors get their products into the marketplace and to help manufacturers find new products to make and market. Brings together inventors and manufacturers for mutual benefit. Maintains speakers' bureau, 1500 volume library, small museum, and hall of fame. **Members:** 8875. **Telecommunication Access:** Electronic bulletin board, (818)246-6546. **Founded:** 1964.

Inventor Associates of Georgia

637 Linwood Ave. N.W.   (404) 656-5361
Atlanta, GA 30306   Tom J. Sutor, Contact

Inventors & Entrepreneurs Association of Austin

6714 Spicewood Springs Rd.   (512) 452-9043
Austin, TX 78759   Veronique Berardino, Contact

Inventors & Entrepreneurs Society of Indiana, Inc.

Box 2224   (219) 989-2354
Hammond, IN 46323   Prof. Daniel J. Yovich, Contact
Provides for communication among inventors, manufacturers, invention developers, patent attorneys, and venture capitalists and arranges meeting among them. Primarily a provider of education. **Members:** 200, mainly inventors and entrepreneurs. **Publications:** Monthly newsletter. Affiliated with National Congress of Inventor Organizations.

Inventors Assistance League

345 West Cypress   (818) 246-6540
Glendale, CA 91204   Arthur Ryan, Contact
  Contact: Ted DeBoer.

Inventors Association of America (IAA)

P.O. Box 1531   (714) 980-6446
Rancho Cucamonga, CA 91730   L. Troy Hall, President
Private inventors. Seeks to assist members in patenting, producing, and marketing their ideas. Maintains a Board of Evaluators to judge the quality and potential of members' creations; aims to thwart efforts by large corporations to improve or manufacture members' inventions and thereby obtain patents. Fulfills commissions to manufacture inventions through parent company, the Corporation of Inventive Minded People. Sponsors educational program in conjunction with schools on inventions and the inventive process. Conducts research and educational seminars. Maintains library. Also sponsors compeitions; bestows awards. Although currently active only in the Los Angeles, CA, area, plans to expand nationally. **Members:** 191. **Publications:** Inventor's Gazette, monthly; makes available audio- and videocassette tapes. **Meetings:** Annual; also holds quarterly meeting. **Founded:** 1985.

Inventors Association of Canada

P.O. Box 281   (306) 773-7762
Swift Current, SK, Canada S9H 3V6   Phyllis Tengum, Secretary
Provides services to inventors in Canada, including disseminating information about inventor organizations and manufacturers in Canada and the U.S. Also conducts patent searches through the services of patent agents in Ottawa.

Inventors Association of Connecticut

9 Sylvan Rd. South   (203) 226-9621
Westport, CT 06880   Murray Schiffman, President
Goal is to establish a community with skills and resources in innovation and invention, prototyping, patenting, commercializing, manufacturing, marketing, and financing to bring products and processes to the market through self-help and association. Works to provide a forum for networking and group innovation. Affiliated with the National Congress of Inventor Organizations. Members include inventors and innovators, business people, executives, consultants, engineers, patent attorneys, entrepreneurs, marketing people, and venture capitalists.

Ask questions before you join a national organization. Compare membership offers. Don't be mislead by slick brochures, certificates, directories, travel and insurance discounts, nonprofit status, press release services, and book clubs. **The bottom line is what can the organization do for you.** Ask for member and business references. Talk to corporate R&D executives who express faith in the organization. Money can shotgun press releases and newsletters out to a range of diversified manufacturers; it does not guarantee results.

According to Dr. Martin Bernard of Argonne National Laboratory, the best organizations are those that have a full-time executive director and some clerical staff. "As a rule of thumb, the executive director should not be an inventor," he suggests.

The Sharper Image Company holds monthly inventor's fairs in San Francisco for those interested in having inventions included in their catalog or retail stores, according to the Inventors Association of New England IANE newsletter. Inventors may write The Sharper Image, 650 Davis St., San Francisco, CA 94111 for further information.

Bobby Toole, executive director of the Inventor's Association of St. Louis, reports that when she finishes counseling an inventor, he or she leaves with two pages of people to contact.

Alexander T. Marinaccio, president of Inventors Clubs of America, says that the typical inventor club lasts only two years. "Everybody joins, wants to get rich and then they all drop out."

Inventors Association of Georgia
241 Freyer Dr. N.E.   (404) 427-8024
Marietta, GA 30060   Hal Stribling, Contact
    Provides education, self-help, and volunteer assistance to inventors. Affiliated with National Congress of Inventors Organizations. Members include inventors and other persons interested in intellectual property. **Publications:** Monthly newsletter.

Inventors Association of Indiana (IAI)
612 Ironwood Dr.   (317) 745-5597
Plainfield, IN 46168   Randall N. Redelman, President
    Provides technical, business, and educational information to members. **Members:** 51. **Publications:** Newsletter, monthly **Founded:** 1979.

Inventors Association of Metro Detroit
19813 E. Nine Mile Rd.   (303) 772-7888
St. Clair Shores, MI 48080   Peter P. Ruppe, Jr., Contact

Inventors Association of New England
P.O. Box 335   (617) 862-5008
Lexington, MA 02173   Dr. Donald Job, Contact
    Provides education, support and networking in eastern New England. Affiliated with National Congress of Inventor Organizations. Members include inventors, manufacturers, patent attorneys, invention developers, and venture capitalists. **Publications:** IANE Newsletter, monthly. In workshops uses The Inventors' Guide: Getting Beyond the Invention, and Guide to Developing a Business Plan, both by Dr. Job, Haley Publications.

Inventors Association of New England, Connecticut Chapter
9 Sylvan Rd. South   (203) 226-9155
Westport, CT 06880   Murray Schiffman, Contact

Inventors Association of St. Louis
P.O. Box 16544   (314) 534-2677
St. Louis, MO 63105   Roberta Toole, Director
    Promotes and encourages invention and innovation through a network by which mutual support and encouragement can be given and received. Regional in scope, with members from six states. Provides for communication among inventors, manufacturers, invention developers, patent attorneys, and venture capitalists and arranges meetings among them through an extensive referral system. Affiliated with Missouri Small Business Development Center and St. Louis University Entrepreneurial Studies Program. Members include inventors, authors of artistic works, manufacturers, investors, attorneys, and accountants; open to anyone who deals with patents, trademarks, or copyrights. **Publications:** The Big Idea, a monthly newsletter; Advanced Technology Guide to the St. Louis Metropolitan Region.

Inventors Association of Washington
P.O. Box 1725   (206) 455-5520
Bellevue, WA 98009   David V. Sires, Contact

The Inventors Club
Rte. 11, Box 379   (904) 433-5619
Pensacola, FL 32514   Bill Bowman, Contact
    Encourages and supports inventors in development and commercialization of new products. Affiliated with Florida Entrepreneurship Program. Membership is primarily inventors. **Contact:** Alternative Address: WSRE-TV, 1000 College Blvd. Pensacola, FL 32504; telephone (904) 478-5800, ext. 1224

Inventors Club of America (ICA)
P.O. Box 450261   (404) 938-5089
Atlanta, GA 30345   Alexander T. Marinaccio, President
    Clubs of inventors, scientists, manufacturers, and others involved in problem-solving and inventing. Stimulates inventiveness and helps inventors in all phases of their work, including patenting, development, manufacturing, marketing, and advertising. Seeks to prevent abuses to the individual inventor, such as theft of ideas. Works to create new industry and lower the tax rate. Conducts research program and competitions. Sponsors children's services, seminars, and professional training; presents awards. Resources include hall of fame, biographical archives, 2000 volume library, and museum. Operates charitable program and speakers' bureau. **Members:** 6000. **Publications:** Inventors News, monthly; also publishes How to Shape an Idea Into an Invention and How to Protect an Idea Before PUitent, and produces educational books. **Meetings:** annual conference (with exhibits) - always second Friday in November. 1989 Atlanta, GA; 1990 Orlando, FL. **Founded:** 1935.

Inventors Club of Greater Cincinnati (ICGC)
18 Gambier Circle   (513) 825-1222
Cincinnati, OH 45218   William M. Selenke, Secretary
    Seeks to further education and professional contact among members. Serves Cincinnati area, including northern Kentucky and southeastern Indiana. Affiliated with National Congress of Inventor Organizations. Members include inventors, manufacturers, patent attorneys, and venture capitalists. **Members:** 40. **Publications:** Newsletter, nine yearly. **Contact:** Alternate telephone (513) 922-9462. **Founded:** 1983.

Inventors Club of Minnesota

Inventors & Technology Transfer
  Corporation
P.O. Box 14-235
St. Paul, MN 55114

(612) 379-7387

Inventors Connection of Greater Cleveland

P.O. Box 46254   (216) 226-9681
Bedford, OH 44146   Bob Abernethy, President

> Serves state of Ohio, with emphasis on the Greater Cleveland area. Provides assistance to the inventor, from the idea to the marketplace. Affiliated with Ohio Inventors Association. Members, 20 percent of whom are retirees, are predominantly inventors. **Publications:** Monthly newsletter.

Inventors' Council

53 W. Jackson, Suite 1041   (312) 939-3329
Chicago, IL 60604   Don Moyer, President

> Provides liaison between inventors and industries in U.S. and Canada. Members include inventors and manufacturers. **Publications:** Opportunity Letter, sent periodically to mailing list of 2,000 inventors. Publicizes opportunities to invent for manufacturers with invention needs.

Inventors Council of Dayton

140 E. Monument Ave.   (513) 439-4497
Dayton, OH 45402   Leonard E. Smith, Contact

> Serves inventors in the Greater Dayton area. Affiliated with National Congress of Inventor Organizations. **Contact:** Ron Versic, P.O. Box 77, Dayton, OH 45409.

Inventors Council of Greater Lorain County

Lorain County Community College   (216) 365-7771
1005 N. Abbe Rd.
Elyria, OH 44035   Hank Ferguson, ., Secretary/Treasurer

> Provides information, motivation, and assistance to inventors and innovators. Affiliated with Ohio Inventors Association. Members include inventors, manufacturers, and patent agents and attorneys. **Publications:** Monthly newsletter.

Inventors Council of Hawaii

P.O. Box 27844   (808) 595-4296
Honolulu, HI 96827   George K.C. Lee, President

> Promotes and develops the innovative and inventive talent of the people of the state of Hawaii and the Pacific Basin. Affiliated with National Congress of Inventor Organizations. Serves as a Patent Information Center for Hawaii. Members are mainly inventors. **Publications:** The Hawaii Inventor, a monthly newsletter.

Inventors Council of Illinois (ICI)

c/o Donald F. Mayer   (312) 922-5616
53 Jackson Blvd.
Chicago, IL 60604   Donald F. Mayer, Executive Officer

Inventors Council of Michigan (INCOM)

c/o University of Michigan, Industrial   (313) 764-5260
  Development Division
Institute of Science and Technology, 220
  Bonisteel Blvd.
Ann Arbor, MI 48109   J. Downs Herold, Chairperson

> Inventors, entrepreneurs, and business, education, and government leaders. Seeks to: establish support network for inventors; aid members with marketing strategies; facilitate the exchange of ideas between business and industry. **Members:** 927. **Publications:** INCOM Newsletter, monthly. **Meetings:** monthly - always fourth Wednesday of the month. **Founded:** 1983.

Inventors Education Network

P.O. Box 14775   (612) 379-7387
Minneapolis, MN 55414   Marge Braddock, Contact

Inventors League

403 Longfield Rd.
Philadelphia, PA 19118

Inventors Network of Columbus

Business Technology Center   (614) 291-7900
1445 Summit St.
Columbus, OH 43201   E.D. Young, Director

> I have known the co-founder of the Inventors Network, E.D. Young, for many years and include a profile of his organization as an example of how much a top flight regional club can offer. Young is quick to point out that to be a member one does not have to be a genius, engineer, or have a patent. The club membership stands at 85 and includes people from diverse backgrounds and levels of expertise. "We have everyone from doctors to professional inventors to blue collar workers," says Young. Half of the members are inventors; other members are manufacturers, patent agents and attorneys, venture capitalists, physicians, retirees, and homemakers.

The Inventors Council of Dayton has developed "inVenture," a program designed to bring out the inventor in junior high students. To support the program, the Council has written two publications (for teachers and students) and produced a two-hour videotape.

The Inventors Network of Columbus is a nonprofit, member-supported organization that is dedicated to the "furthering of invention and creativity in the Central Ohio area through education, motivation, promotion and networking," according to the Network.

Young Inventor's Award Program
The Inventors Network originated and sponsors a progressive school awards and recognition program. It has been successful in introducing rewards and encouragement for young people to exercise their creativity within the bounds of formal education.

E.*D. Young,
director of the
Inventors Network,
says that inventor
organizations help
members meet the right
people. "Networking
with others eliminates
the need to reinvent
the wheel, thereby
shortening the journey
from concept to
product."*

O*ver and above
dues, find out
whether the
organization you're
considering expects a
piece of your
invention in return
for its assistance.*
**Make sure that it is
not an invention
marketing company
in disguise.** *Query
the salary of full-
time organization
executives and
relevant budget
information. As a
potential member,
this kind of
information should
be made available to
you.*

The Inventors Network supplies new members with a package that includes literature on idea protection, prototyping, marketing, financing, business start-up, manufacturing and licensing. It operates a monthly speaker series featuring corporate presidents, senior marketing executives, and leading patent authorities. The club provides for communication among inventors, manufacturers, invention developers, patent attorneys, and venture capitalists and arranges meetings between individual inventors and manufacturers at monthly meetings or via its message center.

Twice a week, IN produces and promotes a 30-minute program on Channel 21 (regional cable) called "You Are an Inventor." The show instructs viewers how to take inventions from the workbench to commercialization.

The Inventors Network meets the second Tuesday of each month at 7:30 in the evening in Columbus. After a two-hour session, the group adjourns to a local restaurant for what E.D. Young calls, "attitude refinement."

Membership is $36 per year.

Inventors of California

215 Rheem Blvd.　　　　　　　　　　　(415) 376-7541
Moraga, CA 94556　　　　　　　　　　Norman C. Parrish, President

Assists the individual inventor through the entire process, focusing on California, national, and international issues. Affiliated with the National Congress of Inventor Organizations. Members include inventors, manufacturers with an active interest in new concepts, patent attorneys, and invention developers on a restricted basis. **Publications:** 1) The California Inventor (monthly) newsletter; 2) Successful Inventing, by Norman C. Parrish, 1988. **Contact:** Alternate address: P.O. Box 158, Rheem Valley, CA 94570.

Inventors Workshop International Education Foundation (Tarzana) (IWIEF)

P.O. Box 251　　　　　　　　　　　　(818) 344-3375
Tarzana, CA 91356　　　　　　　　　Alan Arthur Tratner, Executive Director

Amateur and professional inventors. To provide for inventors and creative persons instruction, assistance, and guidance in such areas as: the steps needed to get patent protection for inventions, including how to make patent searches; how to offer inventions for sale; how to get inventions and products designed, production engineered, and manufactured; how to choose the experts who may be required for these services; how inventors may perform as many of these vital actions as capabilities and resources provide. Conducts research, seminars, and semiannual programs on invention promotion and "Reduction to Practice." Maintains exhibit at Los Angles County Museum of Science and Industry in California. Works in cooperation with Inventors Licensing and Marketing Agency which was formed to help inventor/members sell or license their products. Operates library of 500 volumes; maintains speakers' bureau; sponsors competitions and bestows awards. **Members:** 15,000. **Publications:** Lightbulb (magazine), bimonthly; also publishes Complete Guide to Making Money With Your Inventions; Little Inventions That Made Big Money; Inventor's Guidebook; Inventor's Journal; Science of Creating Ideas for Industry; and Simplicity: The Key to Success. **Meetings:** Annual Invention Convention (with exhibits). **Contact:** Maggie Weisberg. **Founded:** 1971.

Inventors Workshop International (Newbury Park)

3537 Old Conejo Rd., Suite 120　　　(805) 499-1626
Newbury Park, CA 91320-2157　　　　Melvin L. Fuller, Contact

Inventrepreneurs' Forum (IF)

Five Riverside Dr.　　　　　　　　　　(212) 874-7362
New York, NY 10023　　　　　　　　　David A. Lee, Co-Founder

Individual inventors and entrepreneurs. To promote the success of emerging proprietary technologies. Serves as a referral service, pairing inventors with entrepreneurs who provide financial backing for new ideas and inventions. **Founded:** 1986.

Kansas Association of Inventors

2015 Lakin　　　　　　　　　　　　　(316) 792-1375
Great Bend, KS 67530　　　　　　　　Clayton Williamson, President

Functions as a support group and information agency for innovative persons, primarily in Kansas but also has members in most other states and Canada. Affiliated with United Association of Manufacturers Sales Represenatives, Western Kansas Manufacturers Association, Mid America World Trade Center. Members include inventors, manufacturers, patent agents and attorneys, invention developers, venture capitalists, and manufacturing representatives having a serious interest in inventing, having an idea, or wanting to help the inventive community. **Publications:** KAI Resource Directory (for members only); quarterly newsletter.

Kearney Inventors Association

c/o Kearney Development Council
2001 Ave. A, Box 607　　　　　　　　(308) 237-3101
Kearney, NE 68847　　　　　　　　　Steve Buttress, President

Assists inventors in the commercial development of their ideas, serving Kearney and surrounding region. Member of a loose network of inventor organizations fostered by the U.S. Small Business Administration. Supported by Kearney Development Council; provides referrals to Nebraska Business Development Center for management assistance, or to Nebraska Technical Assistance Center for engineering. Members include inventors, manufacturers, and venture capitalists. **Publications:** Monthly newsletter.

Licensing Executives Society (LES)

71 East Ave., Suite S　　　　　　　　(203) 852-7168
Norwalk, CT 06851　　　　　　　　　James E. Menge, Executive Director
　　U.S. and foreign businessmen, scientists, engineers, and lawyers having direct responsibility
　　for the transfer of technology. Maintains placement service. Sponsors quarterly technology
　　transfer seminar. **Members:** 5200. **Publications:** (1) LES Nouvelles (journal), quarterly; (2)
　　Membership Directory, annual; (3) International Technology Transfer Directory, biennial; has
　　also published Law and Business of Licensing. **Meetings:** Annual - always October. 1989 Maui,
　　HI; 1990 Homestead, WV; 1991 San Diego, CA; 1992 Boca Raton, FL. **Founded:** 1966.

Lincoln Inventors Association

P.O. Box 94666　　　　　　　　　　(402) 471-3782
Lincoln, NE 68509　　　　　　　　　Steve Williams, Contact

Los Angeles Copyright Society (LACS)

3000 W. Alameda Ave.　　　　　　　(818) 840-3508
Burbank, CA 91523　　　　　　　　　Donald L. Zachary, Executive Officer
　　Sponsors lectures and discussions on all phases of entertainment law, both domestic and
　　international. **Members:** 300. **Meetings:** Monthly (September through May). **Founded:** 1952.

Michigan Biotechnology Institute

P.O. Box 27609　　　　　　　　　　(517) 355-2277
Lansing, MI 48909　　　　　　　　　Dr. Jack H. Pincus, Director
　　Promotes the commercialization of biotechnology in the state. Provides in-house research and
　　provides technology transfer.

Michigan Energy and Resource Research Association

328 Executive Plaza　　　　　　　　(313) 964-5030
1200 6th St.
Detroit, MI 48226　　　　　　　　　Todd Anuskiewics, Director
　　Serves as a statewide partnership between industry, university, and government, promoting
　　energy-resource-technology research and working to bring public and private research and
　　development grants and contracts into the state. Provides information about the Small Business
　　Innovation Research (SBIR) program.

Michigan Patent Law Association (MPLA)

Brook and Kushman　　　　　　　　(313) 358-4400
2000 Town Center, Suite 200
Southfield, MI 48075　　　　　　　　James A. Kushman, President
　　Patent attorneys and agents organized for the promotion of high ethical standards, the exchange
　　of information concerning the industry, and fellowship among members. Sponsors annual
　　judges banquet. **Members:** 265. **Publications:** Newsletter, monthly **Founded:** 1913.

Midwest Inventors Group (MIG)

P.O. Box 518　　　　　　　　　　　(715) 723-5061
Chippewa Falls, WI 54729　　　　　　Steve Henry, Executive Officer

Minnesota Entrepreneurs Club

511 11th Ave. S.
Minneapolis, MN 55415

Minnesota Inventors Congress

P.O. Box 71　　　　　　　　　　　(507) 637-2344
Redwood Falls, MN 56283　　　　　　Penny Becker, Coordinator
　　Objective is to meet the needs of inventors through education, promotion, and referral. Serves
　　inventors and entrepreneurs from any geographic location. Membership is open to any person,
　　corporation, association, or organization that pays the membership dues and agrees to be
　　bound by the by-laws and articles of incorporation. **Publications:** MIC Memo Newsletter,
　　quarterly. **Telecommunication Access:** Toll-free in Minnesota: 800-INVENT-1

Minnesota Project Innovation, Inc.

Hazeltine Gates Office Bldg.　　　　(612) 448-8826
1107 Hazeltine Blvd.
Chaska, MN 55318　　　　　　　　　James Swiderski, Executive Director
　　Promotes and provides technical assistance with a key federal program for small business
　　innovation and development. Affiliated with Minnesota Department of Energy and Economic
　　Development; U.S. Small Business Administration; and various private companies. Services
　　available to the public at large. **Publications:** Minnesota Small Business R&D Funding Report.
　　Telecommunication Access: Toll-free in Minnesota only is 800-247-0864.

Mississippi Inventors Workshop

4729 Kings Hwy.　　　　　　　　　(601) 366-3661
Jackson, MS 39206　　　　　　　　　Karl Rabe, President

Mississippi Research and Development Center

3825 Ridgewood Rd.　　　　　　　　(601) 982-6425
Jackson, MS 39211-6453　　　　　　R.W. Parkin, Marketing Consultant

The Licensing Executives Society is a professional organization whose members are "engaged in the domestic and international licensing and other transfer of technology and intellectual property rights." For the most part, members in the U.S. are attorneys or licensing professionals from large firms. But LES does have a Small Business Committee and publishes a list of brokers and consultants who have agreed to provide an initial consultation at no cost.

The Missouri Inventors Association is developing a computerized database for inventors, industry, and investors which will match state inventors with industries that want new products and "invention to order." The database will also track investment opportunities, agents and brokers in particular industries, legitimate licensing opportunities, and technology exchange and promotion.

Mississippi Society of Scientists
508 Cindy Lane
Pearl, MS 39208

Mississippi Society of Scientists and Inventors
Box 2244
Jackson, MS 39205

Missouri Inventors Association
204 E. High St.   (314) 636-2026
Jefferson City, MO 65101   Roberta Toole, Executive Director
    Provides services, information, networking, programs, and contacts to the five inventors' associations in Missouri. Affiliated with Inventor Association of St. Louis and inventors' associations in Rolla, Springfield, Columbia/Jefferson City, and Kansas City. Membership is restricted to inventor groups in Missouri. **Publications:** Monthly newsletter.

National Association of Black Women Entrepreneurs (NABWE)
P.O. Box 1375   (313) 341-7400
Detroit, MI 48231   Marilyn French-Hubbard, Founder
    Black women who own and operate their own businesses; black women interested in starting businesses; organizations and companies desiring mailing lists. Acts as a national support system for black businesswomen in the U.S. and focuses on the unique problems they face. Objective is to enhance business, professional, and technical development of both present and future black businesswomen. Maintains speakers' bureau and national networking program. Offers symposia, workshops, and forums aimed at increasing the business awareness of black women. Shares resources, lobbies, provides placement service, and bestows annual Black Woman Entrepreneur of the Year Award. **Members:** 3000. **Publications:** (1) Making Success Happen Newsletter, bimonthly; (2) Membership Directory, annual. **Meetings:** Annual conference (with exhibits). **Founded:** 1979.

National Association of Minority Entrepreneurs (Balch Springs) (NAME)
3300 Shepherd Ln., #E   (214) 286-9705
Balch Springs, TX 75180   Leon Johnson, President
    Minority professionals. Seeks to: develop a minority professional organization and form alliances with other professional groups; improve and increase the minority entrepreneurial community. Works to educate members in areas such as management, marketing, finance, and networking. Serves as a referral service and offers users reduced rates for the services of members. Holds workshops and seminars. Plans to: act as a clearinghouse for information; establish a credit union and an entrepreneurial institute; encourage and negotiate joint ventures between majority and minority firms; offer moral and financial support to minority businesses. Also plans to provide a telex service and hold annual convention. **Members:** 150. **Publications:** (1) Focus, bimonthly; (2) NAME Directory, annual. **Founded:** 1985.

National Association of Minority Entrepreneurs (New York) (NAME)
271 W. 125th St., Rm. 302   (212) 316-3706
New York, NY 10027   William J. Hampton, Executive Director
    Minority individuals working for economic development in minority communities. **Members:** 500. **Founded:** 1972.

National Association of Plant Patent Owners (NAPPO)
1250 I St., N.W., Suite 500   (202) 789-2900
Washington, DC 20005   Robert F. Lederer, Executive V.P.
    Owners of patents on newly propagated trees, shrubs, fruits, and other plants. Seeks to keep members informed of plant patents issued, provisions of patent laws, changes in practice, and new legislation. **Members:** 60. **Publications:** Membership Roster, annual; (2) Bulletin, periodic. **Meetings:** Annual - always July. **Founded:** 1940.

National Business Incubation Association (NBIA)
One President St.   (614) 593-4331
Athens, OH 45701   Dinah Adkins, Executive Director
    Incubator owners and operators; real estate developers; venture capital investors; economic development professionals. (Incubators are facilities where office space, consultants, receptionists, and other related services are shared.) Helps keep newly formed businesses operating, in part, by lowering business expenses; provides information to members. Educates businesses and investors on incubator benefits; offers specialized training in incubator formation. Conducts research and referral services. Maintains speakers' bureau; bestows awards. **Members:** 400. **Publications:** (1) NBIA Review, quarterly; (2) National Directory of Incubators, quarterly; (3) Newsletter, quarterly; also distributes books and sponsors publications on incubators. **Meetings:** Annual conference (with exhibits). **Founded:** 1985.

National Congress of Inventor Organizations (NCIO)
215 Rheem Blvd.   (415) 376-7541
Moraga, CA 94556   Norman C. Parrish, President
    Inventors' groups. Coordinates information relating to inventor education and programs such as wanted and available inventions and credible organizations offering development and marketing assistance. Conducts National Innovation Workshops in cooperation with the National Bureau of Standards. Sponsors competitions; operates children's services; maintains speakers' bureau and library of books and tapes relating to invention, innovation, marketing, and idea development and protection. **Members:** 32. **Publications:** (1) NCIO Newsletter, monthly; (2) NCIO

Membership Directory, annual; (3) NCIO Proceedings, annual; (4) NCIO Bulletin, periodic; (5) Inventors Source Book. **Meetings:** Bimonthly conference. **Founded:** 1981.

National Council for Industrial Innovation (NCII)

105 Charles St., Suite 530
Boston, MA 02114
(617) 367-0072
Harry G. Pars, Ph.D., Chairperson
Encourages the abilities of individuals and small innovative enterprises to create new products, thus stimulating the economy. **Meetings:** Annual. **Founded:** 1970.

National Council of Patent Law Associations (NCPLA)

c/o Office of Public Affairs, U.S. Patent
and Trade Office
2021 Jefferson Davis Hwy., Crystal Plaza
2, Room 1A05
Arlington, VA 22202
(703) 557-3341
To inform member associations of matters of interest in patent, trademark, and copyright fields and to facilitate exchange of information among member associations. Annually inducts inventors into National Inventors Hall of Fame. Maintains numerous committees. Cosponsors, with the Patent and Trademark Office, the National Inventors Hall of Fame Foundation. Bestows awards; maintains speakers' bureau; compiles statistics. **Members:** 53. **Publications:** (1) Chairman's Letter, monthly; (2) Legislative Letter, monthly; (3) Newsletter, quarterly. **Meetings:** Quarterly. **Founded:** 1934.

National Inventors Cooperative Association

P.O. Box 6585
Denver, CO 80206
(303) 756-0034
Morton J. Levand, Contact

National Inventors Foundation (NIF)

345 W. Cypress St.
Glendale, CA 91204
(818) 246-6540
Ted DeBoer, Executive Director
Independent inventors united to educate individuals with regard to the protection and promotion of inventions and new products. Instructs potential inventors on how to protect their inventions through the use of methods developed by the foundation and patent laws. Has assisted individuals throughout the U.S. and in 44 other countries. Maintains speakers' bureau, hall of fame, and museum. **Members:** 8089. **Telecommunication Access:** 24-hour on-line bulletin board with information for potential inventors. **Founded:** 1963.

National Patent Council (NPC)

Crystal Plaza One
2001 Jefferson Davis Hwy., Suite 301
Arlington, VA 22202
(703) 521-1669
Eric P. Schellin, Executive V.P.
Manufacturers, patent attorneys, inventors, and others interested in patents. **Publications:** Patent Trends, monthly. Presently inactive. **Founded:** 1945.

National Society of Inventors

539 Laurel Place
South Orange, NJ 07079
(201) 596-3322
Lawrence Jay Schmerzler, President
National nonprofit organization providing a forum for inventors to help one another and network. Members include inventors, entrepreneurs, patent attorneys, educators, and other individuals interested in innovation. **Publications:** Monthly newsletter. **Contact:** Frank Sowa, P.O. Box 434, Cranford, NJ 07016; (201) 276-0213. Alternate phone: (201) 763-6197.

National Venture Capital Association (NVCA)

1655 N. Fort Myer Dr., Suite 700
Arlington, VA 22209
(202) 528-4370
Daniel T. Kingsley, Executive Director
Venture capital organizations, corporate financiers, and individual venture capitalists who are responsible for investing private capital in young companies on a professional basis. Organized to foster a broader understanding of the importance of venture capital to the vitality of the U.S. economy and to stimulate the free flow of capital to young companies. Seeks to improve communications among venture capitalists throughout the country and to improve the general level of knowledge of the venturing process in government, in the universities, and in the business community. Maintains speakers' bureau. **Members:** 223. **Publications:** Annual Membership Directory. **Meetings:** Annual - always Washington, DC. **Founded:** 1973.

Network of American Inventors and Entrepreneurs

402 Pierce, Suite 300
Houston, TX 77002
(713) 650-0645
Wessie Cramer, President
A networking organization; purposes include the simulation of creativity and sharing of knowledge in the Houston area. Members include inventors and, as associate members, providers of services to inventors. **Publications:** NAIE News, a quarterly newsletter.

Nevada Inventors Association

c/o Institute for Business & Industry,
Truckee Meadows Comm. College
4001 S. Virginia St.
Reno, NV 89502
(702) 322-9636

Don Coster, President

The National Invention Center is a nonprofit national organization which is constructing the National Inventors Hall of Fame. The Hall of Fame will be a hands-on science and technology museum that promotes invention and creative thought. The Center is located at Suite 206, 146 High St., Akron, OH 44308; (216) 762-4463. Dr. Edwin J.C. Sobey is the Executive Director.

Don Coster, president of the newly formed Nevada Inventors Association, a nonprofit, self-help group in Reno, finds that the value of inventor organizations "lies mainly in the education of the members and the public against scams and rip-offs and people that take advantage."

Frederick L. Jones, a member of the Palm Beach Society of American Inventors, says his club has approximately 150 members. How many of his colleagues make a success at inventing? "Practically none," he admits. "We're all dreamers. Inventors are all dreamers. If we weren't we wouldn't be inventors." Typical of most independent inventors in his club, Fred Jones holds nine patents and has yet to license one. Why? "I've been too busy inventing," he confesses.

A Cincinnati, Ohio inventor group calls itself The Salmon Club, named for the fish that, like inventors, always swim upstream. The Salmon Club is large enough to run occasional seminars for its members and guests. One recent seminar was a two-day event with speakers presenting oral and written information on topics encompassing invention evaluation, licensing, financing, basic patent law, incubators, and marketing.

Nevada Innovation & Technology Council
c/o U.S. Dept. of Commerce - ITA   (702) 784-5203
1755 E. Plumb Ln., Rm. 152
Reno, NV 89502-3680   J.J. Jeremy, Contact

New York Society of Professional Inventors
State University of New York at   (516) 420-2397
    Farmingdale
Lupton Hall
Farmingdale, NY 11735   J.E. Manuel, Contact
     Contact: Phillip Knapp, 116 Stuart Ave., Amityville, NY 11701 (516) 598-0036 or (516) 331-3606.

Ohio Inventors Association (OIA)
10595 Sand Ridge Rd.   (614) 797-4434
Millfield, OH 45761   Ron Docie, President
     Goal is to help local inventor organizations and to solve problems common to all inventors. Affiliated with Akron/Youngstown Inventors Organization, Inventors Club of Greater Cincinnati, Inventors Connection of Greater Cleveland, Inventors Network of Columbus, Inventors Council of Dayton, and Inventors Council for Greater Lorain County. **Publications:** Prepared the Ohio Inventors Resource Guide for publication and distribution by the state of Ohio. Sponsors the Ohio Inventors Contest.

Oklahoma Inventors Congress
P.O. Box 54625   (405) 848-1991
Oklahoma City, OK 73154   Albert Janko, Jr., Contact
     Helps inventors promote their inventions to the manufacturing stage, with jobs for Oklahomans a priority. Serves Oklahoma and surrounding states. Affiliated with the Office of the Governor, Department of Commerce; National Congress of Inventor Organizations; and has chapters in Oklahoma City, Tulsa, Bartlesville, Lawton, and McAlester. Members include inventors, service providers, and people who want to help inventors and help promote the art of invention. **Publications:** The Oklahoma Inventor, a monthly newsletter.

Omaha Inventors Club
c/o U.S. Small Business Administration   (402) 221-3604
11145 Mill Valley Rd.
Omaha, NE 68145   Bob Simon, Contact

Palm Beach Society of American Inventors
P.O. Box 26   (407) 736-6594
Palm Beach, FL 33480   Robert E. White, Contact
     Supports inventors and educates about organizations that prey on inventors. Affiliated with Florida Entrepreneurship Program, National Congress of Inventor Organizations, Boynton Beach Chamber of Commerce. Members are mainly inventors, but anyone interested in inventions may join. **Publications:** Monthly newsletter. **Contact:** Kiki Shapero (305) 655-0536.

Pan Hellenic Society Inventors of Greece in U.S.A. (PHSIG)
2053 Narwood Ave.   (516) 223-5958
South Merrick, NY 11566   Dr. Kimon M. Louvaris, Executive
         Director
     Inventors of Greek-American descent whose inventions are patented in both the U.S. and Greece; individuals with patents pending in the U.S., South America, or in Athens, Greece. To assist the Greek-American individual with the patenting, protection, and marketing of his/her invention; to promote the marketing of inventions which contribute to health. Operates biographical archives and museum. Compiles statistics. Conducts triennial seminars in Athens and research programs. **Members:** 150. **Publications:** Newsletter, monthly; also publishes Inventors Greek-English Guide and technical books. **Meetings:** Annual. **Founded:** 1969.

Patent and Trademark Office Society (PTOS)
P.O. Box 2089   (703) 557-6511
Arlington, VA 22202   Richard T. Stouffer, Pres.
     Professional patent examiners in the U.S. Patent and Trademark Office (850); patent and trademark attorneys outside the patent office and former examiners (450). Purposes are educational, social, and legislative. Legislative program concerns federal patent and trademark legislation. Social program is for members and their families and patent and trademark bench and bar members. Educational programs are designed for the public and for continuing professional education. Supplies judges for annual International Science Fair Competition and local fairs. Bestows annual Rossman Award for the best article in the Journal. **Members:** 1300. **Publications:** (1) Journal, monthly; (2) Unofficial Gazette, monthly; also publishes monographs. **Meetings:** Annual. **Founded:** 1917.

Patent Office Professional Association (POPA)
P.O. Box 2745   (703) 557-2975
Arlington, VA 22202   Randall P. Myers, Sec.
     Professional, nontrademark, and nonmanagement employees of the U.S. Patent and Trademark Office. To establish better working conditions and professionalism in the U.S. Patent and Trademark Office. Maintains library of labor reports and legal texts. **Members:** 1500. **Publications:** Newsletter, monthly. **Meetings:** Annual - always first Thursday in December, Arlington, VA. **Founded:** 1962.

Rocky Mountain Inventors Congress
P.O. Box 4365   (303) 231-7724
Boulder, CO 80204-0365   Ken Richardson, Contact

Rolling Inventors Education Network
P.O. Box 14775
Minneapolis, MN 55414   Marge Braddock
    Sponsors mini-workshops on topics of interest to inventors in small communities. Affiliated with Inventor and Technology Transfer Society.

Silicon Valley Entrepreneurs Club, Inc.
TECHMART Bldg., Ste. 360   (408) 562-6040
5201 Great America Parkway
Santa Clara, CA 95054   Robert Hanson, President
    Facilitates the creation and management of team-driven start-up and emerging high-growth companies. Affiliated with Resource Network of Northern California. Members includes inventors, patent agents, sources of capital, entrepreneurial teams, attorneys, and accountants. **Publications:** The Achiever (monthly) newsletter; The Calendar (monthly) listing of events.

Society for Inventors and Entrepreneurs
306 Georgetown Dr.   (305) 859-4855
Casselberry, FL 32707   Dr. Frank Dumont, Contact

Society for the Encouragement of Research and Invention (SERI)
P.O. Box 412   (201) 273-1088
100 Summit Ave.
Summit, NJ 07901   Dr. J. A. M. LeDuc, Exec.Dir.
    Independent organization devoted exclusively to the advancement of research and invention. Principal function is to recognize persons of diverse activities, of all ages, at different times in their lives, who have demonstrated significant achievements. Encourages researchers and scientists of all disciplines and fields and honors those who have distinguished themselves by their activities and contributions furthering the evolution and growth of research and invention. Also rewards authors of scientific research, inventions, and technical realizations. Cooperates in the development of the national economy by providing the atmosphere and means to facilitate achievements in research and invention. Fosters international understanding, relationships, and interchange of ideas. Has established an Invention Center for the advancement of research, invention, and innovation. Compiles statistics; bestows annual awards. **Publications:** Newsletter, bimonthly. **Meetings:** Annual; also holds annual general meeting - always October. **Founded:** 1976.

Society of Minnesota Inventors
20231 Basalt St., N.W.   (612) 753-2766
Anoka, MN 55303   Paul Paris, Contact
    Provides a forum for the discussion of problems common to private inventors. Members are inventors and others who are interested in helping inventors achieve success. Functions as a clearinghouse for information and acts as liaison with other inventor organizations. **Contact:** Helen Saatzer, P.O. Box 355, St. Cloud, MN 56302; (612) 253-2537.

Society of Mississippi Inventors
P.O. Box 5111   (601) 984-6047
Jackson, MS 39296   Eric Rommerdale, President
    Promotes economic development and assists individuals, whether members or not. Receives advice from the Mississippi Department of Economic Development and refers inventors who wish to commercialize their own work or are in need of financial assistance to this department. Membership is primarily inventors, with a few manufacturers, a patent attorney, and a representative of the DED. **Publications:** Monthly newsletter.

Society of University Patent Administrators (SUPA)
c/o Spencer L. Blaylock   (515) 294-4740
315 Beardshear Hall
Iowa State Univ.
Ames, IA 50511   Spencer L. Blaylock, Pres.
    Patent administrators from institutions of higher learning and affiliated hospitals; attorneys; foundations. Assists administrators in the licensing of technology and reporting of inventions. Recommends more effective technology transfer procedures. Sponsors national workshops and regional information sessions. **Members:** 520. **Publications:** (1) Newsletter, quarterly; (2) SUPA Journal, annual; (3) SUPA Membership Directory, annual. **Meetings:** Semiannual conference. **Founded:** 1975.

South Dakota Inventors Congress
Watertown Area Chamber of Commerce   (605) 886-5814
P.O. Box 1113
Watertown, SD 57201   Barry Wilfahrt, Executive V.P.
    Conducts biannual Inventors Congress. Provides referrals to regional experts who can assist inventors.

> "**A**nyone having difficulty moving an invention from conception to the marketplace should get in contact with the closest local inventor organization," advises Dr. Martin Bernard of the Argonne National Laboratory. "Local inventor organizations are fonts of information, have diverse skills and expertise in the membership, and have extensive networks."

The Texas Innovation Information Network System provides a computerized clearinghouse containing information on current research, engineering developments, technology management, and education in Texas. The database also describes the capabilities and needs of the state's scientists, engineers, entrepreneurs, advanced technology companies, professional services, and educational institutions.

Spirit of the Future Creative Institute (SFCI)
3308 1/2 Mission St., Suite 300
San Francisco, CA 94110   Gary Marchi, Founder & Dir.

"Idea creators, pioneers, innovators, inventors, and futurists"; research and development groups, educators, consultants, and researchers. Acts as a "creative innovation center" whose purpose is to explore, discover, create, and utilize new ideas and concepts, technologies, systems, processes, and applications of existing technologies. Seeks to facilitate the "process of positive potential" and assist "emerging growth industries." Primary areas of interest include: future science and space technology; economics and the free enterprise system; the U.S. Constitution; mental development, such as applied logic, creativity, and learning improvement systems; natural health and fitness; conservation; historical "re-discoveries" and mysterious phenomena. Provides advisory services on future planning, evaluations, forecasts, and applied research. Produces Future Consumer radio and television documentary series. Holds seminars and workshops; sponsors internship program. Maintains speakers' bureau. Operates library of 425 volumes, 100 cassette tapes, biographical archives, and clippings. Compiles statistics. **Members:** 20. **Founded:** 1976.

Tampa Bay Inventor's Council
805 W. 118th Ave.   (813) 933-9124
Tampa, FL 33612   F. MacNeill MacKay, Contact

Encourages and supports inventors in development and commercialization of new products; provides a network for services and information in the Tampa Bay area and Florida. Affiliated with Florida Entrepreneurship Program, National Congress of Inventor Organizations, Intellectual Property Owners, and World Intellectual Property Organization. Members include inventors, manufacturers, patent attorneys, venture capitalists, and invention developers that have a good record, and others who can help promote America ingenuity; must have successful review by membership committee and subscribe to the Council's code of ethics. **Publications:** Florida Inventors News Digest, a monthly newsletter. **Contact:** Alternative address: P.O. Box 2254, Largo, FL 34294-2254; phone (813) 937-4783 or 596-3384 (Stan Stawski).

Tennessee Inventors Association
P.O. Box 11225   (615) 690-3109
Knoxville, TN 37939-1225   Igor Alexeff, President

Goal is to advance technology throughout Tennessee by providing information, guidance, and encouragement to those involved in innovation. Affiliated with National Congress of Inventor Organizations; open to inventors, small business developers, engineers, and scientists. **Publications:** Monthly newsletter.

Texas Innovation Information Network System
INFOMART   (214) 746-5140
1950 Stemmons Freeway
P.O. Box 471
Dallas, TX 75207   John Rodman, Executive Director

Goal is to develop a system in Texas to improve the scientific and engineering information transfer between universities, small high-technology businesses, and other sectors in order to increase the number of Texas innovations, improve the state's research competitiveness, and foster economic growth for the state. Open to inventors, entrepreneurs, small high-technology businesses, and universities.

Texas Inventors Association
4000 Rock Creek Dr., #100   (214) 528-8050
Dallas, TX 75204   Tom E. Workman, President

Assists inventors in learning about the activities of inventing, protecting ideas, and commercializing ideas. Open to inventors, manufacturers, patent attorneys, invention developers, and others in the Dallas-Fort Worth area. Affiliated with National Congress of Inventor Organizations. **Publications:** Quarterly newsletter. **Contact:** Alternate phone number is (214) 256-1540.

Trademark Society (TMS)
P.O. Box 2631, Eads Station   (703) 557-3277
Arlington, VA 22202   Jerry Price, Pres.

Labor union of trademark attorneys in the U.S. Department of Commerce. **Members:** 100.

United States Trademark Association (USTA)
Six E. 45th St.   (212) 986-5880
New York, NY 10017   Robin A. Rolfe, Exec.Dir.

Trademark owners; associate members are lawyers, law firms, advertising agencies, designers, market researchers, and public relations practitioners. Seeks to: protect the interests of the public in the use of trademarks and trade names; promote the interests of members and of trademark owners generally in the use of their trademarks and trade names; disseminate information concerning the use, registration, and protection of trademarks in the United States, its territories, and in foreign countries. Maintains informal placement service, speakers' bureau, and library of 1750 volumes. Presents essay award. **Members:** 1750. **Publications:** (1) Bulletins, weekly; (2) Trademark Reporter, bimonthly; (3) Executive Newsletter, quarterly; (4) Trademark Stylesheets, semiannual; (5) Roster of Members, annual; also publishes books about trademarks. **Meetings:** Annual (with exhibits) - always April and May. **Founded:** 1878.

U.S. Patent Model Foundation (USPMF)

1331 Pennsylvania Ave., N.W., Suite 903   (202) 737-1836
Washington, DC 20004   Nancy Metz, Exec.Dir.

Foundation established to stimulate American inventiveness and productivity. Seeks to enhance public awareness of American inventions, both past and present. Works to recover many patent models of the 19th century, most of which were sold at public auction by the U.S. Patent Office in 1925; plans to donate these models to the Smithsonian Institution. (Patent models are working replicas of devices for which patents are sought.) Sponsors Invent America Program, an educational competition for elementary school students; provides training seminars for teachers involved in the program; bestows awards and grants to the winning students, teachers, and schools. Plans to develop a resource center to provide educators and potential inventors with information on the patent process;develops curriculum materials. Compiles statistics. **Publications:** Newsletter, quarterly; also publishes Invent America! Creative Resource Guide; plans to publish a quarterly newsletter for the IAP. **Meetings:** annual conference (with exhibits). **Founded:** 1984.

Venture Exchange Forum

P.O. Box 23184   (615) 694-6772
Knoxville, TN 37933-1184   David Patterson, Contact

Provides an environment in which aspiring entrepreneurs can interact on a personal basis with successful entrepreneurs and with professionals who provide necessary support services for start-up businesses. Serves Knoxville, Oak Ridge, Blount County area. Members include inventors, entrepreneurs, patent agents and attorneys, venture capitalists, and other service providers. **Publications:** VEF Bulletin, a monthly newsletter; Directory.

Women Entrepreneurs (WE)

1275 Market St., Suite 1300   (415) 929-0129
San Francisco, CA 94103   Sharon Cannon, Pres.

Offers the woman business owner support, recognition, and access to vital information and resources. Has participated in government studies and in the 1980 White House Conference on Small Business. Conducts monthly programs featuring speakers and technical assistance educational seminars and workshops; sponsors Advice Forum, providing business and problem-solving information; bestows Appreciation Awards. **Members:** 150. **Publications:** (1) Prospectus (newsletter), monthly; (2) Membership Roster, annual. **Founded:** 1974.

Women Inventors Project, Inc.

22 King St. S., Suite 500   (519) 746-3443
Waterloo, ON, Canada N2J 1N8 Dr. Shelly Beauchamp, Project Co-Director

Develops workshop manuals, informatin kits, and audiovisual materials for women inventors, potential women inventors, and younger school-aged women in Canada. Affiliated with National Congress of Inventor Organizations. Membership open to inventors, patent attorneys, and interested individuals. **Publications:** Focus, a newsletter; The Book for Women Who Invent or Want To.

Worcester Area Inventors USA

132 Sterling St.   (617) 835-6435
Worcester, MA 01583   Barbara N. Wyatt, Contact

Young Entrepreneurs Organization (YEO)

Campus Box 147   (316) 689-3000
Wichita State Univ.
Wichita, KS 67208   Douglas K. Melligan, Dir.

Entrepreneurs under the age of 30. Facilitates communication among members; sponsors YEO 100 listing, which notes the top 100 young entrepreneurs worldwide based on gross revenues. Cooperates with the Association of Collegiate Entrepreneurs (see separate entry). Maintains speakers' bureau; bestows awards; compiles statistics. **Members:** 750. **Publications:** ACE-YEO Newsletter, monthly. **Meetings:** Annual conference. **Telecommunication Access:** Electronic news network. **Founded:** 1985.

Women Inventors Project, Inc. developed a traveling exhibit, Inventing Women, which highlighted 28 women inventors. The exhibit traveled Canada in 1989-90.

Inventors can receive guidance and support from myriad sources, e.g. patent lawyers, government agencies, universities, reference books, business and technical resource centers, etc. But often the most sincere information, albeit not always the most useful, comes from colleague inventors, peers who understand and also pursue the "quest for fire."

University Innovation Research Centers

Listed here are more than 300 principal university research institutes, technology transfer centers and research parks whose purpose it is to promote innovation, invention and product development. Nearly every state has at least one college or university that can provide research and development facilities to technically oriented companies or individuals.

Advanced Science & Technology Institute
University of Oregon
319 Hendricks Hall
Eugene, OR 97403

(503) 686-3189
Dr. Robert S. McQuate, Executive
Director

Bridges university research activities and resources with the corporate community. Encourages cooperative research projects with industry through its Industrial Associates Program Serves as a broker for patenting and licensing arrangements between industry and Oregon State University and University of Oregon. **Publications:** Newsletter; also a research directory of Oregon State University and University of Oregon researchers. Sponsors technical seminars, workshops, and conferences.

Alabama Small Business Development Center
Medical Towers Bldg., Univ. of Alabama/
  Birmingham
1717 11th Ave. South, Suite 419
Birmingham, AL 35294

(205) 934-7260

Dr. Jeff D. Gibbs, State Director

Ann Arbor Technology Park
Richard Wood & Company, Inc.
230 Huron View Boulevard
Ann Arbor, MI 48103

(313) 769-6030

David R. Tyler, Vice President

820-acre site providing tenants access to research, educational, and computing resources at the University.

Applied Information Technologies Research Center
1880 Mackenzie Drive
Suite 111
Columbus, OH 43220

(614) 442-1955

George Minot, President

Information environment, information representation, and user interfaces and search and retrieval technologies. Performs contract reseach and development for member and nonmember companies and assists entrepreneurial ventures and start-up companies in the area of information product and/or service development.

Argonne National Laboratory Technology Transfer Center
9700 South Cass Avenue
Argonne, IL 60439

(312) 972-4929
Dr. Brian Frost, Director

Facilitates the exchange of Argonne research resources and inventions with industry on the state and national level and develops industrywide partnerships for the Laboratory. Collaborates with Argonne/University of Chicago Development Corporation to oversee patenting and licensing activities.

Arizona Small Business Development Center
108 North 40th St., Suite 150
Phoenix, AZ 85034

(602) 392-5224
Dave Smith, State Director

**Arizona State University
Engineering Excellence Program**
Center of Research
College of Engineering and Applied
  Science
Tempe, AZ 85287

(602) 965-1725

Charles Bachus, Director

Partnership of local industries, universities, and state government striving to promote economic development by attracting new businesses to the state. Provides general advice and lobbying in the state legislature and promotes economic development.

Research parks are planned groupings of technology companies, often near universities, that encourage university/private partnerships. They draw industry to a particular location and provide incubator services.

Arizona State University Research Park

2049 East ASU Circle   (602) 752-1000
Arizona State University
Tempe, AZ 85284   Michael S. Ammann, Executive Director

Seeks to link technological results of university and individual research with private industry. The Park provides tenants with leased land for construction of research and development facilities, laboratories, offices, pilot plants, facilities for production or assembly of prototype products, and University and government research facilities.

Association of University Related Research Parks

4500 South Lakeshore Drive   (602) 752-2002
Suite 336
Tempe, AZ 85282-7055   Chris Boettcher, Executive Director

Serves as forum for the exchange of information on planning, construction, marketing, and managing of university-related research parks, particularly information on university-industry relations, innovation, and technology transfer to the private sector. Monitors legislative and regulatory actions affecting the development and operation of research parks. Acts as a clearinghouse for career opportunities. **Publications:** The Research Park Forum (quarterly newsletter).

Auburn University
Auburn Technical Assistance Center

111 Drake Center   (205) 826-4659
Auburn University, AL 36849-5350   Henry Burdg, Contact

Provides management assistance and applied research to private sector organizations.

Ball State University
Center for Entrepreneurial Resources

Carmichael Hall, 2nd Floor   (317) 285-1588
Muncie, IN 47306   Dr. B.J. Bischoff Whittaker, Acting
       Director

Entrepreneurship and development of existing businesses, including corporate training, executive development, intrapreneurship, adult literacy, technology transfer, employee honesty testing, strategic planning, human resource development, needs assessment, creativity, program evaluation, consultant selection, computer software, new product development, and sales training. **Publications:** Practical Guides for Professions (annually).

SBIR Proposal Assistance

Office of Research   (317) 285-1600
1825 Riverside Avenue
Muncie, IN 47306

Offers assistance on submissions to the Federal Small Business Innovation Research (SBIR) program.

Battelle Memorial Institute

505 King Avenue   (614) 424-6424
Columbus, OH 43201   Dr. Douglas E. Olesen, President and
       Chief Executive

Advanced materials, biological and chemical sciences, electronic and defense systems, and engineering and manufacturing technology and information systems. Activities include research and development, commercialization of technology, and management services.

Baylor University
Center for Private Enterprise

Hankamer School of Business   (817) 755-3766
Waco, TX 76798   Dr. Calvin A. Kent, Director

Entrepreneurship and private enterprise, including studies on characteristics of entrepreneurs, student and teacher attitudes toward the private enterprise system, business taxation, and women entrepreneurs.

Biomedical Research Zone

740 Court Street   (901) 526-1165
Memphis, TN 38105   Daniel S. Beasley, Ph.D., Acting Director

Serves as a development center for companies involved with the health care industry and as an international headquarters for pharmaceutical, biomedical, dental, and veterinary firms and professional health care societies. **Publications:** Newsletter (quarterly).

Boise State University
Idaho Business and Economic Development Center

1901 University Drive   (208) 385-1511
Boise, ID 83725   Ronald R. Hall, Director

Based in the College of Business. Manages the outreach functions of the College. Research Unit provides high technology support and research capabilities to private companies and functions as local affiliate to the National Aeronautics and Space Administration Industrial Applications Center (Boise), which promotes technology transfer. Operates the Idaho Small Business Development Center and Idaho Economic Development Center, which provide business assistance.

Bradley University
Technology Commercialization Center
Lovelace Technology Center   (309) 677-2263
Peoria, IL 61625   Dr. William M. Hammond, Director
    Multidisciplinary organization assisting in the development and commercialization of new products and in the transfer of new technologies and manufacturing processes. Activities include product and material testing, prototype development, microelectronic design layout, very-large-scale-intergrated circuit design, medical instrumentation, accelerated corrosive testing, printed circuit board layout, sports medicine instrumentation, optical electronics investigations, and microprocessory control system design.

British Columbia Research Corporation
3650 Wesbrook Mall   (604) 224-4331
Vancouver, BC, Canada V6S 2L2   Dr. T.E. Howard, President
    Conducts cost-accountable research, development, and other technical work, including industrial development, operations management, educational planning, and industrial assistance through technology transfer. Research services to industry include product development, process development, and systems and operational analysis. **Publications:** Annual Report.

California University of Pennsylvania
Mon Valley Renaissance
Box 62   (412) 938-5938
California, PA 15419   Richard H. Webb, Executive Director
    Multiprogram consortium with a focus on applied research and economic development services to business and industry.

Canada Centre for Mineral and Energy Technology, Office of Technology Transfer
555 Booth Street   (613) 995-4267
Ottawa, ON, Canada K1A 0G1   J. Kuryllowicz, Director
    Addresses technology transfer issues. Develops guidelines to aid at all stages of research and development, including planning, bench-scale, pilot plant, and demonstration phases, obtaining financial assistance, and commercial applications. Also assists with patenting procedures and handles intellectual property, including licensing and legal matters arising from research and development contracts.

Canadian Industrial Innovation Centre/Waterloo
156 Columbia Street West   (519) 885-5870
Waterloo, ON, Canada N2L 3L3   Gordon F. Cummer, Chief Operating Officer
    Processes involved in invention, innovation, and entrepreneurship. Projects include an evaluation and implementation model for new product innovations, development of research and development strategies in high technology companies, innovation in small and medium sized firms (lessons from Japan), and the role of University of Waterloo as a technology growth pole in a regional development strategy. **Publications:** Innovation Showcase (quarterly); The Canadian Inventors Newsletter (quarterly); Eureka!.

Carnegie Mellon University
Biotechnology Center
4400 Fifth Avenue   (412) 268-3188
Mellon Institute
Pittsburgh, PA 15213-2683   Dr. Edwin Minkley, Director
    Transfers molecular biology technology into viable commercial processes. Activities emphasize large-sale production of recombinantly engineeered proteins, from cost-efficient expression systems through in-house commercial production.

Center for Education and Research with Industry
213 Northcott Hall   (304) 696-3367
Huntington, WV 25701   William A. Edwards, Director
    Facilitates joint ventures between the academia, business, and government by linking Marshall University campus resources and faculty expertise with technology transfer activities around the state. Administers joint research programs between academia and industry.

Center for Entrepreneurial Development
120 South Whitfield Street   (412) 621-0700
Pittsburgh, PA 15206   Prof. Dwight M. Baumann, Executive Director
    Entrepreneurial activities, including small business and industrial projects as a vehicle for research and experimentation in teaching, advancement of technology and management sources, innovation and entrepreneurship, and transfer of university-based technology in these fields. Aids potential businessmen in overcoming barriers in moving projects from their conception to realization within the business community, assists professionals in starting their own businesses, and provides technical advice and introductions to lending institutions, potential clients, and suppliers.

Idea evaluation is the first major step after a concrete, detailed idea has been developed. This is a critical phase since every following phase requires the investment of more time and money. University research centers are an excellent place to have an invention evaluated to determine its overall technical and commercial feasibility.

Center for Entrepreneurial Studies and Development, Inc.

West Virginia University   (304) 293-4607
College of Engineering
P.O. Box 6101
Morgantown, WV 26506   Dr. Jack Byrd, Executive Director

Business operations improvement, employee relations, management, and systems. Operations improvement studies focus on materials handling systems, cost reduction, work standards development, training manual development, facilities utilization and planning, work methods, inventory control systems, production automation, computer applications, quality control, and facility relocation models. Employee relations studies focus on job incentive systems, wage payments, job enrichment, employee motivation programs, job evaluation, labor/management relations, and employee staffing. Management studies focus on development programs, organization development, management incentives, succession planning, policy manual development, and small business organizations. Systems studies focus on financial planning, computer systems, sales forecasting, strategic planning, office systems improvements, competition, insurance policies, and business plan development.

Center for Innovation & Business Development

Box 8103   (701) 777-3132
University Station
Grand Forks, ND 58202   Bruce Gjovig, Director

Works with applied research personnel at the University of North Dakota and other institutions to facilitate the commercialization of new technologies by providing research support services to entrepreneurs, inventors, and small manufacturers in the areas of invention evaluation, technology transfer, SBIR applications, technical development, licensing, and business development, including market feasibility studies and marketing and business plans. Acts as the University of North Dakota National Aeronautics and Space Administration Industrial Applications Center. **Publications:** Annual Report; Entrepreneur Kit; Business Plan for Start-ups; Marketing Plan for Start-ups.

Center for Innovative Technology

Hallmark Building   (703) 689-3000
Suite 201
13873 Park Center Road
Herndon, VA 22071   Dr. Edward M. Davis, President

Facilitates the transfer of technology for commonwealth universities to industry in biotechnology, computer-aided engineering, information technology, and materials science engineering.

Center for the New West

Suite 1700 South Tower   (303) 623-9298
600 Seventeenth Street
Denver, CO 80202   Philip M. Burgess, President

Seeks to improve the quality and usefulness of information on the new economy, to promote the economic growth and diversification of the western United States, and to improve the competitiveness of western enterprise in a changing global economy. Focuses on national and global trends and their impact on the economic vitality of the western regions. Emphasizes enterprise development, including the formation, retention, expansion, and revitalization of traditional industries, especially agriculture, mining, and forestry; capital formation, including the formation, availability, and cooperative use of public and private capital; technology and innovation development, diffusion, and use; expanded international commerce; human capital, including demographic trends and their impact on economic growth and productivity; and area and regional development, including infrastructure requirements and innovative institutional arrangements. **Publications:** Profile of the West (annual compendium of 200 demographic, economic, and social indicators on the western states).

Central Florida Research Park

12424 Research Parkway   (407) 282-3944
Suite 100
Orlando, FL 32826   Joe Wallace, Director of Marketing

1,027-acre site zoned for commercial and light manufacturing adjacent to the University of Central Florida. Established to create an environment which promotes and fosters relationships between industry and the University.

Centre for Advanced Technology

P.O. Box 483   (303) 482-2916
601 South Howes Street
Fort Collins, CO 80522   Robert B. Hutchinson, III, Contact

235-acre site that facilitates the exchange of Colorado State University research resources with Park tenants.

Chicago Technology Park

2201 West Campbell Park Drive   (312) 829-7252
Chicago, IL 60612   Nina M. Klarich, President

Seeks to coordinate industry, university, and government partnerships to stimulate the formation of science-based companies and economic development in the Chicago area. Provides access to university and hospital resources, offers assistance in the creation of new venture companies, and provides space in an incubator building.

Clemson University
Emerging Technology Development and Marketing Center
338 University Square   (803) 656-4237
P.O. Box 5703
Clemson, SC 29634-5703

Seeks to enhance economic development in the state. Provides technical and marketing assistance to new technology-oriented product manufacturing business firms; seeks to stimulate the transfer of technology and new business concepts to existing companies; enters research and development partnerships with existing businesses that intend to manufacture in the state; and provides assistance to University faculty, staff, and students with products, patents, and ideas that may be commercialized and manufactured.

College of DuPage
Technology Commercialization Center
Business and Professional Institute   (312) 858-6870
22nd and Lambert Road
Glen Ellyn, IL 60137-6599   Nancy Pfahl, Manager

Links high technology businesses to university and other resources to assist in the production and commercialization of new ideas and products and to enhance the transfer of technologies from University laboratories into the marketplace.

College of St. Thomas
Entrepreneurial Enterprise Center
1107 Hazeltine Boulevard   (612) 448-8800
Chaska, MN 55318

Incubator facility.

Colorado School of Mines
Table Mountain Research Center
5930 McIntyre Street   (303) 279-2581
Golden, CO 80403   Gary E. Butts, Director

Provides laboratory, pilot-plant, and office space to engineers, scientists, and researchers in the fields of geology, mining, and ore processing for scientific and technological discoveries, processes, and inventions.

Columbia University
Center for Law and Economic Studies
435 West 116 Street, Box E-2   (212) 854-3739
School of Law
New York, NY 10027   Prof. Jeffrey Gordon, Codirector

Fundamental economic and legal problems of the modern industrial society, including studies on takeovers and the market for corporate control, new directions in law and economics, international taxation, competition in international business, the impact of the modern corporation, and the relationship between administrative law and political economy. Also studies contracts, torts, nuisance, takeovers and the market for corporate control, regulation, antitrust, intellectual property, and the economics of the legal profession.

Center for Studies in Innovation and Entrepreneurship
315 Uris Hall   (212) 280-2830
New York, NY 10027   Michael Tushman, Director

Designs for innovation, entrepreneurship, and technology and organizations.

Office of Science and Technology Development
411 Low Memorial Library   (212) 280-8444
New York, NY 10027   Jack M. Granowitc, Director

University technology transfer office specializing in biotechnology.

Competitive Enterprise Institute
611 Pennsylvania Avenue S.E.   (202) 547-1010
Washington, DC 20003   Fred L. Smith, Jr., President

Domestic economic policy issues, including tax reform, deregulation of industry, deficit reduction, privatization, antitrust, free trade, free-market environmentalism, intellectual property, transportation deregulation, and risk and insurance. **Publications:** Washington Antitrust Report (quarterly newsletter); CEI Update (monthly newsletter on Institute events).

Cornell Research Foundation
Office of Patents and Licensing
East Hill Plaza   (607) 255-7367
Ithaca, NY 14850   Walter Haeussler, Director

University-affiliated technology transfer office specializing in agriculture and engineering.

Cornell University
Cornell Business and Technology Park
102 Langmuir Lab   (607) 255-5341
Ithaca, NY 14853   Richard E. deVito, Marketing Manager

The Park serves as a conduit between Cornell University and business, especially in electronics and biotechnology, and is a home for the independent development of technologies resulting from efforts by Cornell researchers. Leases space for business incubator activities and venture start-up companies.

Independent inventor Shawn Tyler Brown turned to the University of Portland in Oregon when he needed to test his invention, Magna Ears, a strap-on hearing enhancer. "Through the university's testing, we discovered that Magna Ears really worked," he says. How much did this exercise cost Brown? "Nothing. The university did it free of charge because they were interested in the product and what it could do for their patients," Brown reports. Ultimately, through the university, the State of Oregon became involved and sponsored Dr. David Lilly, who is associated with NASA, to test the product. Today Brown is manufacturing and marketing Magna Ears.

*T*ed Erikson, inventor of POC Stix (a.k.a Puzzles of Chaos) was assisted in his work by Purdue University. He feels that if university programs are operated effectively they can be very helpful. The holder of U.S. Patent No. 4,763,902, Erikson has invested more than $10,000 of his own money to date in development and inventory.

Corporation for Enterprise Development

1725 K Street, N.W.   (202) 293-7963
Suite 1401
Washington, DC 20006   Robert E. Friedman, President

Research, development, and dissemination of entrepreneurial policy initiatives at the local, state, and federal levels. Specific studies include economic climate of the states, transfer payment investment, pension funds, job creation, seed capital assessment, international exchange, flexible manufacturing networks, and state capital market analysis. Emphasizes economic empowerment of economically disadvataged populations. **Publications:** The Entrepreneurial Economy (monthly); Making the Grade: The Development Report Card for the States; State Enterprise Development Implementation Packets (paper series); State Strategy Memoranda (paper series); Investing In ... (technical report series); Business Climate Tax Index.

Cummings Research Park

Chamber of Commerce   (205) 535-2008
P.O. Box 408
Huntsville, AL 35804-0408   Jim Reichardt, Vice President

3,600-acre site linking University research and educational resources with Park tenants, particularly in the areas of aerospace, missile research and development, aplied optics, artificial intelligence, software development, and data communications.

Dandini Research Park

DRI Research Foundation   (702) 673-7315
P.O. Box 60220
Reno, NV 89506   Dale F. Schulke, Contact

470-acre site that links the research and development activities of Park tenants with the Desert Research Institute's technological equipment, personnel, laboratories, and training programs. Instrument design, environmental testing, and research and development services may be done in cooperation with DRI staff.

Discovery Foundation

220-3700 Gilmore Way   (604) 430-3533
Burnaby, BC, Canada V5G 4M1   F.C. Hodges, General Manager

Spearheads joint efforts by industry, government, and higher education in the province. Activities are carried out through: Discovery Parks Incorporated, which manages the industrial research parks adjacent to University of British Columbia, Simon Fraser University, British Columbia Institute of Technology, and University of Victoria; Discovery Enterprises Inc., which provides seed money to high-technology enterprises, products, and processes through the commercial stage of development; and Discovery Innovation Office, which offers counseling and referral services to innovators. **Publications:** BC-R&D (bimonthly); Hi-TECH (monthly); BC Technology Directory.

Discovery Parks Incorporated

220-3700 Gilmore Way   (604) 430-3533
Burnaby, BC, Canada V5G 4M1   F. Hodges, Vice President

Manages 55-acre park at University of British Columbia, 75-acre park at Simon Fraser University, 80-acre park at British Columbia Institute of Technology, and 10-acre park at University of Victoria that link university research resources with technological and research companies in the parks. Offers office and laboratory space to encourage exchange between University and tenant researchers. Serves as a network of investment and business contacts for new companies. Provides serviced land for construction of corporate owned facilities.

East Carolina University
Center for Applied Technology

Greenville, NC 27834   (919) 757-6708

Provides business planning, information, management, and technical assitance. Acts as the coordinating agency between the business community and the University.

Eastern Michigan University
Center for Entrepreneurship

121 Pearl Street   (313) 487-0225
Ypsilanti, MI 48197   Dr. Patricia B. Weber, Director

Applied research toward the development of entrepreneurship and growth management. Focuses on the vital transistion from start-up to sustained long-term growth and stability. A major project involves following the progress of 150 companies during their first year. **Publications:** Working Paper Series.

Edmonton Research Park

203 Advanced Technology Centre   (403) 462-2121
9650-20 Avenue
Edmonton, AB, Canada T6N 1G1   Glenn A. Mitchell, General Manager

Serves as a link between University of Alberta research resources and tenants involved in basic, applied, and developmental research, product development, light production, advanced technology activities, and related support services.

First Coast Technology Park

4567 St. John's Bluff Road South   (904) 646-2500
Jacksonville, FL 32216   William A. Ingram, Executive Director

    Facilitates cooperative research and development activities between Park tenants and the University of North Florida community.

Florida Atlantic Research Park

Office of Academic Affairs   (407) 393-3066
P.O. Box 3091   Dr. J.S. Tennant, Associate Vice
Boca Raton, FL 33431-0991   President

    Serves as a bridge between the research interests of the tenant companies and research activities of the University of South Florida community. The Park features an innovation center to aid in transferring technology and houses University functions such as the Small Business Development Center and National Aeronautics and Space Administration/Southern Technology Application Center.

Florida State University
Florida Economic Development Center

335 College of Business   (904) 644-1044
Tallahassee, FL 32306-1007   Roy Thompson, Director

    Business planning, venture capital, insurance, purchasing, workman's compensation, community development, and management. Conducts target industry studies, area business analysis studies, and downtown business surveys. **Publications:** Special Reports/Studies; Small Business News and Views (weekly column); Florida Venture Capital Handbook; Special Sources of Credit; Business Planning Guide; Florida Entrepreneurial Network (quarterly leaflets announcing all small business workshops in Florida).

Florida Entrepreneurial Network

Florida Economic Development   (904) 644-1044
  Center
College of Business
Room 335
Tallahassee, FL 32306-1007   Roy Thompson, Contact

    Provides information sharing, networking, and business development. **Publications:** North Florida Entrepreneurial Network and South Florida Entrepreneurial Network.

Fox Valley Technical College
Technical Research Innovation Park

1825 North Bluemound Drive   (414) 735-5600
P.O. Box 2277
Appleton, WI 54913-2277   Stanley Spanbauer, President

    Seeks to encourage the growth of business and industry in the region through the organization and establishment of regional industry and association headquarters. Encourages marketing ventures, supports inventors and entrepreneurs, fosters research and development for the paper industries, and promotes new technology industries as tenants in the Park. Plans include establishing the D.J. Burdini Technological Innovation Center of the College, which will include a technical library, economic development center, communications network area, product development service center, flexography laboratory, high technology demonstration laboratories, and facilities for conferences and classes.

George Mason University
Center for the Productive Use of Technology

3401 North Fairfax Drive   (703) 841-2675
Suite 322
Arlington, VA 22201-4498   David S. Bushnell, Director

    Technology transfer, work group dynamics and motivation, production measurement, human factors and sociotechnical systems design, and management development. Assists public and private sector organizations in planning, implementing, and evaluating alternative strategies for productivity improvement. **Publications:** Monographs.

George Mason Entrepreneur Center

4400 University Drive   (703) 323-2568
Fairfax, VA 22030

    Assists businesses through financial contacts and provides access to the Commonwealth Technology Information Service, a catalog of the state's technology and research resources.

Institute for Advanced Study in the Integrative Sciences

Thompson Hall, Room 219   (703) 425-3998
4400 University Drive
Fairfax, VA 22030   Dr. John N. Warfield, Director

    Acts to stimulate on-campus research activity by encouraging cooperation between the University and high-technology industry. Principal investigators specialize in systems research, communications research, systems design, and computer science.

State governments are heavily involved with university research through cooperative programs and support of research parks and technology transfer programs. Known at some universities as "Advanced Technology Centers" or "Centers of Excellence," these programs are designed to increase cooperation between academic institutions and state-based industries. They assist in the creation of new firms through the development of new technology, attracting new business to the state and increasing competition.

George Washington University
Center for International Science and Technology Policy
Gelman Library, Suite 714   (202) 676-7292
2130 H Street, Northwest
Washington, DC 20052   Robert W. Rycroft, Director
  Interdisciplinary research and policy analysis. Program includes such disciplines as public administration, physics, political science, sociology, economics, international affairs, urban planning, and environmental resources for application to science and technology policy, international science policy, technology transfer, research and development policy, risk analysis and management, regulatory process, institutional analysis, public perception assessment, telecommunications policy, space policy, environmental quality, disaster recovery, and emerging technologies.

Georgia Institute of Technology
Advanced Technology Development Center
430 Tenth Street, N.W.   (404) 894-3575
Suite N-116
Atlanta, GA 30318   Dr. Richard Meyer, Director
  Serves as a conduit to the University system's research programs, faculty, and facilities for developing technology businesses, particularly in the areas of aerospace vehicles and equipment, biotechnology products, telecommunications equipment, computers and peripheral devices, computer software, electronic equipment, medical devices, instrumentation and test equipment, pharmaceuticals, new materials, and robotics.

Economic Development Laboratory
Georgia Tech Research Institute   (404) 894-3841
Atlanta, GA 30332   Dr. David S. Clifton, Jr., Director
  Agricultural technology, industrial energy conservation, hazardous waste management, safety engineering, industrial hygiene, asbestos abatement, market planning and research, target industry analysis, cost-benefit analysis, energy modelling, technology transfer, industrial training, productivity improvement, small business assistance, analytical chemistry, and indoor air pollution. Supports 12 regional offices throughout Georgia to assist local industries. **Publications:** Engineering Reviews (occasionally); Technical Briefs (occasionally). Also publishes five quarterly newsletters: The Industrial Advisor; Industrial Energy Conserver; Environmental Spectrum; Poultry and Egg Computing; and Poultry Engineering Progress.

Patent Assistance Program
Department of Microfilms   (404) 894-4508
Atlanta, GA 30332-0999   Jean Kirkland, Head

Technology Policy and Assessment Center
Office of Interdisciplinary Programs   (404) 894-2375
Atlanta, GA 30332   Dr. Frederick A. Rossini, Director
  Policy and societal aspects of science and technology, both domestic and international. Studies include technology and impact assessment, technological innovation and diffusion of innovations, cost benefit analysis, socioeconomic development, research and development policy and management, energy and environmental policy, and the interdisciplinary research process.

Georgia State University
International Center for Entrepreneurship
College of Business Administration   (404) 651-3782
University Plaza
Atlanta, GA 30303   Dr. Francis W. Rushing, Director
  Conducts and facilitates studies in entrepreneurship. Serves as a resource center.

Gulf South Research Institute, Technology Management Division
P.O. Box 14787   (504) 766-3300
Baton Rouge, LA 70898   James H. Clinton, President
  Performs technology documentation, financial planning, technology protection coordination, preliminary commercial assessment, technology development, market analysis and planning, preproduction development, and technology transfer.

Harvard Medical School
Office for Technology, Licensing, and Industry-Sponsored Research
221 Longwood Avenue, Room 202   (617) 732-0920
Boston, MA 02115
  University technology transfer office.

Harvard University
Office for Patents, Copyrights, and Licensing
1350 Massachusetts Avenue   (617) 495-3067
Holyoke Center, Room 499
Cambridge, MA 02138   Joyce Brinton, Director
  University technology transfer office specializing in applied sciences, biotechnology, chemistry, material science, software, courseware, biomedical technology, and research products.

Hawaii Ocean Science and Technology Park (HOST)
220 South King Street, Suite 840   (808) 548-8996
Honolulu, HI 96813   William M. Bass, Executive Director
     Facilitates cooperative research and development activities between Park tenants and University of Hawaii community, particularly in the areas of marine microbiology, oceanography, and alternative energy production and other forms of ocean-related high technology. **Publications:** Hawaii High Tech Journal (quarterly).

Idaho State University Research Park
Box 8044   (208) 236-2430
Pocatello, ID 83209-0009   T. Les Purce, Director
     Facilitates the exchange of University research community with Park tenants, particularly in the areas of health professions, life sciences, pharmacy, engineering, electronics, and business.

Illinois Institute of Technology
Manufacturing Productivity Center
IIT Research Institute   (312) 567-4800
10 West 35th Street
Chicago, IL 60616
     Provides information on new manufacturing techniques and maintains a list of manufacturing technology centers.

Technology Commercialization Center
     Stuart School of Business Bldg.,   (312) 567-5115
       Room 229B
     Chicago, IL 60616   Stephen J. Fraenkel, Director
       Provides assistance and support to small businesses, entrepreneurs, and inventors through technology transfer, including technical and business feasibility studies.

Illinois State University
Technology Commercialization Center
Media Services Bldg., Room 215   (309) 438-7127
Normal, IL 61761   Jerry W. Abner, Director
     General technology commercialization. Provides innovation evaluation service for new product and business proposals, including complete research and development assistance for those proposals which meet the economic development objectives of the I-TEC Illinois program. **Publications:** Annual Report; Quarterly Report (both distributed to I-TEC).

Indiana Institute of Technology
McMillen Productivity and Design Center
1600 East Washington Boulevard   (219) 422-5561
Fort Wayne, IN 46803
     Offers technical assistance in industrial production hardware and software and on-campus leasing.

Indiana University
Entrepreneur in Residence
School of Business, Room 640D   (812) 335-9200
Bloomington, IN 47405
     Offers business planning, management, and networking through contacts with venture capitalists, bankers, and investors.

Indiana University-Purdue University Indianapolis
Commercial/Industrial Liaison
355 North Lansing   (317) 264-8285
Indianapolis, IN 46202
     Helps businesses seeking assistance in identifying faculty expertise. Also helps commercial firms develop university innovation.

Economic Development Administration University Center
     611 North Capitol Avenue   (317) 262-5083
     Indianapolis, IN 46204   Lucinda Pile, Contact
       Provides technical assistance and technology transfer services with affiliation to National Aeronautics and Space Administration/Indianapolis Center for Advanced Research.

Tech Net
     611 North Capitol Avenue   (317) 262-5003
     Indianapolis, IN 46204
       State-wide referral service for scientific, engineering, and technology needs of business.

Industrial Technology Institute
P.O. Box 1485   (313) 769-4000
2901 Hubbard Road
Ann Arbor, MI 48106   George Kuper, President
     Design and application of advanced manufacturing technologies and processes and the human issues new technologies spawn. Specific areas include computer-aided design and engineering, computer-aided manufacturing (flexible machining, assembly, inspection, and materials handling), manufacturing information systems, artificial intelligence, computer communication networks, computer-based information and technology transfer products and

services, and economic, organizational, and social impacts. **Publications:** Gateway: The MAP/TOP Reporter (bimonthly newsletter); Modern Michigan (quarterly manufacturing journal).

Innovation and Productivity Strategies Research Program
606 Hill Hall   (201) 648-5837
360 Dr. Martin Luther King, Jr.
  Boulevard
Rutgers University
Newark, NJ 07102   Robert Wharton, Codirector
Innovation and productivity improvement strategies in U.S. and international industrial establishments. Studies include productivity strategies for invention and management of new product commercialization from an international perspective, strategic partnerships between large firms and small hi-tech firms, entrepreneurship strategies of small high-tech industries, intrapreneurship strategies for corporations, employee attitudes and behaviors, quality management techniques, and joint ventures and technology transfer strategies. Collaborates with universities in the U.S. and overseas to carry out international projects.

Innovation Park
1673 West Dirac Drive   (904) 575-6381
Tallahassee, FL 32304   1977 Mike Lea, Contact
Fosters research partnerships between Florida A&M University, Florida State University, and industry tenants.

Innovation Place
15 Innovation Boulevard   (306) 933-6258
Saskatoon, SK, Canada S7N 2X8   Doug Tastad, Manager
120-acre research and development park providing office, industrial, and research space for lease to tenants interested in accessing research resources at the University of Saskatchewan.

Institute for New Enterprise Development
Box 360   (617) 491-0203
Cambridge, MA 02238   Dr. Stewart E. Perry, President
Assesses specialized resources for community economic development; analyzes problems and potential problems of minorities, younger workers, women, and the elderly in relation to local economic revitalization; analyzes the differences in entrepreneurial patterns related to the ethnicity of a community; develops community based ventures, especially in the field of energy; and evaluates federal, state, and municipal policies for economic development.

Inventors Workshop International Education Foundation (Camarillo)
HQ, Inventor Center USA   (805) 484-9786
3201 Corte Malpaso, #304-A
Camarillo, CA 93010   Alan Arthur Tratner, President
New product development and market research focusing on technology-oriented inventions. Also studies patent classifications. **Publications:** Invent! (magazine).

Iowa State University (CIRAS)
Center for Industrial Research and Service
205 Engineering Annex   (515) 294-3420
Ames, IA 50011   David H. Swanson, Director
Problem areas of business, manufacturing, technology transfer, productivity, new product design, manufacturing processes, marketing, and related topics. Acts as a problem-handling facility and a clearinghouse for efforts to help Iowa's industry grow through studies highlighting not only production and management problems but also markets, marketing, and profit potential of possible new developments.

Iowa State University Research Park
125 Beardshear Hall   (515) 294-5121
Ames, IA 50011   Leonard C. Goldman, President
195-acre site on the University's South Campus facilitating interaction between corporate research laboratories and the University research community.

Jacksonville State University
Center for Economic Development & Business Research
College of Commerce and Business   (205) 231-5324
  Administration
114 Merrill Hall
Jacksonville, AL 36265   Pat W. Shaddix
Conducts industrial strategy studies and provides management assistance to business community in northeast Alabama.

Johns Hopkins University
Bayview Research Campus
Dome Corporation   (301) 955-7724
550 North Broadway   David Hash, Director, Property
Baltimore, MD 21205   Development
Biomedical research park established to provide government and industry with space for offices and laboratory facilities.

Whenever you require specialized engineering skills on a project, especially to build complicated prototypes, one of the first places to check is the talent pool available at local university research centers. The costs are reasonable and there is no better place to locate fresh, young minds to work on a resolution to a particular problem.

Universities do have specialities, so it is best to check in advance before going out to the campus. For example, some universities concentrate on biomedical and life sciences research while others handle aerospace engineering and like disciplines.

Kansas State University
Engineering Extension
Ward Hall 133   (913) 532-6026
Manhattan, KS 66506
Promotes technology transfer in Kansas.

Kansas State University Foundation
1408 Denison   (913) 532-6266
Manhattan, KS 66502   Arthur Loub, President
Provides financial assistance to research and educational activities of the University. Owns and operates the TechniPark, a research and development site.

Knowledge Transfer Institute
1308 4th Street S.W.   (202) 554-9434
Washington, DC 20024   Dr. Ronald G. Havelock, Director
Knowledge dissemination, transfer, and use, including scientific knowledge and technology transfer and use in education, medicine, and other areas. Conducts studies of networks involving schools and universities, the role of external innovations agents in dissemination and use of educational innovations, transfer of new cancer technologies into routine patient care, the effects of the mandating of legislation on use of new knowledge, and utilization of new technologies by managers.

Laval University
Industrial Organization Research Group
Faculte des sciences de l'administration   (418) 656-7973
Cite universitaire
Ste-Foy, PQ, Canada G1K 7P4   The-Hiep Nguyen, Ph.D., Contact
Industrial organization, including theory and methodology, technological innovations, technology transfer, investment strategies, and economic policy.

Small Business and Entrepreneurship Research Group
Faculte des sciences de   (418) 656-7960
l'administration
Cite universitaire
Ste-Foy, PQ, Canada G1K 7P4   Yvon Gasse, Ph.D., Contact
Small and medium-sized businesses and entrepreneurship, including growth strategies and technological innovations to promote development.

Lehigh University
Center for Innovation Management Studies (CIMS)
Johnson Hall 36   (215) 758-4819
Bethlehem, PA 18015   Alden S. Bean, Director
Management of technological innovation. Seeks to understand the reasons for the success or failure of technological innovation emphasizing the role of management in improving industrial innovation. Specific projects include comparison of management of external versus internal technological ventures, analysis of the response of the stock market to corporate financial decisions concerning research and development, examination of the productivity and creativity of research and development teams, success and failure of high-tech innovations, termination decisions in monitoring research and development projects, technological cycles and industrial innovation, leadership and productivity in innovation activities, cooperative ventures in large and small firms, boundary management in innovation groups, and creation of a database on technological entrepreneurship.

Office of Research and Sponsored Programs
Bethlehem, PA 18015   (215) 758-3020
     Dr. Richard B. Streeter, Director
Administers and coordinates, as administrative agency of the University, sponsored and cooperative research supported by government agencies, industry, and technical associations, including studies in physical, natural, social, and engineering sciences, and the humanities. Assists faculty and students in unsponsored research and scholarly efforts.

Small Business Development Center
301 Broadway   (215) 758-3980
Bethlehem, PA 18015   Dr. John W. Bonge, Director
Problems faced by small businesses, the impact of the general economy on the formation and operation of small business, and characteristics on entrepreneurs. **Publications:** Lehigh University Small Business Reporter (semiannually); Financing Guide for Northampton, Lehigh and Berks County; Market Planning Guide; Export Planning Guide; Lehigh Valley Business Support Services; Financing Your Business.

Lorain County Community College
Advanced Technologies Center
1005 North Abbe Road   (216) 366-6618
Elyria, OH 44035   Eugene Voda, Director
Serves as an application center in the transfer of technological information to the design, manufacture, and marketing of systems in robotics, flexible manufacturing, microelectronics, computer system maintenance and repair, computer-aided design, injection molding simulation, and computer numerical control products.

Technokitsch, or the Art of Inventing the Unnecessary

A recent U.S. News & World Report reported on electronic inventions that "answer questions nobody asked." Chief among them:

The electronic fish detector, a $70 device that fastens to your fishing pole and provides invaluable digitized information. It beeps when a fish strikes, translates the difficulty of the fish's struggle into a numeric readout and confirms reality by emitting a three-second tone when a fish is landed. As reporter Vic Sussman noted, "No more looking to see whether the fish is actually flopping on the deck."

The 1990 Inventor of the Year Award, presented by Intellectual Property Owners Inc., was shared by three Detroit-based Chrysler engineers who developed an electronic automobile transmission.

Maurice B. Leising, Howard L. Benford and Gerald L. Holbrook split a $5,000 award. The transmission, which has one-third fewer parts than normal automatic transmissions, first was available on 1989-model cars.

Louisiana State University
Office of Technology Transfer
60 University Lakeshore Drive   (504) 388-6941
Baton Rouge, LA 70803   Ted Kohn, Director

Identifies, protects, and transfers technology that originates from University research activities. Activities include patent work, finding state or national licensees, and starting new entrepreneurial projects. Projects include work in such areas as computer chips, enzymes, instrumentation, genetic engineering, drugs, and aquaculture. **Publications:** Newsletter (occasionally).

Maine Technology Park
Stillwater Ave. at I-95   (207) 942-6380
Orono, ME 04469

Associated with the University of Maine. Offers technical consultants, computer and laboratory access, and financial assistance to businesses.

Management Advisory Institute
University of Alberta   (403) 432-2225
222 Faculty of Business Building
Edmonton, AB, Canada T6G 2R6   Dr. Charles A. Lee, Executive Director

Technological innovation, new ventures and entrepreneurship, managing technological change, and student/business interface.

Marquette University
Center for the Study of Entrepreneurship
College of Business Administration   (414) 224-5578
Milwaukee, WI 53233   William J. Gleeson, Director

New business formation, including studies on nature of the entrepreneur, his background and training, opportunities for independent venture, social, economic, and legal climates conducive to new firm formation, business failures, and incidence of new firm formation by region, industry type, and growth of industry type.

Massachusetts Biotechnology Research Park
373 Plantation Street   (508) 755-2230
Worcester, MA 01605   Raymond L. Quinlan, Executive Director

Research and development space designed for biotechnology. Fosters the exchange between the research communities of the affiliated institutions and park tenants.

Massachusetts Institute of Technology
MIT Research Program on the Management of Technology
50 Memorial Drive   (617) 253-4934
E52-535
Cambridge, MA 02139   Prof. Edward B. Roberts, Chairman

Managerial research on technology-based innovation, with emphasis on industry and government. Studies focus on organizations in the United States, Europe, Asia, and Latin America.

Technology Licensing Office
28 Carleton Street, Room E32-300   (617) 253-6966
Cambridge, MA 02139   John Preston, Director

University technology transfer office specializing in biotechnology, biomedical science, ceramics, chemistry, computers, electro-optics, integrated circuits, polymers, metallurgy, optics, semiconductors, sensors, signal processing, and software.

University Park
77 Massachusetts Avenue   (617) 253-5278
Cambridge, MA 02139   Walter Milne, Contact

Fosters interaction between the Institute's research community and park tenants.

Maui Research and Technology Park
Maui Economic Development Board, Inc.   (808) 871-6802
P.O. Box 187
Kahului Maui, HI 97632   Donald G. Malcolm, President

300-acre research park fostering research activities between the academic sector and Park tenants, particularly in the areas of optic systems, electronics design and assembly, information systems and telecommunications, biotechnology, and alternate energy. **Publications:** Newsletter; Proceedings of Symposium.

Medical College of Wisconsin
MCW Research Foundation
8701 Watertown Plank Road   (414) 257-8219
Milwaukee, WI 53226   D.H. Westermann, Executive Vice President

Creates intellectual property from ideas and processes of College researchers for license of products and services to the marketplace primarily in the areas of biophysics, biomedical engineering, biochemistry, microbiology, pharmacology, medical instrumentation, medical imaging, magnetics, computer software design, and genetic engineering, and immunology.

Memorial University of Newfoundland
P.J. Gardiner Institute for Small Business Studies
Faculty of Business
St. John's, NF, Canada A1B 3X5

(709) 737-8855
Garfield Pynn, Director

Small businesses and entrepreneurship in Newfoundland and Labrador. Conducts feasibility studies, break-even analyses, market analyses and market planning, and marketing research studies. Specific projects include an international comparison of the response of small business to offshore development, a survey of 2,800 small businesses in Western Newfoundland, and a survey of 1,250 business alumni and 800 engineering alumni on entrepreneurial activity.

MetroTech
333 Jay Street
Brooklyn, NY 11201

(718) 260-3665
Dr. Seymour Scher, President

10-block, 16-acre urban research park linking Polytechnic University research resources with information technology industries.

Miami Valley Research Institute
1850 Kettering Tower
Dayton, OH 45423

(513) 228-7987
John F. Torley, President

Facilitates the transfer of basic and applied scientific technological research from Miami Valley Research Park tenants and member institutions to production, manufacture, and marketing of materials and services. Solicits public and private grants for research personnel and facilities of member institutions and recruits tenants for the Park.

Miami Valley Research Park (MVRI)
P.O. Box 20026
Dayton, OH 45420

(513) 224-5930
Peter H. Forster, Chairman

Located in close proximity to area universities, Park tenants are offered availability to the research capabilities and expertise of MVRI institutions, especially in the areas of computer and information science, materials, biomedical and human factors engineering, biomedicine, environmental systems, earth resources and energy utilization, applied mathematics, and aeronautical, astronautical, and allied sciences.

Michigan State University
Center for the Redevelopment of Industrialized States
403 Olds Hall
College of Social Science
East Lansing, MI 48824-1047

(517) 353-3255

Dr. Jack H. Knott, Director

Economic and social changes affecting the state of Michigan and ways to diversify and improve the state's business climate. Addresses issues such as technological innovation and strategic human resources management, modeling municipal expenditure patterns in Michigan, determinants of state industrial policy expenditures, implementing new technologies, organizational training practices and the facilitation of technological change, models of plant and investment decisions, and implications of changes in world automobile production for local communities. **Publications:** Newsletter; Reprints.

Michigan Technological University
Bureau of Industrial Development
Houghton, MI 49931

(906) 487-2470
Richard E. Tieder, Director

Small business, business analysis and operations research, natural resource economics, and technology transfer; provides a broad base of knowledge for developing the resources, industries, markets, and communities of Michigan's Upper Peninsula through research, service, and academic activities.

Microelectronics Center of North Carolina
P.O. Box 12889
Research Triangle Park, NC 27709

(919) 248-1800

A nonprofit institution that works in cooperation with the Research Triangle Institute, North Carolina State University, North Carolina Agricultural and Technical, Duke University, and the University of North Carolina (Chapel Hill and Charlotte campuses). Helps the state's universities to educate more students in fields relating to microelectronics by providing complete research and training facilities.

Midwest Technology Development Institute
Suite 815 BTC
245 East 6th Street
St. Paul, MN 55101-1940

(612) 297-6300

Wm. C. Morris, President/Chairman

Seeks to enhance competitiveness and economic growth in the Midwest through the establishment of cooperative research and development ventures involving industry, universities, and government. Institute ventures include the Rural Enterprise Partnership, which conducts case studies of midwestern farmers successfully using new tillage techniques and crops and livestock management systems; and the Advanced Ceramics and Composites Partnership, a coalition to promote start-up programs in materials. Also seeks to facilitate the equitable transfer of technology among industry, universities, and domestic and foreign governments. **Publications:** Technology Advance.

The man who gave the world the Butterball turkey and designer butter pats has a new invention: the Fun-Bun. Leo Peters' "no-mess hot dog bun" resembles a tiny canoe. It's closed at both ends to prevent relish, mustard and anything else you put on your dog from slipping out the sides or bottom. Peters, of Grand Rapids, Michigan, is a multi-millionaire inventor who developed a plumper, heartier breed of turkey in the 1950s that became known as the Butterball. An exclusive licensing agreement was granted to Swift Co. to market the birds. Peters also patented a process for coloring margarine yellow and a process for manufacturing butter pats in a variety of designs. The butter pats became a hit with hotels and food chains.

Minnesota Technology Corridor
Midland Square Building
331 Second Avenue South
Suite 700
Minneapolis, MN 55401
(612) 348-7140

Judy Cedar, Project Coordinator

128-acre site established to foster technology transfer, research and development, and prototype manufacturing activities between the University of Minnesota and related businesses.

Mississippi Research and Technology Park
P.O. Box 2740
Starkville, MS 39759
(601) 324-3219

Dr. W.M. Bost, President

Operates laboratories, incubator and multitenant facilities, professional and business services, and office space to foster high technology research collaboration between Mississippi State University and industry tenants.

Montana State University Entrepreneurship Center
412 Reid Hall
Bozeman, MT 59717
(406) 994-4423

Provides business training, research, and technical assistance to small businesses.

Morgantown Industrial/Research Park
1000 DuPont Road, Bldg. 510
Morgantown, WV 26505-9654
(304) 292-9453

John R. Snider, Executive Vice President

600-acre site linking West Virginia University research resources with industrial research and development activities. The 200-acre Industrial Park provides pilot plants for industrial research tenants to test findings for entry into commercial markets and further generate research in the 400-acre Research Park.

National Aeronautics and Space Administration/Indianapolis Center for Advanced Research
611 North Capitol Avenue
Indianapolis, IN 46204
(317) 262-5000

Dr. Thomas D. Franklin, Jr., President

Diagnostic and therapeutic applications of ultrasound, urban technology, technology transfer, computer engineering, software development for engineering applications, medical instrumentation, automated manufacturing, and advanced electronics. **Publications:** Newsletter (quarterly).

National Research Council of Canada (NRC)
Montreal Road, Building M58
Ottawa, ON, Canada K1A 0R6
(613) 993-9101

Dr. Larkin Kerwin, President

Conducts basic and directed research programs aimed at improving Canada's economic competitiveness. Provides Canadian industry with services, facilities, and technology transfer programs and collaborative research opportunities. Maintains Industrial Research Assistance Program, a technology network. **Publications:** Bulletins (quarterly); Annual Report; NRC Directory of Research Activities; Fact Sheets.

New Jersey Institute of Technology Center for Information Age Technology
Newark, NJ 07102
(201) 596-3035

William R. Kennedy, Executive Director

Serves as a bridge between the Institute and industry to transfer new and existing computer and information technologies from university research laboratories to state and local governments and to small- and medium-sized businesses.

New Mexico Research Park
New Mexico Institute of Mining and Technology
Socorro, NM 87801
(505) 835-5600

Laurence Lattman, President

Facilitates the exchange of research resources between the Institute's research community and Park tenants.

New Mexico State University Arrowhead Research Park
Box 30001-Dept. 3RED
Las Cruces, NM 88003
(505) 646-2022

Dr. Averett S. Tombes, V.P., Res. & Econ. Development

Fosters opportunities for exchange between Park tenants and the University community. Multidisciplinary activities include heavy metal recovery from waste streams, flower production, hearing disorder studies, computer telemetry data, fiber optic imaging, and water purification.

New York University Center for Entrepreneurial Studies (CES)
Leonard N. Stern School of Business
90 Trinity Place, Room 421
New York, NY 10006
(212) 285-6150

Prof. William D. Guth, Director

Conducts and funds research on the factors that promote entrepreneurship and lead to the creation of new wealth and business revenues and on business venturing within established firms. Topics include the major pitfalls and obstacles to start-ups, securing of venture capital, psychology and sociology of entrepreneurship, management of new ventures, innovation and innovative problem-solving, organizing for corporate venturing, control and reward systems, and cross-cultural environments that stimulate entrepreneurship. Also operates a program focusing on entrepreneurship in nonprofit corporations. **Publications:** CES Reports (semiannually); Journal of Business Venturing (copublished with Wharton, quarterly); INE Reports (for nonprofit corporations).

Initiatives for Not-for-Profit Entrepreneurship (INE)
Graduate School of Business (212) 285-6548
 Administration
90 Trinity Place
New York, NY 10006 Laura Landy, Director
 Works with those who are seeking entrepreneurial solutions to the problem of revenue generation and assists and supports them in their efforts. Promotes approaches that enhance success and reduce the risk of venturing, including the correlation between planning, investment, management, size, and venture success in not-for-profit corporations and competition between not-for-profit corporations. **Publications:** INE Reports (semiannual newsletter); Research Reports.

Northeast Louisiana University
Louisiana Small Business Development Center
College of Business Administration (318) 342-2464
Monroe, LA 71209 Dr. John Baker, Director
 Offers business planning, information, management, and technical assistance to the small business community in Louisiana.

Northeastern Texas Small Business Development Center
Dallas County Community College (214) 746-0555
302 N. Market Street, Suite 300
Dallas, TX 75202-1806 Norbert Dettmann, Director
 Telecommunication Access: FAX (214) 746-2475.

Northern Advanced Technologies Corporation (NATCO)
Van Housen Hall (315) 265-2194
P.O. Box 72
State University College at Potsdam
Potsdam, NY 13676 Steven C. Hychkano, Executive Director
 Provides support for technology transfer and for research and development in the areas of nondestructive testing, software development, computer applications, materials processing, and high technology. Clients are offered building space, general assistance, and seed money for a variety of projects to commercialize new technology.

Northern Illinois University
Technology Commercialization Center
Dekalb, IL 60115-2874 (815) 753-1238
 Dr. Larry Sill, Director
 Assists inventors, entrepreneurs, and small businesses with product research and development, technical and commercial assessments, patent applications, and licensing, particularly in the areas of basic sciences, engineering, computer science, and business.

Northern Kentucky University Foundation Research/Technology Park
Highland Heights, KY 41076 (606) 572-5126
 Paul A. Gibson, Contact
 Facilitates the exchange of research resources between the University and park tenants.

Northwestern University
Technology Innovation Center
906 University Place (312) 491-3750
Evanston, IL 60201 Dr. Jack L. Bishop, Director
 Matches University resources with the needs of state business by developing small business innovation research programs, linking businesses to share technologies, developing Japan-Illinois technology cooperatives, commercializing University technology, and providing business planning activities for entrepreneurs.

Northwestern University/Evanston Research Park
1710 Orrington Avenue (312) 475-7170
Evanston, IL 60201 Ron Kysiak, Executive Director
 Encourages the exchange of research activities between the University and Park tenants and transfer technological advances to basic industry. Current tenants include the Basic Industry Research Laboratory, which is funded by the U.S. Department of Energy and owned and staffed by Northwestern University; the Lab focuses on manufacturing and applied materials research, including ceramics, coatings, friction, harsh environments, and energy-efficiency. The Park also provides a small-business incubator system to support newly-developing, high-technology companies, and well as technical assistance, and a seed capital fund of one million dollars.

In 1970, Edward Zimmer and two colleagues started their own company, Ann Arbor Terminals Inc. Starting with $15,000 in capital, the engineers began designing and producing state-of-the-art computer display terminals.

"We totally boot-strapped it," he says of the firm. "Within three years, we were up to 50 employees and $5 million in sales. We never had an unprofitable quarter." An experience with expansion ended after less than a year. "It wasn't fun anymore. And we decided we'd go back to whatever we could comfortably do and have fun, which turned out to be about $5 million sales annually." Zimmer says the experience taught him that finding new products and technical expertise can be a big problem for a small firm. "It would always seem that six to nine months after I was looking for something here I would find it."

Oakland Technology Park
Schostak Brothers & Co., Inc. (313) 262-1000
First Center Office Plaza
26913 Northwestern Highway
Southfield, MI 48034 Philip J. Houdek, Vice President
> Facilitates technology transfer between Oakland University, Oakland Community College, and Park tenants, enhancing the research strength of the affiliated institutions. Facilities are oriented toward robotics, engineering, automation, computer technology, and advanced manufacturing applications.

Ohio State University Research Park
104 Research Center (614) 292-9250
1314 Kinnear Road
Columbus, OH 43212
> 200-acre park offering research-oriented companies a site for their administrative, research, and development facilities to foster exchange between the University and industry. Seeks to enhance the University's teaching and research capabilities through stimulating the exchange of ideas and sharing of resources between University and tenant researchers. Coordinates University and tenant resources in developing commercial applications for new discoveries and technologies.

Ohio University
Innovation Center and Research Park
One President Street (614) 593-1818
Athens, OH 45701-2979 Dr. Wilfred R. Konneker, Director
> Created to foster entrepreneurial activities and to provide technical and business assistance to new and expanding companies. Facilitates consulting, product testing, and technical assistance between tenants and the University.

Oklahoma State University
Technology Transfer Center
OSU District Office (405) 332-4100
P.O. Box 1378
Ada, OK 74820
> Provides evaluations of new technology.

Oregon Graduate Center Science Park
1600 N.W. Compton Drive (503) 690-1025
Suite 300 Bert Gredvig, Executive Vice President/
Beaverton, OR 97006 COO
> Science Park serves as a site for interaction between tenants and Center faculty, students, and facilities, including joint scientific research ventures and internship opportunities for Center students.

Oregon State University
Office of Vice President for Research, Graduate Studies, and International Programs
Corvallis, OR 97331 (503) 754-3437
 Dr. George H. Keller, Vice President
> Coordinates research activities of the University, including individual projects in various academic schools and special research organizations. Also administers the Technology Transfer Program and international activities of the University.

Pan American University
Center for Entrepreneurship and Economic Development (CEED)
School of Business Building, Room 124 (512) 381-3361
1201 West University Drive
Edinburg, TX 78539-2999 Dr. J. Michael Patrick, Director
> South Texas business assistance and economic research, focusing on business plans, economic area profiles, economic impact studies, economic development planning, market feasibility studies, international trade, urban and rural commercial revitalization, industrial park development and feasibility, and land use surveys.

Pennsylvania State University
Technology Transfer Office
101 George Building (814) 865-6277
306 West College Avenue
University Park, PA 16801 Dr. Kenneth J. Yost, Liaison
> University technology transfer office specializing in electronics materials, structural ceramics, ploymers, and manufacturing.

Pittsburg State University
Center for Technology Transfer
School of Technology and Applied (316) 237-7000
  Science
Pittsburg, KS 66762 Dr. Victor Sullivan, Dean
> Development, introduction, and transfer of technology to Kansas regional industries, particularly to wood and plastics industries. Emphasizes design, testing, and development of

products and processing methods, including the applications of computer-aided design, computer numerical control, and robotics to manufacturing.

Institute for Economic Development
Pittsburg, KS 66762 (316) 231-7000

Provides one-stop managerial, financial, and technical assistance to new and expanding businesses in southeast Kansas. Services include financial packaging, business plan development, small business counseling, training and skill development, and research on business operations and markets.

Princeton University
Princeton Forrestal Center
105 College Road East (609) 452-7720
Princeton, NJ 08540 David H. Knights, Jr., Director of Marketing

The Center is designed as a planned multiuse development area creating an interdependent mix of academia and business enterprise in the Princeton area.

Progress Center: University of Florida Research and Technology Park
One Progress Boulevard, Box 25 (904) 462-4040
Alachua, FL 32615 Rick Finholt, Director, Development

200-acre research and technology park open to both public and private research and manufacturing organizations. High-technology development is emphasized, including the areas of electronics, biotechnology, advanced materials, pharmacology, and agriculture. Provides a link between University researchers and industry and is designed to transfer new technologies from the laboratory to the marketplace. Mainframe computer resources available.

PTC Research Foundation
Franklin Pierce Law Center (603) 228-1541
2 White Street
Concord, NH 03301 Robert Shaw, Director

Patents, trademarks, copyrights, invention, and legal and practical systems for dealing with industrial and intellectual property, both in the U.S. and worldwide. **Publications:** IDEA (a journal of law and technology); Monographs; Project Reports.

Purdue Industrial Research Park
Purdue Research Foundation (317) 494-8642
Frederick L. Hovde Hall of
   Administration
Purdue University Winfield F. Hentschel, V.P., Purdue
West Lafayette, IN 47907 Research Foundati

Provides facilities and University research and technical support services to industrial tenants.

R&D Village
Montgomery County (301) 217-2345
Office of Economic Development
101 Monroe Street, Suite 1500
Rockville, MD 20850 Dyan Lingle, Director

1,200-acre site linking biomedical research and development activities between government, park tenants, and academia. Houses the Center for Advanced Research in Biotechnology, a joint research venture of the National Institute of Standards and Technology, University of Maryland, and Montgomery county government; also houses a Johns Hopkins University facility focusing on advanced study programs in computer science, electrical engineering, and technical management. The Village encompasses the Shady Grove Life Sciences Center, Shady Grove Executive Center, and The Washingtonian, a mixed use development, and branch facilities of University of Maryland and Johns Hopkins University. **Publications:** Economic Focus Newsletter.

Rensselaer Polytechnic Institute
George M. Low Center for Industrial Innovation
Troy, NY 12180 (518) 276-6023
Dr. Christopher W. LeMaistre, Director

Consists of five component centers—Center for Interactive Computer Graphics, Center for Manufacturing Productivity and Technology Transfer, Center for Integrated Electronics, Decision Sciences and Engineering Systems Department, and Automation and Robotics Department. Coordinates research related to high technology business and industrial productivity and quality, including solid geometric modeling, computer integrated manufacturing, vision systems, and computer-aided design of very-large-scale-integrated circuits.

Rensselaer Technology Park
100 Jordan Road (518) 283-7102
Troy, NY 12180 Michael Wacholder, Director

Serves as a conduit for joint research activities, consultancies, refresher studies, associate programs, and human interactions between Park tenants and the Institute.

More on Ed Zimmer: he thinks it's getting ever tougher for small companies to find the new products they need to survive. "Back in the 1950s a company doing $2 million to $4 million annual sales could easily afford its own internal research and development because engineers were making $5,000 a year," he says. "Engineers now are making $50,000 a year. That means now a company needs $20 million to $40 million in annual sales to afford that same R&D they need for their survival."

Research Corporation Technologies

6840 East Broadway Boulevard
Tucson, AZ 85710-2815

(602) 296-6400
Dr. John P. Schaefer, President

Evaluation, patent, development, and commericialization of inventions from colleges, universities, medical research organizations, and other nonprofit laboratories. Provides incentives for invention disclosure, funds applied research and new start-up and joint ventures, and develops commercialization strategies.

Research Institute for the Management of Technology (RIMTech)

215 North Marengo, 3rd Floor
Pasadena, CA 91101

(818) 584-9139
Dr. Steven M. Panzer, President

Management issues, technology policy, and the international competitiveness of technology industries. Develops programs facilitating the commercialization of technology from national research facilities, particularly the National Aeronautics and Space Administration (NASA) and its Jet Propulsion Laboratory (JPL), operated by the California Institute of Technology. Projects focus on the commercialization of environmental technology.

Research Triangle Park

2 Hanes Drive
P.O. Box 12255
Research Triangle Park, NC 27709

(919) 549-8181

James O. Roberson, President

Facilitates interaction between industrial and governmental research and development organizations with the research communities of North Carolina State University, University of North Carolina at Chapel Hill, and Duke University.

Rio Grande Research Corridor

Pinion Building, Suite 358
1220 St. Francis Drive
Sante Fe, NM 87501

(505) 827-5964
Ponziano Ferarracio, Senior Program Officer

Links University research resources with government laboratories and private research facilities, particularly in the areas of non-invasive diagnosis, materials, explosives technology, plant genetic engineering, computer research applications, and commercial product development.

RiverBend

2501 Gravel Drive
Fort Worth, TX 76118-6999

(817) 284-5555
David Newell, Contact

Fosters interaction between the University of Texas at Austin research community and industry, particularly in the areas of robotics and automation.

Riverfront Research Park

University Planning Office
1295 Franklin Boulevard
Eugene, OR 97403

(503) 686-5566

Diane K. Wiley, Representative

67-acre site that provides opportunities for research interaction between university and Park tenants engaged in such activities as industrial research and development, biotechnology, materials science, environmental technology, computer software development, and business, educational, and governmental research and consulting services. Seeks to assist in the diversification of the economic base of the state.

Rochester Institute of Technology Research and Business Park

One Lomb Memorial Drive
Rochester, NY 14623-3435

(716) 475-5316
Eric Hardy, Director

1,300-acre site linking business and industry tenants with research resources of the Institute, particularly in the areas of optical and imaging sciences, photolithography, printing systems, laser, and microchip production. Operates research, office, and industrial facilities.

Rose-Hulman Institute of Technology
Innovators Forum

5500 Wabash Avenue
Terre Haute, IN 47803

(812) 877-1511
Russell Holcomb, Contact

Offers inventors technical screening and access to faculty expertise.

Rutgers University
Technical Assistance Program

180 University Avenue
Newark, NJ 07102

(201) 648-5891
Patricia Johnson, Contact

Provides managerial and technical assistance to small business community.

Sangamon State University
Entrepreneurship and Enterprise Development Center

Springfield, IL 62708

(217) 786-6571
Richard J. Judd, Ph.D., Director

Performs the following activities for private and nonprofit organizations: marketing research (including customer/client surveys), economic analyses, and organizational analyses and development. **Publications:** Economic Business Review (quarterly). Offers training to private and nonprofit organizations.

Science and Technology Research Center
P.O. Box 12235   (919) 549-0671
Research Triangle Park, NC 27709-2235   J. Graves Vann, Director
Provides information, technical assistance, referrals, workshops, and seminars on subjects of concern to busines and industry. Provides access to ongoing research in federal laboratories through affiliation with NASA.

Science Park
Five Science Park   (203) 786-5000
New Haven, CT 06511   William W. Ginsberg, President and
           C.E.O.
80-acre technology and light industrial site providing scientific facilities, services, and assistance for tenants to interface with university research resources. Houses the New Enterprise Center, an incubator facility.

South Carolina Research Authority
P.O. Box 12025   (803) 799-4070
Columbia, SC 29211   Robert E. Henderson, Ph.D., Director
Promotes cooperative research and development activities between the University of South Carolina at Columbia and Clemson University and park tenants. The Authority serves as the contractor and manager of the American Manufacturing Research Consortium, comprised of four companies and research institutions involved in computer-integrated manufacturing. The Consortium develops factory plans that employ technologies to reduce overhead costs in manufacturing and provide low cost products, including factory design and construction based on RAMP (rapid acquisition of manufactured parts) technologies for small machined parts and printed wire assemblies for the U.S. Navy.

Southeastern Oklahoma State University
Central Industrial Applications Center
Station A, Box 2584   (405) 924-6822
Durant, OK 74701-2584   Dr. Dickie Deel, Director
Data searches are provided for technology transfer to state businesses.

Southern Illinois University at Carbondale
Technology Commercialization Center
Washington Square C   (618) 536-7551
Carbondale, IL 62901-6706   Martha Cropper, Director
Facilitates the production, transfer, and commercialization of new technology in agriculture, forestry, engineering, materials, mining, and biomedicine by investigating technical feasibility, commercial development, marketing, and cash flow. **Publications:** Connections (ten times per year).

Southern Illinois University at Edwardsville
Technology Commercialization Center
Box 1108   (618) 692-2166
Edwardsville, IL 62026-1108   James W. Macer, Director
Provides assistance to inventors, entrepreneurs, and businesses in the commercialization of ideas and products, particularly in the areas of robotics, CAD/CAM, optical coatings, laser communications, automatic inspection, inferfacing microcomputers to instrumentation, computer simulation, analysis of variance, statistical quality control, statistical process control, plant operations, materials planning, linear and nonlinear programming, radioactive biochemicals, human resources, and marketing.

Southern Willamette Research Corridor
72 West Broadway   (503) 687-5033
Eugene, OR 97401   Bill Barrons, Chairman
40-mile research, development, and specialized manufacturing site facilitating cooperative ventures between participating colleges and universities, industry, and local government.

Southwest State University
Science & Technology Resource Center
Marshall, MN 56258   (800) 642-0684
           James Babcock, Contact

Stanford University
Office of Technology Licensing
350 Cambridge Avenue, Suite 250   (415) 723-0651
Palo Alto, CA 94306   Niels Reimers, Director
University technology transfer office specializing in biotechnology, electronics, and lasers.

Stanford Research Park
105 Encina Hall   (415) 725-6886
Stanford, CA 94305-6080   Zera Murphy, Managing Director
655-acre site linking research resources of the University with Park tenants, particularly in the areas of electronics, space, publishing, pharmaceutics, and chemistry. Activities between park tenants and the University community include cooperative research ventures, instruction, and consulting.

203

According to Harry Hawkins, the director of the Center for Innovative Technology Transfer (CITT) at State University of New York (SUNY), CITT has provided its services to researchers at the College and at other universities in the state as well as to central New York businesses and independent inventors. For example, among its clients have been a marine biologist from Cornell University for whom several designs for a contraption to entrap trout eggs spawned in Lake Ontario were tested, and a local building contractor who required a design for a plastic "Easy Spacer" to facilitate the building of wood decking.

State University of New York at Binghamton
Office of Sponsored Program Development
Administration Building, Room 242   (607) 777-6136
Binghamton, NY 13901   Stephen A. Gilje, Associate Vice Provost
                                                     for Res
      Coordinates preparation and submission of grant and contract proposals for sponsored programs. Assists faculty and administrators in developing preliminary and formal proposals locating potential sponsors, processing and transmitting applications, and negotiating budgets. Negotiates grant and contract awards and assists faculty in technology transfer. **Publications:** Graduate Studies Research News (monthly).

State University of New York at Oswego
Center for Innovative Technology Transfer
209 Park Hall   (315) 341-2128
Oswego, NY 13126   Harry Hawkins, Director

State University of New York at Plattsburgh
Economic Development and Technical Assistance Center
Plattsburgh, NY 12901   (518) 564-2214
                                            Stephen Hyde, Contact
      Provides business planning and information, including management, marketing, financial packaging, and feasibility analysis services.

Sunset Research Park
P.O. Box 809   (503) 929-2477
Corvallis, OR 97339   B. Bond Starker, General Manager
      Facilitates the exchange of research resources and expertise between Park tenants and faculty at Oregon State University; special emphasis in areas of biotechnology, instrumentation, superconductivity, food science, and natural resources.

Syracuse University
Technology and Information Policy Program (TIPP)
103 College Place   (315) 443-1890
Syracuse, NY 13244-1120   Barry Bozeman, Director
      Nexus of technology, information, and public policy, particularly computer-based technological forecasting and assessment, management of scientific and technical information flows in organizations, technological innovation studies, and the relationship of management information systems, information technology, and organization design.

Tampa Bay Area Research & Development Authority
University of South Florida   (813) 974-2890
Administration 275
Tampa, FL 33620   John Hennessey, Executive Director
      Research and development park leasing land to industries needing facilities for research and related scientific manufacturing in medicine, engineering, and natural sciences. Tenants share University of South Florida research facilities and services, use faculty as consultants, and utilize graduate students as part-time work force. Oversees development of university-related research parks in the Tampa Bay area, in particular University Center Research and Development Park.

Tennessee Center for Research and Development
Tennessee Technology Foundation   (615) 675-9505
P.O. Box 23184
Knoxville, TN 37933-1184   David A. Patterson, President
      Promotes economic development in Tennessee through two components: a research division to conduct basic and applied studies for industrial development and to perform technology transfer activities, and a for-profit subsidiary to provide management support services and temporary laboratory facilities to new companies. Serves as a resource center for public information on the effects of science and technology on the environment.

Tennessee Technology Corridor
10915 Hardin Valley Road   (615) 694-6772
P.O. Box 23184
Knoxville, TN 37933-1184   Dr. David A. Patterson, President
      Research park linking the research resources of the University of Tennessee, Oak Ridge National Laboratory, and Tennessee Valley Authority with tenants, particularly in the areas of advanced materials, biotechnology, information sciences, waste technology, measurements and control, and electrotechnology. Fosters the transfer of technology from research and development resources in the Corridor and in Tennessee.

Texas A&M University
Technology Business Development

Texas Engineering Experiment Station   (409) 845-0538
Suite 310
Wisenbaker Engineering Research
  Center
College Station, TX 77843-3369   Helen Baca Dorsey, Director

Statewide mechanism to commercialize innovative technologies developed in Tecas. Facilitates the start-up of a limited number of commercial operations each year involving university research in engineering and licenses university-developed technology. Provides assistance to the University, Texas entrepreneurs, and small companies in technical research and development, marketing and pricing studies, technical evaluations, and production and financial planning of unique ideas. **Publications:** Diversification Report (quarterly newsletter).

Texas A&M University Research Park

College Station, TX 77843   (409) 845-7275
    Dr. Mark L. Money, Vice Chancellor

Located on 434 acres west of main campus, the Park serves to assist private utilization of University resources, to promote closer ties between industry engaged in research and the University, to improve the quality and productivity of University research activities, and to accelerate the dissemination of new knowledge and the transfer of new technologies.

Texas Engineering Experiment Station (TEES)

College Station, TX 77843   (409) 845-5510
    Dr. Herbert H. Richardson, Director

Engineering and related sciences. Major research thrusts include biotechnical engineering, electrooptics and telecommunications, artificial intelligence and expert systems, electrochemical engineering, engineering toxicology, and manufacturing systems. Assists technology transfer, commercialization, and economic development of University research through the Technology Business Development Division. **Publications:** Windows (four per year); Research Report; Annual Report; Engineering Issues.

Texas Research and Technology Foundation

8207 Callaghan Road, Suite 345   (512) 342-6063
San Antonio, TX 78230   John F. D'Aprix, President and CEO

Supports basic and applied science and new advanced technology enterprises in the San Antonio area by providing research and development facilities, equipment, endowment, and business and scientific expertise. Manages the 1,500-acre Texas Research Park providing industry with access to the resources of several educational and research organizations. Activities include creation of research centers conducting studies in areas such as connective tissue, human performance, neuroscience, battlefield casualty management, toxicology, and clinical trials. Also conducts an inventory and analysis of scientific and technical resources in San Antonio.

Texas Research Park

Texas Research and Technology   (512) 342-6063
  Foundation
8207 Callaghan Road, Suite 345
San Antonio, TX 78230   Chris Harness, Vice President

Supports technology development efforts of private industry by providing facilities for basic and applied research and access to the resources of local educational and research organizations. 1,300 acres of the Park are available to private firms through sale or long-term lease; 200 acres are reserved for nonprofit research and development. The Institute of Biotechnology, supporting basic and applied research in the biosciences, the Institute of Applied Sciences, providing technology transfer to industry and government, and the Invention and Investment Institute, offering business and technical services in support of advanced technology business ventures, are collaborative programs based at the Park.

Troy State University
Center for Business and Economic Services

Sorrell College of Business   (205) 566-3000
Troy State University, AL 36082-0001   Joseph W. Creek

Supports applied research in business and government throughout Alabama. Topics include small business management, government funds management, income tax, salesmanship, venture capital, and Small Business Administration workshops.

Tulane University
U.S.-Japan Biomedical Research Laboratories

Herbert Research Center   (504) 394-7199
Belle Chase, LA 70037

A biomedical research center being developed as an international research park.

University Center Research and Development Park

c/o Vantage Companies   (813) 882-0601
7650 West Courtney Campbell Causeway
Suite 1100
Tampa, FL 33607-1432   Daniel Woodward, Project Manager

Park provides an interface between research and development tenants and University resources, particularly in the areas of medical technology and engineering sciences. Tenants

University research centers can offer independent inventors many services. If you have a well-developed design completed, the center may be capable of prototyping and testing it, while offering design modifications to facilitate its commercialization. Or if your design is very rough or merely a definition of function, a center might be able to find the solution through noodling sessions between teachers and students charged with the challenge. According to Inventor-Assistance Program News, a typical fee is $50 per hour for shop time plus the cost of materials. The university covers the facilities and cost of faculty. Contracts with independent inventors are typically very flexible. Confidentiality is practiced.

Other kinds of services that are available at many university research centers include innovation evaluation, invention testing, counseling on the start-up of new high-technology businesses, counseling regarding patents, trademarks, and copyrights, and assistance in identifying sources of government contracts for goods or services needed. They also may provide assistance in identifying potential research funding from the federal government under programs such as the Small Business Innovation Research (SBIR) Grants and the Energy Related Inventions Program (ERIP).

access University of South Florida research facilities and services, faculty consulting, and graduate student assistance in research and development and prototype assembly activities.

University of Alabama in Huntsville
Alabama High Technology Assistance Center
101 Morton Hall (205) 895-6409
Huntsville, AL 35899 M. Carl Ziemke, Director

Offers technology transfer, research, analysis, counseling, education, and information services to small businesses in Alabama, focusing on technical and high technology assistance. Projects include assistance with product prototypes to determine patentability and producibility.

University of Alaska
Alaska Economic Development Center (AEDC)
School of Business and Public (907) 789-4402
  Administration
1108 F St.
Juneau, AK 99801 Dr. Henry Kohl, Science Advisor

Provides technical assistance state and local governments, community and civic organizations, nonprofit corportations, and Alaska Native Claims Settlement Act corporations involved in economic development activities.

University of Alberta
Office of Research Services
Room 1-3 University Hall (403) 432-5360
Edmonton, AB, Canada T6G 2J9 Robert E. Armit, Director

Provides assistance in obtaining grants and contracts. Facilitates the transfer of technology from the University to industry. Administers University patent policy and markets research capabilities of the University to industry and government. Operates University of Alberta PRIME (Principal Researcher's Interests and Major Expertise) database, open to the public.

University of Arkansas
Arkansas Center for Technology Transfer
Engineering Experiment Station (501) 575-3747
Fayetteville, AR 72701 William H. Rader, Director

Facilitates research and technology transfer in the areas of robotics and automation, interactive technology, productivity and industrial efficiency, and business incubation and analysis. Specific projects include bar-code reading of trucks at weigh stations, scales to weigh-in-motion, tactile sensing, and advanced brakes, clutches, and transmissions.

Entrepreneurial Service Center
Bureau of Business and Economic (510) 575-4151
  Research
College of Business Administration
Fayetteville, AR 72701 Dr. Phillip Taylor, Director

Operates as a consulting service to Arkansas entrepreneurs. Provides business plans, capital location, problem-solving for new businesses, marketing plans, seminars, and other services tailored to the needs of individual businesses.

University of Arkansas at Little Rock
Technology Center
100 S. Main, Suite 401 (501) 371-5381
Little Rock, AR 72201 Paul McGinnis, Director

University of Calgary
Office of Technology Transfer
2500 University Drive, N.W. (403) 220-7220
Calgary, AB, Canada T2N 1N4 Dr. E.L. Jessop, Acting Director

Offers contract research and tranfers new technologies, inventions, and products conceived by University and hospital researchers to industry. Coordinates University and hospital inventions, products, patents, expertise, facilities, and technologies with research and development needs of industry. Adminsters license and joint venture programs.

University of California, Berkeley
Patent, Trademark and Copyright Office
1250 Shattuck Avenue (415) 642-5000
Berkeley, CA 94720 Roger Ditzel, Director

University technology transfer office specializing in biotechnology.

University of California, Los Angeles
Microelectronics Innovation and Computer Research Opportunities (MICRO)
7514 Boelter Hall (213) 206-6814
Los Angeles, CA 90024 Prof. C.R. Viswanathan, MICRO
Executive Committee

Conducts research in microelectronics in collaboration with industry. Seeks to increase interaction with university researchers to enhance the transfer of technology.

University of Colorado Foundation Inc.
Office of Patents and Licensing
Box 1140   (303) 492-8134

Boulder, CO 80306   John P. Holloway, Patent Officer

    University technology transfer office specializing in optoelectronics, molecular biology, pharmacy, chemistry, and medical technology.

University of Hawaii
Pacific Business Center Program
College of Business Administration   (808) 948-6286

2404 Maile Way

Honolulu, HI 96822

    Provides direct counseling and referral services as well as access to the faculty and physical resources of the University to small business owners.

University of Illinois
Bureau of Economic and Business Research
428 Commerce West   (217) 333-2330

1206 South Sixth Street

Champaign, IL 61820   Marvin Frankel, Acting Director

    Economics and business, including studies in business expectations, forecasting and planning, public utilities, innovation, entrepreneurship, consumer behavior, poverty problems, small business operations and problems, investment and growth, productivity, and research methodology. **Publications:** Project Reports; Monographs; Bulletins; Illinois Business Review (six times a year); Illinois Economic Outlook (annually); Illinois Statistical Abstract; Quarterly Review of Economics and Business; Research Projects and Publications (annual report of the faculty of the College of Commerce and Business Administration).

Business Development Service
109 Coble Hall   (217) 333-8357

801 South Wright Street

Champaign, IL 61820   Arthur H. Perkins, Director

    Business development aimed at University entrepreneurs. Produces and commercializes new ideas and products and promotes technology transfer from laboratories into the marketplace.

ILLITECH
Office of Federal and Corporate   (217) 333-8634
   Relations

363 Administration Building

506 South Wright Street

Urbana, IL 61801   Ms. Sunny Christensen, Director

    Produces and commercializes new ideas and products and promotes technology transfer from laboratories to the marketplace. Stimulates the development and commercialization of University technologies through a partnership with the State of Illinois, the University, and Illinois business and industry.

University of Illinois at Chicago
Technology Commercialization Program
P.O. Box 4348   (312) 996-9131

M/C 345

Chicago, IL 60680   Dr. L.F. Barry Barrington, Director

    Produces and commercializes new ideas and products and promotes technology transfer in the fields of biotechnology, robotics, mechanical devices, engineering software, and data processing hardware components. Arranges assistance for qualified clients' requirements for testing, prototyping, demonstration, market surveys, and business planning. **Publications:** Workshops on Business Plans; Workshops on Technology Transfer (semiannually).

University of Illinois at Urbana-Champaign
Office of the Vice Chancellor for Research
601 E. John Street   (217) 333-7862

Swanlund Building, 4th Floor

Champaign, IL 61820   Dillon Mapother, Assoc. Vice Chancellor

    University technology transfer office specializing in engineering and computing.

University of Kansas
Space Technology Center
Raymond Nichols Hall   (913) 864-4775

Lawrence, KS 66045   Prof. B.G. Barr, Director

    Supports development of new knowledge, concepts, and technology for surveying earth resources and evaluating environmental quality, including remote sensing, technology transfer, flight research, economic and business research, microprocessor control, computer integrated manufacturing, mineral resource surveys, geochronology, and energy research. Staff also works with industry and government in applying newly developed technology.

University of Kansas Center for Research, Inc. (CRINC)
2291 Irving Hill Drive   (913) 864-3441
Lawrence, KS 66045   Dr. Carl E. Locke, Ph.D., Director
> Remote sensing technology (development, analysis, and applications), energy, environmental quality, aircraft performance improvement, stress analysis, communications systems, technology transfer, microprocessing, transportation, biomedical engineering, isotope geochemistry, and radiation physics. **Publications:** Annual Report; Focus Newsletter (semiannually).

University of Kentucky
National Aeronautics and Space Administration/University of Kentucky
Technology Applications Program
109 Kinkead Hall   (606) 257-6322
Lexington, KY 40506-0057   William R. Strong, Acting Director
> Provides access to scientific and technical information for use in problem-solving by government and industry. Promotes secondary use of NASA technology and stimulates technology transfer through its information services and technical assistance. Program is also a member of the Federal Laboratory Consortium Technology Transfer Network.

University of Louisville
Center for Entrepreneurship and Technology
School of Business   (502) 588-7854
Louisvill, KY 40292   Lou Dickie, Director
> Provides business counseling and management consulting services to technology-based businesses, including entrepreneurs and inventors.

University of Manitoba
Office of Industrial Research (OIR)
230 Engineering Building   (204) 474-9463
Winnipeg, MB, Canada R3T 2N2   Prof. R.E. Chant, Director
> Industrial problems, including studies related to product development, agriculture, energy, energy conservation, laser holographic techniques, product evaluation and testing, wind power development, home economics, policy and planning analysis, management, and small enterprises. Also provides collaboration on innovative research and development projects.

University of Maryland
Engineering Research Center (ERC)
College of Engineering   (301) 455-7941
Wind Tunnel Building
Room 2104
College Park, MD 20742   Dr. David Barbe, Executive Director
> Engineering and scientific disciplines. Activities include a Technology Extension Service (TES), which offers technical assistance to the Maryland business community; a Technology Advancement Program (TAP), an incubator for start-up companies engaging in the development of technically oriented products and services; a Technology Initiatives Program (TIP) to support technological capabilities within the University; and the Maryland Industrial Partnerships (MIPS) program, a matching fund for industry sponsored research. **Publications:** ERC Update (quarterly newsletter).

University of Maryland Science and Technology Center
7505 Greenway Center Drive   (301) 982-9400
Suite 203
Greenbelt, MD 20770   Dr. Jon K. Hutchison, Project Director
> 466-acre research and development park focusing in the areas of math, physics, computer sciences, electrical engineering, and mechanical engineering. Researchers are provided access on a contract basis to University of Maryland faculty and equipment. The Supercomputing Research Center is housed at the Center.

University of Miami
Innovation and Entrepreneurship Institute
P.O. Box 249117   (305) 284-4692
Coral Gables, FL 33124   Jeanie L. McGuire, Director
> Entrepreneurship and innovation in Florida, including studies on high-technology ventures and black and Cuban-American entrepreneurs. Promotes interaction of entrepreneurs and capital and service providers. **Publications:** Research Report Series; Friends of the Forum Directory.

University of Michigan
Technology Transfer Center, Special Projects Division
College of Engineering   (313) 763-9000
2200 Bonisteel Blvd.
Ann Arbor, MI 48109   Larry Crockett, Director
> University technology transfer office specializing in automated engineering.

University of Minnesota
Center for the Development of Technological Leadership
107 Lind Hall   (612) 624-2006
Institute of Technology
Minneapolis, MN 55455   W.T. Sackett, Acting Director
> Technology transfer and technology leadership, including technology transfer methods and pilot experiments.

Office of Patents and Licensing
1919 University Avenue   (612) 624-0550
St. Paul, MN 55104   John Thuente, Director
> University technology transfer office specializing in health sciences and engineering.

Office of Research and Technology Transfer Administration
1919 University Avenue   (612) 624-1648
St. Paul, MN 55104   A.R. Potami, Assistant Vice President
> Serves as the research support unit for University of Minnesota faculty members by administering non-programmatic aspects of all research, training, and public service projects funded by external sources. Reviews and processes all proposals and awards for research, training, and public service projects and is responsible for financial management, cash receipt, and financial reporting of project funds. Works with faculty to stimulate the disclosure of patentable discoveries resulting from University research. Assists faculty in locating potential sources of support for research, training, and public service programs.

University of Missouri
Missouri Research Park
215 University Hall   (314) 882-3397
Columbia, MO 65211   Dr. Duane Stucky, Executive Director
> A 240-acre research park designed to be a link between academia and industry and being developed as a center for research and development in such fields as agriculture, chemicals, medicine, computers, engineering, and food processing. Also planned site for an international, policy-level think tank in the field of agribusiness.

University of Missouri—Rolla
Center for Technology Transfer and Economic Development
One Ngogami Terrace   (314) 341-4559
Rolla, MO 65401   H. Dean Keith, Director
> Technological innovation.

University of Moncton
Center of Research in Administrative Sciences (CRAS)
Moncton, NB, Canada E1A 3E9   (506) 858-4555
                          Dr. Jean Cadieux, Director
> Regional development, small businesses, capital venture, and impact of new technologies on the Atlantic region of Canada. **Publications:** Cahiers du CRSA (annually).

University of Nebraska, Lincoln
Nebraska Technical Assistance Center
W191 Nebraska Hall   (402) 472-5600
Lincoln, NE 68588-0635   Herbert Hoover, Director
> Provides information about new technologies, nonlegal assistance in patent searches and the patent process, and educational activities.

University of New Hampshire
Industrial Research and Consulting Center
Thompson Hall   (603) 862-3750
Durham, NH 03824   Dr. Donald C. Sundberg, Director
> Promotes research and development relationships between the private sector and the University. Organizes problem solving teams and makes available University instrumentation and computer facilities and research and development laboratories. Assists business and industry with product development, process development, long-range research and planning, modeling, software development, technical troubleshooting, feasibility studies, development of laboratory testing procedures, market analysis, risk analysis, and planning and educational programs. Develops patents and licenses technology and other intellectual property of the University.

University of New Mexico
Technology Application Center
Albuquerque, NM 87131   (505) 277-3622
                          Dr. Stanley A. Morain, Director
> Retrieves, processes, and analyzes satellite and aerial data for earth resources and develops geographic information systems (GIS). Image processing and GIS activities include mineral exploration, cover type mapping, habitat mapping and modeling, and surveys of archeological locations. Photo search and retrieval services include satellite images, aerial photos, maps, digital data, LANDSAT data, and photos from Gemini, Apollo, Apollo-Soyuz, Skylab, and Space Shuttle missions. Designated as University of New Mexico National Aeronautics and Space Administration Industrial Applications Center to assist in commercializing the national program for space especially in the areas of remote sensing and image dissemination. Offers national and international visiting scientist programs providing customized technical assistance and

The federal government spends about $60 billion per year on R&D; $40 billion is awarded to private industry (companies ranging from IBM and Ford to small companies and universities). $20 billion is spent at 1,100 federal laboratories. The government labs employ more than 100,000 scientists and engineers, or one-sixth of the nation's total.

training in remote sensing and image processing. **Publications:** Remote Sensing of Natural Resources: A Quarterly Literature Review.

Technology Innovation Center
Albuquerque, NM 87131   (505) 277-5934
     Gary Smith, Program Director

University of North Carolina at Charlotte
Engineering Research and Industrial Development Office
Charlotte, NC 28223   (704) 547-4150
     Dr. J.M. Roblin, Contact

Seeks to assist business in the region to compete more effectively by providing industrial development facilities, equipment, applied research, and consulting resources for solving practical industrial problems.

University of Oklahoma
Swearingen Research Park
1700 Lexington Avenue   (405) 325-7233
Norman, OK 73069   George Hargett, Airpark Administrator

Provides on a 900-acre tract facilities and services for research administered by the University's Office of Research Administration, as well as U.S. and state government, industry, and University of Oklahoma laboratories located in the Park.

University of Pennsylvania
Office of Corporate Programs and Technology
133 South 36th Street, Suite 300   (215) 898-7293
Philadelphia, PA 19104   George C. Farnbach, Director

University technology transfer office specializing in biomedical technology.

Office of Sponsored Programs
1335 South 36th Street   (215) 898-7293
Philadelphia, PA 19104-3246   Anthony Merritt, Director

Administers extramurally sponsored research for all departments and research units of the University and handles processing of research applications. Responsible for licensing of inventions and other technology transfer activities.

Sol C. Snider Entrepreneurial Center
Vance Hall, Suite 400   (215) 898-4856
Philadelphia, PA 19104   Dr. Ian C. MacMillan, Director

Entrepreneurship in private sector, public sector, and nonprofit organizations, including venture capital, corporate venturing, development of internal corporate entrepreneurship, technology transfer, and university spin-offs.

University of Pittsburgh
Center for Applied Science and Technology
100 Loeffler Building   (412) 624-2024
Pittsburgh, PA 15260   Richard K. Olson, Director

Fosters collaborative research efforts between the University and industry for the transfer of technology.

National Aeronautics and Space Administration Industrial Applications Center
823 William Pitt Union   (412) 648-7000
Pittsburgh, PA 15260   Paul A. McWilliams, Ph.D., Executive Director

Works with business and industrial clients to transfer technology developed by the National Aeronautics and Space Administration (NASA) and other government agencies. Activities center on new product identification and testing, engineering analyses, corporate resource redeployment initiatives, literature searching and document procurement, database system design and electronic publishing, and science curriculum enhancements for grades K-12. The Center is hard-wired to the NASA Recon database and it maintains direct link with researchers in NASA field centers, federal research laboratories, and academic centers. **Publications:** Economic Books: Current Selections (quarterly); United States Political Science Documents (annually).

University of Puerto Rico
Center for Business Research
College of Business Administration   (809) 764-0000
Department of Management & Business
Rio Piedras, PR 00931   Prof. Alvin J. Martinez, Director

Commerce and economics, including studies on consumer analysis, marketing, budget analysis, transfer of managerial technology, economic and business development, marketing, budget analysis, and problems of private enterprise. Compiles statistical data on personnel, employment and unemployment, health and nutritional problems management training and proficiency, entrepreneurship, international commerce, economic problems, finance, and accounting in Puerto Rico. **Publications:** Annual Report; Revista de Ciencias Comerciales (semiannually).

University of Quebec at Hull
Organizational Efficiency Research Group
283 Tache Boulevard   (819) 595-2318
P.O. Box 1250
Hull, PQ, Canada J8X 3X7   Prof. Alain Albert, Director
 Office automation and management and high-technology and venture capital.

University of Rhode Island
Business Assistance Programs
Kingston, RI 02881   (407) 792-2337
                                             Robert Comerford, Director
 Provides assistance through Rhode Island Innovation Center, which evaluates potential of
 inventions and advises inventors on commercialization strategies.

University of Scranton
Technology Center
Scranton, PA 18510-2192   (717) 961-4050
                                             Jerome P. DeSanto, Executive Director
 Works with business and industry to find practical solutions in such areas as computer-aided
 design, applied research, chemistry and biotechnology, software engineering, computer
 networking, telecommunications, and business planning and management to aid the economic
 development of northeastern Pennsylvania. **Publications:** Newsletter.

University of South Florida
Small Business Development Center
College of Business Administration   (813) 974-4274
Tampa, FL 33620   Bill Manck, Director
 Small business operations, entrepreneurship, and success and failure factors for small
 business development and management, including developing business plans, marketing
 strategy, and loan packages. **Publications:** Speaking of Small Business (monthly newsletter).

University of Southern California
Office of Patent and Copyright Administration
376 South Hope Street, Suite 200   (213) 743-4926
Los Angeles, CA 90007   L. Kenneth Rosenthal, Assistant Director
 University technology transfer office specializing in medicine and chemistry.

   Urban University Center, Western Research Applications Center
   3716 South Hope Street   (213) 743-2371
   Los Angeles, CA 90007-4344
    Supports the Small Business Innovation Research program through seminars, workshops,
    conferences, and one-to-one counseling activities.

University of Southern Maine
Center for Business and Economic Research
118 Bedford Street   (207) 780-4187
Portland, ME 04103   Richard J. Clarey, Dean of SBEM
 Business expansion in Maine, including management and small business development.
 Publications: Maine Business Indicators (quarterly); Reports.

   The New Enterprise Institute
   Center for Research and Advanced   (207) 780-4420
     Study
   246 Deering Avenue
   Portland, ME 04102   Dr. Richard J. Clarey, Director
    Creates, tests, and implements economic development mechanisms. Provides technical
    assistance, feasibility studies, applied research, and productivity enhancement analysis to
    all economic sectors of the state. Also acts as a connection between users of
    entrepreneurial and technology-related information.

University of Texas
Center for Technology Development and Transfer
Ernest Cockrell Jr., Hall 2:516   (512) 471-3700
College of Engineering
Austin, TX 78712-1080   Dr. Stephen A. Szygenda, Director
 Links university researchers with industrial needs and facilitates the commercialization of
 academic research, university-generated technology, scientific information, and other
 intellectual property. Activities range from serving as a clearinghouse for technology to product
 development and the formation of businesses. Also serves as the hub for a statewide technology
 development and transfer network.

University of Texas at Arlington
Technology Enterprise Development Center
UTA Station Box 19029   (817) 273-2559
Arlington, TX 76019   Dr. John Troutman, Director

The U.S. government applies for some 1,200 patents per year; 500 of them issue and 700 remain as applications. There is a database of 28,000 U.S. government patents. These patents are in 394 categories.

University of Texas at Austin
IC2 Institute
2815 San Gabriel
Austin, TX 78705

(512) 478-4081
Dr. George Kozmetsky, Director

Management of technology, creative and innovative management, measurement of the state of society, dynamic business development and entrepreneurship, new methods of economic analysis, and the evaluation of attitudes, concerns, and opinions on key issues. **Publications:** Newsletter (quarterly).

University of Texas Health Science Center at Houston
Institute for Technology Development and Assessment
P.O. Box 20334
6901 Bertner
Houston, TX 77225

(713) 792-4609

R.W. Butcher, Ph.D., Director

Established to evaluate, develop, and promote cost-effective new technologies relating to health care, preventative medicine, and biomedical research. Acts as an administrative umbrella for multidisciplinary research at the Health Science Center.

University of Utah
University of Utah Research Park
505 Wakara Way
Salt Lake City, UT 84108

(801) 581-8133
Charles A. Evans, Director

300-acre tract designed to facilitate scientific research projects between government agencies, industrial organizations, and the University. Facilities include general and technical libraries, scientific and research equipment, and computer center. The Park is the site of 22 buildings, including facilities for the Utah Innovation Center, Inc.

University of Washington
Office of Technology Transfer
201 Administration Building, AG/10
Seattle, WA 98195

(206) 543-5900
Donald Baldwin, Director

University technology transfer office specializing in biotechnology, bioengineering, engineering, and medical applications.

University of Wisconsin—Madison
The Wisconsin Alumni Research Foundation
Box 7365
Madison, WI 53707

(608) 263-2500
Thomas Hinkes, Director of Licensing

University technology transfer office specializing in pharmacy and life sciences.

University of Wisconsin—Milwaukee
Office of Industrial Research and Technology Transfer
P.O. Box 340
Milwaukee, WI 53201

(414) 229-5000
Irving D. Ross, Jr., Director

Serves as a catalyst in developing university/industry research programs and facilitates the transfer of technology between the University and industry. Provides assistance on patents, copyrights, and proprietary agreements to faculty.

University of Wisconsin—Stout
Center for Innovation and Development
Menomonie, WI 54751

(715) 232-1252
Dr. John F. Entorf, Director

Manufacturing productivity, electronics, and process evaluation. Evaluates inventions and new products, develops prototype models of innovative products, and evaluates materials for fabrication of new products.

University of Wisconsin—Whitewater
Wisconsin Innovation Service Center
402 McCutchan Hall
Whitewater, WI 53190-1790

(414) 472-1365
Debra Knox-Malewicki, Program Manager

University Research Park (Charlotte)
Suite 1980, Two First Union Plaza
Charlotte, NC 28282

(704) 375-6220
Seddon Goode, Jr., President

2,700-acre site providing companies located in the park with research and educational interaction with University of North Carolina at Charlotte.

University Research Park (Madison)
946 WARF Building
610 Walnut Street
Madison, WI 53705

(608) 262-3677

Wayne McGown, Director

The 180-acre Park facilitates technology transfer between the research produced on the University of Wisconsin—Madison campus and applied research of industry and provides a long-term endowment income to the University. Leases land to private companies and agencies, particularly those involved with microelectronics, forestry economics, genetic engineering, microbiology, computer research and software development, environmental research, food research, medical instrumentation, robotics, energy conservation, biotechnology, remote sensing, space sciences, manufacturing systems, materials science, polymers, toxicology, and hazardous waste. Utilizes University resources and encourages entrepreneurial innovation among faculty.

University Technology Incorporated

Research Building #1   (216) 368-5514
11000 Cedar Avenue
Cleveland, OH 44106   William Carlson, President

Established to strengthen regional business and industry, stimulate research, and provide a channel for the commercialization of Case Western University campus technologies. Identifies, evaluates, and implements development strategies for technologies, including intellectual property and protection; patent strategy; defining business and marketing opportunities; designing business structures; and forming business, commercial, and financial relationships. Licenses proprietary rights to companies prepared to invest in further on-campus research and development of technology, including established businesses and newly formed ventures.

Utah State University
Utah State University Research and Technology Park

1780 North Research Park Way   (801) 750-6924
Suite 104
North Logan, UT 84321   Wayne Watkins, Director

Fosters the interaction between the University research community and Park tenants. Administers a technology transfer program and provides incubator services to start-up companies.

Virginia Polytechnic Institute and State University
Virginia Tech Corporate Research Park

Development and University Relations   (703) 961-7676
315 Burruss Hall
Blacksburg, VA 24061   Charles M. Forbes, Vice-President

120 acres adjacent to the main campus and university airport providing building sites for lease to companies interested in developing or expanding a research relationship with the University. Houses an innovation center to provide facilities for start-up companies requiring support of University programs.

Washington State University
Innovation Assessment Center

Small Business Development Center   (206) 464-5450
Pullman, WA 99164   Jim Van Orsow, Director

Research and Technology Park

N.E. 1615 East Gate Boulevard   (509) 335-5526
Pullman, WA 99164   John R. Schade, Director

Encourages research interaction between Washington State University and Universiy of Idaho and industry. Seeks to promote economic development in the 3,000 square miles of Idaho's Panhandle Region and southeastern Washington. The Park leases land to industries engaged in research, development, and light manufacturing. Industrial tenants share University research facilities and services, use faculty as consultants, and utilize graduate students as a part-time work force. Major research areas include agriculture, forestry, mines, veterinary medicine, plant and animal biotechnology, and engineering. **Publications:** The Research Connection; Quarterly Report.

Washington University
Industrial Contracts and Licensing

Campus Box 8013   (314) 362-5866
St. Louis, MO 63130   H.S. Leahy, Director

University technology transfer office specializing in biotechnology.

Wayne State University
Technology Transfer Center

5050 Cass Avenue, Room 242   (313) 577-2788
Detroit, MI 48202   Nate R. Borofsky, Director

Assists manufacturers, entrepreneurs, and inventors in gaining access to technical resources at the University.

Western Illinois University
Center for Business and Economic Research

College of Business   (309) 298-1594
Macomb, IL 61455   Dr. Richard E. Hattwick, Director

Business and economics, with emphasis on labor markets, occupational education, regional economic analysis, and entrepreneurial case studies. **Publications:** Monographs; Working Papers; Journal of Behavioral Economics; Illinois Business Leader (newspaper); American Business Leader (newspaper).

Technology Commercialization Center

Room 212 Seal   (309) 298-2211
Macomb, IL 61455   Daniel Voorhis, Director

Produces and commercializes new ideas and products. Promotes the transfer of technology from the laboratory to the marketplace. Provides assistance to develop technology-based

---

Celebrity Inventors

Magician Harry Houdini: an open ocean diver's suit

Actor Danny Kaye: a child's toy noisemaker

Actress/dancer Julie Newmar: panty hose design

Actress Hedy Lamarr: an RF torpedo control system

Operatic soprano Lillian Russell: a dresser trunk

Ventriloquist Paul Winchell: an artificial heart

Bandleader Fred Waring: a food/ beverage blender

Singer Edie Adams: cigarette and cigar holders

Comic Herbert "Zeppo" Marx: a wristwatch heart monitor

Writer John Dos Passos: a soap bubble gun

Bandleader Lawrence Welk: lunch box designs

products and businesses through technical and commercial assessments and design assistance.

Western Michigan University
Institute for Technological Studies
Kalamazoo, MI 49008   (616) 387-4022

Robert M. Wygant, Director

Provides administrative support for internal and external research of the College. Also links industry with University resources by providing technical assessments, information, referrals, testing, and instruction.

Wichita State University
Center for Entrepreneurship
008 Clinton Hall—Campus Box 147   (316) 689-3000

Wichita, KS 67208   Prof. Fran Jabara, Director

Entrepreneurship, particularly the effect entrepreneurial education has on business start-ups and success rates, and profile studies of entrepreneurs. **Publications:** Center Report (twice yearly).

Center for Productivity Enhancement
Engineering, Room 203   (316) 689-3525

Wichita, KS 67208-1595   Dr. Richard Graham, Contact

Technology transfer in the areas of computer-integrated manufacturing and advanced composite materials.

Wisconsin for Research, Inc.
210 North Bassett Street   (608) 258-7070

Madison, WI 53703   Reed Coleman, President

Seeks to encourage economic development by promoting the transfer of University of Wisconsin research and technology to businesses in the state. Operates the Madison Business Incubator to generate economic development in Dane County by providing office space, support services, and business assistance. Administers a seed capital fund.

Worcester Polytechnic Institute
Management of Advanced Automation Technology Center
Department of Management   (617) 793-5000

Worcester, MA 01609   Arthur Gerstenfeld, Director

Management of technological innovations in the areas of computer-aided manufacturing, decision support systems, office automation, group technology, manufacturing communication networks, automation assembly process, application of bar coding, user friendly shop floor control, and manufacturing strategy. Conducts applied research for the specific technological needs of industrial clients. **Publications:** Newsletter (quarterly).

Yale University
Office of Cooperative Research
252 JWG, Box 6666   (203) 432-3003

New Haven, CT 06511   Dr. R.K. Bickerton, Director

University technology transfer office specializing in biotechnology, computer science, and engineering.

Inventor Support: State & Regional Opportunities

Alabama

Alabama Small Business Development Consortium
1717 11th Ave., S., Suite 419   (205) 934-7260
Medical Towers Bldg.
Birmingham, AL 35294   Dr. Jeff D. Gibbs, Director
　　Nonprofit consortium provides managerial and technical assistance and training through twelve
　　small business development centers, the Alabama International Trade Center, Alabama High
　　Technology Assistance Center, Alabama Small Business Procurement System, and six affiliated
　　centers.

Alabama State Department of Economic and Community Affairs
Planning and Economic Development
3465 Norman Bridge Rd.   (205) 261-3572
P.O. Box 2939
Montgomery, AL 36105-0939   Dr. Don Hines, Director
　　Assists small and developing businesses through information dissemination for federal grants.
　　Also provides consulting services.

U.S. Small Business Administration
Birmingham District Office
2121 Eighth Ave., N., Suite 200   (205) 731-1338
Birmingham, AL 35203-2398

Alaska

Alaska Small Business Development Center
Anchorage Community College   (907) 274-7232
430 W. Seventh Ave., Suite 115
Anchorage, AK 99501   Janet Nye, Director
　　Serves the Alaskan businessperson or a person considering going into business in Alaska.
　　Counsels inventors on commercialization and patent processes. Arranges meetings between
　　individual inventors and manufacturers and other interested paties. **Publications:** Quarterly
　　newsletter.

Alaska State Department of Commerce and Economic Development
P.O. Box D   (907) 465-2500
Juneau, AK 99811
　　Maintains a Business Development Division and an Office of Enterprise. **Telecommunication**
　　Access: Alternate telephone (907) 465-2017.

U.S. Small Business Administration
Anchorage District Office
701 C St., Room 1068   (907) 271-4022
Anchorage, AK 99513

The SBA supports a network of local offices that provide training and assistance to entrepreneurs. In this chapter I list many state-supported programs as well as more than 100 regional SBA offices.

Also listed here are a number of state-supported organizations providing referrals or other services related to obtaining venture capital. For a listing of private venture capital firms, see the chapter, Venture Capitalists.

Arizona

Arizona Enterprise Development Corporation

Arizona State Department of Commerce,   (602) 255-1782
  Development Finance Program
State Capitol Tower, 1700 W.
  Washington, Fifth Floor
Phoenix, AZ 85007

A private, nonprofit corporation that offers loans to small businesses through the U.S. Small Business Administration's 504 Program. Grants such loans for the purchase of land, buildings, machinery, and equipment, for construction, for renovation/leasehold improvements, and for related professional fees. Loans are available to for-profit businesses that generated an average net profit below $2 million for the preceding two years and that have a net worth below $6 million. (Businesses in Phoenix and Tucson do not qualify but have similar programs.)
Telecommunication Access: Toll-free/Additional Phone Number: (602)255-5705.

Arizona State Department of Commerce

State Capitol Tower   (602) 255-5371
1700 W. Washington, Fourth Floor
Phoenix, AZ 85007

Offers business counseling and provides information about establishing a business, licensing, taxation, funding, labor and environmental regulations, and foreign trade zones. Participates in the National Small Business Revitalization Program. Administers three federal financing programs for Arizona's small businesses. Also refers loan applicants to other sources such as the loan guarantee programs of the U.S. Small Business Administration, the Economic Development Administration, and the Farmers Home Administration. Publications: Guide to Establishing a Business in Arizona.

Revolving Loans Program

State Capitol Tower   (602) 255-5705
1700 W. Washington
Phoenix, AZ 85007

A federally funded program that provides loans for economic development projects of small for-profit and nonprofit businesses in Arizona, except Maricopa and Pima counties. Offers below-market interest rates—with negotiable terms—for site/facility acquisition and improvements, construction, machinery and equipment, building rehabilitation, leasehold improvements, and sometimes working capital.

Urban Development Action Grants   (602) 255-5705

Grants subordinated loans to new and expanding businesses for site improvements, industrial and commercial rehabilitation, machinery and equipment, and like expenses—except working capital, debt refinancing, and consolidation.

U.S. Small Business Administration
Phoenix District Office

2005 N. Central Ave., Fifth Floor   (602) 261-3732
Phoenix, AZ 85004

Tucson District Office

300 W. Congress St., Room 3V   (602) 629-6715
Tucson, AZ 85701

Arkansas

Arkansas Capital Corporation/Arkansas Capital Development Corporation

Governor's Office
State Capitol
Little Rock, AR 72201

Utilizes both state and private funds dedicated to the growth and stimulation of economic activity within the state. The Arkansas Capital Corporation is authorized to borrow $20 million from the interest earned on the state's average daily balances; the ACC works in concert with financial institutions and economic/industrial development agencies. The Arkansas Capital Development Corporation is a for-profit, risk-taking partner of the ACC. It extends financial assistance in the forms of loans, loans with options to purchase equity interest, and the direct acquisition of equity interest. The ACDC serves as a supplement to the ACC and other financial devices within the public and private sectors. The ACDC has the ability to engage in venture capital-type projects.

Arkansas Industrial Development Commission

One Capitol State Mall   (501) 371-1121
Little Rock, AR 72201

Serves as an information and referral source for small business concerns.

Arkansas Science and Technology Authority
200 Main St., Suite 450   (501) 371-3554
Little Rock, AR 72201   Dr. John W. Ahlen, Executive Director
> Administers an investment fund to provide seed capital in the form of loans, royalty agreements, and the purchase of limited amounts of stock for new and developing technology-based companies. **Contact:** Alan Gumbel, SBIR Assistance.

Arkansas Small Business Development Center
University of Arkansas at Little Rock   (501) 371-5381
Little Rock Tech Center
100 S. Main, Suite 401
Little Rock, AR 72201   Paul McGinnis, State Director

U.S. Small Business Administration
Little Rock District Office
320 W. Capitol Ave., Suite 601   (501) 378-5871
Little Rock, AR 72201

California

California Small Business Development Center Program
Department of Commerce   (916) 324-9234
1121 L St., Suite 600
Sacramento, CA 95814   Edward Kawahara, Director

California State Department of Commerce
Office of Small Business
1121 L St., Suite 600   (916) 445-6545
Sacramento, CA 95814
> Provides management and technical assistance to small business professionals. Helps entrepreneurs develop business skills and assists them in problem solving. Directs entrepreneurs to specific financing resources, including sources of seed money and venture capital.

U.S. Small Business Administration
Fresno District Office
2202 Monterey St., Suite 108   (209) 487-5189
Fresno, CA 93721

> **Los Angeles District Office**
> 350 S. Figueroa St., Sixth Floor   (213) 894-2956
> Los Angeles, CA 90071

> **Sacramento District Office**
> 660 J St., Suite 215   (916) 551-1445
> Sacramento, CA 95814

> **San Diego District Office**
> 880 Front St.   (619) 557-7269
> San Diego, CA 92188

> **San Francisco District Office**
> 211 Main St., Fourth Floor   (415) 974-0642
> San Francisco, CA 94105

U.S. Small Business Administration (Region Nine)
450 Golden Gate Ave., Box 36044   (415) 556-7487
San Francisco, CA 94102

Colorado

Colorado Housing Finance Authority
500 E. Eighth Ave.   (303) 861-8962
Denver, CO 80203
> Sponsors the Quality Investment Capital (QIC) Program to provide fixed-rate financing for small business loans guaranteed by the U.S. Small Business Administration. Also administers the ACCESS Program—a small business financing tool for fixed assets. Has created the Flex Fund Program to provide a secondary market for the purchase of up to 75 percent participation in loans originated under locally administered economic development loan programs. Also offers export credit insurance.

For more information on the Association of Small Business Centers and its programs, contact John C. Sandefur, membership services director, 1313 Farnam on the Mall, Suite 132, Omaha, NE 68182-0472 or telephone (402) 595-2387.

Colorado Small Business Development Center
600 Grant St., Suite 505   (303) 894-2422
Denver, CO 80203   Dr. Richard Wilson, Director

Colorado State Department of Economic Development
Small Business Office
1625 Broadway, Suite 1710   (303) 892-3840
Denver, CO 80202
    Operates an information referral service for small business professionals. Maintains data bases. Sponsors a hotline for small business-related inquiries. Toll-free/Additional Phone Number: 800-323-7798 (hotline).

Colorado State Division of Commerce and Development
1313 Sherman St., Room 523   (303) 833-2205
Denver, CO 80203
    Provides capital formation assistance.

CU Business Advancement Centers
1690 38th St., #101   (303) 444-5723
Boulder, CO 80301
    Offers business consulting, technology transfer, information retrieval, and information on selling products and services to the U.S. Government. Centers are located in Boulder, Grand Junction, Durango, Colorado Springs, Trinidad, and Burlington.

U.S. Small Business Administration
Denver District Office
721 19th St., Room 407   (303) 844-2607
Denver, CO 80202-2599

U.S. Small Business Administration (Region Eight)
999 18th St., Suite 701   (303) 294-7033
Denver, CO 80202

Connecticut

Connecticut Business Development Corporation
Connecticut State Department of   (203) 566-4051
   Economic Development
210 Washington St.
Hartford, CT 06106
    For businesses whose net worths are less than $6 million and whose after-tax profits of the last two years are less than $2 million. Arranges federally guaranteed debenture financing for fixed-asset projects on a second-mortgage basis. (Second mortgages may not exceed $500,000 or 40 percent of the cost of the project.)

Connecticut Development Authority
217 Washington St.   (203) 522-3730
Hartford, CT 06106
    Helps companies undertake new capital expansions. Offers assistance to manufacturing, research and development, distribution, warehouse, and office facilities and for pollution control and energy conservation projects. Provides financing to cover the cost of land, the purchase and installation of machinery and equipment, and the construction, purchase, improvement, or expansion of buildings. Sponsors the Umbrella Bond/Direct Loan Program for projects up to $850,000, the Self-Sustaining Industrial Revenue Bond Program for projects up to $10 million, and the Mortgage Insurance Program for projects up to $10 million.

Connecticut Product Development Corporation
93 Oak St.   (203) 566-2920
Hartford, CT 06106
    Offers financing for new product development. Provides up to 60 percent of the eligible development costs of specific projects—from initial concept through fabrication of the production-ready prototype. Also offers low-cost loans of up to $200,000. Does not request a share in company ownership or management; requires only a limited royalty on the sales of sponsored products.

Connecticut Small Business Development Center
University of Connecticut, School of   (203) 486-4135
   Business Administration
368 Fairfield Rd., Room 422, Box U-41
Storrs, CT 06268   John O'Connor, Director

Connecticut State Department of Economic Development
Corner Loan Program
210 Washington St.   (203) 566-3322
Hartford, CT 06106

Offers financing for working capital or fixed assets to manufacturers and wholesale distributors whose gross sales for their most recently completed fiscal years were less than $10 million and that have been doing business in the region for at least one year. Also offers financing to manufacturers and wholesale distributors who have been doing business in the region for at least two months (but less than one year) and whose gross revenues did not exceed an average of $600,000 per calendar month during that period.

Naugatuck Valley Revolving Loan Fund   (203) 566-3322
Offers financing for fixed assets such as land or building purchases, construction, renovation, rehabilitation, and the purchase and installation of machinery and equipment; for start-up capital; and for working capital. Funds are available to manufacturers and to wholesale distributors (no gross sales limit).

Sales Contact Centers   (203) 566-4051
Helps businesses generate new sales by matching manufacturers with the purchasing executives of large corporations. Also assists corporations with locating their own suppliers.

Set-Aside Program   (203) 566-4051
A procurement program for small businesses and for minority- and women-owned firms. Requires that small businesses provide 15 percent of all goods and services purchased by the state and that 25 percent of this amount be supplied by women- and minority-owned businesses. Eligible companies must have been doing business in Connecticut for at least one year, with annual gross revenues not exceeding $1.5 million.

Small Business Services   (203) 566-4051
Offers managerial, job training, and employment assistance to small businesses. Sponsors low-cost financing programs for small businesses. Administers a procurement program for small and minority- and women-owned businesses. Conducts workshops and seminars.

Small Manufacturers Loan Program   (203) 566-3322
For companies whose gross revenues do not exceed $5 million. Offers sole source direct loans for fixed-asset financing. Makes available working capital direct loans on a matching basis.

Urban Enterprise Zone Program   (203) 566-3322
Offers investment incentives for manufacturers, commercial businesses, retailers, and residential property owners undertaking new capital investments in six designated cities. Special financing programs include low-cost working capital, venture capital, and small business loans. Incentives include a seven-year graduated deferral of any increase in property taxes attributable to new improvements. Added incentives for manufacturers and research-and-development facilities include 80 percent property tax abatements, 50 percent state corporate business tax reductions, and $1000 grants for each new job created. Urban enterprise zones are located in Bridgeport, Hartford, New Britain, New Haven, New London, and Norwalk.

Urban Jobs Program   (203) 566-3322
Assists firms undertaking new capital investments in key urban areas.

U.S. Small Business Administration
Hartford District Office
330 Main St., Second Floor   (203) 240-4670
Hartford, CT 06106

Delaware

Delaware Development Office
99 Kings Hwy.   (302) 736-4271
P.O. Box 1401
Dover, DE 19903   Dr. Dale E. Wolf, Director

Administers the U.S. Small Business Administration's 503 loan program in the state. **Contact:** Donna Murray, Business Development Specialist (302) 736-4271.

Delaware Small Business Development Center
University of Delaware, Purnell Hall,   (302) 451-2747
  Suite 005
Newark, DE 19716   Linda Fayerweather, Acting Director

Delaware State Chamber of Commerce
One Commerce Center, Suite 200   (302) 655-7221
Wilmington, DE 19801

Sponsors a series of special programs on small business financing.

Although each federal agency makes its own awards using contracts, grants, or cooperative agreements, the Small Business Administration is charged with the formulation and issues policy direction for the government-wide Small Business Innovation Research (SBIR) programs. For further information, contact Raymond Marchakitus, Small Business Administration, SBIR, 1441 L St., N.W., Premier Building #414, Washington, D.C. 20416.

Delaware State Development Office
99 Kings Hwy.   (302) 736-4271
P.O. Box 1401
Dover, DE 19903

> Encourages business development by maintaining contacts with existing industries to stimulate growth and expansion and by providing economic information to new industrial prospects. Offers several special financing programs to assist businesses that will add to Delaware's employment base. Helps small businesses obtain financing. Provides assistance to firms experiencing regulatory problems with governmental agencies. Recommends improvements in state programs that affect small business. Collects, analyzes, and distributes statistical data on the state's economy and business climate. Coordinates employment and training programs to meet the needs of business and industry. Publications: Small Business Start-up Guide; Selling to the State of Delaware: A Guide to Procurement Opportunities.

New Castle County Economic Development Corporation
12th and Market Sts.   (302) 656-5050
Wilmington, DE 19801

> Offers industrial revenue bonds and financing through the U.S. Small Business Administration's 503 Program.

U.S. Small Business Administration
Wilmington District Office
844 King St., Room 5207   (302) 573-6295
Wilmington, DE 19801

Wilmington Department of Commerce
800 French St., City/County Bldg., Ninth   (302) 571-4610
  Floor
Wilmington, DE 19801

> Maintains an industrial revenue bond and other financing programs.

Wilmington Economic Development Corporation
Two E. Seventh St.   (302) 571-9088
Wilmington, DE 19801

> Provides financing through the U.S. Small Business Administration's 503 Program and through other government mechanisms.

District of Columbia

District of Columbia Office of Business and Economic Development
Revolving Loan Fund
1111 E St., N.W.   (202) 727-6600
Washington, DC 20004

> Offers direct loans to be used in conjunction with private funds. Provides relatively short-term, gap financing. Also provides direct, below-market, long-term loans for property acquisition and capital equipment.

District of Columbia Small Business Development Center
Howard University   (202) 636-5150
Sixth and Fairmount St., N.W., Room 128
Washington, DC 20059   Nancy Flake, Director

U.S. Small Business Administration
Washington, D.C., District Office
1111 18th St., N.W., Sixth Floor   (202) 634-4950
Washington, DC 20036

Florida

Florida Small Business Development Center
University of West Florida, College of   (904) 474-3016
  Business
Building 38, Room 107 #245
Pensacola, FL 32514-5750   Jerry Cartwright, Coodinator
   Telecommunication Access: FAX (904) 474-2030.

Florida State Department of Commerce
Bureau of Business Assistance
107 W. Gaines St., Room G26　　　　(904) 487-1314
Tallahassee, FL 32301　　　　　　　Murry Hagerman, Development
　　　　　　　　　　　　　　　　　　Representative
　　　Assists and counsels small businesses. **Telecommunication Access:** Alternate telephone
　　　(904) 487-1314.

　　　Entrepreneurship Program
　　　107 West Gains St.　　　　　　(904) 488-9357
　　　Tallahassee, FL 32399-2000　　James Hosler, Director
　　　　　Encourages the creation and development of local organizations that support
　　　　　entrepreneurship.

Product Innovation Center
The Progress Center, One Progress　　(904) 462-3942
　Blvd., Box 7
Alachua, FL 32615　　　　　　　　　Pamela H. Riddle, Director
　　　Provides small business inventors and entrepreneurs with technical and market feasibility
　　　assessments.

U.S. Small Business Administration
Coral Gables District Office
1320 S. Dixie Hwy., S. 501　　　　(305) 536-5533
Coral Gables, FL 33136

　　　Jacksonville District Office
　　　400 W. Bay St., Room 261　　(904) 791-3782
　　　Jacksonville, FL 32202

　　　Tampa District Office
　　　700 Twiggs St., Room 607　　(813) 228-2594
　　　Tampa, FL 33602

　　　West Palm Beach District Office
　　　3500 45th St., Suite 6　　　　(305) 689-2223
　　　West Palm Beach, FL 33407

Georgia

Georgia Small Business Development Center
University of Georgia　　　　　　　(404) 542-5760
Chicopee Complex, 1180 E. Broad St.
Athens, GA 30602　　　　　　　　　James McGovern, Director

Georgia State Department of Community Affairs
Small Business Revitalization Program
40 Marietta St., N.W., Eighth Floor　(404) 656-6200
Atlanta, GA 30303
　　　Offers assistance in securing long-term financing.

U.S. Small Business Administration
Atlanta District Office
1720 Peachtree Rd., N.W., Sixth Floor　(404) 347-4749
Atlanta, GA 30309

　　　Statesboro District Office
　　　52 N. Main St., Room 225　　(912) 489-8719
　　　Statesboro, GA 30458

U.S. Small Business Administration (Region Four)
1375 Peachtree St., N.E., Fifth Floor　(404) 347-4999
Atlanta, GA 30367

Guam

U.S. Small Business Administration
Agana District Office
Pacific News Bldg., Room 508　　　(671) 472-7277
238 Archbishop F. C. Flores St.
Agana, GU 96910

The ten regional SBA offices can provide you with information on the Small Business Innovation Research (SBIR) program and updates on what's happening at the participating agencies and departments.

Small Business Answer Desk: The SBA operates a toll-free number. Dial 1-800-368-5855.

Every year the federal government spends more than one billion dollars on research and development. For more information on the federal government's participation in funding R&D, see **R&D Funding: Federal Sources.**

Hawaii

Chamber of Commerce of Hawaii
Small Business Center
735 Bishop St., Suite 225   (808) 531-4111
Honolulu, HI 96813
> Serves as a resource center offering information, referrals, and consulting services. Provides business financial management and/or loan packaging assistance to meet business needs. Conducts special entrepreneur training programs.

Hawaii State Department of Planning and Economic Development
Small Business Information Service
P.O. Box 2359   (808) 548-8741
Honolulu, HI 96814   Carl Swanholm, Science & Technology
     Officer
> Offers information and referrals on permits and licensing, market data, business plan writing, consulting, government procurement, alternative financing, and business classes. Maintain the Small Business Information Service, a one-stop clearinghouse of small business resources. Publications: 1) Hawaii's Business Regulations; 2) Starting a Business in Hawaii.

U.S. Small Business Administration
Honolulu District Office
300 Ala Moana, Room 2213   (808) 541-2990
Honolulu, HI 96850

Idaho

The Idaho Company
P.O. Box 6812   (208) 334-6308
Boise, ID 83707
> Offers assistance in obtaining venture capital.

Idaho First National Bank
Economic Action Council
Communications Department #1-5011   (208) 383-7265
P.O. Box 8247
Boise, ID 83733
> Administers a program for economic action fund grants.

Idaho Small Business Development Center
Boise State University   (208) 385-1640
College of Business, 1910 University Dr.
Boise, ID 83725   Ronald R. Hall, Director

Idaho State Department of Commerce
Statehouse, Room 108   (208) 334-2470
Boise, ID 83720
> Offers assistance with all types of business inquiries. Provides financial information about community development block grants, urban development action grants, industrial revenue bonds, and Idaho Travel Council grants. Also offers information about regulations, permits, and licenses; state procurement programs; travel and tourism; and international trade. Maintains a census data center that provides economic and demographic data.

U.S. Small Business Administration
Boise District Office
1020 Main St., Second Floor   (208) 334-1696
Boise, ID 83702

Illinois

Build Illinois Small Business Development Program
Equity Investment Fund
Illinois State Department of Commerce   (217) 782-7500
     and Community Affairs
620 E. Adams St.
Springfield, IL 62701
> Designed to stimulate the development of technology-based companies. Provides equity financing to companies having significant potential for job creation. Offers up to one-third of anticipated project costs, with a maximum of $250,000. Program funding may be used for the

purchase of real estate, machinery, and equipment, for working capital, for research-and-development costs, and for organizational fees.

Illinois Development Finance Authority
Direct Loan Program
Two N. LaSalle St., Suite 980   (312) 793-5586
Chicago, IL 60602

Offers supplemental financing to small and medium-sized businesses undertaking fixed-asset projects that will increase employment opportunities. Covers no more than 30 percent of a project.

Illinois Small Business Development Center
Illinois State Department of Commerce   (217) 524-5856
   and Community Affairs
620 E. Adams St., Fifth Floor
Springfield, IL 62701   Jeff Mitchell, Director

Illinois State Department of Commerce and Community Affairs
Business Innovation Fund
620 E. Adams St.   (217) 782-7500
Springfield, IL 62701

Provides funding for projects that may not otherwise attract traditional lenders or venture capitalists. Offers financial aid to businesses using a university's assistance to advance technology, to create new products, or to improve manufacturing processes. Businesses may qualify for up to $100,000 in financial assistance for projects matched with private resources. A royalty agreement reimburses the state when the product is developed and sold in the marketplace.

Small Business Advocacy   (217) 782-7500
Responsible for the representation, coordination, legislative, and ombudsman services to small businesses and related organizations throughout the state. Refers small business owners to technical, management, and special financial assistance programs. The department also sponsors a program for entrepreneurial and employee education.

Small Business Assistance Bureau   (217) 782-7500
Provides information about all state government forms and applications for new and existing businesses. Maintains a one-stop permit center that offers a business start-up kit featuring all state permit forms and license applications. Toll-free/Additional Phone Number: 800-252-2923 (Small Business Hotline).

Small Business Development Program   (217) 782-7500
Provides—along with a participating financial institution—direct financing to small businesses for expansion and subsequent job creation or retention. Offers long-term, fixed-rate, low-interest loans.

Small Business Financing Program   (217) 782-7500
Helps finance economic development/job creation projects by combining dollars from the U.S. Small Business Administration, private capital, and public block grants. Other funding categories for community improvement activities include a Competitive Economic Development Program, a Public Facilities and Housing Program, and Set-Aside Funds.

Small Business Fixed-Rate Financing Fund   (217) 782-7500
Purpose is to provide affordable start-up and expansion financing to small businesses and to create jobs for low- and moderate-income people. Offers long-term, fixed-rate financing to new or expanding companies for fixed assets and working capital.

Small Business Micro Loan Program   (217) 782-7500
Offers direct financing to small businesses at below-market interest rates in cooperation with private sector lenders. Funds may be used for acquisition of land or buildings; construction, renovation, or leasehold improvements; purchase of machinery or equipment; or inventory and working capital.

Illinois State Department of Commerce and Community Development
Procurement Assistance Program
620 E. Adams St.   (217) 782-7500
Springfield, IL 62701

Helps businesses participate in government markets. Sponsors procurement assistance centers that help firms seeking state and federal government contracts.

Illinois Venture Capital Fund
Illinois Development Finance Authority   (312) 793-5586
Two N. LaSalle St., Suite 980
Chicago, IL 60602

Funded by the state of Illinois and the Frontenac Venture Company. Provides seed capital for enterprises seeking to develop and test new products, processes, technologies, and inventions.

U.S. Small Business Administration
Chicago District Office
219 S. Dearborn St., Room 437   (312) 353-4528
Chicago, IL 60604-1779

Springfield District Office
Four N. Old State Capitol Plaza   (217) 492-4416
Springfield, IL 62701

U.S. Small Business Administration (Region Five)
230 S. Dearborn St., Room 510   (312) 353-0359
Chicago, IL 60604-1593

Indiana

Corporation for Innovation Development
One N. Capitol, Suite 520   (317) 635-7325
Indianapolis, IN 46204   Mr. Marion C. Dietrich, President &
                                                C.E.O.
Encourages capital investment, with the primary objective of providing additional employment through the expansion of business and industry—particularly new and existing small business enterprises. Invests in small business investment companies as well as directly in new and existing businesses. Includes common or preferred stock equity investments and various forms of debt, with warrants or other convertible features.

Indiana Corporation for Science and Technology
One N. Capitol, Ninth Floor   (317) 635-3058
Indianapolis, IN 46204
Administers funds set aside for research and development of new products.

Indiana Institute for New Business Ventures
One N. Capitol, Suite 420   (317) 634-8418
Indianapolis, IN 46204   Robert Cummings, President
Encourages and supports the creation and development of emerging enterprises that are expected to provide growth opportunities for Indiana's economy. Introduces businesses to appropriate financing sources. Offers management, educational, technical, and financial resources to assist in starting and operating small enterprises. Provides referral services and small business counsellors. Sponsors conferences and educational programs.

Indiana Seed Capital Network
Indiana Institute for New Business   (317) 634-8418
  Ventures
One N. Capitol, Suite 420
Indianapolis, IN 46204   David C. Clegg, Director
A computerized matching service for investors and entrepreneurs. Identifies investment opportunities in entrepreneurial ventures as well as informal investors. Acts as a referral service to entrepreneurs and investors.

Indiana Small Business Development Center
Indiana Economic Development Council,   (219) 282-4350
  Inc.
One N. Capitol, Suite 200
South Bend, IN 46601   Stephen Thrash, Director

Indiana State Department of Commerce
One N. Capitol   (317) 232-8800
Indianapolis, IN 46204
Administers financial assistance programs for small businesses.

U.S. Small Business Administration
Indianapolis District Office
575 N. Pennsylvania St., Room 578   (317) 269-7272
Indianapolis, IN 46204-1584

Iowa

Iowa Business Development Credit Corporation
901 Insurance Exchange Bldg.   (515) 282-2164
Des Moines, IA 50309
Stimulates economic development through loans to new or established firms in conjunction with banks, insurance companies, savings and loan associations, and other financial institutions. Loan proceeds may be used to purchase land, to purchase or construct buildings, machinery, equipment, or inventory, or as working capital. A portion may be used to retire debt. Makes available loans up to $500,000—plus the amount of bank participation. (Most loans range from $100,000 to $200,000.) Nonprofit enterprises, lending or financial institutions, and those firms

able to acquire funds at reasonable rates from other sources are ineligible for assistance. Also offers counseling in financial, marketing, and managerial areas.

Iowa Business Growth Company

901 Insurance Exchange Bldg.   (515) 282-2164
Des Moines, IA 50309

Maintains the U.S. Small Business Administration's 503/504 loan programs for businesses in every Iowa community. Helps stimulate growth and expansion of small businesses by providing long-term, fixed-asset financing at a fixed rate of interest slightly below the market rate. Applicants must demonstrate that—as a result of the loan—they will hire additional employees, fill a needed service in the community, or help upgrade a deprived area. May finance up to 45 percent of the cost of plant construction, conversion, or expansion, including the acquisition of land, existing buildings, leasehold improvements, and equipment (if the equipment has a useful life of at least ten years). May lend up to $500,000 for 15, 20, or 25 years on a second-mortage basis. The small business must provide a minimum of 10 percent of the project costs. The remaining project funds are provided by a conventional first-mortgage lender for at least ten years at market interest rates.

Iowa Small Business Development Center

Iowa State University, College of   (515) 292-6351
  Business Administration
Chamberlynn Bldg., 137 Lynn Ave.
Ames, IA 50010   Ronald A. Manning, Director
Telecommunication Access: FAX (515) 292-2009.

Iowa State Department of Economic Development
Community Economic Betterment Account

200 E. Grand Ave.   (515) 281-3415
Des Moines, IA 50309

Established to invest proceeds of the Iowa Lottery in local economic development projects. Cities, counties, or merged area schools are eligible to apply on behalf of local enterprises that are expanding, modernizing, or locating. Key criteria for approval of applications are the number of jobs created, the cost per job for the funds involved in the project, and significant community involvement and interest. Funds are used to buy-down principal or interest on business loans to acquire land or buildings for construction or reconstruction, to purchase equipment, to prepare industrial sites, and to assist in other business financings.

Iowa Procurement Outreach Center

c/o Kirkwood Community College   (319) 398-5665
6301 Kirkwood Blvd., S.W.
Cedar Rapids, IA 52406

A service center for Iowa small businesses interested in selling goods and services to the federal government. Uses existing resources and programs to identify interested businesses and assists them in bidding for and receiving government contracts. Toll-free/Additional Phone Number: 800-458-4465 (in Iowa); (319)398-5666.

Iowa Product Development Corportation

200 E. Grand Ave.   (515) 281-5459
Des Moines, IA 50309   Glen Burmeister, Deputy Director

Provides risk capital to feasible ventures that have the potential to create new jobs or to develop new products for manufacture in Iowa. The return on investments received by the IPDC is reinvested in new ventures.

Iowa Small Business Advisory Council   (515) 281-7252

Works to determine the problems and priorities of Iowa's business owners. Makes recommendations on legislative matters affecting businesses. Serves as an advocate for small businesses in the state. Toll-free/Additional Phone Number: 800-532-1216 (in Iowa).

Iowa Small Business Loan Program   (515) 281-4058

Assists in the development and expansion of small businesses in Iowa through the sale of bonds and notes that are exempt from federal income tax and through the use of the proceeds to provide limited types of financing for new or existing small businesses. The loans may be used for purchasing land, construction, building improvements, or equipment. Funds cannot be used for working capital inventory or operating purposes. The maximum loan is $10 million. Rates vary with the level of risk.

Iowa Small Business Vendor Application Program   (515) 281-8310

An aid to small businesses competing for state government contracts. Application forms are available to vendors wanting their names placed on lists for use by state purchasing agencies soliciting bids. Toll-free/Additional Phone Number: 800-532-1216 (in Iowa).

Self-Employment Loan Program   (515) 281-3600

Assists low-income entrepreneurs by providing low-interest loans for new or expanding small businesses. The loans shall not exceed $5000 or be issued at a rate to exceed 5 percent simple interest per annum.

Targeted Small Business Loan Guarantee Program   (515) 281-4058

Provides guarantees to lenders for loans made to certified targeted small businesses in the state of Iowa. Provides up to a 75 percent loan guarantee for private lenders making loans to targeted small businesses.

U.S. Small Business Administration
Cedar Rapids District Office
373 Collins Rd., N.E., Room 100   (319) 399-2571
Cedar Rapids, IA 52402-3118

> **Des Moines District Office**
> 210 Walnut St., Room 749   (515) 284-4567
> Des Moines, IA 50309

Kansas

Avenue Area, Inc.
New Brotherhood Bldg., Seventh and   (913) 371-0065
Minnesota
Kansas City, KS 66101
> Provides financial packaging services for U.S. Small Business Administration and other financing programs to eligible small businesses in Wyandotte County.

Big Lakes Certified Development Company
104 S. Fourth   (913) 776-0417
Manhattan, KS 66502-6110
> Provides financial packaging services for U.S. Small Business Administration and other financing programs to eligible small businesses located in Clay, Geary, Pottawatomie, Marshall, and Riley counties.

Citywide Development Corporation of Kansas City, Kansas, Inc.
Kansas State Department of Economic   (913) 573-5730
Development
One Civic Plaza, Municipal Office Bldg.
Kansas City, KS 66101
> Provides financial packaging services for U.S. Small Business Administration and other financing programs to eligible small businesses in Kansas City, Kansas.

Four Rivers Development, Inc.
108 E. Main   (913) 738-2210
P.O. Box 365
Beloit, KS 67420
> Provides financial packaging services for U.S. Small Business Administration and other financing programs to eligible small businesses in Cloud, Dickinson, Ellsworth, Jewell, Lincoln, Mitchell, Ottawa, Republic, Saline, and Washington counties.

Great Plains Development, Inc.
1111 Kansas Plaza   (316) 275-9176
Garden City, KS 67846
> Provides financial packaging services for U.S. Small Business Administration and other financing programs to eligible small businesses in Barber, Barton, Clark, Cloud, Comanche, Edwards, Finney, Ford, Grant, Gray, Greeley, Hamilton, Haskell, Hodgeman, Kearney, Kiowa, Lane, Meade, Morton, Ness, Pawnee, Pratt, Rush, Scott, Seward, Stafford, Stanton, Stevens, and Wichita counties. Maintains an office at 407 S. Main, P.O. Box 8776, Pratt, KS 67124. Toll-free/Additional Phone Number: (316)672-9421 (Pratt office).

Kansas Small Business Development Center
Wichita State University, College of   (316) 689-3193
Business Administration
021 Clinton Hall, Campus Box 48
Wichita, KS 67208   Tom Hull, State Director
> **Telecommunication Access:** FAX (316) 689-3770.

Kansas State Department of Administration
Small Business Procurement Program
Landon State Office Bldg., Room 102   (913) 296-2376
Topeka, KS 66612
> The Kansas Small Business Procurement Act of 1978 establishes a set-aside program for small businesses. It is the policy of the state to ensure that a fair proportion of the purchase of and contracts for property and services for the state be placed with qualified small business contractors. Program includes supplies, materials, equipment, maintenance, contractual services, repair services, and construction.

Kansas State Department of Commerce
Kansas Enterprise Zone Act
400 W. Eighth St., Fifth Floor   (913) 296-3485
Topeka, KS 66603-3957
> Allows qualified businesses and industries located in an approved enterprise zone to take enhanced job expansion and investment income tax credits and Kansas sales tax exemptions on specified capital improvements.

One-Stop Permitting   (913) 296-5298

Serves as a central point of contact for general information needed to establish or operate a business in Kansas. Provides necessary state applications and forms required by agencies that license, regulate, or tax businesses. Answers questions about starting or expanding a business in Kansas.

Kansas State Procurement Technical Assistance Program

Wichita State University   (316) 689-3193
Kansas Small Business Development
  Centers
021 Clinton Hall, Campus Box 148
Wichita, KS 67208-1595

Assists Kansas businesses in getting their fair share of government contracts. Offered through the Kansas State Small Business Development Center system, which provides one-on-one counseling assistance and procurement training seminars. (Counseling assistance is extended to any Kansas business free of charge.)

Kansas Technology Enterprise Corporation

Kansas State Department of Commerce   (913) 296-5272
400 W. Eighth St., Fifth Floor
Topeka, KS 66603-3957   Kevin Carr, Director

A nonprofit corporation designed to foster innovation, with programs ranging from research-and-development grants to equity investments. Offers research matching grants, which fund 40 percent of the cost of industry research-and-development projects that lead to job creation in Kansas. Also maintains a seed capital fund to provide equity financing for high-tech product development. Matching funds are provided for the Small Business Innovation Research Program. Also conducts an annual high-tech expo.

Lenexa Development Company, Inc.

8700 Monrovia, Suite 300   (913) 888-1826
P.O. Box 14244
Lenexa, KS 66215

Provides financial packaging services for U.S. Small Business Administration and other financing programs to eligible small businesses in the city of Lenexa.

McPherson County Small Business Development Association

101 S. Main   (316) 241-2100
P.O. Box 1032
McPherson, KS 67460

Provides financial packaging services for U.S. Small Business Administration and other financing programs to eligible small businesses in McPherson County.

Mid-America, Inc.

1715 Corning   (316) 421-6350
P.O. Box 708
Parsons, KS 67357

Provides financial packaging services for U.S. Small Business Administration and other financing programs to eligible small businesses in Allen, Anderson, Bourbon, Cherokee, Crawford, Labette, Montgomery, Neosho, Wilson, and Woodson counties.

Neosho Basin Development Company

Emporia State University   (316) 342-7041
1200 Commercial, Box 46
Emporia, KS 66801

Provides financial packaging services for U.S. Small Business Administration and other financing programs to eligible small businesses in Chase, Coffey, Lyon, Morris, Osage, and Wabaunsee counties.

Pioneer Country Development, Inc.

317 N. Pomeroy Ave.   (913) 674-3488
P.O. Box 248
Hill City, KS 67642

Provides financial packaging services for U.S. Small Business Administration and other financing programs to eligible small businesses in Cheyenne, Decatur, Ellis, Gove, Graham, Logan, Norton, Osborne, Phillips, Rawlins, Rooks, Russell, Sheridan, Sherman, Smith, Thomas, Trego, and Wallace counties.

South Central Kansas Economic Development District

727 N. Waco, River Park Place, Suite 565   (316) 262-5246
Wichita, KS 67203

Provides financial packaging services for U.S. Small Business Administration and other financing programs to eligible small businesses in Butler, Chautauqua, Cowley, Elk, Greenwood, Harper, Harvey, Kingman, Marion, Reno, Rice, Sedgwick, and Sumner counties.

Topeka/Shawnee County Development, Inc.
820 S.E. Quincy, Suite 501   (913) 234-0072
Topeka, KS 66612
  Provides financial packaging services for U.S. Small Business Administration and other financing programs to eligible small businesses in Shawnee County.

U.S. Small Business Administration
Wichita District Office
110 E. Waterman St.   (316) 269-6571
Wichita, KS 67202

Wakarusa Valley Development, Inc.
1321 Wakarusa Dr., Suite 2102   (913) 841-7120
P.O. Box 1732
Lawrence, KS 66046
  Provides financial packaging services for U.S. Small Business Administration and other financing programs to eligible small businesses in Douglas County.

Wichita Area Development, Inc.
350 W. Douglas   (316) 265-7771
Wichita, KS 67202
  Provides financial packaging services for U.S. Small Business Administration and other financing programs to eligible small businesses in the city of Wichita.

Kentucky

Kentucky Development Finance Authority
Kentucky State Commerce Cabinet   (502) 564-4886
2400 Capital Plaza Tower
Frankfort, KY 40601
  Encourages economic development, business expansion, and job creation by providing financial support to manufacturing, agribusiness, and tourism projects. Issues industrial revenue bonds for eligible projects and second-mortgage loans to private firms in participation with other lenders. Houses the Commonwealth Small Business Development Corporation. Maintains the Kentucky Community Development Block Grant Program. **Telecommunication Access:** Alternate telephone (502) 564-4554.

Kentucky Small Business Development Center
University of Kentucky, 20 Porter Bldg.   (606) 257-7668
Lexington, KY 40506-0205   Jerry Owen, Director

Kentucky State Commerce Cabinet
Business Information Clearinghouse
Capital Plaza Tower, 22nd Floor   (502) 564-4252
Frankfort, KY 40601
  Provides new and existing businesses with a centralized information source on business regulation. Assists businesses in securing necessary licenses, permits, and other endorsements, including an ombudsman service with regulatory agencies. Acts as a referral service for government financial and management assistance programs. Serves as a regulatory reform advocate for business. Toll-free/Additional Phone Number: 800-626-2250.

  Small Business Division
  Capital Plaza Tower, 22nd Floor   (502) 564-4252
  Frankfort, KY 40601   Patsy Wallace, Acting Director
    Encourages small business development in the state. Assists existing small businesses in Kentucky through referrals and information. Operates a computerized procurement service. Functions as a small business development center in conjunction with other such facilities across the state. Toll-free/Additional Phone Number: 800-626-2250.

U.S. Small Business Administration
Louisville District Office
600 Federal Place, Room 188   (502) 582-5971
Louisville, KY 40202

Louisiana

Louisiana Small Business Development Center
Northeast Louisiana University, College   (318) 342-2464
  of Business Administration
700 University Ave.
Monroe, LA 71209   Dr. John Baker, Director

Louisiana Small Business Equity Corporation
4521 Jamestown Ave., Suite 9   (504) 925-4112
Baton Rouge, LA 70808
  Administers financial assistance programs for small businesses. **Telecommunication Access:** Alternate telephone (504) 342-9213.

Louisiana State Department of Commerce and Industry
Small Business Development Corporation
P.O. Box 94185   (504) 342-5361
Baton Rouge, LA 70804   Mr. Nadia Goodman, Acting Executive
                                      Director
  Maintains a variety of programs to assist owners and operators of small businesses.

U.S. Small Business Administration
New Orleans District Office
1661 Canal St., Suite 2000   (504) 589-6685
New Orleans, LA 70112

  Shreveport District Office
  500 Fannin St., Room 8A08   (318) 226-5196
  Shreveport, LA 71101

Maine

Finance Authority of Maine
83 Western Ave.   (207) 623-3263
P.O. Box 949
Augusta, ME 04330
  An economic development agency that administers all state-operated business financing programs, including a mortgage insurance program, small business loan guarantees, veterans' loans, a natural resources entrants program, and several industrial development bond programs.

Maine Small Business Development Center
University of Southern Maine, 246   (207) 780-4420
  Deering Ave.
Portland, ME 04102   Robert H. Hird, Director
  Telecommunication Access: FAX (207) 780-4417.

Maine State Development Office
Financial Programs
State House Station #59   (207) 289-2656
Augusta, ME 04333
  Administers a variety of financing programs, including job and investment tax credits, revolving loan funds, a development fund, and tax increment financing.

  Maine Growth Program   (207) 289-2656
    Offers financial packaging to healthy small businesses. Funds may be used for building new plants, expanding existing facilities, or purchasing new equipment.

U.S. Small Business Administration
Augusta District Office
40 Western Ave., Room 512   (207) 622-8378
Augusta, ME 04330

Maryland

Maryland Small Business Development Center
217 E. Redwood St., Room 1006   (301) 333-6996
Department of Economic & Employment
  Development
Baltimore, MD 21202   Elliott Rittenhouse, Director

Maryland Small Business Development Financing Authority
World Trade Center, 401 E. Pratt St. (301) 659-4270
Baltimore, MD 21202
  Assists socially and economically disadvantaged business persons in the development of their businesses. Acts as a funds precipitator and a risk taker. Minority businesses are the primary recipients of assistance, although nonminorities may seek surety bond guarantees. Provides working capital and funds for equipment acquisition through direct loans and loan guarantees for the completion of projects that are under federal, state, and local government and public utility

contracts. Offers long-term loan guarantees to lending financial institutions for property improvements and for working capital and equipment financing (up to a ten-year period and 80 percent of the loan request). Provides up to a 4 percent interest rate subsidy on long-term loan guarantees. Insures up to a maximum of 90 percent loss on a $1 million face contract for both minority and nonminority applicants seeking surety bonding. Provides equity participation franchise financing up to a maximum of $100,000 per franchisee or 45 percent of costs.

Maryland State Department of Economic and Community Development
Maryland Business Assistance Center
45 Calvert St.   (301) 974-2945
Annapolis, MD 21401

Provides expertise in industrial and community development, financial management, and technical assistance to the private sector and to local governments. Offers information on financing programs, facility location, state-funded employee training, procurement, licensing and permits, and starting a business. Toll-free/Additional Phone Number: 800-OK-GREEN (hotline).

U.S. Small Business Administration
Baltimore District Office
Ten N. Calvert St.   (301) 962-2054
Baltimore, MD 21202

Massachusetts

Massachusetts Business Development Corporation
One Boston Place   (617) 723-7515
Boston, MA 02108

Provides loans for small and medium-sized businesses that cannot obtain all of their financial requirements from conventional sources. Makes loans for working capital, leveraged buy outs, second mortgages, government guaranteed loans, U.S. Small Business Administration 503 loans, and long-term loans for new equipment or energy conversion. Lending terms are flexible.

Massachusetts Capital Resource Company
545 Boylston St.   (617) 536-3900
Boston, MA 02116

Supports the maintenance and growth of Massachusetts businesses by investing in traditional and technology-based industries, high-risk start-up companies, expanding businesses, management buy outs, and turnaround situations. Structures its investments to fit the specific needs of its portfolio companies and to increase employment opportunities within Massachusetts.

Massachusetts Community Development Finance Corporation
131 State St., Suite 600   (617) 742-0366
Boston, MA 02109

Provides flexible financing for working capital needs and real estate development projects when there is some clear public benefit. Makes investments in conjunction with community development corporations, which are organized to promote economic development in targeted areas of Massachusetts. Financing is available to businesses that are able to provide good employment opportunities, that are unable to meet their capital needs in the traditional markets, and that have the sponsorship of an eligible community development corporation. Includes venture capital investment, community development investment, and small loan guarantee programs.

The Massachusetts Government Land Bank
Six Beacon St., Suite 900   (617) 727-8257
Boston, MA 02108

A quasi-public state agency functioning as a financier and public developer. Offers below-market mortgage financing to qualifying public-purpose development projects that lack sufficient public/private investment. Can acquire, rehabilitate, prepare, construct, demolish, and dispose of eligible properties. Has access to $40 million in Massachusetts general obligation bonds, half of which are industrial development bonds. Project investments generally range from $200,000 to $3 million.

Massachusetts Industrial Finance Agency
125 Pearl St.   (617) 451-2477
Boston, MA 02110

Promotes employment growth through incentives that stimulate business investment in Massachusetts. Incentives for industrial companies include industrial revenue bonds, loan guarantees, and pollution control bonds. Financing is available for commercial real estate if these projects are located in locally identified commercial area revitalization districts.

Massachusetts Product Development Corporation

Industrial Services Program, 12 Marshall   (617) 727-8158
  St.
Boston, MA 02108

> Promotes new product development in Massachusetts firms. Offers priority in funding decisions to firms that will experience job loss because of foreign competition, declining markets, or shifts in industry technologies. Investments may be used for product prototypes, marketing efforts, distribution, or new production processes. Firms must detail the number and types of jobs created or maintained in order to receive capital.

Massachusetts Small Business Development Center

University of Massachusetts, 205 School   (413) 549-4930
  of Management
Amherst, MA 01003   John Ciccarelli, Director

Massachusetts State Department of Commerce
Massachusetts Suppliers and Manufacturers Matching Service

100 Cambridge St.   (617) 727-3206
Boston, MA 02202

> Serves as a liaison between manufacturers and suppliers in Massachusetts to meet their contractual purchasing needs of parts, raw materials, or components. Maintains the Massachusetts Matching Service, a computerized listing of suppliers.

Office of Financial Development   (617) 727-2932

> Assists businesses and individuals attempting to utilize the many federal, state, and local finance programs established to help businesses expand. Sponsors the Massachusetts Venture Capital Fair. Publications: 1) Massachusetts Financial Resources Directory; 2) Massachusetts Venture Capital Directory.

Small Business Assistance Division   (617) 727-4005

> Provides technical assistance and programs to support small business development. Serves as an advocate within state government for small business concerns. Provides information on starting a business, developing business plans, doing business with the state, and identifying sources of financing, management, marketing, or other business assistance. Advises firms interested in exporting. Administers and monitors the Small Business Purchasing Program.

Massachusetts State Department of Revenue
Corporations Bureau

215 First St., Third Floor   (617) 727-4264
Cambridge, MA 02142

> Administers a variety of financing programs to assists businesses, including loss carry over for new corporations, local tax exemption on tangible property, expensing research and development and other federal tax incentives, and dividend deductions.

Massachusetts State Executive Office of Consumer Affairs and Business Regulation
Thrift Institution Fund for Economic Development

One Ashburton Place   (617) 727-7755
Boston, MA 02108

> A $100 million lending pool to be invested over a ten-year period for a variety of economic development and job-generating purposes, including small business development.

Massachusetts State Office of Facilities Management
Division of Capital Planning and Operations

One Ashburton Place, 15th Floor   (617) 727-4028
Boston, MA 02108

> Offers financial assistance and tax reductions for pollution control facilities and alternative energy sources.

Massachusetts Technology Development Corporation

84 State St., Suite 500   (617) 423-9293
Boston, MA 02109

> Provides capital to new and expanding high-technology companies that have the capacity to generate significant employment growth in Massachusetts and that offer other public benefits.

Massachusetts Technology Park Corporation

Westborough Executive Park   (617) 870-0312
P.O. Box 663
Westborough, MA 01581   Ms. Christine Sheroff, Public Relations

> Will establish educational centers such as the Massachusetts Microelectronics Center. Centers will contain design, fabrication, and testing facilities and equipment for postsecondary academic and practical training programs required to satisfy the education and employment needs of the state's businesses and industries.

U.S. Small Business Administration
Boston District Office

Ten Causeway St., Room 265   (617) 565-5561
Boston, MA 02222-1093

U.S. Small Business Administration (Region One)
60 Batterymarch St., Tenth Floor   (617) 223-3204
Boston, MA 02110

Michigan

Michigan Investment Fund
Michigan State Department of Treasury,   (517) 373-4330
    Venture Capital Division
P.O. Box 15128
Lansing, MI 48901
> Invests in businesses with strong management that show a substantially above-average potential for growth, profitability, and equity appreciation. Will consider both high-technology and existing technology businesses, but the company must have a proprietary competitive edge. Will also consider joint investment with other financial institutions. Real estate investments and requests involving working capital alone will not be considered. A typical investment is $1 million or more. Additional requirements include an annual return on the investment of at least 35 percent and a business location or planned location with at least half of the assets and personnel in Michigan.

Michigan Small Business Development Center
Wayne State University, 2727 Second   (313) 577-4848
    Ave.
Detroit, MI 48201   Norman J. Schlafmann, Director
> **Telecommunication Access:** FAX (313) 963-7606.

Michigan State Business Ombudsman
P.O. Box 30107   (517) 373-6241
Lansing, MI 48909
> Serves as a one-stop center for business permits. Utilizes a computerized process to determine all permits and licenses required for a specific business. Applications for licenses and permits also are available through this office, as well as a business start-up kit and an employer packet, which include all necessary forms, posters, and applicable information on Michigan laws concerning employer rights and responsibilities. Toll-free/Additional Phone Number: 800-232-2727.

Michigan State Department of Commerce
Procurement Assistance
Law Bldg., Fourth Floor   (517) 373-9626
P.O. Box 30004
Lansing, MI 48909
> Helps businesses tap into government and corporate contracts by training companies to use standard procurement procedures.

Technology Transfer Network
P.O. Box 30225   (517) 335-2139
Lansing, MI 48909   Sharon Woollard, Director
> Seeks to link Michigan businesses with needed technical innovations and inventions. Provides technical assistance, faculty counsulting, research and development, joint research, specialized equipment and facilities, and seminars and workshops through programs at affiliated universities, which include University of Michigan, Michigan State University, Michigan Technological University, Wayne State University, and Western Michigan University.

Michigan State Department of Transportation
Small Business Liaison
425 W. Ottawa   (517) 373-2090
P.O. Box 30050
Lansing, MI 48909
> Enforces set-aside procurement programs. Publishes contract notices. Accepts bids for surface transportation projects.

Michigan Strategic Fund
Seed Capital Companies
Law Bldg.   (517) 373-6378
P.O. Box 30234   Dr. James Kenworthy, Manager R&D
Lansing, MI 48909   Technology
> Focuses on addressing the financing gap that exists in the early stages of business development to finance pre-start-up activities such as developing a working prototype, preparing a business plan, initial market analysis, and assembling a management team. Also administers the Capital Access Program, which provides financing based on a risk pooling concept for small and medium-sized businesses. Toll-free/Additional Phone Number: (517)373-0349 (Capital Access Program).

U.S. Small Business Administration
Detroit District Office
477 Michigan Ave., Room 515   (313) 226-6075
Detroit, MI 48226

   Marquette District Office
   300 S. Front St.   (906) 225-1108
   Marquette, MI 49885

Minnesota

Indian Affairs Council
Indian Business Loan Program
1819 Bemidji Ave.   (218) 755-3825
Bemidji, MN 56601
   Provides Minnesota-based Indians with loans to establish or expand a business in the state.
   Loan funds cannot be used to repay or consolidate existing debt. Also provides management and
   technical assistance resources.

Minnesota Science and Technologies Office
150 E. Kellogg Blvd., 900 American   (612) 297-1554
   Center Bldg.
St. Paul, MN 55101
   Provides business access to NASA technologies. Publishes a directory of state high-technology
   companies and software companies and a guide to business assistance for high-tech
   companies.

Minnesota Small Business Development Center
College of St. Thomas   (612) 448-8810
1107 Haxeltine Blvd., Suite 245
Chaska, MN 55318   Jerry Cartwright, Director
   Telecommunication Access: FAX (612) 448-8832.

Minnesota State Department of Administration
Division of Procurement
50 Sherburne Ave., 112 Administration (612) 296-3871
   Bldg.
St. Paul, MN 55155
   Publishes Selling Your Product to the State of Minnesota. Toll-free/Additional Phone Number:
   (612)296-6949.

Minnesota State Department of Energy and Economic Development
Community Development Corporation Program
150 E. Kellogg Blvd., 900 American   (612) 297-1304
   Center Bldg.
St. Paul, MN 55101
   An annual grant cycle whereby community development corporations make applications to the
   authority for funds for the administrative costs of their operations and for assistance to ventures
   within the defined community area through direct loan participation.

   Energy Development Loan Program   (612) 297-1940
      Finances loans through the sale of tax-exempt or taxable industrial development bonds,
      providing a lower interest rate to the borrower than is available commercially. Loans may be
      for up to 90 percent of project costs, the actual percentage determined by credit and project
      analysis.

   Energy Loan Insurance Program   (612) 297-1940
      Operates in partnership with private financial institutions. Applicants and participating
      financial institutions develop a loan package together, based on the lending institution's
      credit analysis and determination of appropriate financing. The program then insures up to
      90 percent of the loan. (The actual percentage insured is determined by the credit analysis
      and loan package.) The maximum insurable amount is $2.5 million. Interest rates are
      negotiated between the borrower and the lender, but may not exceed three percentage
      points above the prime rate.

   Minnesota Fund   (612) 297-3547
      Provides below-market-interest-rate, fixed-asset financing for new and existing
      manufacturers and other industrial enterprises. Eligible firms are for-profit manufacturing
      or industrial corporations, partnerships, or sole proprietorships; are independently owned
      and operated; and are not dominant in their fields of business. Loan funds may be used to
      purchase land to construct a new manufacturing or industrial facility; for construction of the
      new facility; to purchase an existing facility; to expand an existing facility; and to purchase
      machinery and equipment. Loan funds may not be used for debt refinancing, interim
      financing, or working capital.

According to a recent report, State Technology Programs in the United States, issued by the Minnesota Office of Science and Technology, more than $550 million was allocated for state science and technology initiatives for the 1988 fiscal year. The largest amount of funding (41.2%) was allocated to technology or research centers to promote research and development. However, states supported a wide variety of programs, including technology offices, research grants, incubators, technology transfer programs, seed/venture capital programs, clearinghouses, and equity/royalty investment programs.

Minnesota State Small Business Assistance Office   (612) 296-3871

Provides information and assistance to businesses in areas of start-up, operation, and expansion. Maintains the Bureau of Business Licenses, which offers information on the number and kind of licenses required for a business venture, the agencies that issue them, and the affirmative burdens imposed on the applicant. Also operates the Bureau of Small Business, which serves as a focal point within state government for information and resources available to small businesses. Sponsors workshops and seminars on small business issues throughout the year. Publications: Guide to Starting a Business in Minnesota; State of Minnesota Directory of Licenses and Permits. Toll-free/Additional Phone Number: 800-652-9747 (in Minnesota).

Office of Project Management   (612) 296-5021

Responsible for assisting businesses with financial packages utilizing a variety of resources. Assists in developing business plans and in obtaining financing needed for businesses to start-up or expand.

Small Business Development Loan Program   (612) 297-3547

Targeted to existing manufacturing and industrial firms that want to expand in Minnesota. Funds for business loans are raised through the sale of revenue bonds issued by the authority, the interest on which is generally exempt from state and federal taxes. These funds generally take the form of long-term, fixed-rate loans for land, buildings, and capital equipment. By pooling bonds and inducing the investment of private capital in connection with the business loans, the program facilitates a partnership between the public and private sectors. Funds may be used for interim or long-term financing for certain capital expenditures, including land and/or building acquisition costs; site preparation; construction costs; engineering costs; underwriting or placement fees; and other fees and costs.

Opportunities Minnesota Incorporated

Minnesota State Department of Energy   (612) 296-0582
  and Economic Development
150 E. Kellogg Blvd., 900 American
  Center Bldg.
St. Paul, MN 55101

Provides subordinated financing through the issuance of debentures for businesses that are purchasing buildings or capital assets with useful lives greater than 15 years. Offers financing for fixed assets, including land acquisition, building construction, leasehold improvements, renovation and modernization, and machinery and equipment.

U.S. Small Business Administration
Minneapolis District Office

100 N. Sixth St., Suite 610   (612) 349-3530
Minneapolis, MN 55403

Mississippi

Mississippi Small Business Development Center

University of Mississippi, School of   (601) 982-6760
  Business Administration
3825 Ridgewood Rd.
Jackson, MS 39211   Dr. Robert D. Smith, Director

Telecommunication Access: FAX (601) 982-6761.

Mississippi State Department of Economic Development
Finance Division

P.O. Box 849   (601) 359-3449
Jackson, MS 39205

Administers a variety of financing programs, including enterprise zones, incentives, and industrial revenue bonds. Assists in securing loans through federal, state, and private sources.

Mississippi State Research and Development Center

3825 Ridgewood Rd.   (601) 982-6606
Jackson, MS 39211   Dr. James Meredith, Director

Provides management and technical assistance to business professionals.

U.S. Small Business Administration
Gulfport District Office

One Hancock Plaza, Suite 1001   (601) 863-4449
Gulfport, MS 39501-7758

Jackson District Office

100 W. Capitol St., Suite 322   (601) 965-4378
Jackson, MS 39269

Missouri

Missouri Small Business Development Center
St. Louis University   (314) 534-7204
3674 Lindell Blvd., Room 007
St. Louis, MO 63108   Bob Brockhaus, Acting Director

Missouri State Department of Economic Development
High Technology Program
Economic Development Programs, P.O.   (314) 751-4241
  Box 118
Jefferson City, MO 65102   John S. Johnson, Director
    Markets the state to outside business firms and entrepreneurs and lends assistance to state businesses and entrepreneurs regarding the high-technology process, including federal and state research programs, research and development parks, innovation centers, venture capital, patents, licenses, and trademarks.

    Missouri Corporation for Science and Technology
    P.O. Box 118   (314) 751-3906
    Jefferson City, MO 65102   John Johnson, Director
      Establishes innovation centers as well as managerial and financial assistance to new firms.

    Small and Existing Business Development
    P.O. Box 118   (314) 751-4982
    Jefferson City, MO 65102
      Offers business skill training and technical assistance to small business professionals.

MO-KAN Development, Inc.
1302 Faraon   (816) 233-8485
St. Joseph, MO 64501
    Provides financial packaging services for U.S. Small Business Administration and other financing programs to eligible small businesses in Atchison, Brown, Doniphan, Jackson, Jefferson, and Nemaha counties.

St. Louis Technology Center
5050 Oakland   (314) 534-2600
St. Louis, MO 63110

U.S. Small Business Administration
Kansas City District Office
1103 Grand Ave., Sixth Floor   (816) 374-3416
Kansas City, MO 64106

    St. Louis District Office
    815 Olive St., Room 242   (314) 425-6600
    St. Louis, MO 63101

U.S. Small Business Administration (Region Seven)
911 Walnut St., 13th Floor   (816) 374-5288
Kansas City, MO 64106

Montana

Montana Small Business Development Center
1424 9th Ave.   (406) 444-4780
Helena, MT 59620   Robert A. Heffner, Director

Montana State Department of Commerce
Business Advocacy and Licensing Assistance
1424 Ninth Ave., Room 1   (406) 444-3494
Helena, MT 59620
    Maintains a Small Business Advocate that serves as a contact for individuals starting businesses and acts as an ombudsman for businesses with questions or complaints regarding state agencies.

    Business Development Specialist   (406) 444-3494
    Provides information and technical assistance to small businesses—particularly manufacturers—on such topics as employee training opportunities, U.S. government contract leads, and sources of loan or grant funds. Maintains a small business advocate. Also offers a wide range of services to prospective, new, and existing businesses. Assists in site selection and development. Provides product promotion and marketing assistance. Helps small businesses wishing to enter foreign markets. Operates the Census and Economic Information Center, which supplies population and economic data for research, planning, and decision making.

R*obert E. Bernier, state director of the Nebraska Small Business Development Center and president of the Association of Small Business Development Centers, tells the story of Larry Anderson of Fremont, Nebreska. Mr. Anderson came to the SBCD office in Omaha back in January of 1987 clutching a scrap of paper. It was his prototype, described as a laminated paper that rolls into a cone with built-in handles to form a small megaphone: the FANOZIE. Through SBCD, Anderson's dream became a reality.*

FANOZIE is only available right now at Omaha Lancers' hockey games where it sells for a dollar a copy, according to Bernier. SBCD is working with the inventor to see if his item can be made less expensively. "Who knows? Maybe someday a Georgia Bulldog fan or an Iowa Hawkeye fan or even a Nebraska Cornhusker will get a kick out of yelling for his favorite team through a FANOIZE," adds Bernier.

Montana Economic Development Board

1520 E. Sixth Ave., Lee Metcalf   (406) 444-2090
  Bldg., Room 050
Helena, MT 59620-0505

Makes investments in qualifying Montana businesses. Offers businesses long-term, fixed-rate financing at competitive interest rates for a variety of needs. Preference is given to small and medium-sized businesses, locally owned enterprises, employee-owned enterprises, and businesses that provide jobs for Montanans, that improve the environment, or that promote Montana's agricultural products. The Department of Commerce also offers development finance technical assistance to businesses. Finance specialists assist in analysis and planning, in preparing loan and bond applications, and with other financing programs available to businesses.

Montana Science and Technology Alliance

46 North Last Chance Gulch, Suite   (406) 449-2778
  2B
Helena, MT 59620   Frank Culver, Director

Established to promote public/private sector partnership. Assists in financing the establishment of technology-intensive businesses in the state. Components include seed capital investment, applied research, technical assistance and technology transfer, and research capability development.

Montana State Department of Natural Resources
Renewable Energy and Conservation Program

1520 E. Sixth Ave., Grant & Loan Section   (406) 444-6774
Helena, MT 59620   Greg Mills, Director

Promotes research, development, and demonstration of energy conservation. Offers grants and contracts with individuals and organizations.

U.S. Small Business Administration
Helena District Office

301 S. Park Ave., Room 528   (406) 449-5381
Helena, MT 59626

Nebraska

Nebraska Investment Finance Authority

1033 O St., Gold's Galleria, Suite 304   (402) 477-4406
Lincoln, NE 68508

Provides lower-cost financing for manufacturing facilities, certain farm properties, and health care and residential development. Has established a Small Industrial Development Bond Program for small businesses as well as an Industrial Development Revenue Bond Program.

Nebraska Secretary of State

State Capitol, Suite 2300   (402) 471-2554
P.O. Box 94608
Lincoln, NE 68509-1608

Registers corporations, trademarks, and service marks. Also responsible for business and occupational licenses.

Nebraska Small Business Development Center

University of Nebraska at Omaha   (402) 554-2521
1313 Farnam-on-the-Mall, Suite 132
Omaha, NE 68182-0248   Robert Bernier, Director

Nebraska State Department of Economic Development
Research and Development Authority

301 Centennial Mall, S.   (402) 471-2593
P.O. Box 94666
Lincoln, NE 68509

Provides seed capital for commercially viable ideas resulting from basic research-and-development phases. Does not make grants but invests in the idea by taking an equity position in the venture.

Small Business Division   (402) 471-4668

Advocates and supports the creation, expansion, and retention of businesses. Provides export sales assistance through planning, workshops and seminars, trade missions and trade shows, a network of honorary commercial attaches, and a trade leads advisory service. Assists in the design and implementation of industrial training programs. Publications: 1) Buy Nebraska—matches buyers and sellers of Nebraska products and services; 2) Directory of Nebraska Manufacturers (every two years); 3) Industry Census Findings Report—summarizes data collected about industry in Nebraska; 4) Nebraska International Trade Directory—lists Nebraska firms interested in exporting; 5) Nebraska Statistical Handbook—a fact book for business planning and analysis.

Nebraska State Energy Office
Nebraska Energy Fund
State Capitol, P.O. Box 95085   (402) 471-2867
Lincoln, NE 68509
    Makes revolving loans available to Nebraskans for energy efficiency improvements.

Nebraska State Ethanol Authority Development Board
301 Centennial Mall, S.   (402) 471-2941
P.O. Box 95108
Lincoln, NE 68509
    Funds projects for the expanded use of Nebraska's agricultural products; for efficient, less
    polluting energy sources; and for the development of protein that will be stored more efficiently.
    Financing is provided by way of grants, loans, or loan guarantees.

U.S. Small Business Administration
Omaha District Office
11145 Mill Valley Rd.   (402) 221-4691
Omaha, NE 68154

Nevada

Nevada Small Business Development Center
University of Nevada—Reno, College of   (702) 784-1717
   Business Administration, Room 411
Reno, NV 89557-0016   Samuel Males, Director

Nevada State Commission on Economic Development
Capitol Complex   (702) 885-4325
Carson City, NV 89710
    Assists businesses in various areas. Operates programs that promote community development
    activities. Encourages commercial energy conservation. Aids businesses seeking government
    contracts. Toll-free/Additional Phone Number: 800-992-0900, extension 4420.

Nevada State Office of Community Services
Procurement Outreach Program
1100 E. William, Capitol Complex, Suite   (702) 885-4420
   117
Carson City, NV 89710
    Assists businesses interested in bidding for government contracts. Maintains an office in Las
    Vegas at Temple Plaza, 3017 W. Charleston Blvd., Suite 20, Las Vegas, NV 89102. Toll-free/
    Additional Phone Number: 800-992-0900, extension 4420; (702)468-6174 (Las Vegas office).

   Small Business Revitalization Program
   1100 E. William St., Capitol   (702) 885-4602
      Complex, Suite 117
   Carson City, NV 89710
       Assists small businesses by reviewing proposed expansion projects, by recommending the
       most practical financing for those projects, and by structuring loan proposals that will meet
       the needs of various funding sources.

U.S. Small Business Administration
Las Vegas District Office
301 E. Stewart   (702) 388-6611
Las Vegas, NV 89125

   Reno District Office
   50 S. Virginia St., Room 238   (702) 784-5268
   Reno, NV 89505

New Hampshire

New Hampshire Small Business Development Center
University of New Hampshire   (603) 625-4522
400 Commercial St., University Center,
   Room 311
Manchester, NH 03101   James E. Bean, Director
   Telecommunication Access: Toll-free telephone 800-322-0390; FAX (603) 624-6658.

**New Hampshire State Department of Resources and Economic Development
Industrial Development Office**
105 Loudon Rd.   (603) 271-2591
P.O. Box 856
Concord, NH 03301   Paul Gilderson, Director
    Sponsors programs to encourage economic development. Offers market, site location, and import/export information.

**U.S. Small Business Administration
Concord District Office**
55 Pleasant St., Room 210   (603) 225-4400
Concord, NH 03301

New Jersey

Corporation for Business Assistance in New Jersey
200 S. Warren St., Suite 600   (609) 633-7737
Trenton, NJ 08608
    Provides businesses with long-term, fixed-asset financing for the acquisition of land and buildings, machinery and equipment, construction, renovation, and restoration.

New Jersey Commission on Science and Technology
122 West State Street, CN832   (609) 984-1671
Trenton, NJ 08625   Edward Cohen, Director
    Awards "bridge" grants of up to $40,000 to small companies that have received seed money from the Federal Small Business Innovation Research (SBIR) program. Grants are awarded on a competitive basis in the following areas: biotechnology, hazardous and toxic waste management, materials science, food technology, fisheries and aquaculture, and plastics recycling.

New Jersey Economic Development Authority
200 S. Warren St., Capitol Place One,   (609) 292-1800
    CN990
Trenton, NJ 08625
    Offers low-interest loans for capital costs (industrial development bonds), loan guarantees, and direct loans. Maintains an Urban Center Small Loans Program to assist existing retail and commercial businesses located in commercial districts of designated municipalities. Provides long-term, fixed-asset financing through the U.S. Small Business Administration's 504 Program. Administers programs for low-interest, fixed-asset loans (local development financing funds) and for recycling loans. Maintains an Urban Industrial Parks Program. Offers technical assistance to manufacturers impacted by imports.

New Jersey Small Business Development Center
Rutgers University Graduate School of   (201) 648-5950
    Management
Ackerson Hall, Third Floor, 180
    University St.
Newark, NJ 07102   Janet Holloway, Director
    Telecommunication Access: FAX (201) 648-5889.

**New Jersey State Department of Commerce and Economic Development
New Jersey Set-Aside Program**
One W. State St., CN823   (609) 984-4442
Trenton, NJ 08625
    Maintains a list of vendors wishing to participate in the state's set-aside program for small, minority-owned, and women-owned firms.

Office of Small Business Assistance
CN 823   (609) 984-4442
Trenton, NJ 08625-0823
    Advises and encourages small, women-owned, and minority-owned businesses in matters related to establishing and operating a small business in the state. General managerial, financial, and economic development assistance are offered to companies with 500 or fewer employees.

**U.S. Small Business Administration
Camden District Office**
2600 Mt. Ephrain   (609) 757-5183
Camden, NJ 08104

Newark District Office
60 Park Place, Fourth Floor   (201) 645-3580
Newark, NJ 07102

New Mexico

New Mexico Research and Development Institute
1220 S. St. Francis Dr., Pinon Bldg.,   (505) 827-5886
  Suite 358
Santa Fe, NM 87501
    Offers a seed capital program to finance the research-and-development stages of projects by
    start-up and small companies in New Mexico. Projects should accelerate the commercialization
    of advanced technology-based products, processes, or services.

New Mexico Small Business Development Center
Santa Fe Community College, P.O. Box   (505) 471-8200
  4187
Sante Fe, NM 87502-4187   Randy Grissom, Acting Coordinator

New Mexico State Purchasing Division
1100 St. Francis Dr., Joseph M. Montoya   (505) 827-0472
  Bldg., Room 2016
Santa Fe, NM 87503
    Maintains a special minority and small business development procurement program. Toll-free/
    Additional Phone Number: (505)827-0425 (Minority and Small Business Development Program).

U.S. Small Business Administration
Albuquerque District Office
5000 Marble Ave., N.E., Patio Plaza   (505) 262-6171
  Bldg., Room 320
Albuquerque, NM 87100

New York

New York Small Business Development Center
State University of New York (SUNY)   (518) 442-5398
State University Plaza, Room S523
Albany, NY 12246   James L. King, Director
    This small business development center services both upstate and downstate New York and is
    considered two separate development centers. **Telecommunication Access:** FAX (518) 465-4992.

New York Small Business Innovation Research Promotion Program
99 Washington Ave., Suite 1730   (518) 474-4349
Albany, NY 12210   Tab Wilkins, Director
    Encourages technology companies to participate in the SBIR program.

New York State Department of Commerce
Division for Small Business
230 Park Ave.   (212) 309-0400
New York, NY 10169
    Responsible for developing programs, providing services, and undertaking other initiatives that
    are responsive to the special needs of the state's small businesses. Offers small businesses an
    array of programs and services aimed at helping them prosper and grow. Maintains a Small
    Business Advocacy Program, Procurement Assistance Unit, a Small Business Advisory Board,
    and an Interagency Small Business Task Force. Also sponsors a Business Services
    Ombudsman Program to assist businesses in resolving red-tape difficulties that they may
    encounter in their interactions with all levels of government. Also offers one-on-one counseling
    services to both start-up and existing small businesses. **Telecommunication Access:** Alternate
    telephone (212) 309-0460.

New York State Science and Technology Foundation
Centers for Advanced Technology Program
99 Washington Ave., Suite 1730   (518) 474-4347
Albany, NY 12210   Vernon Ozarow, Director
    Provides a network of technological centers within the state's research institutes. Designed to
    meet the needs for increased investment in applied research and development, professional and
    technical expertise, and links with business and university sectors.

Corporation for Innovation Development
99 Washington Ave., Room 924   (518) 474-4349
Albany, NY 12210   Mr. H. Graham Jones, Executive Director
    Administers financing programs for businesses.

Industrial Innovation Extension Service
99 Washington Ave., Suite 1730   (518) 474-4349
Albany, NY 12210   Tab Wilkins, Director
    Program objectives include dissemination and educational service based on current
    research and development strengths in the area of manufacturing technology, assisting

business retention and expansion, and improving competitiveness and market share of New York state industries through increased knowledge of new technologies and other innovations.

Regional Technology Development Corporation
99 Washington Ave., Suite 1730   (518) 474-4349
Albany, NY 12210   Mark Tebbano, Director
    Assists emerging science and technology-oriented businesses, disseminate information, and assist in the development of small business incubator facilities.

U.S. Small Business Administration
Albany District Office
445 Broadway, Room 242   (518) 472-6300
Albany, NY 12207

    Buffalo District Office
    111 W. Huron St., Room 1311   (716) 846-4301
    Buffalo, NY 14202

    Elmira District Office
    333 E. Water St., Fourth Floor   (607) 734-6610
    Elmira, NY 14901

    Melville District Office
    35 Pinelawn Rd., Room 102E   (516) 454-0764
    Melville, NY 11747

    New York City District Office
    26 Federal Plaza, Room 3100   (212) 264-1318
    New York, NY 10278

    Rochester District Office
    100 State St., Room 601   (716) 263-6700
    Rochester, NY 14614

    Syracuse District Office
    100 S. Clinton St., Room 1071   (315) 423-5371
    Syracuse, NY 13260

U.S. Small Business Administration (Region Two)
26 Federal Plaza, Room 29-118   (212) 264-7772
New York, NY 10278

North Carolina

North Carolina Biotechnology Center
P.O. Box 13547   (919) 541-9366
79 Alexander Dr., 5401 Bldg.   W. Steven Burke, Director of Public
Research Triangle Park, NC 27709-3547   Affiars
    Acts as a catalyst to stimulate further development of the state's biotechnology research community. Facilitates collaboration between universities and industry. Advises the governor and university presidents of developments of long-term plans regarding biotechnology. **Contact:** Barry D. Teater, Communications Specialist.

North Carolina Small Business Development Center
University of North Carolinahapel Hill   (919) 733-4643
4509 Creedmoor Rd., #201
Raleigh, NC 27612   Scott R. Daugherty, Executive Director
    Telecommunication Access: FAX (919) 733-0659.

North Carolina State Department of Commerce
Business Information Referral Center
430 N. Salisbury St., Dobbs Bldg., Room   (919) 733-9013
  2238
Raleigh, NC 27611
    Uses a computer-searchable data base of agencies that regulate, license, certify, issue permits, or provide assistance to North Carolina businesses. Offers information on small business assistance; training programs; market, labor, and facilities data; and technical problems.

    Small Business Development Division
    430 N. Salisbury St., Dobbs Bldg.,   (919) 733-6254
      Room 2019
    Raleigh, NC 27611
      Offers information and assistance to those planning to start a business or to established business professionals. Assists with financing, taxes, regulations, marketing, education, and training. Sponsors workshops and conferences, including the Industrial Buyer/Supplier Exchange. Maintains a business clearinghouse to aid businesses and investors.

North Carolina Technological Development Authority
Incubator Facilities Program
430 N. Salisbury St., Dobbs Bldg., Room   (919) 733-7022
  4216
Raleigh, NC 27611   Julie Tenney, Executive Director
> Provides one-time grants of not more than $200,000, which must be matched in cash or real
> estate, to nonprofit organizations for the establishment of facilities that provide low-rent space,
> shared support services, and basic equipment to resident small businesses. Offers the small
> business community technical, management, and entrepreneurial advice. Includes facilities in
> Haywood County, Ahoskie, Goldsboro, and McDowell County.

Innovation Research Fund
430 N. Salisbury St., Dobbs Bldg.,   (919) 733-7022
  Room 4216
Raleigh, NC 27611
> Grants royalty financing of not more than $50,000 to North Carolina small businesses for
> research leading to the improvement or to the development of new products, processes, or
> services. Enables small businesses to secure technical and management assistance.
> Conducts research leading to the growth or start-up of a firm, thereby creating new jobs.

U.S. Small Business Administration
Charlotte District Office
222 S. Church St., Room 300   (704) 371-6563
Charlotte, NC 28202

North Dakota

North Dakota Small Business Development Center
Liberty Memorial Bldg., Capitol Grounds   (701) 224-2810
Economic Development Committee
Bismark, ND 58505   Carole Bordenkircher, Director
  Telecommunication Access: FAX (701) 223-3081.

North Dakota State Economic Development Commission
Entrepreneurial Assistance
Liberty Memorial Bldg.   (701) 224-2810
Bismarck, ND 58505
> Identifies agencies and individuals that assist inventors and entrepreneurs with technical, legal,
> financial, and marketing issues that must be resolved before launching a new venture. Provides
> training for inventors and entrepreneurs. Helps North Dakota companies develop international
> markets for their products. Maintains a marketing committee for home-based manufactured
> goods. Undertakes special projects.

Federal Procurement   (701) 224-2810
> Assists in-state businesses in obtaining contracts with the federal government for new
> markets for goods and services.

Financial Assistance   (701) 224-2810
> Assists locally referred businesses looking to expand but having difficulty assembling
> financial packages. Works to improve the financial referral service for North Dakota
> businesses and industries.

U.S. Small Business Administration
Fargo District Office
657 Second Ave., N., Room 218   (701) 237-5771
Fargo, ND 58102

Ohio

Ohio Small Business Development Center
Ohio State Department of Development   (614) 466-5111
30 E. Broad Street, 23rd Floor
Columbus, OH 43266-0101   Jack Brown, Director

Ohio State Department of Development
Economic Development Financing Division
P.O. Box 1001   (614) 466-5420
Columbus, OH 43266-0101
> Provides direct loans and loan guarantees to businesses for new fixed-asset financing for land,
> buildings, and equipment. Also issues industrial revenue bonds to provide businesses with 100
> percent of the financing for eligible fixed assets. Interest rates are below current market rates.
> Terms are up to 30 years, often with an option to buy the project financed at maturity.

Minority Development Financing Commission   (614) 462-7709

Helps finance minority business expansion through direct loans and through construction bonds. Direct loans are granted to finance up to 40 percent (or $200,000, whichever is less) of an eligible project for an eligible company, including the purchase and/or improvement of fixed assets such as land, buildings, equipment, and machinery. Issues up to $1 million in construction bonds to a minority business.

Small and Developing Business Division   (614) 466-4945

Offers a variety of programs to small business professionals, including a One-Stop Business Permit Center and a Women's Business Resource Program. The Department of Development also maintains Small Business Enterprise Centers that offer advice and assistance to small businesses. The department's other services include business relocation, site selection, and international trade programs. Publications: International Business Opportunities (monthly)—a newsletter. Toll-free/Additional Phone Number: 800-282-1085 (business and international trade development); (614)466-5700 (Small Business Enterprise Centers).

U.S. Small Business Administration
Cincinnati District Office
550 Main St., Room 5028   (513) 684-2814
Cincinnati, OH 45202

Cleveland District Office
1240 E. Ninth St., Room 317   (216) 522-4182
Cleveland, OH 44199

Columbus District Office
85 Marconi Blvd., Room 512   (614) 469-6860
Columbus, OH 43215

The Withrow Plan of Linked Deposits
c/o Mary Ellen Withrow, Treasurer of the   (800) 228-1102
  State of Ohio
30 E. Broad St., State Office Tower, Ninth
  Floor
Columbus, OH 43215

Allows the treasurer to channel a portion of the state's investment portfolio into reduced-rate investments. These reduced-rate investments are then linked to reduced-rate loans. To qualify, small businesses must document that the proceeds of the loans will help to create or retain jobs.

Oklahoma

Oklahoma Industrial Finance Authority
c/o Oklahoma State Department of   (405) 843-9770
  Commerce
6601 Broadway Extension
Oklahoma City, OK 73116-8214

Has a borrowing capacity of $90 million. Empowered to loan up to 66 2/3 percent of the cost of land and buildings on a secured first mortgage and up to 33 1/3 percent on a second mortgage. Maximum loan amount is $1 million for the state's share of a single project.

Oklahoma Small Business Development Center
Southeastern Oklahoma State University   (405) 924-0277
P.O. Box 4194, Station A
517 W. University
Durant, OK 74701   Dr. Grady Pennington, Director
  Telecommunication Access: Toll-free telephone 800-522-6154.

Oklahoma State Department of Commerce
Business Development Division
6601 Broadway Extension   (405) 843-9770
Oklahoma City, OK 73116-8214

Identifies, counsels, and assists locating new and expanding industries. Helps with site evaluation. Coordinates programs with waterway authorities. Maintains a film office geared toward promoting Oklahoma as a location for film and video productions.

Procurement Section   (405) 843-9770

Maintains a computerized inventory system that allows for the tracking of property and supplies. The department also maintains 21 state bid assistance centers established to assist firms with federal procurement.

Special Services Division   (405) 843-9770

Provides financial package analysis to Oklahoma firms. Assists local communities with state and federal loan or grant contract work. Administers the Industrial Development Bond Fund allocation program. Assists with finance program development, particularly when bond issues are used as part of a package.

U.S. Small Business Administration
Oklahoma City District Office
200 N.W. Fifth St., Suite 670   (405) 231-4301
Oklahoma City, OK 73102

Oregon

Oregon Resource and Technology Development Corporation
Oregon State Economic Development   (503) 373-1200
   Department
595 Cottage St., N.E.
Salem, OR 97310
   Funded by lottery revenues. An independent public corporation established to provide financing
   and other services to existing and start-up businesses, particularly small businesses. **Contact:**
   John Beaulieu, President, One Lincoln Center, Suite 430, 10300 Southwest Greenburg Road,
   Portland, OR 97233 (503) 246-4844.

Oregon Small Business Development Center
99 W. 10th Ave., Suite 216   (503) 726-2250
Eugene, OR 97401   Edward Cutler, Director

Oregon State Department of Commerce
Corporation Division
158 12th St., N.E.   (503) 378-2290
Salem, OR 97310
   Provides forms and registrations required by the state for starting a business. Maintains a
   business registry information center and hotline. Also handles state business and occupation
   licenses and permits. Toll-free/Additional Phone Number: (503)378-4166 (business registry
   information center).

Oregon State Department of Economic Development
Regulation Assistance Service
595 Cottage St., N.E.   (503) 373-1234
Salem, OR 97310
   Refers business owners to appropriate state agencies for licenses, permits, and registrations.
   Maintains a one-stop permit information center. Toll-free/Additional Phone Number: (503)373-
   1999 (one-stop permit information center).

Oregon State Economic Development Department
Business Development Division
595 Cottage St., N.E.   (503) 373-1200
Salem, OR 97310
   Offers services to businesses in such areas as business recruitment, industrial property,
   licenses and permits, business information, financing, and international trade. Publications: 1)
   Directory of Oregon Manufacturers; 2) Exporters Handbook; 3) The Oregon Advantage: Services
   for Oregon Business Firms; 4) Oregon International Trade Directory; 5) Starting a Business in
   Oregon.

   Financial Services Division   (503) 373-1200
      Administers the Industrial Development Revenue Bond Program to promote and develop
      industrial activities, tourism, and international trade in Oregon. Also maintains the
      Umbrella Revenue Bond Program, which provides the benefits of industrial development
      revenue bonds to growing small and medium-sized Oregon businesses. Commercial banks
      provide 10 to 35 percent of the financing for each individual project. Small projects
      ($100,000)—and large projects ($1 million)—may be funded. **Telecommunication Access:**
      Alternate telephone (503) 373-1215.

   Oregon Business Development Fund   (503) 373-1240
      Provides loans for both long-term, fixed-assest financing and working capital. Loans may be
      made only where there is a demonstrated creation of new jobs or retention of existing jobs.
      Program emphasizes businesses with fewer than 50 employees or rural and lagging areas.

U.S. Small Business Administration
Portland District Office
1220 S.W. Third Ave., Room 676   (503) 423-5221
Portland, OR 97204-2882

Pennsylvania

New-Penn-Del Regional Minority Purchasing Council, Inc.
5070 Parkside Ave., Suite 1400   (215) 578-0964
Philadelphia, PA 19131
    Links major corporate buyers with minority business owners who can provide the needed products or services.

Pennsylvania Small Business Development Center
University of Pennsylvania; The Wharton   (215) 898-1219
    School
3201 Steinberg-Dietrich Hall/CC
Philadelphia, PA 19104   Gregory Higgins, Jr., Director
    Telecommunication Access: FAX (215) 898-2400.

Pennsylvania State Department of Commerce
Ben Franklin Partnership
c/o Small Business Action Center, P.O.   (717) 783-5700
    Box 8100
Harrisburg, PA 17105
    Includes the Small Business Incubator Loan Program, which funds the development of incubator facilities that provide new manufacturing or product development companies with the space and business development services they can use to start-up and survive the early years of business growth. Four privately-managed Seed "Venture" Capital Funds, established through a BFP challenge grant program, provide equity financing to new businesses during their earliest stages of growth. Also awards Research "Seed" Grants of up to $35,000 to small businesses seeking to develop or introduce advanced technology into the marketplace.

Business Infrastructure Development Program
467 Forum Bldg.   (717) 787-7120
Harrisburg, PA 17120
    Provides communities with funds for infrastructure improvements that will encourage private firms to locate or expand in Pennsylvania. Locally sponsored manufacturing, industrial, agricultural, and research-and-development enterprises may apply. Eligible projects include energy facilities, fire and safety facilities, sewer systems, waste disposal facilities, water supply systems, drainage systems, and transportation facilities. Job creation requirements vary depending on the amount of the grant or loan.

Office of Minority Business Enterprise
491 Forum Bldg.   (717) 783-1301
Harrisburg, PA 17120
    The primary liaison for minority-owned companies interested in contracting with state government agencies. Helps strengthen Pennsylvania's minority business community by providing state contracting information to minority companies. Serves as a liaison between minority-owned firms and public sector purchasers.

Pennsylvania Capital Loan Fund
405 Forum Bldg.   (717) 783-1768
Harrisburg, PA 17120
    Receives funds from both federal and state appropriations. Loans are targeted to small manufacturing, industrial, and export service businesses with the goal of creating at least one new job for each $15,000 loaned; one job for each $10,000 to apparel manufacturers, which includes shoe and garment manufacturers. Loans are meant to encourage private sector investment in small businesses by subordinating its collateral position. Funds may be used for land and building acquisition and renovation, machinery and equipment, and working capital. Funds may not be used to replace private sector money readily available at reasonable rates; refinance existing debt or pay for previously incurred obligations; or speculate.

Pennsylvania Energy Development Authority
462 Forum Bldg.   (717) 787-6554
Harrisburg, PA 17120
    Promotes development and efficient use of Pennsylvania's energy resources. Provides financial assistance for qualifying energy projects. Funds are appropriated by the General Assembly and deposited in the authority's Energy Development Fund. The amount and type of funds available are tied to the stage of development and include grants for research and feasibility studies; repayable grants (venture capital) for pilot and demonstration projects; and loans for commercial-stage projects. In addition, revenue bond financing may be possible for commercially viable, revenue producing energy projects, depending upon conformance with the Federal Internal Revenue Codes and Internal Revenue Service regulations on tax-exempt bonds and state availability.

Pennsylvania Industrial Development Authority
405 Forum Bldg.   (717) 787-6245
Harrisburg, PA 17120
    Loans are available to businesses engaged in manufacturing or classified as industrial enterprises. An industrial enterprise includes the areas of research and development, food processing and related agribusiness, computer or clerical operation centers, buildings

being used for national and regional headquarters, and certain warehouse and terminal facilities. Excluded are commercial, mercantile, or retail enterprises. Funds may be used for land and building acquisition, new construction, expansion, or renovation in conjunction with an acquistion. Loan applications are first approved by a local industrial development corporation and then reviewed by the authority's board.

Revenue Bond and Mortgage Program
405 Forum Bldg.   (717) 783-1108
Harrisburg, PA 17120

    Financing for projects approved through this program is secured from private sector sources such as banks, insurance companies, individuals, and through public bond issues. These funds are borrowed through a local industrial development authority and can be used for manufacturing projects, and specialized projects such as solid waste disposal. Businesses or investors can use the funds to acquire land, buildings, machinery, and equipment.

Small Business Action Center
P.O. Box 8100   (717) 783-5700
Harrisburg, PA 17105

    Answers general business questions. Publications: 1) Resource Directory for Small Business; 2) Small Business Planning; 3) Starting a Small Business in Pennsylvania.

Pennsylvania State Department of Transportation
Disadvantaged and Women Business Enterprises
Transportation and Safety Bldg., Room   (800) 468-4201
  109
Harrisburg, PA 17120

    Operates a special procurement program for firms owned by women, minorities, or disadvantaged persons. Toll-free/Additional Phone Number: 800-845-3375 (support services).

U.S. Small Business Administration
Harrisburg District Office
100 Chestnut St., Suite 309   (717) 782-3840
Harrisburg, PA 17101

Philadelphia District Office
231 St. Asaphs Rd., Suite 400 East   (215) 596-5801
Philadelphia, PA 19004

Pittsburgh District Office
960 Penn Ave., Fifth Floor   (412) 644-4306
Pittsburgh, PA 15222

Wilkes-Barre District Office
20 N. Pennsylvania Ave., Room 2327   (717) 826-6497
Wilkes-Barre, PA 18701

U.S. Small Business Administration (Region Three)
231 St. Asaphs Rd., Suite 640 West   (215) 596-5962
Philadelphia, PA 19004

Puerto Rico

Puerto Rico Department of Commerce
P.O. Box S-4275   (809) 724-0542
San Juan, PR 00905

Puerto Rico Economic Development Administration
P.O. Box 2350   (809) 765-1303
San Juan, PR 00936

Puerto Rico Small Business Development Center
University of Puerto Rico   (809) 834-3790
College Station, Bldg. B
P.O. Box 5253
Mayaguez, PR 00709   Jose M. Romaguera, Director

U.S. Small Business Administration
Hato Rey District Office
Carlos Chardon Ave., Federico Degatan   (809) 753-4003
  Federal Bldg., Room 691
Hato Rey, PR 00918

Rhode Island

Ocean State Business Development Authority
c/o Rhode Island State Department of   (401) 277-2601
   Economic Development
Seven Jackson Walkway
Providence, RI 02903-3189
> An independent, private, nonprofit corporation established and supported by businesses and government to encourage economic growth in Rhode Island. The authority can provide 90 percent financing on loan requests from $200,000 up to $1,000,000 at interest rates keyed to long term treasury rates established by the U.S. Department of the Treasury. Loan proceeds may be used to purchase land, renovate and construct buildings, and acquire machinery and equipment. Financing is not available for nonprofit organizations, print media such as newspapers and magazines, lending institutions, gambling facilities, recreation facilities that are not open to the public, and real estate development and speculation.

Rhode Island Partnership for Science and Technology
Seven Jackson Walkway   (401) 277-2601
Providence, RI 02903-3189   Bruce Lang, Executive Director
> A nonprofit corporation. Provides state matching grants for applied research that creates a linkage between a private business venture and a Rhode Island nonprofit research facility.

Rhode Island Small Business Development Center
Bryant College   (401) 232-6111
450 Douglas Pike, Route 7
Smithfield, RI 02917   Doug Jobling, Director

Rhode Island State Department of Economic Development
Rhode Island Financial Assistance Programs
Seven Jackson Walkway   (401) 277-2601
Providence, RI 02903-3189
> Provides new or expanding businesses with financing for applied research, real estate, machinery and equipment, and working capital through a variety of programs. The Taxable Industrial Development Revenue Bond Program provides competitive interest rates as well as tax advantages. Tax Exempt Industrial Development Revenue Bonds, available for manufacturing projects up to $10,000,000, offer lower interest rates and other advantages. The Mortgage Insurance Program reduces the capital necessary for new manufacturing facilities, renovation of manufacturing facilities, and the purchase of new machinery and equipment in financing projects up to $5,00,000. Insured Industrial Development Bond Financing reduces the interest rate for smaller firms that otherwise would not be eligible for Industrial Revenue Bond financing. Revolving Loan Fund Program provides eligible Small Business Fixed Asset Loans from $25,000 to a maximum of $150,000 and Working Capital Loans to a maximum of $30,000. Rhode Island Business Investment Fund provides small businesses with loans from $30,000 to a maximum of $500,000. Applied Research Grants—Rhode Island Partnership for Science and Technology provides matching state grants for joint ventures between a Rhode Island nonprofit research facility and a business venture. Administers U.S. Small Business Administration 504 Loans.

U.S. Small Business Administration
Providence District Office
380 Westminster Mall, Fifth Floor   (401) 528-4586
Providence, RI 02903

South Carolina

South Carolina Small Business Development Center
University of South Carolina, College of   (803) 777-4907
   Business Administration
Columbia, SC 29208   W.F. Littlejohn, Director
> **Telecommunication Access:** FAX (803) 252-2435.

South Carolina State Development Board
P.O. Box 927   (803) 734-1400
Columbia, SC 29202   Ron Young, Manager, Entrepreneur Dev.
> Offers financing programs and business assistance services. **Telecommunication Access:** Alternate telephone (803) 737-0400.

Division of Business Assistance and Development
1301 Gervais St.   (803) 734-1400
P.O. Box 927
Columbia, SC 29202

State of South Carolina
Office of Small and Minority Business Assistance
1205 Pendleton St., Edgar A. Brown   (803) 734-0562
  Bldg., Room 305
Columbia, SC 29201
  Offers procurement assistance, management training, and technical assistance to small and
  minority firms.

U.S. Small Business Administration
Columbia District Office
1835 Assembly St., Third Floor   (803) 765-5339
Columbia, SC 29202

South Dakota

South Dakota Small Business Development Center
University of South Dakota   (605) 677-5272
School of Business
414 E. Clark
Vermillion, SD 57069   Donald Greenfield, Director
  Telecommunication Access: FAX (605) 677-5073.

South Dakota State Governor's Office of Economic Development
711 Wells, Capitol Lake Plaza   (605) 773-5032
Pierre, SD 57501   Roland Dolly, Enterprise Initiation
  Offers a variety of financial resources to small businesses.

U.S. Small Business Administration
Sioux Falls District Office
101 S. Main Ave., Suite 101   (605) 336-2980
Sioux Falls, SD 57102-0577

Tennessee

Tennessee Small Business Development Center
Memphis State University   (901) 678-2500
330 DeLoach, FEC Room 220W
Memphis, TN 38152   Dr. Leonard Rosser, Director

Tennessee State Department of Economic and Community Development
Office of Small Business
320 Sixth Ave., N., Rachel Jackson Bldg.,   (615) 741-3282
  Seventh Floor
Nashville, TN 37219
  Telecommunication Access: Toll-free telephone 800-872-7201.

Tennessee Technology Foundation
P.O. Box 32184
Knoxville, TN 37933   Dr. David A. Patterson, Director
  Promotes high-technology economic development in the state. Provides site selection
  assistance; research access to Oak Ridge National Laboratory, Tennessee Valley Authority, and
  University of Tennessee; and supports new product commercialization and support of
  entrepreneurs, including access to venture capital sources.

U.S. Small Business Administration
Nashville District Office
404 James Robertson Pkwy., Suite 1012   (615) 736-5881
Nashville, TN 37219

Texas

Advanced Robotics Research Institute
2501 Gravel Dr.   (817) 284-5555
Fort Worth, TX 76118-6999
  Provides high-technology information on robotics and automation. Institute will be surrounded
  by related technological research and manufacturing businesses.

Texas Small Business Development Center (Houston)
University of Houston   (713) 223-1141
401 Louisiand, 8th Floor
Houston, TX 77002   Jon P. Goodman, Director
    Telecommunication Access: FAX (713) 223-5507.

Texas Small Business Development Center (Lubbock)
Texas Tech University   (806) 744-5343
2005 Broadway
Lubbock, TX 79401   J.E. (Ted) Cadou, Director
    Telecommunication Access: FAX (806) 744-0792.

Texas Small Business Development Center (San Antonio)
University of Texas at San Antonio,   (512) 224-0791
  Center for Economic Development
Hemisphere Plaza Bldg., #448
San Antonio, TX 78285   Henry Travieso, Director

U.S. Small Business Administration
Austin District Office
300 E. Eighth St., Room 520   (512) 482-5288
Austin, TX 78701

    Corpus Christi District Office
    400 Mann St., Suite 403   (512) 888-3331
    Corpus Christi, TX 78401

    Dallas District Office
    1100 Commerce St., Room 3C36   (214) 767-0608
    Dallas, TX 75242

    El Paso District Office
    10737 Gateway West, Suite 320   (915) 541-7676
    El Paso, TX 79935

    Fort Worth District Office
    819 Taylor St.   (817) 334-3777
    Fort Worth, TX 76102

    Harlingen District Office
    222 E. Van Buren St., Room 500   (512) 427-8533
    Harlingen, TX 78550

    Houston District Office
    2525 Murworth, Room 112   (713) 660-4407
    Houston, TX 77054

    Lubbock District Office
    1611 Tenth St., Suite 200   (806) 743-7462
    Lubbock, TX 79401

    Marshall District Office
    505 E. Travis, Room 103   (214) 935-5257
    Marshall, TX 75670

    San Antonio District Office
    727 E. Durango St., Room A-513   (512) 229-6105
    San Antonio, TX 78206

U.S. Small Business Administration (Region Six)
8625 King George Dr., Bldg. C   (214) 767-7643
Dallas, TX 75235-3391

Utah

U.S. Small Business Administration
Salt Lake City District Office
125 S. State St., Room 2237   (801) 524-5804
Salt Lake City, UT 84138-1195

Utah Innovation Center
419 Wakara Way, Research Park, Suite   (801) 584-2500
  206
Salt Lake City, UT 84108   Dr. Gerald L. Davey, Director
    Encourages technical innovation and entrepreneurship. Assists creation of companies, providing legal, administrative, accounting, technical and clerical support.

Utah Small Business Development Center
University of Utah   (801) 581-7905
660 South 200 St., Suite 418
Salt Lake City, UT 84111   Kumen B. Davis, Director

Utah State Department of Community and Economic Development
Division of Economic and Industrial Development
6150 State Office Bldg.   (801) 533-5325
Salt Lake City, UT 84114

Utah Technology Finance Corporation
419 Wakara Way   (801) 583-8832
Salt Lake City, UT 84108

Vermont

U.S. Small Business Administration
Montpelier District Office
87 State St., Room 205   (802) 828-4474
Montpelier, VT 05602

Vermont Industrial Development Authority
58 E. State St.   (802) 223-7226
Montpelier, VT 05602
      Promotes economic growth and increases employment through a variety of financing programs.
      Makes low-interest loans available to businesses for the purchase or construction of land,
      buildings, machinery, and equipment for use in an industrial facility. Issues tax-exempt, low-
      interest bonds to provide funds for the acquisition of land, buildings, and/or machinery and
      equipment for use in a manufacturing facility. Provides loans to nonprofit local development
      corporations for the purchase of land for industrial parks, industrial park planning and
      development, and the construction or improvement of speculative buildings or small business
      incubator facilities. Aids businesses by guaranteeing loans of commercial lending institutions.

Vermont Small Business Development Center
University of Vermont, Extension Service   (802) 656-4479
Small Business Development, Morrill
   Hall
Burlington, VT 05405   Norris A. Elliott, Director

Vermont State Agency of Development and Community Affairs
Department of Economic Development
109 State St., Pavilion Office Bldg.,   (802) 828-3221
   Fourth Floor
Montpelier, VT 05602   Graeme Freeman, Development Spcialist
      Provides advice, information, and referrals to help prospective or operating businesses develop
      business plans and create workable financing packages. Creates a network of entrepreneurs
      and business service providers through entrepreneurship forums and incubator facilities.
      Mobilizes a network of public agencies to help troubled businesses and their employees. Assists
      businesses in obtaining state permits and other approvals. Maintains and shares information
      about new changes in technology affecting business. Disseminates information about Vermont
      companies and their products and services. Helps Vermont companies sell their products by
      researching markets, developing merchandising ideas, and informing them about export
      practices and policies. Helps locate sites and financing for companies who want to find facilities
      in Vermont. Assists in training workers in the specific skills needed by new and expanding
      companies. Organizes international trade missions and trade show presentations.

Vermont State Department of Economic Development
Entrepreneurship Program
State St.   (802) 828-3221
Montpelier, VT 05602   Curt Carter, Director
      Helps develop business incubators.

Virgin Islands

U.S. Small Business Administration
St. Thomas District Office
Veterans Dr., Room 210   (809) 774-8530
St. Thomas, VI 00801

Virgin Islands Small Business Development Agency
P.O. Box 2058   (809) 774-8784
St. Thomas, VI 00801

Virgin Islands Small Business Development Center
College of the Virgin Islands   (809) 776-3206
Grand Hotel Bldg., Annex B
P.O. Box 1087
St. Thomas, VI 00801   Solomon S. Kabuka, Jr., Director

Virginia

U.S. Small Business Administration
Richmond District Office
400 N. Eighth St., Room 3015   (804) 771-2741
Richmond, VA 23240

Virginia State Department of Economic Development
Office of Small Business and Financial Services
1000 Washington   (804) 786-3791
Richmond, VA 23219
    Provides expertise in financial management and technical assistance.

Washington

U.S. Small Business Administration
Seattle District Office
915 Second Ave., Room 1792   (206) 442-5534
Seattle, WA 98174

    Spokane District Office
    W920 Riverside Ave., Room 651   (509) 456-3781
    Spokane, WA 99210

U.S. Small Business Administration (Region Ten)
2615 Fourth Ave., Room 440   (206) 442-5676
Seattle, WA 98121

Washington Small Business Development Center
Washington State University, 441 Todd   (509) 335-1576
  Hall
Pullman, WA 99164-4740   Lyle Anderson, Director
    Telecommunication Access: FAX (509) 335-3421.

Washington State Department of Trade and Economic Development
101 General Administration Bldg.   (206) 753-3065
Olympia, WA 98504
    Offers a variety of financing programs to businesses.

West Virginia

U.S. Small Business Administration
Charleston District Office
550 Eagan St., Room 309   (304) 347-5220
Charleston, WV 25301

    Clarksburg District Office
    168 W. Main St., Fifth Floor   (304) 623-5631
    Clarksburg, WV 26301

West Virginia Industrial Trade Jobs Development Corporation
State Capitol, M-146   (304) 348-0400
Charleston, WV 25305
    Supplements other financial incentive programs to help create jobs. The State Board of Investment makes loans available to the corporation for up to $10 million per project at negotiable rates and terms.

West Virginia Small Business Development Center

Governor's Office of Community and   (304) 348-2960
    Industrial Development
115 Virginia St., E.
Charleston, WV 25305   Eloise Jack, Director

West Virginia State Economic Development Authority

State Capitol, Building 6, Room 525   (304) 348-3650
Charleston, WV 25305

Provides low-interest loans from a revolving fund for land acquisition, building construction, and equipment purchases; loans are geared toward manufacturing firms with an emphasis on new job creation. Issues low-interest industrial revenue development bonds that are exempt from income tax at the federal and state levels; bonds are available to a wide variety of projects with the potential of reducing fees and interest rates.

West Virginia State Office of Community and Industrial Development

State Capitol, M-146   (304) 348-0400
Charleston, WV 25305

Offers low-interest loans and other financial arrangements to businesses. Administers such programs as the West Virginia Certified Development Corporation, which provides long-term fixed-rate loans to small and medium-sized firms, and the Treasurer's Economic Development Deposit Incentive, which offers low-cost financing to businesses that operate exclusively in West Virginia and that employ less than 200 or whose gross receipts total less than $4 million (financing must create or preserve jobs). Maintains contacts with several area and regional banks; helps businesses seeking assistance through federal financing programs.

West Virginia State Small Business Office

State Capitol Complex   (304) 348-2960
Charleston, WV 25305

Provides guidance to existing and prospective businesses. Offers technical assistance, loan packaging, and procurement services.

Wisconsin

Innovation Network Foundation

P.O. Box 71   (608) 256-8348
Madison, WI 53701   Diane Curtz, Director

Nonprofit group seeking to join marketing and sales ideas to those willing to finance them. Also oversees the development of an entrepreneurial resource library newsletter covering training programs, and telephone contacts generally not available to the entrepreneur.

Northwest Wisconsin Business Development Corporation
Northwest Wisconsin Business Development Fund

302 Walnut St.   (715) 635-2197
Spooner, WI 54801

Promotes private sector investment in long-lived assets. Creates jobs by addressing capital gaps in the market for long-term debt. Serves businesses primarily in timber and wood, manufacturing, and tourism industries in northwestern Wisconsin.

State of Wisconsin Investment Board
Private Placements

121 E. Wilson St.   (608) 266-2381
Madison, WI 53702

Meets fiduciary responsibilities to the trusts under the care of the board. Minimum loan is $3 million for Wisconsin-based firms and $10 million for firms based elsewhere.

U.S. Small Business Administration
Eau Claire District Office

500 S. Barstow St., Room 37   (715) 834-9012
Eau Claire, WI 54701

    Madison District Office
    212 E. Washington Ave., Room 213   (608) 264-5268
    Madison, WI 53703

    Milwaukee District Office
    310 W. Wisconsin Ave., Room 400   (414) 291-3941
    Milwaukee, WI 53203

The Wisconsin SBDC has been active. In 1989, it reports that 2,200 people received assistance and 11,000 attended the 700 seminars the organization sponsored.

According to William Pinkovitz, director of the Wisconsin SBDC, a small group of computer specialists who had created a software package that permitted IBM and DEC computers to communicate appealed to the Wisconsin unit for help in organizing a firm. With the assistance of the SBDC and a Small Business Institute student team directed by the SBDC coordinator, an organizational structure and internal control systems were developed. Since then the firm has grown to several million dollars in sales and continues as a leader in the field of specialty software development.

Wisconsin Housing and Economic Development Authority
Linked-Deposit Loan Program
One S. Pinckney St., #500   (608) 266-0976
P.O. Box 1728
Madison, WI 53701
> Offers low-interest loans to improve access to capital for small businesses owned or controlled by women or minorities. Toll-free/Additional Phone Number: (608)266-7884.

Wisconsin Small Business Development Center
University of Wisconsin   (608) 263-7793
432 N. Lake St., Room 423, Ext. Bldg.
Madison, WI 53706   Bill Pinkovitz, Director

Wisconsin State Department of Development
Financing Programs
123 W. Washington Ave.   (608) 266-1018
P.O. Box 7970
Madison, WI 53707
> Offers a variety of financial resources, including tax increment financing, industrial revunue bonds, employee ownership assistance loans, and a technology development fund. Administers the Wisconsin Development Fund-Economic Development Program.

> **Small Business Ombudsman**   (608) 266-0562
> Identifies the problems of individuals or groups of business people. Gathers facts concerning regulations and policies. Rosolves problems between businesses and state agencies or officials. Recommends changes in procedures, regulations, or laws that are determined to be unfair or discriminatory to the stability of small businesses in the state. Toll-free/Additional Phone Number: 800-HELPBUS (Business Hotline).

Wisconsin State Housing and Economic Development Authority
One S. Pinckney St.   (608) 266-0191
Box 1728
Madison, WI 53701
> Provides below-market, fixed-rate loans to manufacturers and first-time farmers for fixed assets. Toll-free/Additional Phone Number: (608)266-0976.

Wisconsin State Rural Development Loan Fund
Route 2, Box 8   (715) 986-4171
Turtle Lake, WI 54889
> Provides loans to create jobs in rural areas of northwestern Wisconsin.

Wyoming

U.S. Small Business Administration
Casper District Office
100 E. B St., Room 4001   (307) 261-5761
Casper, WY 82602-2839

Wyoming Community Development Authority
139 W. Second, Suite E   (307) 265-0603
Casper, WY 82602
> Offers financing for new and expanding businesses through direct loans, leases, guarantees to banks, and interest rate subsidies.

Wyoming Office of the State Treasurer
Small Business Assistance Act
Capitol Bldg.   (307) 777-7408
Cheyenne, WY 82002
> Provides businesses with fixed-asset financing and a five-year interest rate subsidy. Allows the state treasurer to purchase a guarnateed portion of U.S. Small Business Administration and Farmers Home Administration loans.

> **State Link Deposit Plan**   (307) 777-7408
> Provides businesses with a five-year, fixed-rate interest subsidy. State treasurer contracts for deposits with Wyoming financial institutions at a rate up to 3 percent below market rates.

Wyoming Small Business Development Center
Casper College, 130 N. Ash, Suite A   (307) 235-4825
Casper, WY 82601   M.C. "Mac" Bryant, Director

Wyoming State Economic Development and Stabilization Board

Herschler Bldg., Third Floor East   (307) 777-7286
Cheyenne, WY 82002

    Administers the Economic Development Loan Program, which provides Wyoming businesses with loans and loan guarantees at flexible rates. Board's Business/Financing staff administers the Wyoming Economic Development Block Grant Program. This program's funds help communities attract or expand local industry. Provides low-interest loans for low- and moderate-income job creation and retention.

Business Incubators

Arkansas

East Arkansas Business Incubator System, Inc
5501 Krueger Dr.　　　　　　　　　(501) 935-8365
Jonesboro, AR 72401

California

California Business Incubation Network
120 N. Robertson Blvd.　　　　　　(213) 855-8393
Los Angeles, CA 90048　　　　　Frans Verschoor, Contact

Lancaster Economic Development Corporation
104 E. Avenue K4, Suite A　　　　(805) 945-2741
Lancaster, CA 93537

McDowell Industrial Business Center
921 Transport Way　　　　　　　　(707) 762-6341
Petaluma, CA 94952

Southern California Innovation Center
225 Yale Ave., Suite H　　　　　　(714) 624-7161
Claremont, CA 91711

Victor Valley College
Small Business Incubator Project
18422 Bear Valley Rd.　　　　　　(619) 245-4271
Victorville, CA 92392-9699

Colorado

Control Data Business and Technology Center (Pueblo)
Business and Technology Center　　(303) 546-1133
301 N. Main
Pueblo, CO 81002

Connecticut

Bridgeport Innovation Center
955 Connecticut Ave.　　　　　　　(203) 336-8864
Bridgeport, CT 06607

Business incubators are organizations which provide below-market rates to encourage entrepreneurship and minimize obstacles to new business formation and growth, particularly for high-technology firms. In addition, these facilities offer shared support for clerical, reception, and computer services. This chapter presents more than 100 incubators, arranged in a state-by-state format.

A growing number of states are actively involved in facilitating business incubator development. Nine of them have associations which serve, among other duties, as clearinghouses for information on incubator facilities in their states. Listings for these clearinghouses, which are in California, Georgia, Illinois, Michigan, North Carolina, Oklahoma, Pennsylvania, Texas, and Wisconsin, are located under the appropriate state heading.

Science Park Development Corporation
Five Science Park   (203) 786-5000
New Haven, CT 06511

Waterbury Industrial Commons Project
1875 Thomaston Ave.   (203) 574-7704
Waterbury, CT 06704

Georgia

Georgia Industrial Developers Association
c/o Advanced Technology Development   (404) 894-3575
   Center
430 10th St., N.W., Suite N-116   C. Michael Cassidy, Sub-Committee
Atlanta, GA 30318   Chairman

Idaho

Business Center for Innovation and Development (Hayden Lake)
11100 Airport Dr.
Hayden Lake, ID 83835

Idaho Innovation Center
457 Broadway Ave.
Idaho Falls, ID 83402

Illinois

Bradley Industrial Incubator Program
c/o Area Jobs Development Association   (815) 933-2537
P.O. Box 845
Kankakee, IL 60901

Business Center for New Technology (Rockford)
1300 Rock St., P.O. Box 1200   (815) 968-6833
Rockford, IL 61101-1200

Control Data Business and Technology Center (Champaign)
Business and Technology Center   (217) 398-5759
701 Devonshire Dr.
Champaign, IL 61820

The Decatur Industrial Incubator
2121 U.S. Route 51, S.   (217) 423-2832
Decatur, IL 62521

Des Plaines River Valley Enterprise Zone Incubator Project
912 E. Washington St.   (815) 726-0028
Joliet, IL 60433

Electronic Decisions, Inc.
University Microelectronics Center   (217) 367-2600
1706 E. Washington St.
Urbana, IL 61801

Fulton-Carroll Center for Industry
Industrial Council of Northwest Chicago (312) 421-3941
2023 W. Carroll Ave.
Chicago, IL 60612

Galesburg Business and Technology Center
Monmouth Blvd.   (309) 343-1194
Galesburg, IL 61401

Illinois Small Business Incubator Association
Illinois State Department of Commerce   (312) 814-2481
  and Community Affairs
100 W. Randolph, Suite 3-400   Robert Allen, Manager, Incubator
Chicago, IL 60601   Program

Macomb Business and Technology Center
Western Illinois University Campus   (309) 298-2212
Seal Hall
P.O. Box 6070
Macomb, IL 61455

Maple City Business and Technical Center
620 S. Main St.   (309) 734-8544
Monmouth, IL 61462

Shetland Properties of Illinois
1801 S. Lumber   (312) 738-3121
Chicago, IL 60616

Sterling Small Business and Technology Center
1741 Industrial Dr.   (815) 625-5255
Sterling, IL 61081

Indiana

Control Data Business and Technology Center (South Bend)
Business and Technology Center   (219) 282-4340
300 N. Michigan St.
South Bend, IN 46601

Indiana Enterprise Center
2523 Merivale St.   (219) 426-5700
Fort Wayne, IN 46805

Iowa

University of Iowa Technology Innovation Center
University of Iowa   (319) 335-4063
109 TIC
Oakdale Campus
Iowa City, IA 52242

Louisiana

Northeast Louisiana Incubator Center
State Route 594, Swartz School Rd.   (318) 343-2262
Monroe, LA 71203

Maryland

Baltimore Medical Incubator
Baltimore Economic Development   (301) 837-9305
  Corporation
36 S. Charles St., Suite 24
Baltimore, MD 21201

Control Data Business and Technology Center (Baltimore)
Business and Technology Center   (301) 367-1600
2901 Druid Park Dr.
Baltimore, MD 21215

Typically, business incubators provide programs that assist in the development of business plans and marketing strategies, advise firms on personnel, accounting and legal matters, and identify sources of financing. Professionals may assist with the evaluation of product lines and manufacturing processes, advise in the use of state-of-the-art design and manufacturing tools, and identify special expertise at universities and other research centers.

Technology Advancement Program
University of Maryland   (301) 454-8827
Engineering Research Center, Bldg. 335
College Park, MD 20742

Massachusetts

China Trade Center
Chinese Economic Development Council   (617) 482-0111
31 Beach St., Second Floor
Boston, MA 02111

J.B. Blood Building
Lynn Office of Economic Development   (617) 581-9399
One Market St., Suite 4
Lynn, MA 01901

Schaeffer Business Center
130 Centre St.   (617) 777-4605
Danvers, MA 01923

Michigan

Business Service Center (Albion)
Albion Economic Development   (517) 629-3926
  Corporation
1104 Industrial Ave.
Albion, MI 49224

Center for Business Development
c/o Greater Niles Economic Development   (616) 683-1833
  Foundation, Inc.
P.O. Box 585
Niles, MI 49120

Chamber Innovation Center
912 N. Main   (313) 662-0550
Ann Arbor, MI 48104

Delta Market Street Incubator
470 Market St., S.W.
Grand Rapids, MI 49503
    Incubator is operated by Delta Properties, 1300 Four Mile Rd., N.W., Grand Rapids, MI 49503;
    (616)451-2561.

Delta Monroe Street Incubator
820 Monroe St., N.W.
Grand Rapids, MI 49503
    Incubator is operated by Delta Properties, 1300 Four Mile Rd., N.W., Grand Rapids, MI 49503;
    (616)451-2561.

Flint Industrial Village for Enterprise
2717 N. Saginaw St.   (313) 235-5555
Flint, MI 48505

Grand Rapids Terminal Ward
446 Granville Ave., S.W.
Grand Rapids, MI 49503
    Incubator is operated by Delta Properties, 1300 Four Mile Rd., N.W., Grand Rapids, MI 49503;
    (616)451-2561.

Manufacturing Resource and Productivity Center (Big Rapids)
Ferris State College   (616) 796-3100
Big Rapids, MI 49306

Metropolitan Center for High Technology
2727 Second Ave.   (313) 963-0616
Detroit, MI 48201

Michigan Business Incubator Association
c/o MRPC, Ferris State University   (616) 667-0080
10 E. Maple
Big Rapids, MI 40307   Tom Schumann, Contact

The Venture Center
Downriver Community Conference   (313) 283-1289
15100 Northline
Southgate, MI 48195

Minnesota

Control Data Business and Technology Center (Bemidji)
P.O. Box 602   (218) 751-6480
Bemidji, MN 56601

Control Data Business and Technology Center (St. Paul)
Business and Technology Center   (612) 292-2693
245 E. Sixth St.
St. Paul, MN 55101

Minneapolis Business and Technology Center Limited Partnership
Business and Technology Center   (612) 375-8066
511 11th Ave., S.
Minneapolis, MN 55415

St. Paul Small Business Incubator
Department of Planning and Economic   (612) 228-3301
  Development
2325 Endicott
St. Paul, MN 55102

University Technology Center (Minneapolis)
1313 Fifth St., S.E.   (612) 623-7774
Minneapolis, MN 55414

Mississippi

New Building Workspace for Women
121 S. Harvey   (601) 335-3523
Greenville, MS 38701

Missouri

Center for Business Innovation, Inc.
4747 Troost   (816) 561-8567
Kansas City, MO 64110

Missouri Incutech Foundation
Route 4   (314) 364-8570
Box 519
Rolla, MO 65401

Nebraska

Control Data Business and Technology Center (Omaha)
2505 N. 24th St.   (402) 346-8262
Omaha, NE 68110

259

Many incubators are located on or near universities or research and technology parks. For additional related listings, see the chapter entitled University Innovation Research Centers.

New Jersey

Princeton Capital Corporation
P.O. Box 384
Princeton, NJ 08540   (609) 924-7614

New Mexico

Los Alamos Economic Development Corporation
P.O. Box 715   (505) 662-0001
Los Alamos, NM 87544

New Mexico Business Innovation Center, Inc.
3825 Academy Parkway South, N.E.   (505) 345-8668
Albuquerque, NM 87109

Ventures in Progress
134 Rio Rancho Dr.   (505) 892-2161
Rio Rancho, NM 87124

New York

Broome County Industrial Development Agency
P.O. Box 1026   (607) 772-8212
Binghamton, NY 13902

Business Incubator (Brooklyn)
Local Development Corporation of East   (718) 385-6700
  New York
116 Williams Ave.
Brooklyn, NY 11207

Cornell Industry Research Park
Cornell University Industry Research   (607) 256-7315
  Park
Brown Rd., Bldg. 1
Ithaca, NY 14850

Greater Syracuse Business Incubator Center
Greater Syracuse Chamber of   (315) 470-1343
  Commerce
100 E. Onandaga
Syracuse, NY 13202

Incubator Industries Building (Buffalo)
Buffalo Urban Renewal Agency   (716) 855-5056
920 City Hall
Buffalo, NY 14202

190 Willow Avenue Industrial Incubator
c/o South Bronx Development   (212) 402-1300
  Organization
190 Willow Ave.
Bronx, NY 10454

Syracuse Incubator
Cental New York Regional High   (315) 470-1350
  Technology Development Council
c/o Knowledge Systems and Research,
  Inc.
500 S. Salina St., Room 826
Syracuse, NY 13202

Troy Incubator Program
Rensselaer Polytechnic Institute (RPI), J   (518) 276-6658
  Bldg.
Troy, NY 12181

Western New York Technology Development Center
2211 Main St.   (716) 831-3472
Buffalo, NY 14214

North Carolina

Ahoskie Incubator Facility
c/o Tomorrow, Inc.
ECSU Box 962
Elizabeth City, NC 27909

Haywood County Incubator
Smokey Mountain Development   (704) 452-1967
  Corporation
100 Industrial Park
Waynesville, NC 28786

Ohio

Akron Industrial Incubator
One Cascade Plaza   (216) 253-7918
Akron, OH 44308

Athens Innovation Center
Ohio University   (614) 593-1818
One President St.
Athens, OH 45701

Barberton Incubator
576 W. Park Ave.   (216) 753-6611
Barberton, OH 44203

Business and Technology Center (Columbus)
1445 Summit St.   (614) 294-0206
Columbus, OH 43201

Control Data Business and Technology Center (Toledo)
Business and Technology Center   (419) 255-6700
1946 N. 13th St.
Toledo, OH 43624

River East
615 Front St.   (419) 698-2310
Toledo, OH 43605

Springfield Incubator
76 E. High St.   (513) 324-7744
Springfield, OH 45502

Oklahoma

Atoka Industrial Incubator
Kiamichi Area Vo-Tech School   (405) 889-7321
Atoka Campus
P.O. Box 220
Atoka, OK 74525

Durant Industrial Incubator
Rural Enterprises Corporation   (405) 924-5094
Ten Waldron Dr.
Durant, OK 74701

Hugo Industrial Incubator
Kiamichi Area Vo-Tech School   (405) 326-6491
Hugo Campus, 107 S. 15th St.
Hugo, OK 74743

McAlester Industrial Incubator
Kiamichi Area Vo-Tech School   (918) 426-0940
McAlester Campus, Box 308
McAlester, OK 74502

Oklahoma State Incubation Association
c/o McAlester Economic Development   (918) 423-5735
  Corporation
P.O. Box 3190
McAlester, OK 74502   Terry Heilig, Executive Director

Oregon

Cascade Business Center Corporation
Portland Community College   (503) 244-6111
573 N. Killingsworth
Portland, OR 97217

Pennsylvania

Altoona Business Incubator
Sixth Ave. and 45th St.   (814) 949-2030
Altoona, PA 16602

Ben Franklin Advanced Technology Center
115 Research Dr.   (215) 861-0584
Bethlehem, PA 18015

Control Data Business and Technology Center (West Philadelphia)
5070 Parkside Ave.   (215) 879-8500
Philadelphia, PA 19131

East Liberty Incubator
Center for Entreprenuerial Development,   (412) 361-5000
  Inc.
120 S. Whitfield St.
Pittsburgh, PA 15213

Executive Office Network, Inc. (Malvern)
Five Great Valley Pkwy.   (215) 648-3900
Malvern, PA 19335

Greenville Incubator
12 N. Diamond St.   (412) 588-1161
Greenview, PA 16125

Hunting Park West
Southwest Germantown Community   (215) 843-2000
  Development Corporation
5002 Wayne Ave.
Philadelphia, PA 19144

Lansdale Business Center
650 N. Cannon Ave.   (215) 855-6700
Lansdale, PA 19446

Liberty Street Market Place
204 Liberty St., P.O. Box 547   (814) 726-2400
Warren, PA 16365

Matternville Business and Technology Center
Road #1, Box 354   (814) 234-1829
Port Matilda, PA 16870

Meadville Industrial Incubator
Meadville Area Industrial Commission   (814) 724-2975
628 Arch St.
Meadville, PA 16335

Model Works Industrial Commons
Girard Area Industrial Development   (814) 774-9339
   Corporation
227 Hathoway St.
Girard, PA 16417

North East Tier Advanced Technology Center
Ben Franklin Advanced Technology   (215) 758-5200
   Center #125
Lehigh University, 125 Goodman Dr.
Bethlehem, PA 18015

Paoli Technology Enterprise Center
19 E. Central Ave.   (215) 251-0505
Paoli, PA 19301

Pennsylvania Incubator Association
c/o Pennsylvania State Technology and   (215) 889-1325
   Development Center
30 E. Swedesford Rd.
Malvern, PA 19355   Marc Kramer, President

Ridgeway Manufacturing Incubator
North Central Pennsylvania Regional   (814) 772-6901
   Planning and Development
   Commission
P.O. Box 488
Ridgeway, PA 15853

River Bridge Industrial Center
One S. Olive   (215) 872-4469
Media, PA 19063

Southwest Pennsylvania Business Development Center
12300 Perry Hwy., P.O. Box 216   (412) 931-8444
Wexford, PA 15090

Technology Centers International (Montgomeryville)
1060 Route 309   (215) 646-7800
Montgomeryville, PA 18936

University City Science Center
3624 Market St.   (215) 387-2255
Philadelphia, PA 19104

University Technology Development Center I (Pittsburgh)
4516 Henry St.
Pittsburgh, PA 15213

University Technology Development Center II (Pittsburgh)
3400 Forbes Ave.
Pittsburgh, PA 15213

South Carolina

Control Data Business and Technology Center (Charleston)
Business and Technology Center   (803) 722-1219
701 E. Bay St.
Charleston, SC 29403

Florence Incubator
City/County Complex, Drawer FF   (803) 665-3141
Florence, SC 29501

North Augusta Incubator
City of North Augusta, P.O. Box 6400   (803) 278-0816
North Augusta, SC 29841

Rock Hill Incubator
Rock Hill Economic Development   (803) 328-6171
   Corporation
P.O. Box 11706
Rock Hill, SC 29731

For more information on incubators and state incubator associations, contact Ms. Dinah Adkins, Director, National Business Incubation Association, One President St., Athens, OH 45701; (614) 593-4331.

Spartanburg Incubator
City of Spartanburg
P.O. Box 1749
Spartanburg, SC 29304

(803) 596-2072

Tennessee

Knox County Business Incubator
City County Bldg.
Knoxville, TN 37901

(615) 521-2275

Tennessee Innovation Center
P.O. Box 607
Oakridge, TN 37830

(615) 576-3375

Texas

Control Data Business and Technology Center (San Antonio)
301 S. Frio
San Antonio, TX 78207

(512) 270-4500

Vermont

East Creek Center
Star Enterprises, Inc.
38 1/2 Center St.
Rutland, VT 05701

(802) 775-8011

Fellows Complex
Precision Valley Development
  Corporation
100 River St.
Springfield, VT 05156

(802) 885-2138

North Bennington Business Incubator
Bennington County Industrial
  Corporation (BCIC)
P.O. Box 357
North Bennington, VT 05257

(802) 442-8975

Washington

The Center (Everett)
917 134th St., S.W., Suite 100
Everett, WA 98204

(206) 743-9669

Spokane Business and Technology Center
Spokane International Airport Business
  Park
3707 S. Godfrey Blvd.
Spokane, WA 99204

(509) 458-6340

Wisconsin

The Advocap Small Business Center
19 W. First St.
Fond du Lac, WI 54925

(414) 922-7760

Madison Business Incubator
c/o Wisconsin for Research, Inc.
210 N. Bassett St.
Madison, WI 53703

(608) 258-7070

Wisconsin Business Incubator Association
210 N. Bassett St.
Madison, WI 53703

(608) 258-7070
Noel Pratt, President

Venture Capitalists

Alabama

Alabama Capital Corporation
16 Midtown Park, E.   (205) 476-0700
Mobile, AL 36606   David C. Delaney, President
  Minority enterprise small business investment company. Investment Policy: No industry preference.

First SBIC of Alabama
16 Midtown Park, E.   (205) 476-0700
Mobile, AL 36606   David Delaney, President
  Small business investment company. Investment Policy: No industry preference.

Hickory Venture Capital Corporation
699 Gallatin, Suite A2   (205) 539-1931
Huntsville, AL 35801   J. Thomas Noojin, President
  Small business investment company. Investment Policy: No industry preference.

Alaska

Calista Business Investment Corporation
601 W. Fifth Ave, Suite 200   (907) 279-5516
Anchorage, AK 99501   Johnny T. Hawk, President
  Minority enterprise small business investment company. Investment Policy: No industry preference.

Arizona

Navajo Small Business Development Corporation
P.O. Box 545   (602) 697-3572
Kayenta, AZ 86033
  Certified development company. Investment Policy: Inquire.

Norwest Venture Capital Management, Inc. (Scottsdale)
8777 E. Via de Ventura, Suite 335   (602) 483-8940
Scottsdale, AZ 85258   Robert Zicarelly, Chairman of the Board
  Small business investment company. Investment Policy: Prefers to invest in high-technology enterprises. Remarks: An affiliated office of the Norwest Ventur Capital Management headquarters in Minneapolis.

Rocky Mountain Equity Corporation
4530 Central Ave., Suite 3   (602) 274-7534
Phoenix, AZ 85012   Anthony J. Nicoli, President
  Small business investment company. Investment Policy: No industry preference.

Banks are creditors. They're interested in the immediate future, yet most heavily influenced by the past. Loan officers examine the product and market position of the company for assurance that the invention can return a steady flow of sales and generate enough cash to repay the loan and interest.

Venture capital companies are owners. They gamble on the future. Venture capitalists can provide the money and management skills that permit inventors to bring their inventions to fruition and market. Inventors, in turn, give the venture capitalists an opportunity to get what they seek, typically a three-to-five times return on their investment within five to seven years.

Valley National Investors, Inc.
201 N. Central Ave., Suite 900
Phoenix, AZ 85004
(602) 261-1577
J. M. Holliman, III, Managing Director
Small business investment company. Investment Policy: No industry preference.

Arkansas

Capital Management Services, Inc.
1910 N. Grant St., Suite 200
Little Rock, AR 72207
(501) 664-8613
David L. Hale, President
Minority enterprise small business investment company. Investment Policy: No industry preference.

Power Ventures, Inc.
829 Highway 270, N.
Malvern, AR 72104
(501) 332-3695
Dorsey D. Glover, President
Minority enterprise small business investment company. Investment Policy: No industry preference.

California

ABC Capital Corporation
610 E. Live Oak Ave.
Arcadia, CA 91006-5740
(818) 355-3577
Anne B. Cheng, President
Minority enterprise small business investment company. Investment Policy: No industry preference.

Accel Partners (San Francisco)
One Embarcadero Center
San Francisco, CA 94111
(415) 989-5656
Private venture capital firm. Investment Policy: Prefers to invest in telecommunications, software, or biotechnology/medical products industries. **Contact:** Dixon R. Doll; Arthur C. Patterson; James R. Swartz, General Partners.

Adler and Company (Sunnyvale)
2882 Sand Hill Rd., Suite 220
Menlo Park, CA 94025
(415) 854-3660
Private venture capital supplier. Investment Policy: Provides all stages of financing. Remarks: A branch office of Adler and Company, New York, New York.

Allied Business Investors, Inc.
428 S. Atlantic Blvd., Suite 201
Monterey Park, CA 91754
(818) 289-0186
Jack Hong, President
Minority enterprise small business investment company. Investment Policy: No industry preference.

Ally Finance Corporation
9100 Wilshire Blvd., Suite 408
Beverly Hills, CA 90212
(213) 550-8100
Percy P. Lin, President
Minority enterprise small business investment company. Investment Policy: No industry preference.

Asian American Capital Corporation
1251 W. Tennyson Rd., Suite 4
Hayward, CA 94544
(415) 887-6888
Jennie Chien, Manager
Minority enterprise small business investment company. Investment Policy: No industry preference.

Asset Management Company
2275 E. Bayshore, Suite 150
Palo Alto, CA 94303
(415) 494-7400
Venture capital firm. Investment Policy: High-technology industries preferred. **Contact:** John Shoch or Craig Taylor, Partners.

Associates Venture Capital Corporation
300 Montgomery St., Suite 421
San Francisco, CA 94104
(415) 956-1444
Walter P. Stryker, President
Investment Policy: No industry preference.

Bankamerica Ventures, Inc.
555 California St., 12th Floor
San Francisco, CA 94104
(415) 953-3001
Patrick Topolski, President
    Small business investment company. Investment Policy: No industry preference.

Bay Partners
10600 N. De Anza Blvd., Suite 100
Cupertino, CA 95014
(415) 725-2444
    Venture capital supplier. Investment Policy: Provides start-up financing primarily to West Coast technology companies that have highly qualified management teams. Initial investments range from $100,000 to $800,000; where large investments are required, will act as lead investor to bring in additional qualified venture investors. **Contact:** John Freidenrich, W. Charles Hazel, and John Bosch, Partners.

Bay Venture Group
One Embarcadero Center, Suite 3303
San Francisco, CA 94111
(415) 989-7680
William R. Chandler, General Partner
    Small business investment company. Investment Policy: High-technology industries preferred; geographic region limited to the San Francisco Bay area.

Biovest Partners
12520 High Bluff Dr., Suite 250
San Diego, CA 92130
(619) 527-4134
Timothy J. Wollaeger, General Partner
    Venture capital partnership. Investment Policy: Prefers medical and biotechnology industries.

BNP Venture Capital Corporation
3000 Sand Hill Rd.
Building One, Suite 125
Menlo Park, CA 94025
(415) 854-1084

Edgerton Scott, II, President
    Small business investment company. Investment Policy: No industry preference.

Brentwood Associates (Los Angeles)
11661 San Vicente Blvd., Suite 707
Los Angeles, CA 90049
(213) 826-6581
    Venture capital supplier. Investment Policy: Provides start-up and expansion financing to technology-based enterprises specializing in computing and data processing, electronics, communications, materials, energy, industrial automation, and bioengineering and medical equipment. Investments generally range from $1 to $3 million. **Contact:** B. Kipling Hagopian or Frederick J. Warren.

Burr, Egan, Deleage, and Company (San Francisco)
One Embarcadero Center, 25th Floor
San Francisco, CA 94111
(415) 362-4022
Jean Deleage
    Private venture capital supplier. Investment Policy: Invests start-up, expansion, and acquisitions capital nationwide. Principal concerns are strength of the management team; large, rapidly expanding markets; and unique products for services. Past investments have been made in the fields of electronics, health, and communications. Investments range from $750,000 to $5 million. Remarks: Maintains an office in Boston, Massachusetts.

Cable and Howse Ventures (Palo Alto)
435 Tasso St., Suite 115
Palo Alto, CA 94301
(415) 322-8400
    Venture capital supplier. Investment Policy: Provides start-up and early-stage financing to enterprises in the western United States, although a national perspective is maintained. Interest lies in proprietary or patentable technology. Investments range from $500,000 to $2 million. Remarks: A branch office of Cable and Howse Ventures, Bellevue, Washington.

California Capital Investors
11812 San Vicente Blvd., Suite 600
Los Angeles, CA 90049
(213) 820-7222
Arthur H. Bernstein, Managing General Partner
    Small business investment company. Investment Policy: No industry preference.

California Partners Draper Association CGP
3803 E. Bayshore Rd., Suite 125
Palo Alto, CA 94303
(415) 961-6669
Tim Draper, President
    Small business investment company. Investment Policy: No industry preference.

Canaan Venture Partners (Menlo Park)
3000 Sand Hill Rd.
Building One, Suite 205
Menlo Park, CA 94025
(415) 854-8092

Eric A. Young, General Partner
    Venture capital supplier. Investment Policy: Primary concern is strong entrepreneurial management. There are no geographic or industry-specific constraints. Remarks: A branch offic of Canaan Venture Partners in Rowayton, Connecticut.

CFB Venture Capital Corporation (Los Angeles)
Union Venture Corp.
455 Figueroa St.
Los Angeles, CA 90071-1620

(213) 236-4092
Pieter Westerbeek, III, Chief Financial Officer

Small business investment company. Investment Policy: No industry preference. Remarks: Maintains an office in San Francisco, California.

Charterway Investment Corporation
222 S. Hill St., Suite 800
Los Angeles, CA 90012

(213) 687-8539
Harold H. M. Chuang, President

Minority enterprise small business investment company. Investment Policy: No industry preference.

Churchill International
444 Market St., Suite 2501
San Francisco, CA 94111

(415) 328-4401

International venture capital supplier and investment management company. Investment Policy: Provides start-up, growth and development, and bridge and buy out capital to small and medium-sized high-technology companies in the United States, Europe, and Pacific Basin.

Comdisco Venture Lease, Inc. (San Francisco)
101 Calfornia St., 38th Floor
San Francisco, CA 94111

(415) 421-1800

Venture capital subsidiary of operating firm. Investment Policy: Involved in all stages of financing except seed. Has no industry preference, but avoids real estate and oil and gas. Investments range from $500,000 to $5 million. **Contact:** William Tenneson or Terrence Fowler.

Concord Partners (Palo Alto)
435 Tasso St., Suite 305
Palo Alto, CA 94301

(415) 327-2600

Venture capital supplier. Investment Policy: Diversified in terms of stage of development, industry classification, and geographic location. Areas of special interest include computers, telecommunications, health care and medical products, energy, and leveraged buy outs. Preferred investments range from $2 to $3 million, with a $1 million minimum and $10 million maximum.

Continental Investors, Inc.
8781 Seaspray Dr.
Huntington Beach, CA 92646

(714) 964-5207
Lac Thantrong, President

Minority enterprise small business investment company. Investment Policy: No industry preference.

Crosspoint Investment Corp.
One First St.
Los Altos, CA 94022

(415) 948-8300
Max Simpson, President

Small business investment company. Investment Policy: No industry preference. Remarks: A division of Crosspoint Venture Partners, Mountain View, California.

Crosspoint Venture Partners
One First St., Suite 2
Los Altos, CA 94022

(415) 948-8300

Venture capital partnership. Investment Policy: Seeks to invest start-up capital in unique products, services, and/or market opportunities located in the Western United States. Primary interests lie in communication devices and systems; biotechnology; medical products, instruments, and equipment; computer software and productivity tools; computers and computer peripherals; industrial automation and controls; semiconductor devices and equipment; instrumentation; and related service and distribution businesses. Investments range from $50,000 to $3 million. Remarks: Maintains an incubator facility in Palo Alto, California, to support new ventures. Also maintains an office in Newport Beach, California. **Contact:** John Mumford; Roger Barry; Bill Cargile; Bob Hoff; Fred Dotzler, General Partners.

Dougery, Jones & Wilder (Mountain View)
2003 Landings Dr.
Mountain View, CA 94043

(415) 968-4820

Venture capital supplier. Investment Policy: Prefers to invest in small to medium-sized companies headquartered or primarily operating in the West and Southwest. The typical company sought is in the computers/communications or medical/biotechnology industries; is privately owned; has an experienced management team; serves growth markets; and can achieve at least $30 million in sales. Initial investment ranges from $250,000 to $2.5 million; may be expanded by including other venture capital firms. Remarks: Maintains an office in Dallas, Texas. **Contact:** John R. Dougery; David A. Jones; Henry L.B. Wilder; A. Lawson Howard; Gerald R. Schoonhoven.

El Dorado Ventures
Two N. Lake Ave., Suite 480
Pasadena, CA 91101

(818) 793-1936
Brent T. Rider, General Partner

Private venture capital firm. Investment Policy: Prefers to invest in communications, electronics, industrial products and services, and medical/health care industries. Remarks: Affiliated with El Dorado Technology Partners at same address.

Enterprise Partners

5000 Birch St., Suite 6200   (714) 833-3650
Newport Beach, CA 92660
> Venture capital fund. Investment Policy: Prefers electronics, medical technology, and health care ventures in southern California. **Contact:** Charles D. Martin or James H. Berglund, General Partners.

Equitable Capital Corporation

855 Sansome St., Second Floor   (415) 434-4114
San Francisco, CA 94111   James Lee, Vice-President
> Minority enterprise small business investment company. Investment Policy: No industry preference.

First American Capital Funding, Inc.

38 Corporate Park   (714) 660-9288
Irvine, CA 92714   Lou Trankiem, President
> Minority enterprise small business investment company. Investment Policy: No industry preference.

First Century Partners (San Francisco)

350 California St., 21st Floor   (415) 955-1612
San Francisco, CA 94104   C. Sage Givens
> Private venture capital firm. Investment Policy: High-technology, medical, and specialty retail industries are preferred; has no interest in real estate or entertainment investments. Minimum investment is $1 million.

First Interstate Capital, Inc.

5000 Birch St., Suite 10100   (714) 253-4360
Newport Beach, CA 92660   Ronald J. Hall, Managing Director
> Small business investment company. Investment Policy: No industry preference. Provides capital for small and medium-sized companies through participation in private placements of subordinated debt, preferred, and common stock. Offers growth-acquisition and later-stage venture capital.

First SBIC of California (Costa Mesa)

650 Town Center Dr., 17th Floor   (714) 556-1964
Costa Mesa, CA 92626   Brian Jones, Senior Vice-President
> Small business investment company and venture capital company. Investment Policy: No industry preference. Remarks: Maintains offices in Palo Alto and Pasadena, California; Boston, Massachusetts; Washington, Pennsylvania; and London.

First SBIC of California (Palo Alto)

2400 Sand Hill Rd., Suite 100   (415) 424-8011
Menlo Park, CA 94025
> Small business investment company. Investment Policy: No industry preference. Remarks: A branch office of First Small business investment company of California, Costa Mesa, California.

First SBIC of California (Pasadena)

155 N. Lake Ave., Suite 1010   (818) 304-3451
Pasadena, CA 91109
> Small business investment company. Investment Policy: No industry preference. Remarks: A branch office of First Small business investment company of California, Costa Mesa, California.

G C and H Partners

One Maritime Plaza, 20th Floor   (415) 981-5252
San Francisco, CA 94111   James C. Gaither, General Partner
> Small business investment company. Investment Policy: No industry preference.

Glenwood Management

3000 Sand Hill Rd.   (415) 854-8070
Building Four, Suite 230
Menlo Park, CA 94025   Dag Tellefsen
> Venture capital supplier. Investment Policy: Provides early-stage financing to companies with proprietary technology and exceptional management teams. Areas of interest include telecommunications; computer hardware, software, and peripherals; applications of genetic engineering; medical instruments; and semiconductor equipment and devices. Initial investments range from $300,000 to $500,000, with subsequent rounds totalling $800,000 to $1.2 million. Geographic area limited to the Western United States.

Hambrecht & Quist Venture Partners (San Francisco)

One Bush St.   (415) 576-3300
San Francisco, CA 94104   Richard Gorman, Vice-President
> Venture capital firm. Investment Policy: Prefers start-up investments in high-technology and biotechnology industries. Investments range from $500,000 to $5 million.

Hamco Capital Corporation

One Bush St.   (415) 986-5500
San Francisco, CA 94104   William R. Hambrecht, President
> Small business investment company. Investment Policy: High-technology industries preferred.

T heir investment unprotected in the event of failure, most venture capital firms set rigorous policies for venture proposal size, maturity of the seeking company, and requirements and evaluation procedures to reduce risks.

Projects requiring under $250,000 are of limited interest because of the high cost of investigation and administration; however, some venture firms will consider smaller proposals if the investment is intriguing enough. Most venture capitalists live in the $250,000 to $1.5 million atmosphere.

Harvest Ventures, Inc. (Cupertino)

19200 Stevens Creek Blvd., Suite 220   (408) 996-3200
Cupertino, CA 95014

Private venture capital supplier. Investment Policy: Prefers to invest in high-technology, growth-oriented companies with proprietary technology, large market potential, and strong management teams. Specific areas of interest include computers, computer peripherals, semiconductors, telecommunications, factory automation, military electronics, medical products, and health-care services. Provides seed capital of up to $250,000; investments in special high-growth and leveraged buy out situations range from $500,000 to $2 million. Will serve as lead investor and arrange financing in excess of $2 million in association with other groups. Remarks: A branch office of Harvest Ventures, Inc., New York, New York.

Helio Capital, Inc.

5900 S. Eastern Ave., Suite 136   (213) 721-8053
Commerce, CA 90040   Frank Remski, General Manager

Minority enterprise small business investment company. Investment Policy: No industry preference.

Hillman Ventures, Inc.

2200 Sand Hill Rd., Suite 240   (415) 854-4653
Menlo Park, CA 94025   Philip S. Paul, Chairman

Venture capital firm. Investment Policy: Medical, biotechnology, and computer-related electronics industries preferred.

I.K. Capital Loans Ltd.

9460 Wilshire, Suite 608   (213) 276-8547
Beverly Hills, CA 90212   Iraj Kermanshahchi, President

Small business investment company. Investment Policy: No industry preference.

Imperial Ventures, Inc.

P.O. Box 92991   (213) 417-5830
Los Angeles, CA 90009

Small business investment company. Investment Policy: No industry preference. **Contact:** H.Wayne Snavely, President; John H. Upshen, Senior Vice-President.

Institutional Venture Partners III

3000 Sand Hill Rd., Bldg. Two, Suite 290   (415) 854-0132
Menlo Park, CA 94025   Reid W. Dennis, General Partner

Venture capital fund. Investment Policy: Will provide $750,000 to $2 million in early-stage financing.

Interscope Investments, Inc.

10900 Wilshire Blvd., Suite 1400   (213) 208-8525
Los Angeles, CA 90024   Murray Hill, Chief Financial Officer

Venture capital firm. Investment Policy: No industry preference.

Interven II LP (Los Angeles)

333 S. Grand Ave., Suite 4050   (213) 622-1922
Los Angeles, CA 90071   Jonatahn E. Funk, General Partner

Venture capital fund. Investment Policy: Semiconductors, factory automation, telecommunications, health care, and medical technologies preferred. Will invest $1.5 to $2 million in seed and first-round technology companies in Oregon, Washington, Idaho, and southern California. Will also handle buy outs of companies with annual sales of $20 to $50 million.

Interwest Partners (Menlo Park)

3000 Sand Hill Rd.   (415) 854-8585
Building Three, Suite 255
Menlo Park, CA 94025   W. Scott Hedrick, General Partner

Venture capital fund. Investment Policy: Both high-tech and low- or non-technology companies are considered. No oil, gas, real estate, or construction projects.

Investments Orange Nassau, Inc.

Westerly Place   (714) 752-7811
1500 Quail St., Suite 540
Newport Beach, CA 92660

Small business investment company. Investment Policy: Provides all stages of financing to technology-based and oil and gas service industries; no geographic preference. Experience of management team and high-growth market potential are emphasized. Investments range from $500,000 to $5 million. Remarks: A branch office of Orange Nassau Capital Corporation, Boston, Massachusetts.

Ivanhoe Venture Capital Ltd.

737 Pearl St., Suite 201   (619) 454-8882
LaJolla, CA 92037   Alan Toffler, General Partner

Small business investment company. Investment Policy: No industry preference.

Jupiter Partners
600 Montgomery St., 35th Floor   (415) 421-9990
San Francisco, CA 94111   John M. Bryan, President
    Small business investment company. Investment Policy: No industry preference.

Kleiner, Perkins, Caufield, and Byers (San Francisco)
Four Embarcadero Center, Suite 3520   (415) 421-3110
San Francisco, CA 94111
    Investment Policy: Provides seed, start-up, second- and third-round, and bridge financing to companies on the West Coast. Past investments have been made in the following fields: computers and computer peripherals, software, office equipment, medical products and instruments, microbiology, genetic engineering, telecommunications, instrumentation, semiconductors, lasers and optics, and unique consumer products and services. Areas avoided for investment include real estate, motion pictures, solar energy, hotels and motels, restaurants, resort areas, oil and gas, construction, and metallurgy. Investments range from $1 to $5 million. **Contact:** Thomas J. Perkins; Frank J. Caufield; Brook H. Byers, General Partners.

Marwit Capital Corporation
180 Newport Center Dr., Suite 200   (714) 640-6234
Newport Beach, CA 92660   Martin W. Witte, President
    Small business investment company. Investment Policy: No industry preference.

Matrix Partners (Menlo Park)
2500 Sand Hill Rd., Suite 113   (415) 854-3131
Menlo Park, CA 94025
    Private venture capital partnership. Investment Policy: Investments range from $500,000 to $1 million. Remarks: A branch office of Matrix Partners in Boston, Massachusetts.

Mayfield Fund
2200 Sand Hill Rd., Suite 200   (415) 854-5560
Menlo Park, CA 94025
    Venture capital partnership. Investment Policy: Prefers technically oriented companies that have potential to achieve revenues of at least $75 to $100 million in 5 years. Principal interest is experience of management team. Initial investments range from $50,000 to $2 million; for subsequent investments, capital commitment may double or triple the initial investment. **Contact:** Thomas J. Davis, Jr.; A. Grant Heidrich III; F. Gibson Myers, Jr.; Norman A. Fogelsong; Glenn M. Mueller, General Partners.

Menlo Ventures
3000 Sand Hill Rd.   (415) 854-8540
Building Four, Suite 100
Menlo Park, CA 94025
    Venture capital supplier. Investment Policy: Provides start-up and expansion financing to companies with experienced management teams, distinctive product lines, and large growing markets. Primary interest is in technology-oriented, service, consumer products, and distribution companies. Investments range from $500,000 to $3 million; also provides capital for leveraged buy outs. **Contact:** Douglas C. Carlisle; Ken E. Joy; Richard P. Magnuson; H. Dubose Montgomery, General Partners.

Merrill, Pickard, Anderson, and Eyre (Palo Alto)
Two Palo Alto Sq., Suite 425   (415) 856-8880
Palo Alto, CA 94306
    Private venture capital partnership. Investment Policy: Provides start-up and early-stage financing to companies with experienced management teams, with the ability to grow to $50 to $100 million in annual revenues within 4 to 6 years and with distinctive product lines. High-technology industries are preferred; little interest in financial, real estate, or consulting companies. Investments range from $750,000 to $2.5 million. **Contact:** Steven L. Merrill; W. Jeffers Pickard; James C. Anderson; Chris A. Eyre; Stephen E. Coit, General Partners.

Metropolitan Venture Company, Inc.
4021 Rosewood Ave., 3rd Floor   (213) 666-9882
Los Angeles, CA 90004   Rudolph J. Lowy, Chairman of the Board
    Small business investment company. Investment Policy: No industry preference.

MIP Equity Advisors, Inc. (Menlo Park)
3000 Sand Hill Rd., Bldg. 4, Suite 280   (415) 854-2653
Menlo Park, CA 94025   Hans Severiens, President
    International venture capital supplier. Investment Policy: Seeks participants from a variety of business activities that wish to expand into European markets by establishing a base in the Netherlands. High-technology ventures are preferred, but not the the exclusion of other industry groups. Minimum investment is $1 million. Remarks: Company headquarters are located in The Hague, the Netherlands. Another U.S. office is located in Boston, Massachusetts.

Montgomery Securities
600 Montgomery St.   (415) 627-2454
San Francisco, CA 94111   May Leong
    Private venture capital and investment banking firm. Investment Policy: Diversified, but will not invest in real estate or energy-related industries. Involved in both start-up and later-stage financing.

Successful venture capitalists put together marriages between good ideas and good money. These wheeling deal-makers have relationships with a wide variety of potential investors such as corporations, insurance companies, union pension funds, university endowments, and wealthy individuals. Yet, unlike passive investors and traditional lenders, venture capitalists take a hands-on, proactive role in managing the companies in which they invest. In the end, investors invest in the venture capitalist as much as in any opportunity they may represent.

Myriad Capital, Inc.
328 S. Atlantic Blvd., Suite 200-A   (818) 570-4548
Monterey, CA 91754   Kuo Hung Chen, President
    Minority enterprise small business investment company. Investment Policy: No industry preference.

New Enterprise Associates (Menlo Park)
3000 Sand Hill Rd., Bldg. 4, Suite 235   (415) 854-2215
Menlo Park, CA 94025   C. Woodrow Rae, General Partner
    Venture capital supplier. Investment Policy: Concentrates on technology-based industries that have the potential for product innovation, rapid growth, and high profit margins. Investments range from $250,000 to $2 million. Past investments have been made in the following industries: computer software, medical and life sciences, computers and peripherals, communications, semiconductors, specialty retailing, energy and alternative energy, defense electronics, materials, and specialty chemicals. Management must demonstrate intimate knowledge of its marketplace and have a well-defined strategy for achieving strong market penetration. Remarks: An affiliated office of New Enterprises Associates, Baltimore, Maryland.

New Enterprise Associates (San Francisco)
235 Montgomery St., Russ Bldg., Suite   (415) 956-1579
   1025
San Francisco, CA 94104
    Venture capital supplier. Investment Policy: Concentrates in technology-based industries that have the potential for product innovation, rapid growth, and high profit margins. Investments range from $250,000 to $2 million. Past investments have been made in the following industries: computer software, medical and life sciences, computers and computer peripherals, communications, semiconductors, specialty training, energy and alternative energy, defense electronics, materials, and specialty chemicals. Management must demonstrate intimate knowledge of its marketplace and have a well-defined strategy for achieving strong market penetration. **Contact:** C. Richard Kramlich or Thomas C. McConnell.

New Enterprise Associates (Westlake Village)
200 N. Westlake Blvd., Suite 215   (805) 373-8537
Westlake Village, CA 91362   James Cole, Partner
    Venture capital partnership. Investment Policy: Early-stage investors, with initial investments ranging from $500,000 to $2 million. Remarks: An affiliated office of the New Enterprise Associates headquarters in Baltimore.

New Kukje Investment Company
3670 Wilshire St., Suite 418   (213) 389-8679
Los Angeles, CA 90010   George Su Chey, President
    Minority enterprise small business investment company. Investment Policy: No industry preference.

New West Partners (San Diego)
4350 Executive Dr., Suite 206   (619) 457-0722
San Diego, CA 92121
    Small business investment company. Investment Policy: No industry preference. Remarks: A branch office of New West Partners, Newport Beach, California.

Oak Investment Partners (Menlo Park)
3000 Sand Hill Rd., Bldg. Three, Suite   (415) 854-8825
   240
Menlo Parks, CA 94025
    Private venture capital firm. Investment Policy: Computer, biotechnology, and communications industries preferred.

Opportunity Capital Corporation
39650 Liberty St., Suite 425   (415) 651-4412
Fremont, CA 94538   J. Peter Thompson, President
    Minority enterprise small business investment company. Investment Policy: No industry preference.

Oxford Partners (Santa Monica)
233 Wilshire Blvd., Suite 830   (213) 458-3135
Santa Monica, CA 90401
    Independent venture capital partnership. Investment Policy: Prefers to invest in high-technology industries. Initial investments range from $500,000 to $1.5 million; will invest up to $3 million over several later rounds of financing. Remarks: A branch office of Oxford Partners, Stamford, Connecticut.

Pacific Capital Fund, Inc.
675 Mariners' Island Blvd., Suite 103   (415) 574-4747
San Mateo, CA 94404   Eduardo B. CuUnjieng, President
    Minority enterprise small business investment company. Investment Policy: No industry preference.

PBC Venture Capital, Inc.
P.O. Box 6008   (805) 395-3555
Bakersfield, CA 93386   Henry L. Wheeler, Manager
> Small business investment company. Investment Policy: No industry preference. Area limited to California.

PCF Venture Capital Corporation
675 Mariners' Island Blvd., Suite 103   (415) 574-4747
San Mateo, CA 94404   Eduardo B. CuUnjieng, President
> Small business investment company. Investment Policy: No industry preference.

Positive Enterprises, Inc.
1166 Post St., Suite 200   (415) 885-6600
San Francisco, CA 94109   Kwok Szeto, President
> Minority enterprise small business investment company. Investment Policy: No industry preference.

R&D Funding Corporation
3945 Freedom Circle, Suite 800   (408) 980-0990
Santa Clara, CA 95054   Richard E. Moser, President
> Venture capital firm. Investment Policy: Invests in high-growth businesses. Direct investment in research and development.

Ritter Partners
3000 Sand Hill Rd., Suite 190   (415) 854-1555
Metro Park, CA 94025   William C. Edwards, President
> Small business investment company. Investment Policy: No industry preference; retail, hotel, or real estate ventures are not considered.

Robertson, Colman & Stephens
One Embarcadero, Suite 3100   (415) 781-9700
San Francisco, CA 94111   Susan Vican, Associate
> Investment banking firm. Investment Policy: Considers investments in any attractive merging-growth area, including product and service companies. Key preferences include health care, hazardous waste services and technology, biotechnology, software, and information services. Maximum investment is $5 million.

Round Table Capital Corporation
655 Montgomery St.   (415) 392-7500
San Francisco, CA 94111   Richard Dumke, President
> Small business investment company. Investment Policy: No industry preference.

RSC Financial Corporation
223 E. Thousand Oaks Blvd., Suite 310   (805) 646-2925
Thousand Oaks, CA 91360   Frederick K. Bae, President
> Minority enterprise small business investment company. Investment Policy: No industry preference.

San Joaquin Capital Corporation
P.O. Box 2538   (805) 323-7581
Bakersfield, CA 93303   Chester Troudy, President
> Small business investment company. Investment Policy: No industry preference.

Security Pacific Capital Corporation
650 Town Center Dr., 17th Floor   (714) 556-1964
Costa Mesa, CA 92626   Brian Jones, Senior Vice-President
> Small business investment company. Investment Policy: No industry preference. Remarks: Maintains offices in Boston, Palo Alto, Pasadena, Pennsylvania, and London.

Sequoia Capital
3000 Sand Hill Rd., Bldg. 4, Suite 280   (415) 854-3927
Menlo Park, CA 94025
> Private venture capital partnership with $300 million under management. Investment Policy: Provides financing for all stages of development of well-managed companies with exceptional growth prospects in fast-growth industries. Past investments have been made in computers and peripherals, communications, health care, biotechnology, and medical instruments and devices. Investments range from $350,000 for early stage companies to $4 million for late stage accelerates. **Contact:** Pierre Lamond; Walter Baumgartner; Nancy Olson; Doug DeVivo; Gordon Russell; Don Valentine; Michael Moritz.

Sprout Group (Menlo Park)
3000 Sand Hill Rd., Bldg. 1, Suite 285   (415) 854-1550
Menlo Park, CA 94025   Keith Geeflin, Partner
> Venture capital partnership. Investment Policy: No industry preference. Remarks: An affiliated office of the Sprout Group headquarters in New York City.

While the inventor works on product development, the venture capitalist secures the money to keep the pump primed, prepares long-term corporate business programs and marketing plans, and recruits personnel. Most inventors are not skilled at these activities and welcome the assistance.

The inventor's motivation usually involves more than money, i.e. things like ego and pride. Typically, the venture capitalist is in it for money alone, the invention being a vehicle to this end. For the relationship to work, the parties must be highly compatible. The chemistry between inventor and venture capitalist is a major factor in how well their business will develop.

Sutter Hill Ventures
755 Page Mill Rd., Suite 8200 (415) 493-5600
Palo Alto, CA 94304
Venture capital supplier. Investment Policy: Provides seed, start-up, and second-stage financing to companies that manufacture products with some degree of technology and whose markets are potentially worldwide. Investments range from $100,000 to $2 million; larger equity needs may be combined with other leading venture capital firms.

TA Associates (Palo Alto)
435 Tasso St. (415) 328-1210
Palo Alto, CA 94301 Jack Bunce, Associate
Private venture capital firm. Investment Policy: Prefers technology companies and leveraged buy outs. Provides from $1 to $20 million in investments.

Technology Funding, Inc.
2000 Alameda de las Pulgas, Suite 250 (415) 345-2200
San Mateo, CA 94403 M. David Titus
Private venture capital supplier. Investment Policy: Provides primarily late first-stage and early second-stage equity financing. Also offers secured debt with equity participation to venture capital-backed companies. Investments range from $500,000 to $1 million.

Technology Venture Investors
3000 Sand Hill Rd., Bldg. 4, Suite 210 (415) 854-7472
Menlo Park, CA 94025 Mark Wilson, Administrative Partner
Private venture capital partnership. Investment Policy: Primary interest is in technology companies, with minimum investment of $1 million.

3i Ventures (Newport Beach)
450 Newport Center Dr., Suite 250 (714) 720-1421
Newport Beach, CA 92660
Venture capital supplier. Investment Policy: Provides start-up and early-stage financing to companies in high-growth fields such as microelectronics, computers, telecommunications, biotechnology, health sciences, and industrial automation. Investments generally range from $500,000 to $2 million. Remarks: A branch office of 3i Ventures, Boston, Massachusetts.

Trinity Ventures Ltd.
115 Bovet Rd., Stuite 700 (405) 358-9700
San Mateo, CA 94402
Venture capital fund. Investment Policy: Will provided up to $3 million in seed capital to third-round financing for West Coast companies. Emphasis on high technology, with flexibility to enter all areas at all stages of development. Will not consider real estate, oil and gas, or construction ventures. **Contact:** Gerald S. Casilli, Noel J. Fenton, David Nierenbert, General Partners.

Twenty-First Century Venture Partners
Oryx Capital Corporation (415) 956-5097
170 Lombard St.
San Francisco, CA 94111 Thomas J. Sherwood, Director
Venture capital partnership. Investment Policy: Prefers later-stage financing for companies on the verge of explosive growth.

U.S. Venture Partners
2180 Sand Hill Rd, Suite 300 (415) 854-9080
Menlo Park, CA 94025
Venture capital partnership. Investment Policy: Prefers specialty retail, consumer products, technology, and biomedical industry. **Contact:** Nancy Glasser and Steven Krausz, General Partners.

USVP-Schlein Marketing Fund
2180 Sand Hill Rd. (415) 854-9080
Menlo Park, CA 94025
Venture capital fund. Investment Policy: Specialty retailing/consumer products companies preferred. **Contact:** Philip S. Schlein or Nancy E. Glasser.

Venrock Associates (Palo Alto)
755 Page Mill, A-230 (415) 493-5577
Palo Alto, CA 94304
Private venture capital supplier. Remarks: Maintains an office in New York City.

Vista Capital Corporation
5080 Shoreman Place, Suite 202 (619) 453-0780
San Diego, CA 92122 Frederick J. Howden, Jr., Chairman
Small business investment company. Investment Policy: No industry preference.

Walden Capital Partners
750 Battery St., Seventh Floor (415) 391-7225
San Francisco, CA 94111 Arthur S. Berliner, President
Small business investment company. Investment Policy: No industry preference.

Westamco Investment Company

8929 Wilshire Blvd., Suite 400   (213) 652-8288
Beverly Hills, CA 90211   Leonard G. Muskin, President
    Small business investment company. Investment Policy: No industry preference.

Wilshire Capital, Inc.

3932 Wilshire Blvd., Suite 305   (213) 388-1314
Los Angeles, CA 90010
    Minority enterprise small business investment company. Investment Policy: No industry preference.

Wood River Capital (Menlo Park)

300 Sand Hill Rd., Bldg. 1, Suite 280   (415) 854-1000
Menlo Park, CA 94025   Vince Topkin, General Partner
    Small business investment company. Investment Policy: Diversified, with interests in early-stage financing. Prior investment experience in technology-related service businesses, environmental technology, telecommunications, semiconductors, and computer companies.

Colorado

Centennial Business Development Fund Ltd.

1999 Broadway, Suite 2100   (303) 298-9066
Denver, CO 80202   David Bullwinkle, General Partner
    Venture capital fund. Investment Policy: Prefers to invest in later-stage companies. Remarks: Managed by Larimer and Company, Denver, Colorado.

Centennial Fund II Ltd.

1999 Broadway, Suite 2100   (303) 298-9066
Denver, CO 80202   Mark Dubovoy, General Partner
    Venture capital fund. Investment Policy: Prefers to invest in early-stage companies in the Rocky Mountain regin or in the telecommunications industry nationwide. Remarks: Managed by Lorimer and Company, Denver, Colorado.

Centennial Fund III LP

1999 Broadway, Suite 2100   (303) 298-9066
Denver, CO 80202   G. Jackson Tankersley, Jr., General partner
    Venture capital fund. Investment Policy: Prefers to invest in early-stage companies in the Rocky Mountain region. Remarks: Managed by Larimer and Company, Denver, Colorado.

Centennial Fund Ltd.

1999 Broadway, Suite 2100   (303) 298-9066
Denver, CO 80202   Mark Dubovoy, General Partner
    Venture capital fund. Investment Policy: Currently not investing in new companies, but still involved in follow-up investments. Remarks: Managed by Larimer and Company, Denver, Colorado.

Colorado Growth Capital, Inc.

1600 Broadway, Suite 2125   (303) 831-0205
Denver, CO 80202   Nicholas H. C. Davis, President
    Small business investment company. Investment Policy: No industry preference.

William Blair and Company (Denver)

1225 17th St., Suite 2440   (303) 825-1600
Denver, CO 80202
    Investment banker and venture capital supplier. Investment Policy: Provides all stages of financing to growth companies; also deals in leveraged acquisitions. Remarks: A branch office of William Blair and Company, Chicago, Illinois.

Connecticut

Abacus Ventures

411 W. Putnam Ave.   (203) 629-1100
Greenwich, CT 06830
    Venture capital fund. Investment Policy: Will provide $500,000 to $2 million in start-up and first-round financings. **Contact:** Yung Wong; Charles T. Lee; Patrick F. Whelan; Thomas J. Crotty.

Professor Mark A. Spikell of George Mason University labels some venture capitalists "vulture capitalists." "Many companies," he cautions, "use the come-on of venture capital opportunity as a way to merely learn what is being developed in a particular field. It is nothing short of industrial espionage when inventors seeking venture capital are invited to put on a dog and pony show for alleged potential funders. Inventors should approach venture capitalists smartly."

Canaan Venture Partners (Rowayton)
105 Rowayton Ave.
Rowayton, CT 06853

(203) 855-0400
Harry T. Rein, Managing General
Partner

Venture capital supplier. Investment Policy: Primary concern is strong entrepreneurial management. There are no geographic or industry-specific constraints. Remarks: Maintains an office in Menlo Park, California.

Capital Impact Corporation
10 Middle St., 16th Floor
Bridgeport, CT 06604

(203) 384-5984
William D. Starbuck, President

Small business investment company. Investment Policy: No industry preference.

Capital Resource Company of Connecticut, LP
699 Bloomfield Ave.
Bloomfield, CT 06002

(203) 243-1114
I. Martin Fierberg, Partner

Small business investment company. Investment Policy: No industry preference.

Fairfield Venture Partners (Stamford)
1275 Summer St.
Stamford, CT 06905

(203) 358-0255
T. Berman

Venture capital firm. Investment Policy: Diversified. Prefers early-stage financing. Minimum investment is $500,000.

First Connecticut SBIC
177 State St.
Bridgeport, CT 06604

(203) 366-4726
David Engelson, President

Small business investment company. Investment Policy: Real estate industry preferred.

James B. Kobak and Company
774 Hollow Tree Ridge Rd.
Darien, CT 06820

(203) 655-8764
James B. Kobak, Owner

Venture capital supplier and consultant. Investment Policy: Provides assistance to new ventures in the communications field through conceptualization, planning, organization, raising money, and control of actual operations. Special interest is in magazine publishing.

Marcon Capital Corporation
49 Riverside Ave.
Westport, CT 06880

(203) 226-6893
Martin A. Cohen, President

Small business investment company. Investment Policy: Media preferred.

Marketcorp Venture Associates
285 Riverside Ave.
Westport, CT 06880

(203) 222-1000
Buck Griswold, President

Venture capital firm. Investment Policy: Prefers to invest in consumer-market businesses, including the packaged goods, specialty retailing, communications, and consumer electronics industries. Investments range from $500,000 to 2.5 million.

Northeastern Capital Corporation
139 Orange St.
New Haven, CT 06510

(203) 865-4500
Mosha Reiss, President

Small business investment company. Investment Policy: No industry preference.

Oxford Partners (Stamford)
Soundview Plaza
1266 Main St.
Stamford, CT 06902

(203) 964-0592

Kenneth W. Rind, General Partner

Independent venture capital partnership. Investment Policy: Prefers to invest in high-technology industries. Initial investments range from $500,000 to $1.5 million; up to $3 million over several later rounds of financing. Remarks: Maintains an office in Santa Monica, California.

Regional Financial Enterprises (New Canaan)
36 Grove St., Third Floor
New Canaan, CT 06840

(203) 966-2800
Robert M. Williams, Managing Partner

Small business investment company. Investment Policy: Prefers to invest in high-technology companies or nontechnology companies with significant growth prospects, located anywhere in the United States. Investments range from $1 to $5 million. Remarks: Maintains an office in Ann Arbor, Michigan.

Saugatuck Capital Company
One Canterbury Green
Stamford, CT 06901

(203) 348-6669

Private investment partnership. Investment Policy: Seeks to invest in building products, transportation, health care products and services, energy services and products, process control instrumentation, industrial and automation equipment, test and measurement instrumentation, communications, fasteners, filtration equipment and filters, and valves and pumps. Prefers leveraged buy out situations, but will consider start-up financing. Investments range from $1 to $5 million. **Contact:** Norman W. Johnson; Steve Crittfield; John S. Crowley; Frank J. Hawley, Jr.; Alexander H. Dunbar, General Manager.

SBIC of Connecticut, Inc.
1115 Main St.
Bridgeport, CT 06604

(203) 367-3282
Kenneth F. Zarrilli, President

Small business investment company. Investment Policy: No industry preference.

Vista Group (New Canaan)
36 Grove St.
New Canaan, CT 06840

(203) 972-3400
Gerald B. Bay, Managing Partner

Venture capital supplier. Investment Policy: Provides start-up and second-stage financing to technology-related businesses that seek to become major participants in high-growth markets of at least $100 million in annual sales. Areas of investment interest include information systems, communications, computer peripherals, medical products and services, retailing, agrigenetics, biotechnology, low technology, no technology, instrumentation, and genetic engineering. Investments range from $500,000 to $1.5 million. Remarks: A branch office of The Vista Group in Newport Beach, California.

District of Columbia

Allied Investment Corporation (Washington, DC)
1666 K. St., Suite 901
Washington, DC 20006

(202) 331-1112
David Gladstone, President

Small business investment company; Minority enterprise small business investment company. Investment Policy: No industry preference. Remarks: Maintains an office in Fort Lauderdale, Florida.

Allied Venture Partnership
1666 K St., Suite 901
Washington, DC 20006

(202) 331-1112
David Gladstone, President and General Partner

Venture capital fund. Investment Policy: Low-technology businesses, such as radio stations or retail companies, preferred. Will provide $1 million in a second- or third-round financing in the form of convertible debentures.

American Security Capital Corporation, Inc.
730 15th St., N.W., Suite A-5
Washington, DC 20005

(202) 624-4843
Brian K. Mercer, Vice-President and Manager

Small business investment company. Investment Policy: No industry preference.

Broadcast Capital, Inc.
1771 N St., N.W., Fourth Floor
Washington, DC 20036

(202) 429-5393
John E. Oxendine, President

Minority enterprise small business investment company. Investment Policy: Communications media industry preferred.

Consumers United Capital Corporation
2100 M St., N.W.
Washington, DC 20037

(202) 872-5262
Esther M. Carr-Davis, President

Minority enterprise small business investment company. Investment Policy: No industry preference.

DC Bancorp Venture Capital Company
1801 K St., N.W.
Washington, DC 20006

(202) 955-6970
Allen A. Weissburg, President

Small business investment company. Investment Policy: No industry preference; preferred geographic area includes District of Columbia, Maryland, and Virginia.

Fulcrum Venture Capital Corporation
1030 15th St., N.W., Suite 203
Washington, DC 20005

(202) 785-4253
Renate Todd, Vice-President

Minority enterprise small business investment company. Investment Policy: No industry preference.

Minority Broadcast Investment Corporation
1200 18th St., N.W., Suite 705
Washington, DC 20036

(202) 293-1166
Walter Threadgill, President

Minority enterprise small business investment company. Investment Policy: Communications industry preferred.

Syncom Capital Corporation
1030 15th St., N.W., Suite 203
Washington, DC 20005

(202) 293-9428
Herbert P. Wilkins, President

Minority enterprise small business investment company. Investment Policy: Telecommunications media industry only.

Florida

Allied Investment Corporation (Fort Lauderdale)
111 E. Las Olas Blvd. (305) 763-8484
Fort Lauderdale, FL 33301
> Small business investment company. Investment Policy: No industry preference. Remarks: A branch office of Allied Investment Corporation, Washington, District of Columbia.

Gold Coast Capital Corporation
3550 Biscayne Blvd., Room 601 (305) 576-2012
Miami, FL 33137 William I. Gold, President
> Small business investment company. Investment Policy: No industry preference.

Ideal Financial Corporation
780 N.W. 42nd Ave., Suite 501 (305) 442-4665
Miami, FL 33126 Ectore E. Reynaldo, General Manager
> Minority enterprise small business investment company. Investment Policy: No industry preference.

J and D Capital Corporation
12747 Biscayne Blvd. (305) 893-0303
North Miami, FL 33181 Jack Carmel, President
> Small business investment company. Investment Policy: No industry preference.

Market Capital Corporation
1102 N. 28th St. (813) 247-1357
P.O. Box 31667
Tampa, FL 33631 E. E. Eads, President
> Small business investment company. Investment Policy: Grocery industry preferred.

SBAC of Panama City, Florida
2612 W. 15th St. (904) 785-9577
Panama City, FL 32401 Charles Smith, President
> Small business investment company. Investment Policy: Lodging and amusement industries preferred.

South Atlantic Capital Corporation
614 W. Bay St., Suite 200 (813) 253-2500
Tampa, FL 33606/2704 Donald W. Burton, President
> Venture capital supplier. Investment Policy: Provides long-term working capital for privately owned, rapidly growing companies located in the Southeast.

Southeast Venture Capital Limited 1
3250 Miami Center (305) 375-6470
201 South Biscayne Blvd.
Miami, FL 33131 Clement L. Hofmann, Chairman
> Small business investment company. Investment Policy: No industry preference, but prefers to invest in firms located in the southwest.

Universal Financial Services, Inc.
225 N.E. 35th St. (305) 573-1496
Miami, FL 33137 Norman Zipkin, President
> Minority enterprise small business investment company. Investment Policy: No industry preference.

Venture Group, Inc.
5433 Buffalo Ave. (904) 353-7313
Jacksonville, FL 32208 Ellis W. Hitzing, President
> Minority enterprise small business investment company. Investment Policy: Automotive industry preferred.

Western Financial Capital Corporation
1380 Miami Cardens Dr., Suite 225 (305) 949-5900
North Miami Beach, FL 33179 Dr. F. M. Rosemore, President
> Small business investment company. Investment Policy: Medical industry preferred.

Georgia

Investor's Equity, Inc.
P.O. Box 18859 (404) 266-8300
Atlanta, GA 30326 I. Walter Fisher, President
> Small business investment company. Investment Policy: No industry preference.

Noro-Moseley Partners
4200 Northside Pkwy., Building 9   (404) 233-1966
Atlanta, GA 30327   Charles D. Moseley, Jr., General Partner
    Venture capital partnership. Investment Policy: Prefers to invest in private, diversified small and medium-size growth companies located in the southeastern United States.

North Riverside Capital Corporation
50 Technology Park/Atlanta   (404) 446-5556
Norcross, GA 30092   Tom Barry, President
    Small business investment company. Investment Policy: No industry preference.

Hawaii

Bancorp Hawaii SBIC, Inc.
111 S. King St., 3rd Floor   (808) 537-8557
P.O. Box 2900   Thomas T. Triggs, Vice-President and
Honolulu, HI 96813   Manager
    Small business investment company. Investment Policy: No industry preference.

Pacific Venture Capital Ltd.
222 S. Vineyard St., #PH-1   (808) 521-6502
Honolulu, HI 96813/2445   Dexter J. Taniguchi, President
    Minority enterprise small business investment company. Investment Policy: Inquire.

Illinois

Allstate Venture Capital
Allstate Plaza, South Building G5D   (312) 402-5681
Northbrook, IL 60062   Robert L. Lestina, Director
    Venture capital supplier. Investment Policy: Investments are not limited to particular industries or geographical locations. Interest is in unique products or services that address large potential markets and offer great economic benefits; strength of management team is also important. Investments range from $500,000 to $5 million.

Alpha Capital Venture Partners LP
Three First National Plaza, 14th Floor   (312) 372-1556
Chicago, IL 60602   Andrew H. Kalnow, General Partner
    Small business investment company providing venture capital. Investment Policy: No industry preference; however, no real estate, oil and gas, or start-up ventures are considered. All investments are structured to provide equity participation in the business; no straight loans are considered. Minimum investment is $200,000; preferred size of investment is $500,000 to $1 million. Company must be located in the Midwest.

Ameritech Development Corporation
Ten S. Wacker Dr., 21st Floor   (312) 609-6000
Chicago, IL 60606
    Venture capital supplier. Investment Policy: Seeks ideas in high-technology information management that relate to Ameritech's existing and expanding business in telecommunications technologies. Remarks: Ameritech venture teams are assigned to particular projects and are responsible for planning and day-to-day management.

Amoco Venture Capital Company
200 E. Randolph Dr.   (312) 856-6523
Chicago, IL 60601   Gordon E. Stone, President
    Minority enterprise small business investment company. Investment Policy: Technical industires or those that relate to Amoco Corporation needs.

Business Ventures, Inc.
20 N. Wacker Dr., Suite 1741   (312) 346-1580
Chicago, IL 60606   Milton Lefton, President
    Small business investment company. Investment Policy: No industry preference; considers only ventures in the Chicago area.

Capital Strategy Group, Inc.
20 N. Wacker Dr.   (312) 444-1170
Chicago, IL 60606   Eric Von Bauer, President
    Investment banker and venture capital supplier. Investment Policy: Provides financing to start-up and early-stage companies, located in the Midwest, in the manufacturing and service industries. Remarks: Also develops and improves business plans and product strategies, assists with acquisitions, and conducts valuations of closely held companies.

Caterpillar Venture Capital, Inc.
100 N.E. Adams St. (309) 765-5503
Peoria, IL 61629 Robert Powers, President
> Venture capital subsidiary of operating firm. Investment Policy: Prefers to invest in industrial electronics, advanced materials, and environmental-related industries.

Chicago Community Ventures, Inc.
104 S. Michigan Ave., Suites 215-218 (312) 726-6084
Chicago, IL 60603 Phyllis George, President
> Minority enterprise small business investment company. Investment Policy: No industry preference.

Comdisco Venture Lease, Inc. (Rosemont)
6111 N. River Rd. (708) 698-3000
Rosemont, IL 60018 James Labe
> Venture capital subsidiary of operating firm. Investment Policy: Involved in all stages of financing except seed. Has no industry preference but avoids real estate and oil and gas industries. Investments range from $500,000 to $5 million.

Continental Illinois Venture Corporation (CIVC)
231 S. LaSalle St. (312) 828-8021
Chicago, IL 60697
> Small business investment company. Investment Policy: Provides start-up and early-stage financing to growth-oriented companies with capable management teams, proprietary products, and expanding markets. Remarks: CIVC is a wholly owned subsidiary of Continental Illinois Equity Corporation (CIEC), which is a wholly owned subsidiary of Continental Illionis Corporation. CIEC invests in more mature companies, with a special interest in leveraged buy outs. **Contact:** John L. Hines or William Putze.

First Capital Corporation of Chicago
Three First National Plaza (312) 732-5414
Chicago, IL 60670 John A. Canning, Jr., President
> Small business investment company. Investment Policy: No industry preference.

First Chicago Venture Capital (Chicago)
Three First National Plaza, Suite 1330 (312) 732-5400
Chicago, IL 60670-0501 John A. Canning, Jr., President
> Venture capital supplier. Investment Policy: Invests a minimum of $1 million in early-stage situations to a maximum of $25 million in mature growth or buy out situations. Emphasis is placed on a strong management team and unique market opportunity. Remarks: Maintains an office in Boston, Massachusetts.

Frontenac Capital Corporation
208 S. LaSalle St., Room 1900 (312) 368-0047
Chicago, IL 60604 David A. R. Dullum, President
> Small business investment company. Investment Policy: No industry preference.

Golder, Thoma and Cressey
120 S. LaSalle St., Suite 630 (312) 853-3322
Chicago, IL 60603
> Private equity investors. Investment Policy: Provides financing for start-up and leveraged buy out situations. No geographic or industry limitations, with the exception of real estate. Past investments have been made in the health care field and in information services. Investments range from $2 to $10 million. **Contact:** Stanley C. Golder; Carl D. Thoma; Bryan C. Cressey, Founding Partners.

IEG Venture Management, Inc.
Ten Riverside Plaza (312) 644-0890
Chicago, IL 60611 Marian Zamlynski, Operations Manager
> Venture capital supplier. Investment Policy: Provides start-up financing primarily to companies located in the Midwest that focus on productivity-enhancing technologies in the medical, manufacturing, electronic, telecommunications, agricultural, service, chemical and mineral, and metal products industries.

Mesirow Capital Partners SBIC Ltd.
350 N. Clark St. (312) 670-6092
Chicago, IL 60610 James C Tyree
> Small business investment company. Investment Policy: No industry preference.

Neighborhood Fund, Inc.
1950 E. 71st St. (312) 684-8074
Chicago, IL 60649 James Fletcher, President
> Minority enterprise small business investment company. Investment Policy: No industry preference.

Peterson Finance and Investment Company
3300 W. Peterson Ave., Suite A
Chicago, IL 60659
(312) 583-6300
James S. Rhee, President
Minority enterprise small business investment company. Investment Policy: No industry preference.

Seidman Jackson Fisher and Company
233 N. Michigan Ave., Suite 1812
Chicago, IL 60601
(312) 856-1812
David S. Seidman, Managing Partner
Private venture capital supplier. Investment Policy: Provides early-stage and growth-equity financing to companies with proprietary or patented products or services that deal with large and rapidly growing industrial markets; limited interest in consumer markets. Leveraged buy outs and turn-around of mature companies are considered under certain circumstances. Investments range from $200,000 to $2 million.

Tower Ventures, Inc.
Sears Tower, BSC 11-11
Chicago, IL 60684
(312) 875-0571
Robert T. Smith, President
Minority enterprise small business investment company. Investment Policy: No industry preference.

Walnut Capital Corporation
208 S. LaSalle, Suite 1043
Chicago, IL 60604
(312) 346-2033
David L. Bogetz, Vice-President
Small business investment company. Investment Policy: No industry preference.

William Blair and Company (Chicago)
135 S. LaSalle St.
Chicago, IL 60603
(312) 236-1600
Investment banker and venture capital supplier. Investment Policy: Provides all stages of financing to growth companies; also deals in leveraged acquisitions. Remarks: William Blair Venture Partners was formed to meet the needs of new clients. William Blair and Company maintains offices in Denver, Colorado, and in London, England. **Contact:** Samuel B. Guren, General Partner, William Blair Venture Partners.

Indiana

Circle Ventures, Inc.
3228 E. Tenth St.
Indianapolis, IN 46201
(317) 633-7303
Ms. Murray Welch, Investment Manager
Small business investment company. Investment Policy: No industry preference.

First Source Capital Corporation
P.O. Box 1602
South Bend, IN 46634
(219) 236-2180
Eugene L. Cavanaugh, Jr., Vice-President
Small business investment company. Investment Policy: No industry preference.

Heritage Venture Group, Inc.
135 N. Pennsylvania St., Suite 2380
Indianapolis, IN 46204
(317) 635-5696
Arthur A. Angotti, President
Venture capital fund. Investment Policy: Prefers communications industries, especially broadcasting.

Mount Vernon Venture Capital Company
P.O. Box 40177
Indianapolis, IN 46240
(317) 253-3606
Thomas J. Grande, General Manager
Small business investment company. Investment Policy: No industry preference.

White River Capital Corporation
P.O. Box 929
Columbus, IN 47202
(812) 376-1759
Bradley J. Kime, Vice-President
Small business investment company. Investment Policy: No industry preference.

Iowa

InvestAmerica Venture Group, Inc. (Cedar Rapids)
101 Second St., S.E., Suite 800
Cedar Rapids, IA 52401
(319) 363-8249
Venture capital fund management company. Investment Policy: No industry preference. Remarks: Manager of Iowa Venture Capital Fund LP, and MorAmerica Capital Corporation.

Maintains branch offices in Milwaukee, Wisconsin, and Kansas City, Missouri. **Contact:** Donald E. Flynn, President; David R. Schroder, Executive Vice-President; Robert A. Comey, Vice-President.

Kansas

Kansas Venture Capital, Inc.
1030 Bank IV Tower, Suite 825
One Townsite Plaza
Topeka, KS 66603
(913) 233-1368

Rex Wiggins, President
Small business investment company. Investment Policy: No industry preference.

Kentucky

Equal Opportunity Finance, Inc.
420 Hurstbourne Lane, Suite 201
Louisville, KY 40222
(502) 423-1943
David Sattich, Vice-President and
Manager
Minority enterprise small business investment company. Investment Policy: No industry preference; geographic areas limited to Indiana, Kentucky, Ohio, and West Virginia.

Financial Opportunities, Inc.
6060 Dutchman's Lane
P.O. Box 35680
Louisville, KY 40232
(502) 451-3800

Joe P. Peden, President
Small business investment company. Investment Policy: No industry preference.

Mountain Ventures, Inc.
P.O. Box 5070
London, KY 40741
(606) 864-5175
Lloyd R. Moncrief, President
Small business investment company. Investment Policy: No industry preference.

Louisiana

Capital Equity Corporation
1885 Wooddale Blvd.
Baton Rouge, LA 70806
(504) 924-9205
Arthur J. Mitchell, General Manager
Small business investment company. Investment Policy: No industry preference.

Capital for Terrebonne, Inc.
27 Austin Dr.
Houma, LA 70360
(504) 868-3933
Hartwell A. Lewis, President
Small business investment company. Investment Policy: No industry preference.

Dixie Business Investment Company
401-1/2 Lake St.
P.O. Box 588
Lake Providence, LA 71254
(318) 559-1558

L. W. Baker, President
Small business investment company. Investment Policy: No industry preference.

First Southern Capital Corporation
P.O. Box 14418
Baton Rouge, LA 70898-4418
(504) 769-3004
Charest Thibaut, President
Small business investment company. Investment Policy: No industry preference.

Louisiana Equity Capital Corporation
451 Florida St.
Baton Rouge, LA 70821
(504) 389-4421
Melvin L. Rambin, President
Small business investment company. Investment Policy: No industry preference.

SCDF Investment Corporation
1006 Surrey St.
P.O. Box 3005
Lafayette, LA 70502
(318) 232-3769

Marvin Beaulieu, President
Minority enterprise small business investment company. Investment Policy: No industry preference.

Walnut Street Capital Company

2330 Canal
New Orleans, LA 70119

(504) 821-4952
William D. Humphries, Managing
General Partner

Small business investment company. Investment Policy: Inquire.

Maine

Maine Capital Corporation

70 Center St.
Portland, ME 04101

(207) 772-1001
David M. Coit, President

Small business investment company. Investment Policy: No industry preference.

Maryland

Broventure Capital Management

16 W. Madison St.
Baltimore, MD 21201

(301) 727-4520
William M. Gust, Principal

Venture capital partnership. Investment Policy: Provides start-up capital to early-stage companies, expansion capital to companies experiencing rapid growth, and capital for acquisitions. Initial investments range from $400,000 to $750,000.

First Maryland Capital, Inc.

107 W. Jefferson St.
Rockville, MD 20850

(301) 251-6630
Joseph A. Kenary, President

Small business investment company. Investment Policy: No industry preference.

Greater Washington Investors, Inc.

5454 Wisconsin Ave.
Chevy Chase, MD 20815

(301) 656-0626
Don A. Christensen, President

Small business investment company. Investment Policy: Provides financing to small developing companies primarily in computer-related industries but also in the medical and communications areas.

New Enterprise Associates (Baltimore)

1119 St. Paul St.
Baltimore, MD 21202

(301) 244-0115
Frank A. Bonsal, Jr., General Partner

Venture capital supplier. Investment Policy: Concentrates in technology-based industries that have the potential for product innovation, rapid growth, and high profit margins. Investments range from $250,000 to $2 million. Management must demonstrate intimate knowledge of its marketplace and have a well-defined strategy for achieving strong market penetration. Remarks: New Enterprise Associates operates nationwide, with principal offices in San Francisco and Baltimore with affiliated offices in Menlo Park and Westlake Village, California and in New York, New York.

Security Financial and Investment Corporation

7720 Wisconsin Ave., Suite 207
Bethesda, MD 20814

(301) 951-4288
Han Y. Cho, President

Minority enterprise small business investment company. Investment Policy: No industry preference.

T. Rowe Price

100 E. Pratt St.
Baltimore, MD 21202

(301) 547-2000
George J. Collins, President

Venture capital supplier. Investment Policy: Offers specialized investment services to meet the needs of companies in various stages of growth.

Massachusetts

Advanced Technology Ventures

Ten Post Office Square
Boston, MA 02109

(617) 423-4050
Dr. Albert E. Paladino, Managing Partner

Private venture capital firm. Investment Policy: Prefers early-stage financing in high-technology industries.

According to the Small Business Administration (SBA), the typical venture capital company receives more than 1,000 proposals a year. Probably 90 percent will be rejected quickly because they don't fit the established geographical, technical, or market-area policies of the firm, or because they have been poorly presented.

The remaining 10 percent are investigated carefully. The venture capitalist may spend between $2,000 and $3,000 per proposal to hire consultants who will make preliminary investigations. According to the SBA, these investigations narrow the field to 10 or 15 proposals of interest. Second investigations, more thorough and expensive than the first, reduce the number of companies under active consideration to three or four. Eventually, the venture capitalist will invest in one or two of these.

Advent Atlantic Capital Company LP
75 State St. (617) 345-7200
Boston, MA 02109 David D. Croll, Managing Partner
 Small business investment company. Investment Policy: Communications industry preferred.

Advent Industrial Capital Company LP
75 State St. (617) 345-7200
Boston, MA 02109 David D. Croll, Managing Partner
 Small business investment company. Investment Policy: Communications industry preferred.

Advent International Corporation
45 Milk St. (617) 574-8400
Boston, MA 02109 Clinton B. Harris, Vice-President
 Venture capital firm. Investment Policy: Specializes in working with companies that need assistance in accessing international markets and major corporations.

Advent IV Capital Company
75 State St. (617) 345-7200
Boston, MA 02109 David D. Croll, Managing Partner
 Small business investment company. Investment Policy: Communications industry preferred.

Advent V Capital Company
75 State St. (617) 345-7200
Boston, MA 02109 David D. Croll, Managing Partner
 Small business investment company. Investment Policy: Communications industry preferred.

Aeneas Venture Corporation
600 Atlantic Ave. (617) 523-4400
Boston, MA 02210 Scott Sperling, Managing Partner
 Venture capital firm. Investment Policy: Diversified. Minimum investment is $1 million.

American Research and Development
75 State St. (617) 345-7200
Boston, MA 02109 Charles J. Coulter, Managing General Partner
 Independent private venture capital partnership. Investment Policy: All stages of financing; no minimum or maximum investment.

Ampersand Ventures
265 Franklin St., Suite 1501 (617) 239-0700
Boston, MA 02110 Richard A. Charpie, Managing General Partner
 Venture capital supplier. Investment Policy: Provides start-up and early-stage financing to technology-based companies. Investments range from $500,000 to $1 million.

Analog Devices Enterprises
One Technology Way (617) 329-4700
P.O. Box 9106 Robert A. Boole, Director of Venture
Norwood, MA 02062-9106 Analysis
 Venture capital division of Analog Devices, Inc., a supplier of integrated analog circuits. Investment Policy: Prefers to invest in industries involved in analog devices.

Atlas II Capital Corporation
101 Federal St., 4th Floor (617) 951-9420
Boston, MA 02110 Joost E. Tjaden, President and Chief Executive
 Small business investment company. Investment Policy: Provides all stages of financing to technology-based and oil and gas service industries; no geographic preference. Experience of management team and high-growth market potential are emphasized. Investments range from $200,000 to $400,000.

Bain Capital Fund
Two Copley Place (617) 572-3000
Boston, MA 02116
 Private venture capital firm. Investment Policy: No industry preference but avoids investing in high-tech industries. Minimum investment is $500,000.

BancBoston Ventures, Inc.
100 Federal St. (617) 434-2442
Boston, MA 02110 Paul F. Hogan, President
 Small business investment company. Investment Policy: Provides start-up and first- and second-round financing to companies in the communications, computer hardware and software, electronic components and instrumentation, industrial products, and health care industries. Investments range from $500,000 to $2 million.

Bever Capital Corporation

101 Federal St., 4th Fl.
Boston, MA 02110

(617) 951-9420
Joost E. Tjaden, President and Chief Executive

Small business investment company. Investment Policy: Provides all stages of financing to technology-based and oil and gas service industries; no geographic preference. Experience of management team and high-growth market potential are emphasized. Investments range from $200,000 to $400,000.

Boston Capital Ventures LP

Old City Hall
45 School St.
Boston, MA 02108

(617) 227-6550

Venture capital firm. Investment Policy: Prefers health care and high-technology industries. **Contact:** Donald Steiner; A. Dana Callow, Jr.; H.J. vaugh-der-Goltz, General Partners.

Burr, Egan, Deleage, and Company (Boston)

One Post Office Sq., Suite 3800
Boston, MA 02109

(617) 482-8020

Private venture capital supplier. Investment Policy: Invests start-up, expansion, and acquisitions capital nationwide. Principal concerns are strength of the management team; large, rapidly expanding markets; and unique products or services. Past investments have been made in the fields of electronics, health, and communications. Investments range from $750,000 to $5 million. Remarks: A branch office of Burr, Egan, Deleage, and Company, San Francisco, California.

Business Achievement Corporation

1172 Beacon St.
Newton Centre, MA 02161

(617) 965-0550
Michael L. Katzeff, President

Small business investment company. Investment Policy: No industry preference.

Charles River Ventures

67 Battery March St., Suite 600
Boston, MA 02110

(617) 439-0477
Barbara A. Piette, Associate

Venture capital partnership. Investment Policy: Diversified, but no real estate. Minimum investment is $750,000.

Chestnut Capital International II LP

75 State St.
Boston, MA 02109

(617) 345-7200
David D. Croll, Managing Partner

Small business investment company. Investment Policy: Communications industry preferred.

Claflin Capital Management, Inc.

77 Franklin St.
Boston, MA 02110

(617) 426-6505
Thomas M. Claflin, II, President

Private venture capital firm investing its own capital. Investment Policy: No industry preference but prefers early stage companies.

Eastech Management Company

260 Franklin St., Suite 530
Boston, MA 02109

(617) 338-0200
Michael H. Shanahan, Associate

Private venture capital supplier. Investment Policy: Provides start-up and first- and second-stage financing to companies in the following industries: communications, computer-related electronic components and instrumentation, and industrial products and equipment. Will not consider real estate, agriculture, forestry, fishing, finance and insurance, transportation, oil and gas, publishing, entertainment, natural resources, or retail. Investments range from $400,000 to $900,000, with a minimum of $250,000. Eastech prefers that portfolio companies be located in New England or on the East Coast, within two hours of the office.

First Chicago Venture Capital (Boston)

One Financial Center
Boston, MA 02110-2621

(617) 542-9185
Kevin McCafferty

Venture capital supplier. Investment Policy: Invests a minimum of $1 million in later-stage situations to a maximum of $25 million in mature growth or buy out situations. Emphasis is placed on a strong management team and unique market opportunity. Remarks: A branch office of First Chicago Venture Capital, Chicago, Illinois.

First United SBIC, Inc.

135 Will Dr.
Canton, MA 02021

(617) 828-6150
Alfred W. Ferrera, Vice-President

Small business investment company. Investment Policy: Retail grocery stores.

Fleet Venture Partners II

Ten Post Office Sq., Suite 600 N.
Boston, MA 02109

(617) 429-9555
Jim Saulfield, General Partner

Venture capital partnership. Investment Policy: No industry preference. Remarks: A branch office of Fleet Venture Resources, Inc., Providence, Rhode Island.

Greylock Management Corporation (Boston)

One Federal St. (617) 423-5525
Boston, MA 02110

> Private venture capital partnership. Investment Policy: Minimum investment of $250,000; preferred investment size of over $1 million. Will function either as deal originator or investor in deals created by others. Remarks: Maintains an office in Palo Alto, California. **Contact:** William W. Helman and William S. Kaiser, General Partners.

Hambro International Venture Fund (Boston)

160 State St. (617) 523-7767
Boston, MA 02109

> Private venture firm. Investment Policy: Seeks to invest in mature companies as well as in high-technology areas, from start-ups to leveraged buy outs. Investments range from $500,000 to $3 million, with initial investments ranging from $800,000 to $1 million. Remarks: A branch office of Hambro International Venture Fund, New York, New York. **Contact:** Richard A. D'Amore, General Partner; Peter Santeusanio, Associate.

John Hancock Venture Capital Management, Inc.

One Financial Center, 39th Floor (617) 350-4002
Boston, MA 02111 Edward W. Kane, Managing Director

> Venture capital supplier. Investment Policy: Primary interest is in high technology-based emerging growth companies with experienced management teams. Investments range from $500,000 to $1.5 million, with a targeted minimum of $1 million.

Matrix Partners (Boston)

One Post Office Sq. (617) 482-7735
Boston, MA 02109 Paul J. Ferri, General Partner

> Private venture capital partnership. Investment Policy: Investments range from $500,000 to $1 million. Remarks: Maintains an office in Menlo Park, California.

McGowan, Leckinger, Berg

Ten Forbes Rd. (617) 849-0020
Braintree, MA 02184

> Venture capital supplier. Investment Policy: Provides early-stage financing to retail ventures and related businesses whose retail concepts follow the demographic trends of the 1980s. **Contact:** James A. McGowan; Robert T. Leckinger, Principals.

MIP Equity Advisors, Inc. (Boston)

283 Franklin St., Suite 200 (617) 484-0781
Boston, MA 02110 Barney Ussher, Vice-President

> International venture capital supplier. Investment Policy: Seeks participants from a variety of business activities that wish to expand into European markets by establishing a base in the Netherlands. High-technology ventures are preferred, but not to the exclusion of other industry groups. Minimum investment is $1 million. Remarks: Company headquarters are located in The Hague, the Netherlands. Another U.S. office is located in Menlo Park, California.

Morgan, Holland Ventures Corporation

One Liberty Square (617) 423-1765
Boston, MA 02109

> Venture capital partnership. Investment Policy: Provides start-up, early-stage, and expansion financing to companies that are pioneering applications of proven technology; also will consider nontechnology-based companies with strong management teams and with plans for expansion. Investments range from $500,000 to $1 million, with a $6 million maximum. **Contact:** James F. Morgan; Daniel J. Holland; Jay Delahanty; Joseph T. McCullen, Jr.; Edwin M. Kania, Jr., Partners.

New England Capital Corporation

One Washington Mall, Seventh Floor (617) 573-6400
Boston, MA 02108 Z. David Patterson, Executive Vice-President

> Small business investment company. Investment Policy: Provides early-stage, second-stage, third-stage, and expansion financing to manufacturing, communications, assembly, high-technology, and sophisticated service businesses with experienced management teams and high profit markets. Leveraged buy out situations also are considered. Investments range from $200,000 to $2 million; will co-invest with other institutional investors if more capital is needed.

New England MESBIC, Inc.

530 Turnpike St. (617) 688-4326
North Andover, MA 01845 Etang Chen, President

> Minority enterprise small business investment company. Investment Policy: No industry preference.

Northeast Small Business Investment Corporation

16 Cumberland St. (617) 267-3983
Boston, MA 02115 Joseph Mindick, Treasurer

> Small business investment company. Investment Policy: No industry preference.

Orange Nassau Capital Corporation
101 Federal St., 4th Floor
Boston, MA 02110

(617) 951-9420
Joost E. Tjaden, President and Chief Executive

Small business investment company. Investment Policy: Provides all stages of financing to technology-based and oil and gas service industries; no geographic preference. Experience of management team and high-growth market potential are emphasized. Investments range from $200,000 to $400,000. Remarks: Maintains offices in Dallas, Texas, and Newport Beach, California. Orange Nassau is able to establish Asian and European manufacturing facilities and local distribution channels for its portfolio companies through an international network. International affiliates are located in The Hague, the Netherlands, in Paris, France, and in Singapore.

Palmer Service Corporation
300 Unicorn Park Dr.
Woburn, MA 01801

(617) 933-5445

Venture capital partnership. Investment Policy: Provides financing to new enterprises and existing growth companies. Investments range from $100,000 to $1 million. **Contact:** William H. Congleton; John A. Shane; Stephen J. Ricci; Karen S. Camp; William T. Fitzgerald, General Partners.

Sprout Group (Boston)
One Center Plaza
Boston, MA 02108

(617) 570-8720

Venture capital affiliate of Donaldson, Lufkin, and Jenrette. Investment Policy: Provides early-stage financing to companies specializing in retailing, financial services, and high-technology products. Will also invest in selected later-stage, emerging/growing companies and in management buy outs of mature companies. Investments range from $1 to $2 million, with a minimum of $500,000. Remarks: A branch office of Sprout Group, headquartered in New York, New York.

Summit Ventures
One Boston Place, Suite 3420
Boston, MA 02108

(617) 742-5500
Harrison B. Miller, Associate

Venture capital firm. Investment Policy: Prefers to invest in emerging, profitable, growth companies in the electronic technology, environmental services, and health care industries. Investments range from $1 to $4 million.

TA Associates (Boston)
45 Milk St.
Boston, MA 02109

(617) 338-0800
Brian J. Conway, Associate

Private venture capital partnership. Investment Policy: Technology companies, media communications companies, and leveraged buy outs preferred. Will provide from $1 to $20 million in investments.

3i Ventures (Boston)
99 High St., Suite 1530
Boston, MA 02110

(617) 542-8560

Venture capital supplier. Investment Policy: Provides start-up and early-stage financing to companies in high-growth industries such as biotechnology, computers, electronics, health care, industrial automation, and telecommunications. Investments generally range from $500,000 to $2 million. Remarks: Maintains offices in Newport Beach, California and London, England.

Transportation Capital Corporation (Boston)
45 Newbury St., Rm. 117
Boston, MA 02116

(617) 536-0344
Jon Hirsch, Assistant Vice-President

Minority enterprise small business investment company. Investment Policy: No industry preference. Specializes in medallion loans. Remarks: A branch office of Transportation Capital Corporation, New York, New York. Toll-free/Additional Phone Number: (617)333-0858.

Vadus Capital Corporation
101 Federal St.
Boston, MA 02110

(617) 951-9420
Joost E. Tjaden, President and Chief Executive

Small business investment company. Investment Policy: Provides all stages of financing to technology-based and oil and gas service industries; no geographic preference. Experience of management team and high-growth market potential are emphasized. Investments range from $200,000 to $400,000.

Venture Capital Fund of New England II
160 Federal St., 23rd Floor
Boston, MA 02110

(617) 439-4646

Venture capital fund. Investment Policy: Prefers New England high-technology companies that have a commercial prototype or initial product sales. Will provide up to $500,000 in first-round financing. **Contact:** Richard A. Farrell; Harry J. Healer, Managing General Partners.

The entire investigative process by the venture capitalist can take anywhere from three to six months. Before any attempts are made to raise money, the venture capitalist and inventor best know each other well. Annulments are rare. Divorces can be messy.

Venture Founders Corporation

One Cranberry Hill   (617) 863-0900
Lexington, MA 02173   Edward Getchell, General Partner
> Venture capital fund. Investment Policy: CAD/CAM and robotics, materials components, information technology, biotechnology, and medical industries preferred. Will provide $250,000 to $1 million in start-up and first-round financing. Preferred initial investment size is between $100,000 and $750,000.

Vimac Corporation

12 Arlington St.   (617) 267-2785
Boston, MA 02116   Max J. Steinmann, President
> Venture capital supplier. Investment Policy: Provides start-up and early-stage financing to businesses in the computer industry, as well as in other suitable high-technology industries.

Michigan

Demery Seed Capital Fund

3307 W. Maple Rd., Suite 101   (313) 433-1722
Birmingham, MI 48010   Thomas Demery, President
> Seed capital fund. Investment Policy: Diversified, but interested in food processing. Invests in start-up companies in Michigan.

Federated Capital Corporation

30955 Northwestern Hwy., Suite 300   (313) 737-1300
Farmington Hills, MI 48018   Louis P. Ferris, Jr., President
> Small business investment company. Investment Policy: No industry preference.

MBW Management, Inc.

4251 Plymouth Rd.   (313) 747-9401
P.O. Box 986
Ann Arbor, MI 48106   Richard Goff, Vice-President
> Manages an Small business investment company and two venture capital funds. Investment Policy: Prefers high-tech industries and leveraged buy outs for manufacturing companies. No geographic limitations.

Metro-Detroit Investment Company

30777 Northwestern Hwy., Suite 300   (313) 851-6300
Farmington Hills, MI 48018   William J. Fowler, President
> Minority enterprise small business investment company. Investment Policy: Food store industry preferred.

Motor Enterprises, Inc.

3044 W. Grand Blvd.   (313) 556-4273
Detroit, MI 48202   James Kobus, Manager
> Minority enterprise small business investment company. Investment Policy: Prefers manufacturing.

Regional Financial Enterprises (Ann Arbor)

325 E. Eisenhower Pkwy., Suite 103   (313) 769-0941
Ann Arbor, MI 48108
> Manages one small business investment company and two private venture capital firms. Investment Policy: Prefers to invest in high-technology companies or nontechnology companies with significant growth prospects located anywhere in the United States. Investments range from $2 to $3 million. Remarks: A branch office of Regional Financing Enterprises, New Canaan, Connecticut.

Minnesota

Capital Dimensions Ventures Fund, Inc.

Two Apple Tree Sq., Suite 244   (612) 854-3506
Minneapolis, MN 55425-1637   T. F. Hunt, Jr., Vice-President
> Minority enterprise small business investment company. Investment Policy: No industry preference.

Cherry Tree Ventures

1400 Northland Plaza   (612) 893-9012
3800 W. 80th Street
Minneapolis, MN 55431
> Venture capital supplier. Investment Policy: Provides start-up and early-stage financing. Fields of interest include communications, medical devices, health care services, applications and systems software, and microprocessor-based systems for the office and factory. There are no

minimum or maximum investment limitations. **Contact:** Anton J. Christianson; Gordon F. Stofer, General Partners.

FBS Venture Capital Company
First Bank Place East (612) 370-4764
120 S. Sixth Street
Minneapolis, MN 55480
Small business investment company. Investment Policy: Generally invests in high-technology companies, although they are not necessarily preferred.

First Midwest Ventures, Inc.
914 Plymouth Bldg. (612) 339-9391
12 S. Sixth Street
Minneapolis, MN 55402 Alan K. Ruvelson, President
Venture capital consultants.

ITASCA Growth Fund, Inc.
19 N.E. Third Street (218) 326-0754
Grand Rapids, MN 55744
Small business investment company. Investment Policy: No industry preference.

North Star Ventures, Inc.
P.O. Box 59110 (612) 936-4500
Minneapolis, MN 55440 Terrence W. Glarner, President
Small business investment company. Investment Policy: Invests start-up and early-stage capital in all industry segments excluding real estate, with a preference toward companies in high-technology, electronics, and/or medical industries. Investments range from $400,000 to $1 million, with a preferred limit of $1 million.

Northland Capital Corporation
613 Missabe Bldg. (218) 722-0545
Duluth, MN 55802 George G. Barnum, Jr., President
Small business investment company. Investment Policy: No industry preference.

Pathfinder Venture Capital Funds
7300 Metro Blvd., Suite 585 (612) 835-1121
Minneapolis, MN 55439 Andrew J. Greenshields, General
Partner
Venture capital supplier. Investment Policy: Provides start-up and early-stage financing to emerging companies in the medical, computer, pharmaceuticals, and data communications industries. Emphasis is on companies with proprietary technology or market positions and with substantial potential for revenue growth. Midwest location is preferred. Investments range from $75,000 to $1.5 million, with a $2.5 million maximum.

Threshold Ventures, Inc.
15500 Wayzata Blvd. (612) 473-2051
12 Oak Center, Suite 819
Minneapolis, MN 55391 John L. Shannon, President
Small business investment company. Investment Policy: No industry preference.

Mississippi

Sun-Delta Capital Access Center, Inc.
819 Main St. (601) 335-5291
Greenville, MS 38701 Howard Boutte, Jr., Vice-President
Minority enterprise small business investment company. Investment Policy: No industry preference.

Vicksburg SBIC
P.O. Box 1258 (601) 636-4762
Vicksburg, MS 39181 David L. May, President
Small business investment company. Investment Policy: No industry preference.

Missouri

Bankers Capital Corporation
3100 Gillham Rd. (816) 531-1600
Kansas City, MO 64109 Raymond E. Glasnapp, President
Small business investment company. Investment Policy: No industry preference.

*P*rofessor Mark A. Spikell of George Mason University advises those contemplating the venture capital route to think about doing it from day one, "so that you don't make a lot of mistakes that have to be recouped before investors will talk to you." And, he says, make sure you know the rules of the game. Offers Spikell: "Moderately intelligent and bright people, when they know the rules of the game, stand a good chance to develop strategies that will enable them to win. It's only when moderately intelligent and bright people don't know the game they are playing that they lose almost every time."

Capital for Business, Inc. (Kansas City)
1000 Walnut, 18th Floor   (816) 234-2357
Kansas City, MO 64106   Bart S. Bergman, Executive Vice-
President
    Small business investment company. Investment Policy: No industry preference. Remarks: A branch office of the Capital for Business headquarters located in St. Louis.

Capital for Business, Inc. (St. Louis)
11 S. Meramec, Suite 800   (314) 854-7427
St. Louis, MO 63105
    Small business investment company. Investment Policy: Focuses primarliy on expanding manufacturing and on leveraged buy outs. Investments range from $300,000 to $1 million. Remarks: Maintains a branch office in Kansas City, Missouri.

Intercapco West, Inc. (St. Louis)
7800 Bonhomme Ave.   (314) 863-0600
St. Louis, MO 63105   Thomas E. Phelps, President
    Small business investment company. Investment Policy: No industry preference.

InvestAmerica Venture Group, Inc. (Kansas City)
911 Main St., Commerce Tower Bldg.,   (816) 842-0114
  Suite 2724A
Kansas City, MO 64105
    Small business investment company. Investment Policy: No industry preference. Remarks: A branch office of InvestAmerica Venture Group, Inc., Cedar Rapids, Iowa.

MBI Venture Capital Investors, Inc.
850 Main St.   (816) 471-1700
Kansas City, MO 64105   Anthony Sommers, President
    Small business investment company. Investment Policy: No industry preference.

United Missouri Capital Corporation
1010 Grand Ave.   (816) 556-7333
Kansas City, MO 64106   Joe Kessinger, Manager
    Small business investment company. Investment Policy: No industry preference.

Nebraska

Community Equity Corporation of Nebraska
6421 Ames Ave.   (402) 455-7722
Omaha, NE 68104   William C. Moore, President
    Minority enterprise small business investment company. Investment Policy: No industry preference.

United Financial Resources Corporation
P.O. Box 1131   (402) 734-1250
Omaha, NE 68101   Terrence W. Olsen, President
    Small business investment company. Investment Policy: Grocery store industry preferred.

Nevada

Enterprise Finance Capital Development Corporation
One E. First St., Suite 1102   (702) 329-7797
P.O. Box 3597
Reno, NV 89501   Robert S. Russell, Chairman
    Small business investment company. Investment Policy: No industry preference.

New Hampshire

Granite State Capital, Inc.
Seven Islington Rd.   (603) 436-5044
P.O. Box 6564
Portsmouth, NH 03801   Richard S. Carey, Treasurer
    Small business investment company. Investment Policy: No industry preference.

VenCap
1155 Elm St.   (603) 644-6100
Manchester, NH 03101   Richard J. Ash, President
  Small business investment company. Investment Policy: No industry preference.

New Jersey

Accel Partners (Princeton)
One Palmer Square   (609) 683-4500
Princeton, NJ 08542
  Venture capital firm. Investment Policy: Telecommunications industry, software, and biotechnology/medical products preferred. Will provide $100,000 to $5 million in first-round and later-stage financing. Remarks: Maintains an office in San Francisco, California. **Contact:** Dixon R. Doll; Arthur C. Patterson; James R. Swartz, General Partners.

Bradford Associates
22 Chambers St., Fourth Floor (609) 921-3880
Princeton, NJ 08540   Winston J. Churchill, General Partner
  Venture capital firm. Investment Policy: No industry preference.

Bridge Capital Advisors, Inc.
Glen Point Center West   (201) 836-3900
Teaneck, NJ 07666
  Venture capital firm. Investment Policy: Prefers later-stage financing. Prefers small growth companies or companies with $5 to $100 million in annual revenues. ridge Capital Investors is a subsidiary at same address. **Contact:** Donald P. Remey, Managing Director; Hoyt J. Goodrich, Geoffrey Wadsworth; Jay Barton Goodwin, General Partners.

DSV Partners
221 Nassau St.   (609) 924-6420
Princeton, NJ 08542
  Private venture capital supplier. Investment Policy: Provides seed, research and development, start-up, first-stage, and second-stage financing to companies specializing in communications, computers, electronic components and instrumentation, energy/natural resources, genetic engineering, industrial products and equipment, and medical-related products and services. Will not consider real estate. Preferred investment is $750,000 or more, with a minimum of $250,000. **Contact:** James R. Bergman; John K. Clarke; Morton Collins; Robert S. Hillas, Partners.

Edelson Technology Partners
Park 80, W., Plaza 2   (201) 843-4474
Saddle Brook, NJ 07662   Harry Edelson, General Partner
  Venture capital partnership. Investment Policy: Prefers high-tech industries.

Edison Venture Fund
997 Lenox Dr., Suite 3   (609) 896-1900
Lawrenceville, NJ 08648
  Private venture capital firm. Investment Policy: No industry preference. **Contact:** John H. Martinton; James F. Mrazek, General Partners.

Eslo Capital Corporation
212 Wright St.   (201) 242-4488
Newark, NJ 07114   Leo Katz, President
  Small business investment company. Investment Policy: No industry preference.

First Princeton Capital Corporation
Five Garret Mt. Plaza   (201) 278-8111
West Patterson, NJ 07424 Michael Lytell, Chairman
  Small business investment company. Investment Policy: No industry preference.

InnoVen Group
Park 80, W., Plaza 1   (201) 845-4900
Saddle Brook, NJ 07662   Gerald A. Lodge, Chief Executive Officer
  Venture capital firm. Investment Policy: Prefers to invest in high-tech industries. Also prefers second- and third-round financing.

Johnston Associates, Inc.
181 Cherry Valley Rd.   (609) 924-3131
Princeton, NJ 08540   Robert F. Johnston, Managing Director
  Venture capital supplier. Investment Policy: Seeks high-technology, medical, and biologically oriented concepts in order to initiate and guide the establishment of a company by providing seed and start-up financing. Remarks: Will work with a scientist or inventor to recruit a management team.

The best way to check out the capabilities of venture capitalists is to speak with the management of other companies that the venture capitalist has assisted. And don't just look into successful enterprises, but find people who have experienced total flops to see how the venture capitalist handles both success and failure. Any seasoned venture capitalist will have had a share of good and bad deals.

Monmouth Capital Corporation
125 Wyckoff Rd.
P.O. Box 335
Eatontown, NJ 07724

(201) 542-4927
Ralph B. Patterson, Executive Vice-
President
Small business investment company. Investment Policy: No industry preference.

Rutgers Minority Investment Company
180 University
Newark, NJ 07102

(201) 648-5627
Oscar Figueroa, President
Minority enterprise small business investment company. Investment Policy: No industry preference.

Unicorn Ventures II LP
Six Commerce Dr.
Cranford, NJ 07016

(201) 276-7880
Frank P. Diassi, General Partner
Small business investment company. Investment Policy: No industry preference.

New Mexico

Ads Capital Corporation
524 Camino del Monte Sol
Santa Fe, NM 87501

(505) 983-1769
A. David Silver
Venture capital supplier. Investment Policy: Prefers to invest in manufacturing or distribution companies and health science industries. Average investment is $750,000.

Albuquerque SBIC
501 Tijeras Ave., N.W.
P.O. Box 487
Albuquerque, NM 87103

(505) 247-0145
Albert T. Ussery, President
Small business investment company. Investment Policy: No industry preference.

Associated Southwest Investors, Inc.
2400 Louisiana, N.E., Building 4, Suite
225
Albuquerque, NM 87110

(505) 881-0066
John R. Rice, President
Minority enterprise small business investment company. Investment Policy: No industry preference.

Fluid Capital Corporation
6400 Uptown
Albuquerque, NM 87111

(505) 884-3600
George T. Slaughter, President
Small business investment company. Investment Policy: No industry preference.

Fluid Financial Corporation
6400 Uptown
Albuquerque, NM 87111

(505) 884-3600
George T. Slaughter, President
Minority enterprise small business investment company. Investment Policy: No industry preference.

Industrial Development Corporation of Lea County
P.O. Box 1376
Hobbs, NM 88240

(505) 327-2039
Harold Lampe, Executive Director
Certified development company. Investment Policy: Inquire.

Meadow Ventures
1650 University Blvd., N.E., No. 500 A
Albuquerque, NM 87102

(505) 243-7600
Venture capital firm. Investment Policy: Inquire.

Mora San Miguel Guadalupe Development Corporation
131 Bridge St.
Las Vegas, NM 87701

Elroy Aragon
Certified development company. Investment Policy: Inquire.

National Central New Mexico Economic Development District
Development Authority of New Mexico,
Inc.
P.O. Box 5115
Sante Fe, NM 87501

Leo Murphy
Certified development company. Investment Policy: Inquire.

New Mexico Business Development Corp.
6001 Marble, N.E., #6
Albuquerque, NM 87110

(505) 268-1316
Venture capital firm. Investment Policy: Inquire.

SBA 503 Development Company
Roswell Chamber of Commerce
131 W. Second St.
Roswell, NM 88201
    Certified development company. Investment Policy: Inquire.

Southwest Capital Investments, Inc.
515 Niagra, N.E.   (505) 334-4332
3500 E. Comanche Rd., N.E.
Albuquerque, NM 87113   Martin J. Roe, President
    Small business investment company. Investment Policy: No industry preference.

United Mercantile Capital Corporation
P.O. Box 37487   (505) 883-8201
Albuquerque, NM 87176   Joe Justice, General Manager
    Small business investment company. Investment Policy: Manufacturing and distribution preferred.

New York

Adler and Company (New York City)
375 Park Ave., Suite 3303   (212) 759-2800
New York, NY 10152   Frederick R. Adler, General Partner
    Private venture capital supplier. Investment Policy: Provides all stages of financing. Remarks: Maintains an office in Sunnyvale, California.

Alan Patricof Associates, Inc. (New York)
545 Madison Ave., 15th Floor   (212) 753-6300
New York, NY 10022   Alan Patricof, Chairman
    Venture capital firm. Investment Policy: No industry preference, but avoids real estate. Interested in all stages of financing. Also handles turnarounds and leveraged buy outs.

AMEV Capital Corporation
One World Trade Center, Suite 5001   (212) 775-9100
New York, NY 10048-0024   Martin S. Orland, President
    Venture capital supplier. Investment Policy: Diversified with respect to industry, stage of development, and geographic location. Preferred minimum investment is $1 million; able to lead or participate in syndications of up to $10 million or more.

AMEV Venture Management, Inc.
One World Trade Center, 50th Floor   (212) 323-9780
New York, NY 10048   Martin S. Orland, President
    Manages two limited partnerships and one Small business investment company. Investment Policy: Prefers leveraged buy outs and later-stage financing. Investments range from $1 to $3 million.

ASEA—Harvest Partners II
767 Third Ave.   (212) 838-7776
New York, NY 10017   Harvey Wertheim, General Partner
    Small business investment company. Investment Policy: No industry preference.

Avdon Capital Corporation
1413 Avenue J   (718) 692-0950
Brooklyn, NY 11230   A. M. Donner, President
    Minority enterprise small business investment company. Investment Policy: No industry preference.

Bessemer Venture Partners
630 Fifth Ave.   (212) 708-9303
New York, NY 10111   Robert H. Buscher, Partner
    Venture capital partnership. Investment Policy: No industry preference.

Bohlen Capital Corporation
767 Third Ave.   (212) 838-7776
New York, NY 10017   Harvey J. Wertheim, President
    Small business investment company. Investment Policy: No industry preference.

BT Capital Corporation
280 Park Ave.   (212) 850-1916
New York, NY 10015   James G. Hellmuth, Deputy Chairman
    Small business investment company. Investment Policy: No industry preference.

Howard Jay Fleisher, 32, inventor of the Polygonzo sculpture toy, says that the big problem he had when hunting for venture capital was that "people saw my product and company as a one-trick pony. Even though in my business plan I elaborated that I was going to use my product as a springboard for other new and innovative products, most of the investors felt that it was a very high risk." Fleisher was forced by circumstances to use his own money.

"Whenever there is a breakdown there is a breakthrough. And when it seemed like everything was about to fail, somehow another thing was delivered to my doorstep," he says. Fleisher has left his secure job, invested $50,000 of his own money, and already has licensed one company, Nasta International, to manufacture his multi-dimensional puzzle for the toy industry.

Business Loan Center
704 Broadway, 2nd Floor
New York, NY 10003
(212) 979-6688
Henry Fierst
Small business investment company.

Capital Investors and Management Corporation
210 Canal St., Suite 607
New York, NY 10013
(212) 964-2480
Rose Chao, Manager
Minority enterprise small business investment company. Investment Policy: No industry preference.

Chase Manhattan Capital Corporation
One Chase Manhattan Plaza, 7th Floor
New York, NY 10081
(212) 552-6275
Gustav H. Koven, President
Small business investment company. Investment Policy: No industry preference.

Chemical Venture Capital Associates
885 Third Ave.
New York, NY 10028
(212) 230-2280
Steven J. Gilbert, Managing General Partner
Small business investment company. Investment Policy: No industry preference.

Citicorp Venture Capital Ltd. (New York City)
Citicorp Center
399 Park Ave., 6th Floor
New York, NY 10043
(212) 559-1127
Peter G. Gerry, President
Small business investment company. Investment Policy: Invests in the fields of information processing and telecommunications, transportation and energy, and health care, providing financing to companies in all stages of development. Also provides capital for leveraged buy out situations. Remarks: Maintains offices in Palo Alto, California, and Dallas, Texas.

Clinton Capital Corporation
704 Broadway
New York, NY 10003
(212) 979-6688
Mark Scharfman, President
Small business investment company. Investment Policy: No industry preference.

CMNY Capital Company, Inc.
77 Water St.
New York, NY 10005
(212) 437-7078
Robert Davidoff, Vice-President
Small business investment company. Investment Policy: No industry preference.

Columbia Capital Corporation
704 Broadway
New York, NY 10003
(212) 979-6688
Mark Scharfman, President
Minority enterprise small business investment company.

Concord Partners (New York City)
535 Madison Ave.
New York, NY 10022
(212) 906-7000
John P. Birkelund, President
Venture capital partnership. Investment Policy: Diversified in terms of stage of development, industry classification, and geographic location. Areas of special interest include computers, telecommunications, health care and medical products, energy, and leveraged buy outs. Preferred investments range from $2 to $3 million, with a $1 million minimum and $10 million maximum. Remarks: Maintains an office in Palo Alto, California.

CW Group, Inc.
1041 Third Ave.
New York, NY 10021
(212) 308-5266
Venture capital supplier. Investment Policy: Interest is in the health-care field, including diagnostic and therapeutic products, services, and biotechnology. Invests in companies at developing and early stages.

Edwards Capital Company
Two Park Ave.
New York, NY 10016
(212) 686-2568
Edward H. Teitlebaum, Managing Partner
Small business investment company. Investment Policy: Transportation industry preferred.

Elk Associates Funding Corporation
600 Third Ave., 38th Floor
New York, NY 10016
(212) 972-8550
Gary C. Granoff, President
Minority enterprise small business investment company. Investment Policy: Transportation industry preferred.

Elron Technologies, Inc.
850 Third Ave., 10th Floor
New York, NY 10022
(212) 935-3110
Venture capital supplier. Investment Policy: Provides incubation and start-up financing to high-technology companies. Remarks: Elron Technologies is a subsidiary of Elron Electronic Industries Ltd., a diversified high-technology holding company based in Israel.

Equico Capital Corporation
1221 Avenue of the Americas, 32nd Floor   (212) 382-8000
New York, NY 10020   Duane Hill, President
>Minority enterprise small business investment company. Investment Policy: No industry preference.

Euclid Partners Corporation
50 Rockefeller Plaza, Suite 1022   (212) 489-1770
New York, NY 10020   Fred Wilson, Associate
>Venture capital firm. Investment Policy: Prefers early-stage financing in health care and information processing industries.

European Development Corporation
767 Third Ave., Seventh Floor   (212) 838-9721
New York, NY 10017   Harvey J. Wertheim, President
>Small business investment company. Investment Policy: No industry preference.

Everlast Capital Corporation
350 Fifth Ave., Suite 2805   (212) 695-3910
New York, NY 10118   Frank J. Segreto, Vice-President
>Minority enterprise small business investment company. Investment Policy: No industry preference.

Exim Capital Corporation
Nine E. 40th St.   (212) 683-3375
New York, NY 10016   Victor K. Chun, President
>Minority enterprise small business investment company. Investment Policy: No industry preference.

Fair Capital Corporation
Three Pell St., Second Floor   (212) 964-2480
New York, NY 10013   Robert Yet Sen Chen, President
>Minority enterprise small business investment company. Investment Policy: No industry preference.

Ferranti High Technology, Inc.
501 Madison Ave.   (212) 688-9828
New York, NY 10022   Michael R. Simon, Vice-President and
Director
>Small business investment company. Investment Policy: No industry preference.

First Century Partnership (New York City)
1345 Avenue of the Americas   (212) 698-6382
New York, NY 10105   Michael J. Myers
>Private venture capital firm. Investment Policy: Diversified; minimum investment is $1.5 million.

First New York Management Co.
One Pierrepont   (718) 858-0800
300 Cadman Plaza W.
New York, NY 11201   Israel Mindick, General Partner
>Small business investment company. Investment Policy: No industry preference.

Fortieth Street Venture Partners (New York City)
545 Madison Ave., Suite 800   (212) 421-0045
New York, NY 10022   Joseph Mizrachi, General Partner
>Venture capital firm. Investment Policy: No industry preference.

Foster Management Company
437 Madison Ave.   (212) 753-4810
New York, NY 10022   John H. Foster, Chairman and President
>Private venture capital supplier. Investment Policy: Not restricted to specific industries or geographic locations; diversified with investments in the health-care, transportation, broadcasting, communications, energy, and home furnishings industries. Investments range from $2 to $15 million.

Franklin Corporation
767 Fifth Ave.   (212) 486-2323
New York, NY 10153   Norman Stroebel, President
>Small business investment company. Investment Policy: No industry preference; no start-ups.

Fredericks Michael and Company
One World Trade Center, Suite 1509   (212) 466-6620
New York, NY 10048   David Fredericks, Managing Director
>Private venture capital supplier. Investment Policy: Provides start-up and early-stage financing to companies located primarily on the East Coast. Also supplies capital for buy outs and acquisitions.

R*arely is the inventor also a savvy businessperson. If you are not an expert in raising venture capital, find yourself a very good investment banker or someone who knows the ropes and is interested in working with early stage companies to advise you.*

Fresh Start Venture Capital Corporation
313 W. 53th St., Third Floor   (212) 265-2249
New York, NY 10019   Zindel Zelmanovich, President
    Minority enterprise small business investment company. Investment Policy: No industry preference.

Fundex Capital Corporation
525 Northern Blvd.   (516) 466-8551
Great Neck, NY 11021   Howard Sommer, President
    Small business investment company. Investment Policy: No industry preference.

Hambro International Venture Fund (New York)
17 E. 71 St.   (212) 288-7778
New York, NY 10021   Anders K. Brag
    Venture capital supplier. Investment Policy: Seeks to invest in mature companies as well as in high-technology areas from start-ups to leveraged buy outs. Investments range from $100,000 to $3 million, with initial investments ranging from $500,000 to $800,000. Maintains an office in Boston, Massachusetts.

Hanover Capital Corporation
315 E. 62nd St., 6th Floor   (212) 980-9670
New York, NY 10021   Robert Wilson
    Small business investment company. Investment Policy: No industry preference.

Harvest Ventures
767 Third Ave.   (212) 838-7776
New York, NY 10017
    Private venture capital supplier. Investment Policy: Prefers to invest in high-technology, growth-oriented companies with proprietary technology, large market potential, and strong management teams. Specific areas of interest include computers, computer peripherals, semiconductors, telecommunications, factory automation, military electronics, medical products, and health-care services. Provides seed capital of up to $250,000; investments in special high-growth and leveraged buy out situations range from $500,000 to $2 million. Will serve as lead investor and arrange financing in excess of $2 million in association with other groups. Remarks: Maintains an office in Cupertino, California. Harvey J. Wertheim; Harvey P. Mallement, Managing Directors.

Holding Capital Management Corporation
685 Fifth Ave., 14th Floor   (212) 486-6670
New York, NY 10022   Sash A. Spencer, President
    Small business investment company. Investment Policy: No industry preference.

Ibero American Investors Corporation
104 Scio St.   (716) 262-3440
Rochester, NY 14604   Emilio Serrano, President
    Minority enterprise small business investment company. Investment Policy: No industry preference.

Intercontinental Capital Funding Corporation
60 E. 42nd St., Suite 740   (212) 689-2484
New York, NY 10165   James S. Yu, President
    Minority enterprise small business investment company. Investment Policy: No industry preference.

Interstate Capital Company, Inc.
380 Lexington Ave.   (212) 972-3445
New York, NY 10017   David Scharf, President
    Small business investment company. Investment Policy: No industry preference.

ITC Capital Corporation
1290 Avenue of the Americas, 29th Floor   (212) 408-4800
New York, NY 10104   J. Andrew McWethy, President
    Small business investment company; bank holding company fund. Investment Policy: No industry preference.

New Jersey

Japanese American Capital Corporation
716 Jersey Ave.   (201) 798-5000
New Jersey, NJ 07302   Stephen C. Huang, President
    Minority enterprise small business investment company. Investment Policy: No industry preference.

New York

Josephberg, Grosz and Company, Inc.
344 E. 49th St.   (212) 935-1050
New York, NY 10017
> Venture capital firm. Investment Policy: No industry preference. **Contact:** Richard Josephberg, Chairman; Ivan Grosz, President.

Kwiat Capital Corporation
576 Fifth Ave.   (212) 391-2461
New York, NY 10036   Sheldon F. Kwiat, President
> Small business investment company. Investment Policy: No industry preference.

Lawrence, Tyrrell, Ortale, and Smith (New York)
515 Madison Ave., 29th Floor   (212) 826-9080
New York, NY 10022   Brian T. Horey
> Venture capital firm. Investment Policy: No industry preference.

M and T Capital Corporation
One M and T Plaza   (716) 842-5881
Buffalo, NY 14240   William Randon, President
> Small business investment company. Investment Policy: No industry preference.

Manufacturers Hanover Venture Capital Corporation
270 Park Ave.   (212) 270-3220
New York, NY 10017
> Venture capital and leveraged buy out firm. Investment Policy: Invests in leveraged buy outs and growth equity.

Medallion Funding Corporation
205 E. 42nd St., Suite 2020   (212) 682-3300
New York, NY 10017   Alvin Murstein, President
> Minority enterprise small business investment company. Investment Policy: Transportation industry preferred.

Minority Equity Capital Company, Inc.
42 W. 38th St., Suite 604   (212) 786-4240
New York, NY 10018   Donald F. Greene, President
> Minority enterprise small business investment company. Investment Policy: No industry preference.

Monsey Capital Corporation
125 Route 59   (914) 425-2229
Monsey, NY 19052   Shamuel Myski, President
> Minority enterprise small business investment company. Investment Policy: No industry preference.

Nazem and Company
600 Madison Ave., 14th Floor   (212) 644-6433
New York, NY 10022   Frederick F. Nazem, Managing General Partner
> Venture capital fund. Investment Policy: Electronics and medical industries preferred. Will provide seed, first-round, and second-round financing.

New Enterprise Associates (New York City)
119 E. 55th St.   (212) 371-8210
New York, NY 10022
> Venture capital supplier. Investment Policy: Concentrates on technology-based industries that have the potential for product innovation, rapid growth, and high profit margins. Investments range from $250,000 to $2 million.

Noro Capital Ltd.
767 Third Ave., Seventh Floor   (212) 838-9720
New York, NY 10017   Harvey J. Wertheim, President
> Small business investment company. Investment Policy: No industry preference.

Norstar Venture Capital
One Norstar Plaza   (518) 447-4043
Albany, NY 12207-2796   Raymond A. Lancaster, President
> Venture capital supplier. Investment Policy: No industry preference. Typical investment is between $500,000 and $1 million. Remarks: A wholly owned subsidiary of Norstar Bancorp.

North American Funding Corporation
177 Canal St.   (212) 226-0080
New York, NY 10013   Franklin F. Y. Wong, President
> Minority enterprise small business investment company. Investment Policy: No industry preference.

An inventor would be well advised to engage the services of an attorney and / or CPA before signing on with any venture captalist. The money you spend will be money well spent if it gets you a good deal or saves you from a bad one.

NYBDC Capital Corporation
P.O. Box 738
Albany, NY 12201-0738
(518) 463-2268
Robert Lazar, President
  Small business investment company. Investment Policy: No industry preference.

Pan Pac Capital Corporation
121 E. Industry Ct.
Deer Park, NY 11729
(516) 586-7653
Dr. In Ping Jack Lee, President
  Minority enterprise small business investment company. Investment Policy: No industry preference.

Pierre Funding Corporation
605 Third Ave.
New York, NY 10016
(212) 490-9540
Elias Debbas, President
  Minority enterprise small business investment company. Investment Policy: No industry preference.

Prospect Group, Inc.
667 Madison Ave.
New York, NY 10021
(212) 758-8500
Thomas A. Barron, President
  Venture capital supplier. Investment Policy: Investments focus on computer communications and software products, fiber optics, genetic engineering, biotechnology, health care management, and solar energy.

Prudential Venture Capital
717 Fifth Ave., 11th Fl.
New York, NY 10022
(212) 753-0901
  Venture capital fund. Investment Policy: Specialty retailing, medical and health services, communications, and technology companies preferred. Will provide $3 to $7 million in equity financing for later-stage growth companies. **Contact:** Robert Knox, President; William Field, Chairman.

Questech Capital Corporation
320 Park Ave.
New York, NY 10022
(212) 891-7300
John E. Koonce, III, President
  Small business investment company. Investment Policy: No industry preference.

R and R Financial Corporation
1451 Broadway
New York, NY 10036
(212) 790-1400
Imre Rosenthal, President
  Small business investment company. Investment Policy: No industry preference.

Rand SBIC, Inc.
1300 Rand Bldg.
Buffalo, NY 14203
(716) 853-0802
Donald Ross, President
  Small business investment company. Investment Policy: No industry preference.

S and S Venture Associates Ltd.
Seven Penn Plaza, Suite 320
New York, NY 10001
(212) 736-4530
Donald Smith, President
  Small business investment company. Investment Policy: No industry preference.

767 Limited Partnership
767 Third Ave.
New York, NY 10017
(212) 838-7776
  Small business investment company. Investment Policy: No industry preference. **Contact:** H. Wertheim; H. Mallement, General Partners.

Situation Ventures Corporation
1626 52nd St.
Brooklyn, NY 11204
(718) 438-4909
  Investment Policy: Inquire.

Sprout Group (New York City)
140 Broadway, 42nd Fl.
New York, NY 10005
(212) 504-3600
Richard E. Kroon, Managing Partner
  Venture capital affiliate of Donaldson, Lufkin, and Jenrette. Investment Policy: Provides early-stage financing to companies specializing in retailing, financial services, and high-technology products. Will also invest in selected later-stage emerging growth companies and management buy outs of mature companies. Investments range from $1 to $2 million, with a minimum of $500,000. Remarks: Maintains offices in Menlo Park, California, and Boston, Massachusetts.

New Jersey

Taroco Capital Corporation
716 Jersey Ave.   (201) 798-5000
New Jersey, NJ 07302   David R. C. Chang, President
    Minority enterprise small business investment company. Investment Policy: Chinese-Americans preferred.

New York

Telesciences Capital Corporation
26 Broadway, Suite 748   (212) 425-0320
New York, NY 10004   Mike A. Petrozzo
    Small business investment company. Investment Policy: No industry preference.

Tessler and Cloherty, Inc.
155 Main St.   (914) 265-4244
Cold Spring, NY 10516
    Small business investment company. Investment Policy: No industry preference. **Contact:** Patricia Cloherty; Dan Tessler, General Partners.

TLC Funding Corporation
660 White Plains Rd.   (914) 332-5200
Tarrytown, NY 10591   Philip G. Kass, President
    Small business investment company. Investment Policy: No industry preference.

Triad Capital Corporation of New York
960 Southern Blvd.   (212) 589-5000
Bronx, NY 10459   Lorenzo J. Barrera, President
    Minority enterprise small business investment company. Investment Policy: No industry preference.

Vega Capital Corporation
720 White Plains Rd.   (914) 472-8550
Scarsdale, NY 10583   Victor Harz, President
    Small business investment company. Investment Policy: No industry preference.

Venture Opportunities Corporation
110 E. 59th St., 29th Floor   (212) 832-3737
New York, NY 10022   A. Fred March, President
    Minority enterprise small business investment company. Investment Policy: No industry preference.

Venture SBIC, Inc.
249-12 Jericho Tnpk.   (516) 352-0068
Floral Park, NY 11001   Arnold Feldman, President
    Small business investment company. Investment Policy: No industry preference.

Warburg Pincus Ventures, Inc. (New York)
466 Lexington Ave.   (212) 878-0600
New York, NY 10017   Christopher Brody, Managing Director
    Private venture capital firm. Investment Policy: No industry preference.

Watchung Capital Corporation
431 Fifth Ave., Fifth Floor   (212) 889-3466
New York, NY 10016   S. T. Jeng
    Investment Policy: Inquire.

Welsh, Carson, Anderson, and Stowe
200 Liberty, Suite 3601   (212) 945-2000
New York, NY 10281   Patrick Welsh, General Partner
    Venture capital partnership. Investment Policy: High-technology industries preferred. Also interested in leveraged buy outs. Minimum investment is $5 million.

WFG-Harvest Partners Ltd.
767 Third Ave.   (212) 838-7776
New York, NY 10017   Harvey J. Wertheim, General Partner
    Small business investment company. Investment Policy: No industry preference.

Winfield Capital Corporation
237 Mamaroneck Ave.   (914) 949-2600
White Plains, NY 10605   Stanley M. Pechman, President
    Small business investment company. Investment Policy: No industry preference.

Wood River Capital Corporation (New York City)
667 Madison Ave., 12th Floor
New York, NY 10021
(212) 758-8500
Thomas A. Barron, President
Small business investment company. Investment Policy: No industry preference.

Yang Capital Corporation
41-40 Kissena Blvd.
Flushing, NY 11355
(718) 849-1037
Ms. Maysing Yang, President
Minority enterprise small business investment company. Investment Policy: No industry preference.

Yusa Capital Corporation
622 Broadway, Second Floor
New York, NY 10012
(212) 420-1350
Christopher Yeung, Chairman of the Board
Minority enterprise small business investment company. Investment Policy: No industry preference.

North Carolina

Falcon Capital Corporation
400 W. Fifth St.
Greenville, NC 27834
(919) 752-5918
P. S. Prasad, President
Small business investment company. Investment Policy: No industry preference.

Heritage Capital Corporation
Two First Union Center
Charlotte, NC 28282
(704) 334-2867
William R. Starnes, President
Small business investment company. Investment Policy: Diversified industries.

Kitty Hawk Capital Ltd.
1640 Independence Center
Charlotte, NC 28246
(704) 333-3777
Small business investment company. Investment Policy: No industry preference. **Contact:** Walter H. Wilkinson, President; Chris Hegele, Vice-President.

NCNB SBIC Corporation
One NCNB Plaza (T05-2)
Charlotte, NC 28255
(704) 374-5583
Troy S. McCrory, Jr., President
Small business investment company. Investment Policy: No industry preference.

NCNB Venture Company LP
One NCNB Plaza (T05-2)
Charlotte, NC 28255
(704) 374-5723
S. Epes Robinson, General Partner
Small business investment company. Investment Policy: No industry preference.

Southgate Venture Partners (Charlotte)
227 N. Tryon St., Suite 201
Charlotte, NC 28202
(704) 372-1410
Alex Wilkins
Private venture capital firm. Investment Policy: Diversified.

Ohio

A. T. Capital Corporation
900 Euclid Ave., T-18
Cleveland, OH 44101
(216) 737-4970
Small business investment company. Investment Policy: No industry preference. **Contact:** Robert C. Salipante, President; Lisa M. Simecek, Vice-President and Manager.

Banc One Capital Corporation
100 E. Broad St.
Columbus, OH 43271-0251
(614) 248-5832
James E. Kolls, Vice-President
Small business investment company. Investment Policy: No industry preference.

Capital Funds Corporation
800 Superior Ave.
Cleveland, OH 44114
(216) 344-5775
Small business investment company. Investment Policy: Diversified. **Contact:** Carl G. Nelson, Chief Investment Officer; David B. Chilcote, Vice-President.

Cardinal Development Capital Fund I

155 E. Broad   (614) 464-5550
Columbus, OH 43215

    Private venture capital firm. Investment Policy: Provides expansion capital to manufacturing and service firms, particularly in Ohio and other parts of the Midwest. Avoids investments in real estate and resource recovery systems. Minimum investment is $250,000, with a preferred investment of $500,000 or more.

Center City MESBIC, Inc.

40 S. Main St., Suite 762   (513) 461-6164
Dayton, OH 45402   Michael A. Robinson, President

    Minority enterprise small business investment company. Investment Policy: Diversified industries.

Clarion Capital Corporation

1801 E. Ninth, Suite 44114   (216) 687-1096
Cleveland, OH 44114   Morton A. Cohen, President

    Small business investment company. Investment Policy: Specialty chemicals, instrumentation, and health care.

First Ohio Capital Corporation

P.O. Box 1868   (419) 259-7141
Toledo, OH 43603   David J. McMacken, General Manager

    Small business investment company. Investment Policy: No industry preference.

Fortieth Street Venture Partners (Cincinnati)

3712 Carew Tower   (513) 579-0101
Cincinnati, OH 45202   Joseph Mizrachi, General Partner

    Venture capital firm. Investment Policy: No industry preference.

Gries Investment Corporation

Statler Office Tower, Suite 1500   (216) 861-1146
Cleveland, OH 44115   Robert D. Gries, President

    Small business investment company. Investment Policy: No industry preference.

Lubrizol Enterprises, Inc.

29400 Lakeland Blvd.   (216) 943-4200
Wickliffe, OH 44092   Bruce H. Grasser, Senior Vice-President

    Venture capital supplier. Investment Policy: Provides seed capital and later-stage expansion financing to emerging companies in the biological, chemical, and material sciences whose technology is applicabel to and related to the production and marketing of specialty and fine chemicals. Investments range from $250,000 to $15 million.

Miami Valley Capital, Inc.

Talbott Tower, Suite 315   (513) 222-7222
131 N. Ludlow St.
Dayton, OH 45402   Everett F. Telljohann

    Small business investment company. Investment Policy: No industry preference.

Morgenthaler Ventures

700 National City Bank Bldg.   (216) 621-3070
Cleveland, OH 44114   David T. Morgenthaler, Managing
                                       Partner

    Private venture capital supplier. Investment Policy: Provides start-up and later-stage financing to all types of business in North America; prefers not to invest in real estate and mining. Investments range from $500,000 to $3 million.

National City Capital Corporation

1965 E. Sixth St., No. 400   (216) 575-2491
Cleveland, OH 44114   John B. Naylor, President

    Small business investment company. Investment Policy: No industry preference.

Primus Capital Fund

One Cleveland Center, Suite 2700   (216) 621-2185
1375 E. Ninth St.
Cleveland, OH 44114

    Venture capital partnership. Investment Policy: Provides seed, early-stage, and expansion financing to companies located in Ohio and the Midwest, preferring to invest in businesses compatible with the manufacturing, service, and technology base in the Midwest. Industries of interest include industrial automation, medical technologies, and consumer and industrial products with technical content. Investments generally range from $1 to $5 million. Equity capital for leverage management buy outs also is available. **Contact:** Loyal Wilson; James Bartlett, Managing Partners.

"Myth: Complex problems require complex solutions arrived at through complex thinking."
-Eric M. Bienstock, Ph.D.

River Capital Corporation (Cleveland)
796 Huntington Bldg.   (216) 781-3655
Cleveland, OH 44115
> Small business investment company. Investment Policy: No industry preference. Remarks: A branch office of River Capital Corporation, Alexandria, Virginia.

Rubber City Capital Corporation
1144 E. Market St.   (216) 796-9167
Akron, OH 44316   Jesse T. Williams, Sr., President
> Minority enterprise small business investment company. Investment Policy: No industry preference.

SeaGate Venture Management, Inc.
245 Summit St., Suite 1403   (419) 259-8526
Toledo, OH 43603   Charles A. Brown, Vice-President
> Small business investment company. Investment Policy: No industry preference.

Seed One
Park Place   (216) 650-2338
Ten W. Streetsboro St.
Hudson, OH 44236   Burton D. Morgan
> Private venture capital firm. Investment Policy: No industry preference. Equity financing only.

Technology Group Ltd.
Scientific Advances, Inc.   (614) 424-7005
601 W. Fifth Ave.
Columbus, OH 43201   Charles G. James, President
> Venture capital partnership managed by Scientific Advances, Inc. Investment Policy: Seeks small to medium-sized companies with innovative, proven technologies. Minimum investment is $250,000. Remarks: Scientific Advances is a subsidiary of Battelle Memorial Institute, a large, independent research organization that assists in the technical evaluation of investment proposals.

Tomlinson Capital Corporation
13700 Broadway   (216) 587-3400
Garfield Heights, OH 44125   John A. Chernak, President
> Small business investment company. Investment Policy: Miniature supermarket industry preferred.

Oklahoma

Alliance Business Investment Company (Tulsa)
17 E. Second St.   (918) 584-3581
One Williams Center, Suite 2000
Tulsa, OK 74172   Barry Davis, President
> Small business investment company. Investment Policy: No industry preference. Remarks: Maintains an office in Houston, Texas.

Signal Capital Corporaton
One Leadership Sq., Suite 400   (405) 235-4440
Oklahoma City, OK 73102   John Lewis
> Small business investment company. Investment Policy: No industry preference.

Western Venture Capital Corporation
P.O. Box 702680   (918) 749-7000
Tulsa, OK 74170   William Baker, President
> Small business investment company. Investment Policy: Seeks growth companies, turnarounds, and leveraged buy outs in the following areas: oil and gas technology; other natural resources; agribusiness; medical technology and biotechnology; health care; electronics and computer technology; communications and information; leisure and entertainment; and specialty retailing. Investments range from $200,000 to $2 million.

Oregon

Cable and Howse Ventures (Portland)
101 S.W. Main, Suite 1800   (503) 248-9646
Portland, OR 97204
> Venture capital supplier. Investment Policy: Provides start-up and early-stage financing to enterprises in the western United States, although a national perspective is maintained. Interest lies in proprietary or patentable technology. Investments range from $500,000 to $2 million. Remarks: A branch office of Cable and Howse Ventures, Bellevue, Washington.

InterVen II LP (Portland)
227 S.W. Pine St., Suite 200   (503) 223-4334
Portland, OR 97204
> Small business investment company. Investment Policy: No industry preference. Remarks: Affiliated with InterVen II LP, in Los Angeles.

Trendwest Capital Corporation
803 Main St., Suite 404   (503) 882-8059
Klamath Falls, OR 97601   Mark E. Nicol, President
> Small business investment company. Investment Policy: No industry preference.

Pennsylvania

Alliance Enterprise Corporation
1801 Market St., Third Floor   (215) 977-3925
Philadelphia, PA 19103   W. B. Priestley, President
> Minority enterprise small business investment company. Investment Policy: Broadcasting and manufacturing preferred.

Capital Corporation of America
225 S. 15th St., Suite 920   (215) 732-1666
Philadelphia, PA 19102   Martin M. Newman, President
> Small business investment company. Investment Policy: No industry preference.

Core States Enterprise Fund
One Penn Center, Suite 1360   (215) 568-4677
1617 J. F. Kennedy Bldg.
Philadelphia, PA 19103
> Venture capital supplier. Investment Policy: No industry preference. Remarks: A subsidiary of the Philadelphia National Bank. **Contact:** Paul A. Mitchell, President; Michael F. Donoghue, Vice-President.

Enterprise Venture Capital Corporation of Pennsylvania
551 Main St., Suite 303   (814) 535-7597
Johnstown, PA 15901   Donald W. Cowie, Vice-President
> Small business investment company. Investment Policy: No industry preference.

Erie SBIC
32 W. Eighth St., Suite 615   (814) 453-7964
Erie, PA 16501   George R. Heaton, President
> Small business investment company. Investment Policy: No industry preference.

First SBIC of California (Washington)
P.O. Box 512   (412) 223-0707
Washington, PA 15301   Daniel A. Dye
> Small business investment company. Investment Policy: No industry preference.

First Valley Capital Corporation
640 Hamilton Mall   (215) 776-6766
Allentown, PA 18101   Matthew W. Thomas, President
> Small business investment company. Investment Policy: No industry preference.

Greater Philadelphia Venture Capital Corporation, Inc.
225 S. 15th St., Suite 920   (215) 732-3415
Philadelphia, PA 19102   Martin M. Newman, General Manager
> Minority enterprise small business investment company. Investment Policy: No industry preference.

Keystone Venture Capital Management Company
211 S. Broad St.   (215) 985-5519
Philadelphia, PA 19107   G. Kenneth Macrae, General Partner
> Private venture capital partnership. Investment Policy: Provides equity-based expansion financing to companies in the computer, communications, automated equipment, medical, and other industries. Also provides financing for small leveraged buy outs.

Meridian Capital Corporation
Horshaw Business Center   (215) 957-7500
455 Business Center Dr.
Horshaw, PA 19044   Joseph E. Laky, President
> Small business investment company. Investment Policy: No industry preference.

Philadelphia Ventures
1760 Market St.   (215) 751-9444
Philadelphia, PA 19103
   Venture capital partnership. Investment Policy: Provides start-up and early-stage financing to companies offering products or services based on technology or other proprietary capabilities. Industries of particular interest are information processing equipment and services, medical products and services, data communications, and industrial automation. Initial investments range from $500,000 to $2 million. Also supplies capital for buy out situations.

PNC Venture Capital Group
Pittsburgh National Bank Bldg.   (412) 762-8892
Fifth Ave. and Wood, 19th Floor
Pittsburgh, PA 15222   David Hillman, Executive Vice-President
   PNC Venture Capital Group comprises two separate entities; PNC Capital Corporation, an Small business investment company, and PNC Venture Corporation, a venture capital supplier. Investment Policy: Prefers to invest in later-stage and leveraged buy out situations. Interest is in regionally based companies, but will consider investing nationwide. Industry preference is diversified.

Salween Financial Services, Inc.
228 N. Pottstown Pike   (215) 524-1880
Exton, PA 19341   Dr. Ramarao Naidu, President
   Minority enterprise small business investment company. Investment Policy: No industry preference.

Trivest Venture Fund
P.O. Box 136   (412) 741-0754
Ligonier, PA 15658   Thomas W. Courtney, General Partner
   Venture capital firm. Investment Policy: No industry preference.

Puerto Rico

North America Investment Corporation
P.O. Box 1831   (809) 754-6177
Hato Rey, PR 00919   Santigo Ruz Betacourt, President
   Minority enterprise small business investment company. Investment Policy: No industry preference.

Rhode Island

Domestic Capital Corporation
815 Reservoir Ave.   (401) 946-3310
Cranston, RI 02910   Nathaniel B. Baker, President
   Small business investment company. Investment Policy: No industry preference.

Moneta Capital Corporation
285 Governor St.   (401) 861-4600
Providence, RI 02906   Arnold Kilberg, President
   Small business investment company. Investment Policy: No industry preference.

Narragansett Venture Corporation
Fleet Center, Ninth Floor   (401) 751-1000
50 Kennedy Plaza   Robert D. Manchester, Managing
Providence, RI 02903   Director
   Private venture capital supplier. Investment Policy: No industry preference.

South Carolina

Charleston Capital Corporation
111 Church St.   (803) 723-6464
P.O. Box 328
Charleston, SC 29402   Henry Yaschik, President
   Small business investment company. Investment Policy: No industry preference.

Lowcountry Investment Corporation
4401 Daley St.　　　　　　　　　(803) 554-9880
P.O. Box 10447
Charleston, SC 29411　　　　　　Joseph T. Newton, Jr., President
　　Small business investment company. Investment Policy: Grocery industry preferred.

Reedy River Ventures, Inc.
P.O. Box 17526　　　　　　　　　(803) 232-6198
Greenville, SC 29606　　　　　　John M. Sterling, President
　　Small business investment company. Investment Policy: No industry preference.

Tennessee

Chickasaw Capital Corporation
67 Madison Ave.　　　　　　　　(901) 523-6404
P.O. Box 387
Memphis, TN 38147　　　　　　　Tom Moore, President
　　Minority enterprise small business investment company. Investment Policy: No industry
　　preference.

Financial Resources, Inc.
200 Jefferson Ave., Suite 750　　(901) 527-9411
Memphis, TN 38103　　　　　　　Milton Picard, Chairman of the Board
　　Small business investment company. Investment Policy: No industry preference.

International Paper Capital Formation, Inc.
Tower 2, 4th Floor　　　　　　　(901) 763-5951
6400 Poplar Ave., 4-061
Memphis, TN 38197　　　　　　　Richard M. Ludwig, President
　　Minority enterprise small business investment company. Investment Policy: No industry
　　preference.

Lawrence Venture Associates
3401 W. End Ave., Suite 680　　(615) 383-0982
Nashville, TN 37203　　　　　　　Tom Gallagher
　　Private venture capital firm. Investment Policy: Prefers to invest in health care industries.

Massey Burch Investment Group
310 25th Ave., N., Suite 103　　(615) 329-9448
Nashville, TN 37203　　　　　　　Bob Fisher, Associate
　　Venture capital firm. Investment Policy: No industry preference. Investments range from $1 to $3
　　million.

Tennessee Venture Capital Corporation
P.O. Box 2567　　　　　　　　　(615) 244-6935
Nashville, TN 37219　　　　　　　Wendell P. Knox, President
　　Minority enterprise small business investment company. Investment Policy: No industry
　　preference.

Valley Capital Corporation
Krystal Bldg., Suite 212　　　　　(615) 265-1557
100 W. Martin Luther King Blvd.
Chattanooga, TN 37402　　　　　Lamar J. Partridge, President
　　Minority enterprise small business investment company. Investment Policy: No industry
　　preference.

Texas

Acorn Ventures, Inc.
520 Post Oak Blvd., Suite 120　　(713) 622-9595
Houston, TX 77027
　　Investment Policy: No industry preference. Remarks: Maintains an office in Austin, Texas.
　　Contact: Stuart Schube; Walter Cunningham.

Alliance Business Investment Company (Houston)
910 Louisiana　　　　　　　　　(713) 224-8224
3990 One Shell Plaza
Houston, TX 77002
　　Small business investment company. Investment Policy: No industry preference. Remarks: A
　　branch office of Alliance Business Investment Company in Tulsa, Oklahoma.

Allied Bancshares Capital Corporation
P.O. Box 3326
Houston, TX 77253

(713) 224-6611
D. Kent Anderson, Chairman of the
Board
Small business investment company. Investment Policy: No industry preference.

Austin Ventures LP
1300 Norwood Tower
114 W. Seventh St.
Austin, TX 78701

(512) 479-0055

Kenneth P. DeAngelis, General Partner
Austin Ventures administers investments through two funds: Austin Ventures LP, and Rust Ventures LP. Investment Policy: Prefers to invest in start-up/emerging growth companies located in the Southwest and in special situations such as buy outs, acquisitions, and mature companies. No geographic limitations are placed on later-stage investments. Investments range from $1 to $2 million for start-up financing and from $2 to $5 million for special situation companies.

Brittany Capital Company
1525 Elm St., 2424 LTV Tower
Dallas, TX 75201

(214) 954-1515
Robert E. Clements, General Partner
Small business investment company. Investment Policy: No industry preference.

Capital Marketing Corporation
100 Nat Gibbs Dr.
P.O. Box 1000
Keller, TX 76248

(817) 431-7309

Ray Ballard, General Manager
Small business investment company. Investment Policy: Retail grocery industry preferred.

Capital Southwest Corporation
12900 Preston Rd., Suite 700
Dallas, TX 75230

(214) 233-8242
William R. Thomas, President
Small business investment company. Investment Policy: No industry preference.

Charter Venture Group, Inc.
2600 Citadel Plaza Dr., Suite 600
Houston, TX 77008

(713) 863-0704
Winston C. Davis, President
Small business investment company. Investment Policy: No industry preference.

Chen's Financial Group, Inc.
1616 W. Loop, S., Suite 200
Houston, TX 77027

(713) 850-1542
Samuel S. C. Chen, President
Minority enterprise small business investment company. Investment Policy: No industry preference.

Citicorp Venture Capital Ltd. (Dallas)
717 N. Harwood, Suite 2920-LB 87
Dallas, TX 75221

(214) 880-9670
Thomas F. McWilliams
Small business investment company. Investment Policy: No industry preference. Remarks: A branch office of Citicorp Venture Capital Ltd. in New York, New York.

Criterion Investments
1000 Louisiana St., Suite 6200
Houston, TX 77002

(713) 751-2400
David O. Wick, Jr., President
Venture capital fund. Investment Policy: Criterion Investments raises venture capital, seeking companies headquartered in the Sunbelt region.

Energy Capital Corporation
953 Esperson Bldg.
Houston, TX 77002

(713) 236-0006
Herbert F. Poyner, President
Small business investment company. Investment Policy: Specializes in oil and gas energy industries.

Enterprise Capital Corporation
515 Post Oak Blvd., Suite 310
Houston, TX 77027

(713) 621-9444
Fred Zeidman, President
Small business investment company. Investment Policy: No industry preference.

FCA Investment Company
5847 San Felipe, Suite 850
Houston, TX 77057

(713) 781-2857
Robert S. Baker, Jr., Chairman
Small business investment company. Investment Policy: No industry preference; no start-ups.

Idanta Partners
201 Main St., Suite 3200
Fort Worth, TX 76102

(817) 338-2020

Venture capital partnership. Investment Policy: Provides start-up and second-stage financing; will also invest in special situations such as leveraged buy outs. Minimum investment is $250,000. David J. Dunn, Managing Partner; Dev Purkayastha, General Partner.

Mapleleaf Capital Corporation
55 Waugh, Suite 710   (713) 880-4494
Houston, TX 77007   Edward Fink, Managing General Partner
    Small business investment company. Investment Policy: No industry preference.

May Financial Corporation
8333 Douglass Ave., Suite 400   (214) 987-5200
Lock Box 82   Jack McGuire, Head of Corporate
Dallas, TX 75225   Financing
    Brokerage firm working with a venture capital firm. Investment Policy: Prefers food, oil and gas, and electronics industries.

MESBIC Financial Corporation of Dallas
12655 N. Central Exwy., Suite 710   (214) 991-1597
Dallas, TX 75243   Ira D. Harrison, Vice-President
    Minority enterprise small business investment company. Investment Policy: No industry preference.

MESBIC Financial Corporation of Houston
811 Rusk, Suite 201   (713) 228-8321
Houston, TX 77002   Lynn H. Miller, President
    Minority enterprise small business investment company. Investment Policy: No industry preference.

Minority Enterprise Funding, Inc.
17300 El Camino Real, Suite 107B   (713) 488-4919
Houston, TX 77058   Frederick C. Chang, President
    Minority enterprise small business investment company. Investment Policy: No industry preference.

Mventure Corp
1717 Main St., Lower Level 1   (214) 939-3131
Momentum Place
Dallas, TX 75201   J. Wayne Gaylord, President
    Small business investment company. Investment Policy: Diversified; no start-ups.

Omego Capital Corporation
755 S. 11th St.   (409) 832-0221
Beaumont, TX 77704   Theodric E. Moor, Jr., President
    Small business investment company. Investment Policy: No industry preference.

Red River Ventures, Inc.
777 E. 15th St.   (214) 422-4999
Plano, TX 75074   J. R. Heard
    Investment Policy: Inquire.

Retzloff Capital Corporation
P.O. Box 41250   (713) 466-4690
Houston, TX 77240-1250   Steven F. Retzloff, President
    Small business investment company. Investment Policy: No industry preference.

San Antonio Venture Group, Inc.
2300 W. Commerce St., Suite 300   (512) 223-3633
San Antonio, TX 78207   Domingo Bueno, President
    Small business investment company. Investment Policy: No industry preference.

SBI Capital Corporation
P.O. Box 570368   (713) 975-1188
Houston, TX 77257-0368   William E. Wright, President
    Small business investment company. Investment Policy: No industry preference; Texas businesses only.

South Texas SBIC
First Capital Group of Texas   (512) 573-5151
P.O. Box 15616
San Antonio, TX 78212   Kenneth Vickers, President
    Small business investment company. Investment Policy: No industry preference.

Southern Orient Capital Corporation
2419 Fannin, Suite 200   (713) 225-3369
Houston, TX 77002   Min H. Liang, Chairman of the Board
    Minority enterprise small business investment company. Investment Policy: No industry preference.

Southwest Enterprise Associates
Two Lincoln Centre, Suite 1266   (214) 991-1620
5420 LBJ Fwy.
Dallas, TX 75240   C. Vincent Prothro
    Venture capital supplier. Investment Policy: Concentrates on technology-based industries that have the potential for product innovation, rapid growth, and high profit margins. Investments range from $250,000 to $1.5 million. Past investments have been made in the following industries: computer software, medical and life sciences, computers and peripherals, communications, semiconductors, and defense electronics. Management must demonstrate intimate knowledge of its marketplace and have a well-defined strategy for achieving strong market penetration.

Southwest Venture Partnerships
300 Convent, Suite 1400   (512) 227-1010
San Antonio, TX 78205
    Venture capital partnership. Investment Policy: Invests in maturing companies located primarily in the Southwest. Invests in $1 million range.

Southwestern Venture Capital of Texas, Inc.
1336 E. Court St.   (512) 379-0380
P.O. Box 1719
Seguin, TX 78155   James A. Bettersworth, President
    Investment Policy: No industry preference.

SRB Partners Fund Ltd.
Sevin Rosen Management Company   (214) 702-1100
Two Galleria Tower, Suite 1670
Dallas, TX 75240   Jon W. Bayless, General Partner
    Venture capital firm. Investment Policy: Prefers early-stage financing in high-technology, medical, and biotechnology industries.

Sunwestern Capital Corporation
Three Forest Plaza   (214) 239-5650
12221 Merit Dr., Suite 1300
Dallas, TX 75251   Thomas W. Wright, President
    Small business investment company. Investment Policy: No industry preference.

Tenneco Ventures, Inc.
P.O. Box 2511   (713) 757-8229
Houston, TX 77252   Richard L. Wambold, President
    Venture capital supplier. Investment Policy: Provides financing to small, early-stage growth companies. Areas of interest include energy-related technologies, factory automation, biotechnology. and health care services. Prefers to invest in Texas-based companies, but will consider investments elsewhere within the United States. Investments range from $250,000 to $1 million; will commit additional funds over several rounds of financing and will work with other investors to provide larger financing. Remarks: Tenneco Ventures is a subsidiary of Tenneco, Inc., which is headquartered in Houston, Texas.

Texas Commerce Investment Company
712 Main St., Suite 3000   (713) 236-4719
Houston, TX 77002   Fred Lummis, Vice-President
    Small business investment company. Investment Policy: No industry preference.

United Oriental Capital Corporation
908 Town and Country Blvd., Suite 310   (713) 461-3909
Houston, TX 77024   Don J. Wang, President
    Minority enterprise small business investment company. Investment Policy: No industry preference.

Virginia

Atlantic Venture Partners (Alexandria)
801 N. Fairfax St., Suite 404   (703) 548-6026
Alexandria, VA 22314   Wallace L. Bennett, President
    Private venture capital partnership. Investment Policy: Provides all stages of financing primarily to high-technology businesses. Initial investments generally range from $250,000 to $750,000. Under certain circumstances, an investment of less than $100,000 or more than $1 million will be undertaken. May provide financing of $1 to $10 million (or more) in participation with other venture capital firms. Remarks: Maintains an office in Richmond, Virginia: mail address is P.O. Box 1493, Richmond, VA 23212; phone (804) 644-5496.

Basic Investment Corporation
6723 Whittier Ave.   (703) 356-4300
McLean, VA 22101   Frank F. Luwis, President
    Minority enterprise small business investment company. Investment Policy: No industry preference.

Crestar Capital
P.O. Box 1776   (804) 643-7358
Richmond, VA 23214   A. Hugh Ewing, Managing General
                                 Partner
    Small business investment company. Investment Policy: No industry preference.

James River Capital Associates
P.O. Box 1776   (804) 643-7323
Richmond, VA 23214   A. Hugh Ewing, Managing Partner
    Small business investment company. Investment Policy: No industry preference.

Metropolitan Capital Corporation
2550 Huntington Ave.   (703) 960-4698
Alexandria, VA 22303   S. W. Austin, Vice-President
    Small business investment company. Investment Policy: Equity or loans with equity features. No retail or real estate.

River Capital Corporation (Alexandria)
5554 Port Royal Rd., Suite 208   (703) 739-2100
Springfield, VA 22151   Carl L. Schmitz, Vice-President
    Small business investment company. Investment Policy: No industry preference. Remarks: Maintains an office in Cleveland, Ohio.

Sovran Funding Corporation
Sovran Center, Third Floor   (804) 441-4041
One Commercial Place
Norfolk, VA 23510   David A. King, Jr., President
    Small business investment company. Investment Policy: No industry preference.

Tidewater Small Business Investment Corporation (TBBIC)
420 Banks   (804) 622-2312
Norfolk, VA 23510   Diane Newell, Manager
    Small business investment company. Investment Policy: Prefers manufacturing, distribution, and service industries.

Washington

Cable and Howse Ventures (Bellevue)
777 108th Ave., Suite 2300   (206) 646-3030
Bellevue, WA 98004
    Venture capital investor. Investment Policy: Provides start-up and early-stage financing to enterprises in the western United States, although a national perspective is maintained. Interest lies in proprietary or patentable technology. Investments range from $50,000 to $2 million.

Capital Resource Corporation
1001 Logan Bldg.   (206) 623-6550
Seattle, WA 98101   T. Evans Wyckoff, President
    Small business investment company. Investment Policy: No industry preference.

Palms and Company, Inc.
6702 139 Ave., N.E.   (206) 883-3580
Redmond, WA 98052   Peter J. Palms, President
    Private venture capital supplier and investment banker. Investment Policy: Provides all stages of financing to companies located anywhere in the United States and Canada. Investments range from $100,000 to $10 million.

Washington Trust Equity Corporation
Washington Trust Financial Center   (509) 455-3821
P.O. Box 2127
Spokane, WA 99210   John M. Snead, President
    Small business investment company. Investment Policy: No industry preference.

Wisconsin

Bando McGlocklin Capital Corporaton
13555 Bishops Ct., Suite 205   (414) 784-9010
Brookfield, WI 53005   George Schonath, Chief Executive
                                  Officer
    Small business investment company. Investment Policy: Fixed assets-based lender.

Capital Investments, Inc.
744 N. Forth St.   (414) 273-6560
Milwaukee, WI 53203   Robert L. Banner, Vice-president
     Small business investment company. Investment Policy: No industry preference.

Future Value Ventures, Inc.
622 N. Water St., Suite 500   (414) 278-0377
Milwaukee, WI 53202   William P. Beckett, President
     Small business investment company. Investment Policy: Provides financing to companies owned by socially and economically disadvantaged persons. Managers should have proven management ability. Prefers businesses with potential to create jobs. Flexible regarding industry preference.

InvestAmerica Venture Group, Inc. (Milwaukee)
600 E. Mason St.   (414) 276-3839
Milwaukee, WI 53202
     Small business investment company. Investment Policy: Prefers second-stage and leveraged buy out investments of $250,000 to $500,000. Will not consider real estate or retail sales. Remarks: A branch office of InvestAmerica Venture Group, Inc., Cedar Rapids, Iowa.

Lubar and Company, Inc.
3380 First Wisconsin Center   (414) 291-9000
Milwaukee, WI 53202   Sheldon B. Lubar, President
     Private investment and management firm. Investment Policy: Interest is in energy-related products and services, manufacturers of industrial products, and unique or niche businesses. Remarks: Manages the Wisconsin Venture Capital Fund, Inc., which provides financing to Wisconsin-based businesses. Typical investments range from $250,000 to $750,000.

M & I Ventures Corporation
770 N. Water St.   (414) 765-7910
Milwaukee, WI 53202   John T. Byrnes, President
     Small business investment company. Investment Policy: Manufacturing, distribution, electronics, and technology-related industries. Approximately $50 million is available for investments.

Marine Venture Capital, Inc.
111 E. Wisconsin Ave.   (414) 765-2274
Milwaukee, WI 53202   H. Wayne Foreman, President
     Small business investment company. Investment Policy: No industry preference. Remarks: Marine Venture Capital is a subsidiary of a commercial bank.

Venture Investors of Wisconsin, Inc.
565 Science Dr.   (608) 233-3070
Madison, WI 53711
     Venture capital firm. Investment Policy: No industry preference. **Contact:** Roger H. Ganser, President; John Neis, Vice-President.

Wisconsin Community Capital, Inc.
14 W. Mifflin St., Suite 314   (608) 256-3441
Madison, WI 53703   Paul J. Eble, President
     Small business investment company. Investment Policy: No industry preference.

R&D Funding: Federal Sources

The federal government funds almost one half of all U.S. research and development. Yet for most inventors, the federal government is an untapped source of project revenue. You don't have to be a Fortune 500 diversified manufacturer to qualify for financing. More than $1 billion each year goes to small companies for R&D through direct awards.

The 100-plus agencies, listed department by department, in this chapter might be a start to a more rewarding relationship with your government. I have conscientiously read through all federal research and development activity descriptions and selected for this chapter only those agencies and departments that are potential funding sources to independent inventors.

National Aeronautics and Space Administration

Ames Research Center (ARC)

Moffett Field, CA 94035

(415) 694-4044
Lawrence Milov, External Relations Officer

Research and development programs in aeronautics, life sciences, space science, space technology, and flight research. Other efforts are aimed at human factors in space, earth resources study, thermal protection systems for atmosphere entry vehicles, computational physics and chemistry, and artificial intelligence and autonomous system. **Publications:** NASA Technical Reports.

Dryden Flight Research Facility

P.O. Box 273
Edwards Air Force Base, CA 93523

(415) 694-5800
Ms. Carolyn Anderson, Small Business Specialist

Manned flight research, including systems, configuration, and technology developments for high performance aircraft.

George C. Marshall Space Flight Center

Huntsville, AL 35813

(205) 453-2675
Conrad Walker, Small Business Specialist

Serves as a primary center for the design and development of space transportation systems, elements of the Space Station, scientific applications payloads, and other systems for present and future space exploration. Operates the Program Development Directorate, which generates plans for promising new programs; Institutional and Program support Directorate, which provides various supporting services; and Science and Engineering Directorate, which provides technical support and furnishes a research base.

Goddard Space Flight Center (GSFC)

Procurement Analysis Branch
Code 202.3
Greenbelt, MD 20771

(301) 286-5417
Franz W. Hoffman, Small Business Specialist

Research interests include orbital spacecraft development, tracking and data acquisition systems, space physics and astronomy payloads, upper atmospheric research, weather and climate, earth dynamics and resources, information systems, sounding rocket and payload development, planetary science, and sensors and experiments in environmental monitoring and ocean dynamics. Maintains an automated source system comprised of approximately 4,000 firms. The Center encourages businesses to submit a Bidders Mailing List Application (SF129) to be added to this system.

Headquarters Contracts and Grants Division

600 Independence Ave.
Washington, DC 20546

(202) 453-2090
Stuart J. Evans, Assistant Administrator

Responsible for planning, negotiating, awarding, and administering contracts based on procurement requirements, including system engineering services, reliability studies, basic and applied research and development, mobile lecture-demonstration units, exhibits, motion picture services, management analysis surveys and automatic data processing equipment, and software services. The Division also has agency-wide responsibility for negotiating and executing NASA contracts with foreign governments and commercial organizations. **Contact:** Mark E. Kilkenny, Procurement System Branch Chief (202) 453-1840; Eugene D. Rosen, Director,

I have omitted those agencies and departments whose interest is more in reports and studies than in patentable inventions and innovative concepts. Many federal R&D budgets are applied towards scientific research programs and policy research and analysis studies and the like and are inappropriate for our discussion.

This chapter is arranged alphabetically by department, followed by an alphabetic listing of programs.

Small and Disadvantaged Business Utilization (202) 453-2088; James T. Rose, Assistant Administrator, Commercial Programs Office (202) 453-1123.

Jet Propulsion Laboratory (JPL)

California Institute of Technology
4800 Oak Grove Drive
Pasadena, CA 91109

(818) 354-5722

Margo Kuhn, Small Business Specialist

Has the primary responsibility of exploration of the solar system with unmanned spacecraft. Research and development activities include aeronautics and aerospace, communications, computer science and mathematics, earth and space sciences, electronics, and physics. **Contact:** Mr. T. May, Minority Business Specialist (818) 354-2121.

John F. Kennedy Space Center

Kennedy Space Center, FL 32899

(305) 867-7353

Norman Perry, Small Business Specialist

Responsible for assembly, checkout, servicing, launch, recovery, and operational support of space transportation system elements. Principle research activity is the design and development of launch and landing facilities. Technology transfer programs are administered by the Center's Advanced Projects and Technology Office. **Publications:** NASA Tech Briefs.

Langley Research Center

Hampton, VA 23365

(804) 865-3751

Norma Patrick, Small Business Specialist

Research and development in aeronautics and space technology, including acoustics and noise reduction, aerodynamics, aerospace vehicle structures and materials, aerothermodynamics, avionics technology, environmental quality monitoring technology, sensor and data acquisition technology, long-haul aircraft, military support, and advanced space vehicle configuration. Activities also include technology transfer and providing development support to government agencies, industry, and other NASA centers.

Lewis Research Center

21000 Brookpark Road
Cleveland, OH 44135

(216) 433-4000

Mr. J. Liwosz, Jr., Small Business Officer

Responsible for managing and design and development of the power generation, storage, and distribution system for the U.S. space station. Areas of research include air breathing and space propulsion systems; turbomachinery thermodynamics and aerodynamics; fuels and combustion; aero and space propulsion systems; power transmission; tribology; internal engine computational fluid dynamics; high temperature engine instrumentation, space communications; and space and terrestrial energy processes, systems technology, and applications.

Lyndon B. Johnson Space Center (JSC)

Houston, TX 77058

(713) 483-4511

Mr. R.L. Duppstadt, Small Business Specialist

Primary mission is the development of spacecraft for manned space flight programs and the conduct of manned flight operations; selection and training of astronauts; providing management in systems engineering and integration and business and operations management for the Space Shuttle Program; conducting investigations of earth resources technology; and conducting space life sciences research.

National Space Technology Laboratories (NSTL)

NSTL, MS 39529

(601) 688-1636

David Anderson, Small Business Specialist

Primary installation for static test firing of large rocket engines and propulsion systems. Beside NASA, resident agencies include NOAA, and departments of Army, Navy, and Interior; U.S. Coast Guard; as well as groups from various universities. These groups are involved in oceanographic, meteorological, and environmental research and, together with NASA, form a scientific and technical community, each pursuing its own programmatic objectives but collectively producing a scientific base for technology exchange. **Contact:** Mr. P.B. Higdon, Contracting Officer, Earth Resources Laboratory (601) 688-1641.

Office of Space Science and Applications (OSSA)

OSSA Steering Committee
Code EP-3
Washington, DC 20546

(202) 453-1409

Leonard A. Fisk, Associate Administrator

Operationally responsible for scientific research into the nature and origin of the universe and applying space systems and techniques to solve problems of Earth.

Small Business Innovation Research Program

Washington, DC 20546

(202) 453-2848

Harry Johnson, Manager

Areas of soliciting proposals are based on the needs of NASA's programs and missions as described by the roles of each NASA Center.

National Institutes of Health

National Heart, Lung, and Blood Institute (NHLBI)
Contracts Operations Branch   (301) 496-7666
Westwood Bldg., Room 650
Bethesda, MD 20892   Robert Carlsen, Chief
> Supports basic and clinical research, development, and other activities related to the prevention, diagnosis, and treatment of cardiovascular, lung, and blood diseases. The Institute plans and directs research in the development, trial, and evaluation of drugs and devices relating to the prevention such diseases.

National Science Foundation

Division of Policy Research and Analysis
1800 G St., N.W., Room 1233   (202) 357-9689
Washington, DC 20550   Dr. Peter House, Division Director
> Funds policy research, including business firms specializing in science and innovation policy and technology and resource policy.

Industry/University Cooperative Research Centers
Room 1121   (202) 357-7307
1800 G St., N.W.
Washington, DC 20550   Dr. Morris Ojalvo, Program Manager
> Fosters university-industry interaction at university centers by supporting a limited number of research projects. **Contact:** Alex Schwarzkopf, Manager.

Office of Small Business Research and Development
1800 G St., N.W., Room 1250   (202) 357-9666
Washington, DC 20550   Dr. Donald Senich, Director
> Provides information and serves as a referral point for small businesses interested in the Foundation's research or procurement opportunities, or how to submit proposals. Also compiles and publishes information on research grants and contracts awarded by NSF to small business. **Contact:** Linda L. Boutchyard, Program Specialist (202) 357-7464.

Small Business Innovation Research Program
1800 G St., N.W., Room 1250   (202) 357-7527
Washington, DC 20550   Roland Tibbetts, Program Manager
> Acts as the principal NSF research opportunity organization for small businesses. Research topics include physics; chemistry; mathematical sciences; computer sciences; materials research; electrical, communications, and systems engineering; chemical, biochemical, and thermal engineering; mechanics, structures, and materials engineering; cellular biosciences; molecular biosciences; biotic systems and resources; behavioral and neural sciences; social and economic sciences; information sciences; astronomy; atmospheric sciences, earth sciences; ocean sciences; design, manufacturing, and computer engineering; and advanced scientific computing. **Contact:** Mr. Ritchie Coryell, Program Manager (202) 357-7527.

U.S. Department of Agriculture

Agricultural Research Service
Bldg. 005, BARC-W   (301) 344-3084
Beltsville, MD 28705   Dr. Gary R. Evans, Deputy Admin.
> Research interests include physical, biological, and chemical engineering; clothing, housing, and household economics; industrial and food products and processing methods for agricultural commodities; and soil and water, crop, animal husbandry, entomology, agricultural engineering, and energy. Published solicitations are not made by the office, but proposals are received at any time.

Office of Grants and Program Systems (OGPS)
CSRS - USDA   (202) 475-5720
14th & Independence Ave., S.W.
Room 324-A Administration
Washington, DC 20250   Dr. William D. Carlson
> Maintains Competitive Research Grants Office, which publishes its solicitations in the Federal Register. Grants are awarded on a competitive basis. Proposals are requested in the areas of biological nitrogen fixation, biological stress on plants, and human nutrition.

Small Business Innovation Research Program
901 D St., S.W., Room 323-J   (202) 475-7002
Washington, DC 20251   Dr. Charles Cleland

Y ou do not have to be a Fortune 500 diversified manufacturer to qualify for federal financing. Under the Small Business Innovation Development Act, each agency with an extramural R&D budget in excess of $100 million must establish a **Small Business Innovation Research (SBIR) Program**. R&D firms ranging in size from one to 500 employees are encouraged to submit proposals. The following agencies are currently participating in the SBIR program: Department of Agriculture Department of Commerce Department of Defense Department of Education Department of Energy Department of Health and Human Service Department of Transportation Environmental Protection Agency National Aeronautics and Space Administration National Science Foundation Nuclear Regulatory Commission

George P. Lewett, Director of Technology Evaluation and Assessment

Mr. Lewett was recently appointed to head the new Office of Technology Evaluation and Assessment at NIST. He will continue to manage the Energy-Related Inventions Program.

Chief of ERIP since the office was established in 1975, George Lewett spent many years with the Western Electric Company as an industrial and production engineer and as head of the Operations Research Department for the Kearny Works. Following this he worked as a manager and consultant with several R&D firms in the D.C. area, principally in operations research, systems engineering, and computer applications.

U.S. Department of Commerce

The U.S. Department of Commerce's National Institute of Standards and Technology (NIST), formerly the National Bureau of Standards, has been recently reorganized. Its new name reflects a broadened role and area of responsibility. In addition to its traditional functions of providing the measurements, calibrations, data and quality assurance support which are vital to U.S. commerce and industry, NIST has a new purpose: "to assist industry in the development of technology and procedures needed to improve quality, to modernize manufacturing processes, to ensure product reliability, manufacturability, functionality, and cost-effectiveness, and to facilitate the more rapid commercialization of products based on new scientific discoveries."

The Department of Commerce is in the process of centralizing all new technology work under a new Technology Administration, headed by the Under Secretary for Technology. NIST has been directed to develop four new major programs that may be of interest to independent inventors. Its mandate is to enhance the contry's technological competitiveness through the fast and effective transfer of new technologies to U.S. industries. NIST will accomplish this through
1) Regional Centers for the Transfer of Technology;
2) The Boehlert-Rockefeller Technology Extension Program and;
3) The Advanced Technology Program.

NIST has also reorganized and established a separate operating unit called Technology Services, which combines the new programs with traditional extramural NIST programs. Although these new programs and new configurations are still evolving, I tell you of them to give you a jump on events that will be unfolding throughout the next three years. If you are interested in the status of any particular program, contact the designated U.S. government officials.

WHAT ARE REGIONAL CENTERS FOR THE TRANSFER OF MANUFACTURING TECHNOLOGY?

These regional centers are intended to provide direct support to small- and medium-sized manufacturing firms in automating and modernizing their facilities. If you are an inventor who has his or her own manufacturing facility, these centers could help you develop technically sound plans for modernizing production runs.

NIST is creating these centers in partnership with nonprofit organizations established by state and local governments, universities or companies. NIST is authorized to provide up to 50 percent of the operating funds for these centers for their first three years, reducing to zero funding after six years.

The first three organizations that have established regional manufacturing technology centers are: The Cleveland Advanced Manufacturing Program (CAMP) in Cleveland, Ohio; Rensselaer Polytechnic Institute in Troy, New York; and the University of South Carolina in Columbia, South Carolina.

For updated information on the regional centers, contact Dr. Phil Nanzetta, NIST, Gaithersburg, Maryland 20899.

THE BOEHLERT-ROCKEFELLER PROGRAM

NIST is taking an active role in promoting the transfer of federal technology the many existing state and local technology extension services. Such state and local programs usually highlight business advice rather than inventions. The NIST program will help to coordinate the state and local extension services with federal technology transfer programs.

The most exciting part of this activity for the independent inventor will be the technical evaluation of promising inventions that are **not** energy related, a program modeled after NIST's successful 12-year old program for energy-related inventions managed by NIST for the Department of Energy. This part of the program still has no funding. But it will. For updates, contact: George P. Lewett, Chief, Office of Energy-Related Inventions, NIST, Gaithersburg, Maryland 20899; (301) 975-5500.

WHAT IS THE ADVANCED TECHNOLOGY PROGRAM?

NIST will provide money to leverage private investment in specific projects to develop new products and processes. Candidate projects include promoising inventions evaluated under the program mentioned above. The program initially will focus on joint R&D ventures which must provide at least 50 percent of the funds. However, as the program evolves the law would allow direct support for individual businesses. For more information, contact George Uriano, Director of the Advanced Technology Program.

National Institute of Standards and Technology (NIST)

Building 301, Room B128	(301) 975-6353
Gaithersburg, MD 20855	Keith Chandler, Small Business Specialist

NIST is organized into three technical laboratories as follows: (1) National Engineering Laboratory (NEL) furnishes technology and technical services to public and private sectors to address national needs and to solve problems in the public interest. NEL conducts research in engineering and applied science; builds and maintains competence in the necessary disciplines required to carry out this research; and develops engineering data and measurement capabilities.

(2) National Measurement Laboratory (NML) provides the national system of physical, chemical, and materials measurement; coordinates the system with measurement systems of other nations, and furnishes services leading to uniform measurement; and conducts research leading to methods of measurement, standards, and data on the properties of materials.

(3) National Computer and Telecommunication Laboratory (NCTL), which develops computer standards, conducts research, and provides scientific and technical services in the selection, acquisition, application, and use of computer technology; manages a government-wide program for standards development and use, including management of federal participation in voluntary standardization.

National Institute of Standards and Technology, Office of Energy Related Inventions (OERI)

Gaithersburg, MD 20899   (301) 975-5500

George L. Lewett

"We don't have the NIH (Not Invented Here) Syndrome rattling around here," says Don Corrigan, Operations Group Leader at the National Institute of Standards and Technology Office of Energy-Related Inventions. NIST and the Department of Energy, under provisions of the Federal Non-nuclear Energy Research and Development Act of 1974 have combined to offer a marvelous opportunity to inventors of energy-related concepts, devices, products, materials, or industrial processes. Called the Energy-Related Inventions Program (ERIP), it is designed as a process to discover and assist the development of worthy inventions which might otherwise never be commercialized.

From a NIST facility in Gaithersburg just outside the District of Columbia off I-270, the program provides a chance for independent inventors and small businesses with promising energy-related inventions to obtain federal assistance in the development and commercialization of their inventions.

Since its inception OERI has received more than 25,000 evaluation requests, accepted about half of these for evaluation, and recommended 440 for U.S. Government support. Approximately $21.5 million in grants or contracts have been awarded to date on some 300 recommended inventions; the remainder are either in process or have received other than financial assistance.

Invention disclosures evaluated by OERI are categorized as follows:

- Fossil Fuel Production
- Direct Solar
- Other Natural Sources
- Combustion Engines and Components
- Building, Structures and Components
- Industrial Processes
- Miscellaneous

OERI is under the experienced leadership of George P. Lewett. "We currently review about 200 inventions per month for evaluation, Mr. Lewett explains, "and we recommend, fairly steadily, three to four inventions per month." And the system works!

Ask Harrison Robert Woolworth, inventor of a mechanism to preheat scrap metal with waste heat. He submitted his invention to OERI in 1976 and 12 months later was awarded $175,000 by DOE.

Ask Phillip Zacuto and Daniel Ben-Schmuel, co-inventors of a heat extractor. They submitted their invention to OERI in 1977 and by 1978 had landed a DOE award in the amount of $125,000.

Ask Maurice W. Lee, inventor of a new way of cooking hamburgers using the dielectric resistance characteristics of meat. He submitted his invention to OERI in 1984 and three years later received an award of $75,000 from DOE.

Ask F.J. Perhats and James V. Enright, inventors of a device which uses an automobile engine's heat to warm the car's interior after the engine has been turned off. They received $71,000 through the Department of Energy.

OERI SUBMISSION PROCESS

Submissions are handled by the OERI at the Gaithersburg office. The disclosure of an invention should include information required by NIST Form 1019 **(see Evaluation of an Energy-Related Invention in the Forms and Documents section)**, but the format may vary widely depending upon a number of factors. here are some suggestions that you should consider in preparing the description of the invention to be submitted with the Evaluation Request Form 1019.

Make a complete disclosure. The principal requirement in submitting a request is a thorough and complete invention disclosure which describes the invention in detail. It is most important to submit ALL information which is available even if the method of presentation and organization is not professional in nature. Test data and information on how tests were conducted are particularly critical, when applicable, since no testing will be done by NIST as part of its evaluation.

Emphasize the energy relation. The program is interested in all energy-related inventions including both those that involve energy conservation and those that involve alternative sources of energy. This means everything from new methods of recovery to drill bits. Your disclosure should emphasize and document, to the extent possible, the amount of energy saved or made available through an alternate source.

It is only after the invention reaches the commercialization stage that its ultimate contribution to the solution of our energy problem can be realized. It is not necessary to calculate energy savings exactly, but the potential should be indicated.

Time to process your request. Do not expect an immediate response to your request for evaluation. The evaluation process is time-consuming and there are a large number of submittals to consider. Remember that submission of an invention to NIST for evaluation is no guarantee that it will be recommended to the Department of Energy and a recommendation is no guarantee that your invention will be accepted for a DOE grant award.

Describe your competition. Make an effort to find out if there are any similar products on the market. Detail the known competition and document why your innovation is superior technically or from an energy or economic standpoint.

Give the status of your invention. Address the question of what needs to be done to bring your invention closer to fruition. Spell out what you would expect from the U.S. Government in the event of a favorable evaluation.

Highlight the innovation. As you can imagine, NIST has seen all kinds of concepts involving common devices, such as windmills, wave machines, furnaces, carburetors, internal combustion engines, and space heaters. And like so many ideas, many of those submitted are neither new nor innovative. Be sure, therefore, to point out and highlight novel principles or innovations that make yours different, particularly if it falls into a common device class.

More ERIP success stories: Albert B. Csonka, inventor of a new kind of micro-carburetor which is claimed to be fuel-saving and pollution-reducing. He was awarded $193,500 to build a working prototype fit to a late model, standard 350 cubic inch V-8 engine.

Or Donald E. Wise, inventor of a convertible flat/drop trailer. He received a grant of $63,000 to build and test his concept. Then, according to DOE, he successfully licensed his technology to Trail King Company of Nebraska.

And there's Karakian Bedrosian, inventor of a method of preserving fruits and vegetable without refrigeration. It only took him one year to win a DOE award of $97,300. Marketed under the trade name "TomAHtoes," DOE reports that 751,000 25-pound boxes of "TomAHtoes"were shipped in 1987 to the tune of $35 million.

Cost Estimate. With your proposal, provide a detailed estimate of all costs. To help you formalize this, I am including a **Defense Small Business innovation Research (SBIR) Program Cost Proposal** in the Forms and Documents section. Take from it those parts that are most appropriate to your project and situation.

"The amazing thing about these SBIR meetings is that I sit there and meet this endless stream of awesomely bright people," explains *Richard Sparks, program manager for the Defense Technical Information Center (DTIC).*

Be factual and realistic. Your proposal will be evaluated by savvy technical and business-oriented professionals. Prepare your disclosure with that in mind. Do not make outrageous claims that cannot be justified or substantiated by data in your disclosure.

You keep the rights. Uncle Sam keeps your secret. It is the Government's policy that inventors retain patent rights to their inventions and that these rights not be compromised by the OERI evaluation.

OERI maintains the confidentiality of submitted invention disclosures by limiting access to them. All Federal Government personnel involved in the program sign a statement advising them of penalties for disclosure or misuse of information under 18 U.S.C. 1905. Non-government consultants also sign an agreement of nondisclosure which includes agreement not to undertake invention reviews that may subject the consultant to a conflict of interest.

THE EVALUATION PROCESS

The OERI evaluation process, under the direction of Howard Robb, is the formal procedure that will determine, in an objective way, whether the invention you have submitted is appropriate for recommendation to the DOE for support. OERI considers an invention appropriate for DOE consideration if it is technically feasible, will offer a favorable impact on the energy situation and holds potential for commercial success. The focus is on your technology; your credentials or capabilities are not a factor in OERI decision making.

Each disclosure is logged in and tracked through a computerized system that records the number of days spent in each stage of the five-step evaluation process and the number of days required to obtain the opinions of government consultants.

Disclosure Acknowledgment. OERI will immediately acknowledge by letter the receipt of each disclosure received. Every disclosure must be accompanied by a signed evaluation request, NIST Form 1019 (see Forms and Documents section).

Disclosure review and analysis. OERI will review the invention disclosures to determine whether they are complete, readable, technically sufficient, and within the scope of its program. Inventors are notified by letter whether their disclosure is accepted for first-stage evaluation or rejected. This is, in effect, a preliminary screening.

First Stage Evaluation. In this initial screening process, OERI will determine whether an invention is promising enough to warrant an in-depth analysis. OERI evaluators typically seek brief expert opinions (invention reviews) independently from two consultants before deciding whether to place the invention into second-stage evaluation or to reject it. The consultants are asked to comply with the following rules:

Do not transmit information on the invention disclosure to anyone who has not signed an NIST Agreement of Nondisclosure (see Forms and Documents section).

Do not copy the disclosure.

Do not write on or otherwise mark the disclosure.

Return the complete disclosure with the invention report to OERI by certified mail or other secure transmittal method.

OERI evaluators and their consultants are expected to be liberal and highlight the positive aspects of an invention rather than seek reasons for rejection. The inventor is notified by letter of said decision. The actual evaluation paper is not made available, but a summary of its finding is communicated to the inventor.

Second Stage Evaluation. OERI will determine whether to recommend an invention to DOE for support. OERI evaluators obtain a deep analysis of the invention from a consultant before deciding whether to recommend the invention to DOE or to reject it. The consultant, to th extent possible, will try to provide information and references to pertinent literature that may be of assistance to the OERI evaluator and the inventor. The inventor is notified by letter of this decision too, in either case, and is sent a copy of the second-stage consultant's analysis.

Recommendation to DOE. OERI will prepare a transmittal document which includes all pertinent material submitted by the inventor, the second-stage consultant's in-depth analysis, any other pertinent material obtained or developed by OERI during the evaluation, and an OERI evaluator's report outlining the reason for recommending the invention and the appropriate next step to DOE. Inventors are provided a copy of this document and the name of an OERI coordinator whom they may contact.

At each stage in the evaluation process, you may present additional information, which is fully considered before further decisions are made. If your invention is rejected at any stage, you may submit additional information addressing the problems raised by OERI and request reevaluation. This "Open Appeal" feature is considered an intrinsic part of the evaluation process. "We'll consider them from now until Doomsday," says Don Corrigan of OERI. "Fifteen to twenty percent of the inventions we recommend to DOE have been turned down at some point in the evaluation process."

The evaluation process is primarily a selection process which does not necessarily include analysis of every aspect of an invention. **OERI will provide you with any useful information it obtains or develops during the evaluation process of your innovation.** The brief first-stage invention reviews made by consultants are not released, but information contained in them may be included in OERI's correspondence with you.

National Oceanic and Atmospheric Administration (NOAA)

NBOC-1, Room 205   (301) 443-8584
11420 Rockville Pike   Randy Linderholm, Small Business
Rockville, MD 20852   Specialist

Research and development interests in environmental monitoring and prediction include automation of meteorological observations, analysis, and communication; remote sensing; oceanic monitoring; stratospheric measurement; satellite technology and development of ground equipment; weather modification; and modeling of atmospheric and oceanic processes, including climate variation. **Contact:** Edward Tiernan, SBIR Program Manager Chief, NOAA/NESDIS, Orta FB4, Room 3316, Suitland, MD 20233 (202) 763-2418.

U.S. Department of Defense

The Department of Defense presents one of the best opportunities for an individual or small firm to land federal R&D funds and/or sell innovative technologies. The Pentagon has an active and wide-ranging Small Business Innovation Research Program which makes special efforts to identify creative individuals, small businesses, and small disadvantaged businesses with R&D capabilities and/or better ideas. DoD not only has an interest in soliciting business, but it also entertains outside queries and submissions.

Individuals and firms with strong R&D capabilities in science and engineering and other disciplines are encouraged to participate. DoD's SBIR Program goals include stimulating technological innovation in the private sector, strengthening the role of small business in meeting DoD research and development needs, fostering and encouraging participation by minority and disadvantaged persons in technological innovation, and increasing the commercial application of DoD-supported research and development results.

Under the SBIR program, the DoD can be approached several ways.

You can go directly to the different departments to present your concepts and/or to receive information on what they need in terms of outside assistance. The names and addresses of same are provided in this chapter.

Each year the Department of Defense, Office of Small Business Innovation Research, conducts seminars around the country at which it presents "wish lists" and explains the SBIR program. For general information on the where and when of the seminars, contact Bob Wrenn, SBIR Coordinator, OSD/SADBU, U.S. Department of Defense, The Pentagon, Room 2A340, Washington, D.C. 20301-3061; (202) 696-1481.

To register for the SBIR program and receive its DoD program solicitations, there is no charge. The one for FY1989 weighed just under two pounds! The best way to get your name on the list to receive the next program solicitation is via the nearest Small Business Administration (SBA) office. They are listed in your telephone directory under U.S. Government. Or you may contact the Defense Technical Information Center (Att: DTIC-SBIR), Alexandria, VA 22304-6145; (202) 274-6902.

The Defense Technical Information Center (DTIC) is a great place to learn what the Pentagon is looking for in terms of technologies and gadgets. "We exist to help the independent inventors and small businesses cut through the red tape and bureaucracy," says a most helpful and hospitable Richard Sparks, program manager for DTIC and the pointman on the front line with outside queries. He boasts the only toll-free number in the Pentagon (800-368-5211) and he answers his own phone! There are no flacks to block your access. The DTIC is the central storehouse of scientific and technical information resulting from and describing R&D projects that are funded by the Department of Defense. It has on file nearly two million technical reports.

DTIC information is also available via University Research Initiative (URI), a basic DoD research program. Nearly 100 universities throughout the U.S. currently participate in the program, which provides free access to DTIC products and services.

Most DTIC services are free to registered users. The only charges are for paper copy documents, which cost $5 for 1-100 pages and $.07 for each additional page over 100, and microfiche copies, which are $.95 per demand document and $.35 per document supplied under the Automatic Document Distribution.

You can receive an information packet that describes DTIC's services, products, and registration process by contacting the Defense Technical Information Center (ATTN: Registration and Services Section—DTIC-FDRB), Bldg. 5, Cameron Station, Alexandria, VA 22304-6145; (202) 274-6871.

Mr. Sparks says that each year near the end of August, military departments begin to publish SBIR Requests for Proposals (RFPs) in the *Commerce Business Daily* and mail formal Program Solicitations. Then at the beginning of October, the DTIC "opens for business" for the next 13 weeks, assisting people wishing to compete for research and development grants.

While "officially" he only has the budget to handle outside queries during this time period, Mr. Sparks is a very warm, knowledgeable, and helpful individual who will never turn away an inventor in need of guidance. "You never know what you'll find," he says. "One that fascinated me most was a three mile-long magnetic gun for putting things into orbit at low cost without expending fuel. These guys were a couple of whiz kids!"

You may call or visit at the following locations:
Defense Technical Information Center
ATTN: DTIC-SBIR
Building 5, Cameron Station
Alexandria, VA 22304-6145
800-368-5211 or (202) 274-6902 (commercial for VA, AK, HI)
   DTIC Boston On-Line Service Facility
DTIC-BOS
Building 1103, Hanscom AFB,
Bedford, MA 01731-5000
(617) 377-2413
   DTIC Albuquerque Regional Office
AFWL/SUL Building 419
Kirtland AFB, NM 87117-6008
(505)846-6797
   DTIC Los Angeles On-Line Service Facility
Defense Contract Administration Services Region
222 N. Sepulveda Blvd.
El Segundo, CA 90245-4320
(213)355-4170

To show how active the DoD program is, consider the Fiscal Year '88 numbers as an example. Out of the 8,625 proposals received, 1,011 were identified as of September 1st as offering the greatest potential in their field for meeting the research and development needs of DoD, and were subjected to further evaluations and negotiations leading to contract awards.

*T*he Federal Government funds almost one-half of all U.S. research and development. More than one billion dollars each year goes to small companies for R&D through direct awards. But pursuing it is not for everyone. Tapping the pot of green at the end of Uncle Sam's red, white and blue rainbow takes a major commitment of both time and money. And unless you are willing to tolerate a cumbersome and inordinate amount of paperwork, red tape and costly bureaucratic excesses and delays, this exercise may not be for you.

On the other hand, there appears to be something for everyone. Washington's interest ranges from innovations in astronomy, earth, enviromental and marine sciences to chemistry, physics, engineering and materials sciences. It awards money for breakthroughs in robotics, genetics, plastics, adhesives, fusion, optics and artificial intelligence, to name just a few categories.

GUIDE TO UNSOLICITED PROPOSALS TO DOD

You do not have to follow a particular format for the submission of proposals. However, a proposal should, at a minimum, cover the points set forth below in the order indicated. Elaborate proposals or presentations are neither necessary nor desirable. Two copies of your proposal should be submitted.

The DoD follows established competitive procurement procedures for awarding contracts. In other cases, it may issue statments of interest or similar notices to announce various program opportunities in such publications as the *Commerce Business Daily*.

DoD also awards contracts based upon unsolicited proposals.

Normally, competitive procurements will be initiated by the issuance of a formal request for proposal (RFP). The RFP will contain instructions telling you how to prepare the proposal.

A formal unsolicited proposal should be in the form of a detailed document signed by the inventor. This document forms the basis for further technical evaluations and possible contract negotiations. Make sure each proposal includes the following elements as appropriate: cover sheet, abstract, narrative, and cost proposal (should be a document that is separate from original proposal).

Cover Sheet. Each proposal must have a cover sheet providing the information set forth in the sample form entitled **Small Business Innovation Research (SBIR) Program Proposal Cover Sheet**, available for your perusal in the Forms and Documents section.

Abstract. Each proposal should start with a narrative in which you describe the relevance of your invention to the DoD mission. Your personal qualifications should be stated. A sample **Small Business Innovation Research (SBIR) Program Proposal Summary** is offered for your use in the Forms and Documents section.

The Department of Defense is primarily interested in technical competence. The people who'll evaluate your proposals want to know that you understand what the project involves and that you can perform the work required. Once your technical competence has been established, DoD will consider costs.

Be sure to specifically address these points:

Purpose and Objective. State briefly the primary purpose, general objective, and expected results of the proposal, such as:

The problem or problems your invention will contribute to solving and the anticipated contribution of any research to that solution.

Enumerate the specific objective of any additional research and specify the questions that your research will attempt to answer. This will be important if you hope for DoD to finance further research.

State the expected consequences of successful completion of your research, including potential economic and other benefits.

Discuss the existing interest by potential beneficiaries or users.

Previous or Ongoing Related Work. Document any knowledge you may have of other related activities with appropriate references to the literature or currently ongoing R&D. In particular, show how your proposed project relates to these activities and how your invention or research will extend the level of knowledge in the field.

Statement of Work. If your proposal involves additional R&D, give full and complete technical details of the procedures that will be followed throughout the scope of your work. Outline it phase by phase. Give time frames and objectives.

Organization, Facilities and Qualifications. Fully describe who you are, if an individual proposing, or tell what your company does.

Patents. If you believe you are presenting a patentable idea, I suggest that you file, as a protection to yourself and the government, necessary patent applications with the PTO, or otherwise identify your intention to do this. Mark your proposal as containing proprietary data.

Notification of DoD action. It may take up to six months for you to get an answer! Be patient. Your proposal will be acknowledged upon it's receipt.

Security. Always submit unclassified material when possible. When this is not possible, make sure you properly label documents as being classified. Ask the Pentagon for a copy of its *Industrial Security Manual for Safeguarding Classified Information* (DoD 5220.22M). Paragraph 11 of this publication will tell you how to correctly classify your submissions.

Defense Nuclear Agency (DNA)

Office of Contracts
Washington, DC 20305

(202) 325-7658
Mrs. Patricia Brooks, Director, Business Utilization

Seeks research and development from small business firms with strong capabilities in the nuclear weapons effects, including simulation, instrumentation, directed energy, nuclear hardening and survivabilty, security of nuclear weapons, and operational planning. **Contact:** Small Business Innovation Research Program, Attn: OAAM/SBIR, Jim Gerding (202) 325-7018.

U.S. Department of Energy

Role of the Department of Energy in OERI activities. Upon receipt of a recommendation from OERI, an Invention Coordinator from DOE's **Inventions and Innovation Programs Division** analyzes it and opens negotiations on the type and amount of support (if any) to be provided. A letter is written to the inventor. The inventor's response to this letter is treated as a preliminary proposal by DOE.

The OERI recommendation, the inventor's preliminary proposal, and other information available to the Invention Coordinator (such as information concerning other DOE projects in related fields) are considered. Also taken into account by the official are the inventor's business and management capabilities and the resources available to the inventor. The Invention Coordinator then decides whether federal support is warranted and if so, what type of support should be provided.

DOE support under this program may consist of a grant award, a contract award, testing service at a DOE facility, guidance in obtaining private venture capital, or other types of assistance.

In practice, each case is handled on an individual basis. There is no standard operational procedure. Each case is unique.

Inventors, especially those in the early stages of development, often require an amount of money that exceeds DOE's authority and guidelines. Grants have ranged from $20,000 to $200,000, averaging $70,000, with most granted to support technical research, scientific testing, or business planning. DOE cannot support marketing efforts, although it can help inventors obtain market information. As the program evolved, the DOE staff, under the direction of the late Anthony J. "Jack" Vitullo sought to fine-tune financial support to the needs and abilities of individual inventors and their technologies, and to expand the range of nonfinancial services.

Among the many benefits the Program provides to independent inventors is **credibility**. The NIST evaluation has served many inventors as a kind of "Good Housekeeping Seal of Approval" that they have been able to convert into a bankable asset. It was reported that one inventor parlayed the NIST evaluation into an endorsement by, and financial support from, a state agency, then went on to build a business that sold $32 million worth of his product. Another testified that he "couldn't get a venture capitalist to talk to him" until he had the NIST evaluation and DOE support as credentials.

Assistant Secretary for Conservation and Renewable Energy
1000 Independence Ave., S.W.   (202) 586-9275
Washington, DC 20585   Dr. Robert L. SanMartin
Programs in conservation and renewable energy support research that attracts limited or no venture capital because the risks are too high or the payoff too long-range or unpredictable. Renewable energy programs involve research, development, and proof-of-concept experiments to enable industry to bring resource and technology options into the marketplace. Research activities are carried out through the award of grants and contracts.

Assistant Secretary for Defense Programs
1000 Independence Ave., S.W.   (202) 252-1870
Washington, DC 20585   Annie Smith
Directs the DOE's programs for nuclear weapons research, development, testing, production, and surveillance; manages programs for the production of the special materials used by the weapons program within the Department; and manages the defense nuclear waste and by-products program. In addition, awards grants and contracts to universities and other research groups for some defense programs.

Assistant Secretary for Nuclear Energy
1000 Independence Ave., S.W.   (202) 353-4380
Washington, DC 20585   A.S. Lyman
Responsible for programs and projects for nuclear power generation and fuel technology; evaluation of alternative reactor fuel cycle concepts; development of naval nuclear propulsion plants and cores; and nuclear waste technology and remedial action programs. Unsolicited proposals for projects in nuclear energy should be sent to the Unsolicited Proposal Coordinator, Reports and Analysis Branch, Procurement and Assistance Management Directorate, Department of Energy, Washington, DC 20585.

Morgantown Energy Technology Center (METC)
3610 Collins Ferry Rd.   (304) 291-4764
P.O. Box 880
Morgantown, WV 26505-0880   Augustine Petrolo, Director
Fossil fuel research and development laboratory responsible for projects in unconventional gas recovery, fluidized bed combustion, gas stream cleanup, fuel cells, heat engines, component development for coal conversion and utilization devices, surface coal gasification, instrumentation and control, oilshale technology, tar sands, underground goal gasification, and arctic and offshore technologies. **Telecommunication Access:** Alternate telephone (304) 599-4511.

Office of Energy Research (OER)
100 Independence Ave.   (301) 586-5430
Washington, DC 20585   Robert Hunter, Jr., Director-Designate
Manages research programs in the basic energy sciences, high energy physics, and fusion energy; administers DOE programs supporting university researchers; funds research in mathematical and computational sciences critical to the use and development of supercomputers; and administers a financial support program for research and development not funded elsewhere in the Department. The Office also manages the Small Business Innovation Research Program for the Department. **Contact:** Jean Marrow (301) 353-5544.

Pittsburgh Energy Technology Center (PETC)
P.O. Box 10940   (412) 675-6400
Pittsburgh, PA 15236   Dr. Sun W. Chun, Director
Primary mission is to promote the use of coal and its derived synthetic fuels in an environmentally sound manner. Research and development programs cover coal liquefaction, alternate fuels, coal preparation, flue gas cleanup, combustion technology, magnetohydrodynamics, solids transport, and peat and anthracite. The Center is also

Inquiries on the DOE/NIST program may be directed care of the U.S. Department of Energy, CE-12, Washington, DC 20585; (202) 586-1478.

When asked about the benefits of the NIST/DOE combined program, many inventors cited the gain in "credibility" as being equal in importance to the monetary awards.

responsible for handling all unsolicited research proposals pertaining to any part of the DOE's fossil energy program. **Publications:** (1) PETC Quarterly Technical Progress Report; (2) D.P. Databeat; and (3) Energizer (all quarterly). **Telecommunication Access:** Alternate telephone (412) 892-6128.

Small Business Innovation Research Program (SBIR)

Mail Stop ER-16, GTN   (301) 353-5544
Washington, DC 20545   Jean Marrow, Contact

Stimulates technological innovation; uses small business to meet federal research and development needs; fosters and encourages participation by minority and disadvantaged persons in technologcial innovation; and increases private secor commercializiaiton of innovations derived from federal research and development. **Telecommunication Access:** Alternate telephone (301) 353-5867.

U.S. Department of Health and Human Services

National Cancer Institute (NCI)

Westwood Bldg., Room 10A10   (301) 496-5583
Bethesda, MD 20205   Dr. Vincent T. DeVita, Jr., Director

Plans, directs, conducts, and coordinates a national research program on the detectin, diagnosis, cause, prevention, treatment, and palliation of cancers. Conducts and directs research performed in its own laboratories and through contracts; supports construction of facilities necessary for research on cancer; supports demonstration projects on cancer control; and collaborates with cancer research with industrial concerns. **Contact:** Louis P. Greenberg, Acting Chief, Research and Contracts Branch.

National Eye Institute

Research Contract Branch   (301) 496-4487
Division of Contracts and Grants
Bldg. 31, Room 1B44
9000 Rockville Pike
Bethesda, MD 20892   Dave Snight

Supports research on the cause, natural history, prevention, diagnosis, and treatment of disorders of the eye and the visual system, especially glaucoma, retinal disease, corneal diseases, cataract, and sensory-motor disorders. Support is provided for basic studies, clinical trials, and the development of animal models for vision disorders. **Publications:** Planning Report (every five years). **Meetings:** Grand Rounds (weekly) and Neuro-Ophthalmology Seminar (monthly).

National Institute of Allergy and Infectious Diseases

Westwood Bldg., Room 707   (301) 496-7116
5333 Westbard Ave.   Lew Pollack, Contract Management
Bethesda, MD 20205   Branch

Supports research on causes, diagnosis, treatment, and prevention of infections, allergic, and other immunologically mediated diseases. Contract programs are concerned with bacterial and viral vaccines, antiviral substances, transplantation, and viral and allergen reagents.

National Institute of Arthritis and Musculoskeletal and Skin Diseases (NIAMS)

NIH Bldg. 31, Room 9A35   (310) 496-7111
9000 Rockville Pike
Bethesda, MD 20892   Patrick Sullivan

Conducts and supports basic and clinical research on arthritis, osteoarthritis, lupus, muscle disease, psoriasis, acne, ichthyosis, and vitiligo. Offers investigator-initiated grants, reseasrch center grants, individual and institutional research training research awards, career development awards, and contracts to public and private research institutions and organizations. **Contact:** Lawrence E. Shulman, Director (301) 496-4353.

National Institute of Dental Research (NIDR)

Contract Management Section   (301) 496-7311
Westwood Bldg., Room 521
Bethesda, MD 20892   Ms. Marion Blevins

Supports investigations into the causes, means of prevention, diagnosis, and treatment of oral diseases, including dental caries, periodontal diseases, lesions of soft and hard tissue, and oral-facial abnormalities. Also supports biomaterials research, improvements in anesthesia and analgesia, basic epidemiological research, and clinical trials.

National Institute of General Medical Sciences (NIGMS)

Research Contract Branch   (301) 496-4487
Division of Contracts and Grants
Bldg. 31, Room 1B44
9000 Rockville Pike
Bethesda, MD 20892   David Snight

Supports research in the sciences basic to medicine, behavioral sciences, and clinical disciplines. The Institute fosters multidisciplinary approaches to research and employs the full range of support mechanisms, including research project grants, program-project grants, research center grants, career development awards, awards to new investigators, and institutional and individual fellowships.

National Institute of Neurological and Communicative Disorders and Stroke (NINCDS)

Contract Management Branch   (301) 496-9203
Federal Bldg., Room 901
7550 Wisconsin Ave.
Bethesda, MD 20892   Lawrence Fitzgerald

Conducts, coordinates, and supports research concerned with the cause, development, diagnosis, therapy, and prevention of disorders and diseases of the central nervous system and the communicative and sensory systems. Research includes neurological and mental development, infectious diseases, multiple sclerosis, epilepsy, head injury and stroke, biomedical engineering and instrumentation, and neural prostheses.

Office of Human Development Services (HDS)

Room 324B-HHS North Bldg.   (202) 245-1787
200 Independence Ave., S.W.   Cynthia Haile Selassie, Small Business
Washington, DC 20201   Specialist

Oversees programs that contribute to the social and economic well-being of a number of vulnerable populations, including the elderly, children, youth and families, the developmentally disabled, and Native Americans.

U.S. Department of the Air Force

Air Force Systems Command

Room 206, Bldg. 1535   (301) 981-6107
Andrews Air Force Base   Dave McNabb, Executive for Small
Washington, DC 20334   Business

Command includes project divisions that acquire aircraft, missiles, space systems, electronic systems, and conventional weapons; test centers; laboratories that conduct research in the mathematical, physical, engineering, and environmental sciences; and divisions that are involved in medical research and technology, identification of foreign technology, and contract management.

Air Force Systems Command, Aeronautical Systems Division (Eglin Air Force Base)

Eglin Air Force Base, FL 32542-5000   (904) 882-2843
     Ralph K. Frangioni, Jr.

Armament systems research and development, testing, and procurement, including total program management responsibility for Air Force non-nuclear munitions, including guided bombs, mines, fuzes, flares, bomb racks, missiles, aerial targets, and munition handling and transportation equipment.

Air Force Systems Command, Aeronautical Systems Division (Wright-Patterson Air Force Base)

Wright-Patterson Air Force Base   (513) 255-5322
Dayton, OH 45433   Tom Dickman, Small Business Specialist

Development and acquisition of aeronautical systems, including aircraft engines, aircraft wheels and brakes, airborne communication systems, aircraft bombing and navigation systems, and aircraft instruments, and aeronautical reconnaissance systems, and mobile land-based tactical information processing and interpretaton facilities. **Contact:** Jim Beach, Small and Disadvantaged Business Utilization Specialist.

Air Force Systems Command, Space Division

Los Angeles Air Force Station   (213) 643-2855
P.O. Box 92960
Worldway Postal Center   Chuck Willeme, Deputy for Small
Los Angeles, CA 90009   Business

Acquisition and evaluation of space systems and equipment; management of research, tracking, telemetry, and recovery; feasibility studies; acquisition, production, quality assurance, and installation of assigned space and missile systems; and quality assurance and installation of boosters, and aerospace ground equipment to support launch control and recovery.

IMPORTANT! Critically evaluate yourself before submitting a proposal for a government contract. And be equally critical of any idea you submit. The federal government is interested only in the most experienced people with pertinent, novel ideas. Don't waste its time or your own on half-baked ideas or proposals for contracts that you are not qualified to complete.

Proprietary Data. If you do not want your invention or idea disclosed to the public or used by the government for any other purpose than proposal evaluation, clearly label it as containing proprietary data. Don't just do it on the cover sheet, but rather on every page that holds such information.

Note this restriction does not limit the government's right to use or disclose any data contained in your proposal if it is obtainable from another source (or from yourself previously) without restriction.

Arnold Engineering Development Center (AEDC)

Arnold Air Force Base, TN 37389-5000   (615) 454-4407
William Lamb, Small Business Specialist

Conducts development, certification, and qualification testing on aircraft, missile, and space systems. Also conducts research and technology programs to develop advanced testing techniques and instrumentation and to support the design of new test facilities.

Ballistic Missile Office

Attn: BMO/BC   (714) 382-2304
Norton Air Force Base   Mr. Terry Carey, Chief, Small Business
San Bernardino, CA 92409   Office

Responsible for intercontinental ballistic missile development. Office integrates activities of contractors, retaining overall engineering responsibility for a particular system.

Contract Management Division

Kirtland Air Force Base, NM 87117-5000   (505) 844-6644

Acquisition contract management agency for the Air Force. Administers contracts and provides an interface with contractors.

Defense Advanced Research Projects Agency (DARPA)

1400 Wilson Boulevard   (202) 694-1440
Arlington, VA 22209   Dr. Robert C. Duncan, Director

Pursues imaginative and innovative research ideas and concepts offering significant military utility. Role in basic research is to develop selected new ideas from conception to hardware prototype for transfer to development agencies. Programs are conducted through contracts with industrial, university, and nonprofit organizations, focussing on improved strategic, conventional, rapid-development, and sea-power forces; and on scientific investigation into technologies for the future. Unsolicited proposals will be accepted if they are related to any of the current technical programs assigned to DARPA. DARPA invites small businesses to submit proposals under its SBIR program, which has as its objectives stimulating technological innovation in the private sector, strengthening the needs, fostering and encouraging participation by minority and disadvantaged persons in technological innovation, and increasing commerical application of Department of Defense-supported research and development. **Contact:** Bud Durand, SBIR Progam Manager (202) 694-1626.

Directorate of Research and Development Procurement

Aeronautical Systems Division   (513) 255-3825
Wright-Patterson Air Force Base   George R. Laudenslayer, Small Business
Dayton, OH 45433·   Specialist

Procurement of exploratory and advanced research and development in the areas of air breathing, electric, and advanced propulsion; fuels and lubricants; and power gerneration, including molecular electronics, bionics, lasers, vehicle environment, photo materials, position and motion sensing devices, navigation, communications, flight dynamics, materials science, and life support. **Contact:** Ms. Dorothy Muhlhauser, Small and Disadvantaged Business Utilization Specialist.

Eastern Space and Missile Center (ESMC)

Patrick Air Force Base, FL 32935-5000   (305) 494-2207
Don Hoskins, Small Business Specialist

Launches the Titan III space booster and represents Department of Defense interests in Space Shuttle operations. Research, development, and procurement program is confined primarily to range test instrumentation, including radar, telemetry, electro-optics, impact locations, data reduction, range/mission safety and control, weather timing and fireing, and frequency control and analysis. **Telecommunication Access:** Alternate telephone (305) 494-2208.

Electronic Systems Division (ESD)

Hanscom Air Force Base, MA 01730-5000   (617) 861-4441
Al Hart

Development, acquisition, and delivery of electronic systems and equipment for the command, control, and communications function of aerospace foreces. **Telecommunication Access:** Alternate telephone (617) 861-4973/4.

Flight Test Center (AFFTC)

Edwards Air Force Base, CA 93523   (805) 277-2619
James A. Beucherie

Flight testing of aerospace vehicles. Specific activities include: (1) conduct and support of aircraft systems tests; (2) flight evaluation and recovery of aerospace research vehicles and development testing of aerodynamic decelerators; (3) management and operation of USAF Test Pilot School; (4) management and operation of the Utah Test and Training Range and the Edwards Flight Test Range; and (5) support for and participation in agency, foreign, and contractor test and evaluation programs. **Telecommunication Access:** Alternate telephone (617) 861-4974.

Office of Scientific Research (AFOSR)

Building 410   (202) 767-4943
Bolling Air Force Base   Louise Harrison, Small Business
Washington, DC 20332   Specialist

Considers basic research proposals from industrial and small business concerns and nonprofit organizations. Accepts unsolicited proposals. Any scientific investigator may make a preliminary inquiry to obtain advice on the degree of interst in a project, or may submit a specific research proposal. Proposals should indicate the field of the investigation and objectives sought, describing previous work and related grants and contract held, if any; in addition, it should outline the approach planned for the research and should include estimates of the time and cost requirements. The principal investigator should be named and an outline of professional background included. Brochures describing research ares of interest and procedures for submitting proposals may be obtained.

Operational Test and Evaluation Center (AFOTEC)

Office of Small and Disadvantaged   (505) 844-3819
Business Utilization
Air force Contract Management Division
Kirtland Air Force Base, NM 87117-5023   Ms. Rosina Aragon, Chief

Independent agency responsible for testing and evaluation of systems being developed for Air Force and joint service use. Center's focus is on the operation effectiveness and suitability of the Air Force's future weapons and supporting equipment, as well as identifying deficiencies requiring corrective action. **Contact:** Nancy Lindquist, SADBU Specialist.

Rome Air Development Center (RADC)

Griffis Air Force Base, NY 13441   (315) 330-4020
                                     Peter Nicotera, Small Business
                                     Specialist

Provides a technology base for projects that pertain to command, control, communications, and intelligence research and development, including radar, airborne and space-based reconnaissance and sensing systems, vast arrays of communications systems, and computers. Has also been designated as a lead laboratory for research and development of the technology of photonics.

Space Technology Center

Kirtland Air Force Base, NM 87117-6008

Manages three Air Force laboratories, integrating space technology efforts in order to explore military space capabilities and the needs of future space systems. Specific areas of interest include directed energy research, nuclear weapons effects, survivability issues, rocket propulsion, and the earth and space environment.

Strategic Defense Initiative Organization (SDIO)

SDIO/OSD   (202) 693-1527
Washington, DC 20301-7100   Carl Nelson, SBIR Coordinator

Supports programs involved with kinetic energy weapons, and surveillance satellites, and directed energy weapons research. **Contact:** Submission of unsolicited proposals should be directed to Col. Kluter, SDIO/OSD, Washington, DC 20301-7100 (202) 653-0034.

Weapons Laboaratory (AFWL)

Kirtland Air Force Base, NM 87117-6008   (505) 844-9426

Responsible for reseasrch efforts in the areas of survivable systems and directed energy weapons, including particle beam technology, advanced weapons technology, high power microwave technology, and space nuclear power.

U.S. Department of the Army

Aberdeen Proving Ground Installation Support Activity

Attn: STEAP-SB   (301) 278-3878
Aberdeen Proving Ground, MD 21005-   Thomas Rodgers, Small Business
5001   Advisor

Research and development, production, and post-production testing of components and complete items of weapons, systems, ammunition, and combat and support vehicles. Also tests items of individual equipment in use throughout the Army.

Armament Research, Development, and Engineering Center (ARDC)

Attn: AMSMC-SBD (D)   (201) 328-4104
Dover, NJ 07801   Ed Smith, Small Business Advisor

Responsible for management of research, development, life-cycle engineering, and initial acquisition of weapon systems and support equipment.

Aviation Systems Command (AVSCOM)

Attn: AMSAV-V   (314) 263-2200
Building 102E, 4300 Goodfellow Blvd.   Charles Robinson, Small Business
St. Louis, MO 63120-1798   Advisor

Aviation design, research, development, maintenance, engineering, stock and supply control, and technical assistance to users of all Army aviation and aerial delivery equipment. Evaluates prototype hardware for fueling and defueling equipment for use in combat areas and in solving fuel contamintation problems.

Because locating and obtaining federal R&D funding requires a significant commitment of time and resources, the advice of NASA headquarters Small Business Specialist Mark Kilkenny should be remembered. "Concentrate on the few," he recommends, "and don't waste time and paperwork on the unimportant many."

Ballistic Missile Defense Systems Command

P.O. Box 1500   (205) 895-3410
Huntsville, AL 35807   Virginia B. Wright

Research and development in the fields of radar, interceptors, optics, discrimination, and data processing applicable to ballistic missile defense, including analysis of new and novel applications of science and engineering seeking revolutionary approaches to ballistic missile defense.

Ballistic Research Laboratory (BRL)

ARRADCOM Ballistic Research   (301) 671-2309
     Laboratories
Building E4455
Aberdeen Proving Ground, MD 21005-
     5066   Harold Siler, Small Business Specialist

Provides basic and applied research in mathematics, physics, chemistry, biophysics, and the engineering sciences related to the solution of problems in ballistics and vulnerability technology.

Belvior Research, Development, and Engineering Center

Attn: STRBE-V   (703) 664-5134
Fort Belvoir, VA 22060-5606   Susan Irwin, Small Business Advisor

Research, development, engineering, and initial production buys in the areas of mobility/countermobility, survivability, energy, and logistics. Research is carried out by the Combat Engineering Directorate, Logistics Support Directorate, and the Materials, Fuels, and Lubricants Directorate. Related support services are provided by the Advanced Systems Concepts Directorate, Information Management Directorate, Product Assurance and Engineering Directorate, and Resource Management Directorate. **Telecommunication Access:** Alternate telephone (703) 644-2482.

Benet Weapons Laboratory

Attn: SMCWV-SB   (518) 266-5765
Watervliet, NY 12180-5000   A.V. Paparian, Small Business Advisor

Research, development, engineering, and design of mortars, recoilless rifles, and cannons for tanks and towed and self-propelled artillery. Laboratory is a division of U.S. Army Close Combat Armament Center (Picatinny Arsenal, New Jersey 07806-5000), which publishes patent disclosures for inventions.

Chemical Research, Development, and Engineering Center

Building E4455   (301) 671-2309
Aberdeen Proving Ground, MD 21010-
     5423   Harold Siler

Research and development in the fields of chemical, smoke, and flame weapons. Activities also include studies in pollution abatement and environmental control technology.

Cold Regions Research and Engineering Laboratory

P.O. Box 282   (603) 646-4324
Hanover, NH 03755   Raymond F. May, Jr., Small Business
     Specialist

Conducts research pertaining to characterists and events unique to cold regions, especially winter conditions, including design of facilities, structures and equipment, and refining methods for building, traveling, living, and working in cold environments.

Communications-Electronics Command (CECOM)

Attn: DRSEL-SB AMSEL-SB   (201) 532-4511
Fort Monmouth, NJ 07703   John Meschler

Covers full spectrum of services to the U.S. soldier in the field of communications and electronics. The first steps in converting concepts into new military materiel are taken in the CECOM Research and Development Center and its laboratories. These laboratories are responsible for research and development of communications and electronics equipment and systems.

Construction Engineering Research Laboratory (CERL)

Procurement/Supply   (217) 352-6511
P.O. Box 4005
Champaign, IL 61820-1305

Mission is to provide research and development to support Army programs in facility construction, operation, and maintenance. Efforts focus on vertical construction as applied to buildings and structures rather than heavy construction. Laboratory has four main divisions: Engineering and Materials Division, Energy Systems Division, Environmental Division, and Facilities Systems Division. Center emphazises communication and interchange of information with academic, engineering, and construction activities within the Department of Defense, other governmental agencies, and the private sector. **Publications:** CERL Reports (quarterly).

Dugway Proving Ground (DPG)

Attn: STEDP-PR   (801) 522-2102
Dugway, UT 84022-5202   Clyde Harris, Small Business Advisor

Conducts field and laboratory tests to evaluate chemical and radiological weapons and defense systems and materiel. Also conducts biological defense research.

Engineer Topographic Laboratories (USAETL)

Fort Belvoir, VA 22060-5546

(202) 355-2659
Mrs. M.L. Williams, Small Business
Specialist

Research and development in the topographic sciences, including mapping, charting, terrain analysis, geodesy, remote sensing, point positioning, surveying, and land navigation. Provides scientific and technical advisory services to meet environmental design criteria requirements of military materiel developers and in support of geographic intelligence and land environmental resources inventory requirements. Composed of the following laboratories and operational centers: Geographic Sciences Laboratory, Research Institute, Space Programs Laboratory, Terrain Analysis Center, and Topographic Developments Laboratory.

Engineer Waterways Experiment Station (WES)

P.O. Box 631

Vicksburg, MS 39180

(601) 636-3111
Mr. A.J. Breithaupt, Utilization Specialist

Principle research, testing, and development facility of the Corps of Engineers. Operats six laboratories, including Coastal Engineering Research Center, Environmental Laboratory, Geotechnical Laboratory, Hydraulics Laboratory, Information Technology Laboratory, and Structures Laboratory, which conduct research in soil mechanics, concrete, engineering geology, pavements, weapons effects, protective structures, water quality, and dredge materials. **Publications:** List of Publications of the Waterways Experiment Station. **Meetings:** Annual Aquatic Plant Control Research and Operations Review (November), attendance is open to all.

Jefferson Proving Ground (JPG)

STEJP-LD-P

Madison, IN 47250-5100

(812) 273-7226
Ms. Mary N. Gassert, Small Business
Advisor

Processing, assembling, and acceptance testing of ammunition and ammunition components. Receives, stores, maintains and issues assigned industrial stocks, including calibrated components.

Laboratory Command (LABCOM)

Attn: DRDEL-SB
2800 Powder Mill Road
Adelphi, MD 20783-1145

(202) 394-1076

Arthur Wolters, Small Business Advisor

Research, development, engineering, and initial procurement of assigned items in the areas of electronic signal intelligence, electric warfare, atmospheric sciences, target acquisition and combat surveillance, electronic fuzing, radars, sensors, night vision, radar frequency and optical devices, nuclear weapons effects, instrumentation and simulation, and fluidics. Has direct control and management of seven Army laboratories, including Atmospheric Sciences Laboratory, Ballistic Research Laboratory, Electronics Technology and Devices Laboratory, Human Engineering Laboratory, Materials Technology Laboratory, and Army Vulnerability Assessment Laboratory.

Materials Technology Laboratory

Attn: DRXMR-AP AMXMR-K
Watertown, MA 02172

(617) 923-5005
Capt. John Rouse, Small Business
Advisor Dir.

Manages Army's research and development programs relating to materials technology, solid mechanics, lightweight technology, and manufacturing testing technology. Is a center for excellence in the area of corrosion protection. **Meetings:** Sponsors Sagamore Army Materials Research Conference (annually in August) for invited participants and Military Handbook-17 Coordinating Group Meeting (semiannually, in May/November).

Materiel Command (AMC)

Attn: AMCSB
5001 Eisenhower Avenue
Alexandria, VA 22333-0001

(202) 274-8185

Kurt Wussow, Disadvantaged Utilization

Responsible for the life cycle of U.S. Army hardware, including research, development, procurement, production, supply, and maintenance. Provides supervisory, planning, and budetary direction to installations where contracts are executed and administered. **Contact:** John Flakenham, Doris Agnew, Janet Tull. **Telecommunication Access:** Alternate phone number (202) 274-8186.

Medical Research and Development Command

Attn: SGRD-SADBU
Fort Detick
Frederick, MD 21701-5012

(301) 663-2744
Mrs. Audrey L. Wolfe, Assistant Director,
SADBU

Research involves assessment, prevention, diagnosis, and treatment of infectious diseases that would hamper military operations; disease vector surveillance; combat casualty care; health hazards of military materiel; factors limiting soldier effectiveness; prevention of oral disease; and dental materials. Operates research and development laboratories, including Aeromedical Research Laboratory, Biomedical Research and Development Laboratory, Institute of Dental Research, Institute of Surgical Research, Medical Research Institute of Chemical Defense, Medical Research Institute of Infectious Diseases, Research Institute of Environmental Medicine, and Walter Reed Institute of Research. Also supports medical research through contract research awards. Announcements of specific proposals solicited are published in *Commerce Business Daily*.

Missile Command (MICOM)

Redstone Arsenal   (205) 876-2561
Attn: DRSMI-B
Huntsville, AL 35809   Dr. Joe Plano, Small Bus. Advisor Cmdr.

Responsible for design and development; product, production, and maintenance engineering; and new equipment design of training devices in the areas of rockets, guided missiles, targets, air defense, fire control equipment, test equipment, missile launching and ground support equipment, and metrology and calibration equipment. **Telecommunication Access:** Alternate telephone (205) 876-5441.

Natick Research, Development, and Engineering Center

Attn: STRNC-25B   (617) 633-4995
Natick, MA 01760-5008   Ms. Victoria Tangherlini, Small Business Advisor

Research, development, and engineering in advanced systems concepts, aero-mechanical engineering, food engineering, individual protection, and science and advanced technology.

Research and Development Directorate

Attn: DEAN-RDZ   (202) 272-0725
20 Massachusetts Ave., N.W.   George Wischmann, Dir., Small Bus.
Washington, DC 20314-1000   Utilization

Serves as scientific advisor to the Chief of Engineers for Research and Development; directs the activities of all Corps laboratories; exercises responsibility for Army technology-based programs in environmental quality; and maintains responsibility for the planning and budgeting of the Corps research and development program and the management of all resources, including military research, development, test, and evaluaiton; civil works research and development appropriations; and mission support funding.

Small Business Innovation Research Program (SBIR)

Headquarters Laboratory   (301) 394-4602
Code AMSLC-TP-TI
2800 Powder Mill Road
Adelphi, MD 20783   Ruth Morarre, SBIR Coordinator

Will provide funding for nearly 500 topics in fiscal year 1989.

Tank-Automotive Command (TACOM)

Attn: AMSTA-CB   (313) 574-5388
Warren, MI 48297-5000   F.D. Folette, Small Business Officer

Responsible for research, design, development, engineering, test management, modification, product assurance, integrated logistics support, acquisition, and deployment of wheeled and tracked vehicles and associated automotive equipment. Research and development are the functional responsibility of TACOM's Tank-Automotive Research, Development, and Engineering Center. **Contact:** Tom Clynes, Small Business Office (313) 574-5406.

Test and Evaluation Command (TECOM)

Attn: AMSTE-PR   (301) 278-5184
Aberdeen Proving Ground, MD 21005-5055   Edward Snodgrass, Small Business Advisor

Directs research activities, proving grounds, installations, boards, and facilities required to test equipment, weapons, and materiel systems intended for use by U.S. Army.

Water Resources Support Center

Humphreys Engineering Center   (202) 355-2153
Kingman Bldg.
Telegraph and Leaf Rds.
Fort Belvoir, VA 22060-5580   John Carpenter

Research and development in the topographic sciences; implementation of emergency operations to provide the capabilities to transfer data between civil organizations; and remote sensing, water control data systems, and telecommunications research planning. Administers the Waterborne Commerce Statistics Center, which collects and reports on traffic and tonnage.

White Sands Missile Range (WSMR)

Attn: STEWS-PR   (915) 678-1401
White Sands Missile Range, NM 88002-5031   Luis E. Sosa, Small Business Advisor

Conducts testing and evaluation of Army missiles, rockets, warheads, and special weapons. Major directorates include National Range Operations, Army Materiel Test and Evaluation, Instrumentation Directorate, and Directed Energy Directorate. Major tenants are Naval Ordnance Missile Test Station, Air Fore Range Operations Office, U.S. Army Vulnerability Assessment Laboratory, LABCOM, U.S. Army Atmospheric Sciences Laboratory, and U.S. Army TRADOC Analysis Command.

Yuma Proving Ground (YPG)
Attn: STEYP-PC
Yuma Proving Ground, AZ 85364-1530

(602) 328-2825
Mr. Loren Cady, Small Business Advisor

Conducts research and development, production, and post-production testing of components and complete items of weapons, systems, ammunition, and combat and support vehicles in desert environments.

U.S. Department of the Interior

Bureau of Mines
Room 1024
2401 E St., N.W.
Washington, DC 20241

(202) 634-1303

David S. Brown, Acting Director

Conducts research and collects, interprets, and analyzes information involving mineral reserves and the production, consumption, and recycling of mineral materials. **Publications:** (1) Guide for the Submission of Unsolicited Research and Development; (2) Reports of Investigations Information Circulars; (3) Minerals Yearbook; and (4) Mineral Facts and Problems. **Contact:** Chief Mining Engineer (202) 634-1303; Division of Procurement (202) 634-4704.

U.S. Geological Survey, Geologic Division
National Center
12201 Sunrise Valley Dr.
Reston, VA 22092

(703) 648-6600

Benjamin A. Morgan, Chief Geologist

Conducts programs to assess energy and mineral resources, identify and predict geologic hazards, and investigate the effects of climate.

U.S. Department of the Navy

David Taylor Research Center (DTRC)
Code 003
Bethesda, MD 20084-5000

(202) 227-1220
David Tychnan, Small Business Specialist

Navy's principal research, development, test, and evaluation center for naval vehicles. Research information may be obtained from the National Technical Information Service, 5285 Port Royal Road, Springfield, VA 22161.

Marine Corps Headquarters
U.S. Marine Corps (LS)
Washington, DC 20380

(202) 694-1939
Mrs. Sheila D'Agostino, Small Business Representative

Electronic equipment, specialized vehicles, and equipment peculiar to the Marine Corps.

Naval Air Development Center (NADC)
Code 094
Street Road
Warminster, PA 18974-5000

(215) 441-2456
Ms. Janet Koch (Code 094), Small Business Specialist

Navy's principal center for research, development, testing, and evaluation of naval aircraft systems, including simulation, personal safety, aircraft stuctures, aircraft hydraulics and fluids, magnetics, air to air strike, antisubmarine warfare systems, computers, and software. Organized into the following departments: Antisubmarine Warfare Systems Department, Battle Force Systems Department, Tactical Air Systems Department, Air Vehicles and Crew Systems Technology Department, Communication and Navigation Technology Department, Mission Avionics Technology Department, and Systems and Software Technology Department.

Naval Air Engineering Center
Code 00Q, Bldg. 129
Lakehurst, NJ 08733-5028

(201) 323-2064
Constance Hardy, Small Business Specialist

Research, development, test, system evaluation, and engineering in support of aircraft and shipboard interface systems, including launch and recovery systems, support equipment, and visual landing aids.

Naval Air Propulsion Center (NAPC)
Box 7176
Trenton, NJ 08628-0176

(609) 896-5653
Robert Reale, Technical Representative

Technical and engineering support for air breathing propulsion systems, including their accessories and components, fuels, and lubricants. Also provides technical and advisory services and consulting services on matters relating to the development, evaluation, and support of new air breathing propulsion systems.

Naval Air Systems Command (NAVAIR)

Jefferson Plaza, Room 478
Washington, DC 20361-0001

(202) 692-0935
Sarah Cross, Small Business
Representative

Design, development, testing, and evaluation of aircraft, airborne weapon systems, avionics, related photographic and meteorological equipment, ranges, and targets. Principal research and engineering components include the Naval Air Engineering Center, Naval Air Propulsion Center, Naval Air Test Center, Naval Training Systems Center, Naval Weapons Evaluation Facility, and Pacific Missile Test Center. **Contact:** Barbara Williams, Small Business Representative (202)692-0936; Mr. J. A. Johnson, Technical Representative, (AIR 303M), Room 424, Washington, DC 20361.

Naval Avionics Center

6000 East 21st St.
Indianapolis, IN 46218

(317) 353-7009
James F. Wilson, Small Business
Specialist

Research, development, pilot, and limited manufacturing and depot maintenance on avionics and related equipment.

Naval Civil Engineering Laboratory (NCEL)

Naval Construction Battalion Center
   (Code 6641)
Port Hueneme, CA 93043

(805) 982-5992
Mary Lorenzana, Small Business
Representative

Principal Navy research, development, test, and evaluation center for shore and seafloor facilities.

Naval Coastal Systems Center (NCSC)

Panama City, FL 32407

(904) 234-4347
Johnny L. Peace, Small Business
Specialist

Research, development, test, and evaluation center for mine and undersea countermeasures, special warfare, amphibious warfare, diving, and other naval mission that take place primarily in the coastal regions. Hosts several tenant activities, including the Navy Experimental Diving Unit and the Naval Diving and Salvage Training Center. **Contact:** Mr. W.R. Donaldson (904) 234-4862.

Naval Explosive Ordnance Disposal Technology Center

Attn: Code 604
Naval Ordnance ·Station
Indian Head, MD 20640-5070

(301) 743-4530
Edward W. Rice, Technical
Representative

Research and development of specialized equipment, tool, techniques, and procedures required to support operational explosive ordnance disposal units in the location, neutralization, and disposal of surface and underwater explosive ordnance. Also provides support toward the demilitarization of chemical weapons. **Telecommunication Access:** Alternate telephone (301) 743-4841.

Naval Medical Research and Development Command (NMRDC)

National Capital Region
Building 54
Bethesda, MD 20814-5044

(202) 295-0548
Ms. Sandy Shepard, Small Business
Representative

Research, development, test, and evaluation programs in submarine and diving medicine, aviation medicine, fleet occupational health, human performance, combat casualty care, infectious disease, oral and dental health, and electromagnetic radiation. Major components include Naval Aerospace Medical Research Laboratory, Naval Dental Research Institute, Naval Health Research Center, Naval Medical Research Institute, and Naval Submarine Medical Research Laboratory.

Naval Ocean Systems Center (NOSC)

271 Catalina Boulevard
San Diego, CA 92152-5000

(619) 225-2707
Forrest L. Hodges, Small Business
Representative

Research, development, test, and evaluation for command, control, and communications; ocean surveillance, surface- and air-launched weapons systems; and submarine arctic warfare. Also involved in development of technologies that support these activities, including ocean engineering, environmental sciences, marine bioscience, electronics, computer sciences, and atmospheric sciences. **Publications:** (1) NOSC Technical Manuals; (2) Technical Documents; and (3) Technical Reports.

Naval Oceanographic Office (NAVOCEANO)

NSTL Station
NSTL, MS 39522-5001

(601) 688-4166
Don Hutchison

Collects, analyzes, and displays oceanographic data in support of Naval operations and establishment. Activities include oceanographic, hydrographic, magnetic, gravity, navigational, and acoustic surveys in support of mapping, charting, and geodesy. **Publications:** (1) NAVOCEANO's Reference Publications; (2) Special Publications; and (3) Technical Reports.

Naval Ordnance Missile Test Station (NOMTS)

Building N-103
White Sands Missile Range, NM 88002

(915) 678-6115
Mr. H.L. Hendon, Technical
Representative

Supports research, development, test, and evaluation programs in flight testing and guided missiles, rocket gun, and directed energy programs.

Naval Ordnance Station

Code 1141D
Naval Ordnance Station
Indian Head, MD 20640-5070

(301) 743-4404
Ernest K. Tunney, Small Business
Specialist

Research, development, test, and evaluation of ammunition, pyrotechnics, and solid propellant used in missiles, rockets, and guns.

Naval Research Laboratory (NRL)

Code 2490
4555 Overlook Ave., S.W.
Washington, DC 20375

(202) 767-2914
Burt Copson, Small Business
Representative

Multidisciplinary scientific research and technological development directed toward materials, equipment, techniques, systems, and related operational procedures for the Navy. **Publications:** (1) NRL Review; (2) Fact Book (annually).

Naval Sea Systems Command (NAVSEA)

SEA 00311
Washington, DC 20362-5101

(202) 692-7713
Mr. S. Tatigian, Small Business
Representative

Provides material support to the Navy and Marine Corps for ships, submarines, and other sea platforms; shipboard combat systems and components; other surface and undersea warfare and weapons systems; and ordnance expendables. **Contact:** George W. Gatling, Jr., SBIR Coordinator, Crystal Plaza 6, Room 850, Crystal City, VA 20362.

Naval Surface Warfare Center (NSWC)

Attn: Code S095
10901 New Hampshire Avenue
Silver Spring, MD 20910-5000

(703) 663-8391

Hugh H. Snider, Jr.

Principal research, development, test, and evaluation center for surface ship weapons systems, ordnance, mines, and strategic systems support. Current areas of interest include low observables technology, applications of artificial intelligence to naval systems, mission and weapons analysis in support of the Navy's use of space systems, and development of technology for advanced autonomous weapons. **Publications:** (1) NSWC Briefs; (2) On the Surface (both weekly).

Naval Training Systems Center

Code N-005
Orlando, FL 32813-7100

(305) 646-5515
Dr. John H. Rhodes, Utilization Specialist

Responsible for procurement of training aids and equipment for the Army, Navy, Marine Corps, Air Force, and other government activities, including research and development in simulation, training psychology, human factors and human engineering, and design and engineering of training equipment. **Contact:** Wiley V. Dykes, Technical Representitive, Code N-731 (305) 646-4629/5464.

Naval Underwater Systems Center (NUSC)

Newport, RI 02840

(401) 841-2675
Robert Bowerman, Technical
Representative

Research, development, test, and evaluation for submarine warfare and weapons systems, including sonar, electromagnetics combat control, weapon systems, launchers, undersea ranges, and combat systems analysis. **Contact:** Marvin Berger, Technical Representative (New London Laboratory), New London, CT 06320 (203) 440-4811.

Naval Weapons Station

Yorktown, VA 23691

(804) 887-4644
LCDR J. Kinlaw, Small Business
Representative

Development of explosives processing and explosives loading methods for Navy weapons. **Telecommunication Access:** Alternate telephone (804) 887-4744.

Naval Weapons Support Center

Code C-2
Crane, IN 47522

(812) 854-1542
Mrs. Reva Swango, Small Business
Representative

Design, engineering, and inservice engineering, evaluation, and analysis programs required in providng support for ships and craft components, shipboard weapons systems, and assigned expendable and non-expendable ordnance items. **Contact:** James F. Short. Jr., Technical Representative, Code 505, (812) 854-1625/1626.

Naval Weapons Systems Center (NWC)

Code 005   (619) 939-2712
China Lake, CA 93555-6001   Lois Herrington, Small Business Specialist

Research, development, test, and evaluation of air warfare and missile systems, including missile propulsion, warheads, fuzes, avionics and fire control, and missile guidance. Also acts as national range facility for parachute test and evaluation. **Publications:** (1) Naval Weapons Center Technical Publications; (2) Current Technical Events (irregular newsletter). **Contact:** George F. Linsteadt, Technical Representative, Code O1T3 (619) 939-2305.

Office of Naval Research (ONR)

Code 1111MA   (202) 696-4313
800 North Quincy Street
Arlington, VA 22217-5000   Dr. Neal Glassman, Program Manager

Mission is to encourage and promote naval research. Offers broad programs to encourage and assist young scientists, including (1) ONR Naval Science Awards Program, which recognizes achievements of high school science students; (2) ONR Young Investigators Program, which is intended to attract the best young academc researchers at U.S. universities; (3) ONR Graduate Fellowship Program which supports 135 students at about 80 universities in nine fields of science and engineering; (4) ONR University Research Initiatives Block Research Program, which offers interaction resesarch between university and naval laboratory personnel; (5) ONR Summer Faculty Research Program, which allows university researchers to spend the summer working at a naval research activity; and (6) ONR Instrumentation Program, which allows for the purchase of major, high-cost university research equipment. **Contact:** Joseph C. Ely (202) 696-6525.

Pacific Missile Test Center (PMTC)

Code 6009   (805) 989-8432
Point Mugu, CA 93042   Eric D. Duncan, Small Business Specialist

Primary test and evaluation facility for air-launched weapons and airborne electronic warfare systems.

Space and Naval Warfare Systems Command (SPAWAR)

National Center Building 1   (202) 692-6091
Room 1E58   Mrs. Betty Geesey, Asst. Dir. for Small
Washington, DC 20363-5100   Business

Research, development, test, and evaluation for command, control, and communications; underseas and space surveillance; electronic test equipment; and electronic materials, components, and devices. **Contact:** Robert H. Branner, Head of Operational ASW Systems Branch (SPAWAR 661), (202) 433-4729.

U.S. Department of Transportation

Coast Guard Office of Engineering and Development

G-FCP-S   (202) 267-1844
2100 2nd St., N.W.
Washington, DC 20593   Rear Adm. Kenneth G. Wilman, Chief

Expects to spend $19-million in 1989 to support reseach and maintain and improve search and rescue systems, environmental protection, marine safety, and aids to navigation. Majority of the research and development tasks are conducted by private business on contract, and almost all research is applied, aimed at developing improved systems for Coast Guard use.

Federal Aviation Administration

Small Business Coordinator   (202) 426-8230
800 Independence Ave., S.W.
Washington, DC 20591

Expects to spend $165-million in 1989 to conduct engineering and development programs in air traffic control, advanced computer applications, navigation, weather, aviation medicine, aircraft safety, and enviornment. For procurement information on the Federal Aviation Administration Technical Center contact: Procurement Officer, ANA-51B, FAA Technical Center, Atlantic City International Airport, NJ 08405 (609) 484-4000.

Federal Highway Administration (FHWA)

Office of Contracts and Procurement   (202) 366-0650
400 7th St., S.W.
Washington, DC 20509   Robert E. Farris, Administrator

Research on civil, mechanical, geotechnical, chemical, hydraulic, electrical, environmental, and human factors engineering, including direct and contract research and development relating to traffic operations, new construction techniques, and social and environmental aspects of highways and programs. **Contact:** George H. Duffy, Disadvantaged Business Enterprise Division (202) 366-1586.

Federal Railroad Administration

Office of Procurement   (202) 366-0563
400 7th St., S.W., Room 8206
Washington, DC 20590   Joseph Kerner, Contact

> Will obligate $9-million to continue emphasis on safety of train operations, including testing and evaluation, computer modeling, systems engineering, and safety hazard analysis.

National Highway Traffic Safety Adminstration

Office of Contracts and Procurement   (202) 366-0607
400 7th St., S.W.
Washington, DC 20590   Thomas Stafford, Director

> Plans to spend $30-million for motor vehicle research, traffic safety research and demonstration, and other statistical and analytical studies in 1989. Enters into contracts with private industry, educational institutions, nonprofit organizations, and state and local governments for defects investigations, crashworthyness programs, alcohol traffic safety programs, systems operations, emergency medical services, safety manpower development, driver/vehicle interaction, experimental vehicles, occupant packaging testing, biomechanics, passive restraint tests, computer support, management studies, and data acquisition.

Procurement Operations Division

Office of the Secretary of Transportation   (202) 366-4952
Washington, DC 20590

> Anticipates to spend $7-million on broad-based policy research on domestic and international transportation issues of importance to the nation. Procurement Operations Division contracts for studies covering social, economic, environmental, safety and policy-oriented transportation research. Such studies include data collection, modeling, economic, and financial studies and projects to support national transportation policy development and evaluation.

Transportation Systems Center (TSC)

DOT SBIR Program Office, DTS-23   (617) 494-2051
Kendall Square
Cambridge, MA 02142   Dr. George Kovatch

> Industrial-funded research and analysis organization that applies technical skills to national transportation and logistics problems. Expertise includes radionavigation, human factors, telecommunications, structural analysis, railroad inspection technology, industry analysis, emergency and readiness planning, air traffic automation, explosives detection, security and surveillance, and information systems development. Manages the Department of Transportation, Small Business Innovation Research Program. **Publications:** (1) National Transportation Statistics; (2) Transportation Safety Information (annual); and (3) Project Directory.

Hot Press: Newsletters, Magazines & Directories

In the United States, an estimated 12,300 magazines are listed as commercial publications. If you add to this smaller publications, newsletters, and directories, the total jumps to no less than 25,000. I have included in this chapter a representative sampling of periodicals helpful to the inventor, but it is by no means comprehensive. Time-sensitive publications, such as magazines, newsletters, and directories, can provide a wide variety of information and opinion to inventors.

Academic Research and Public Service Centers in California: A Guide
California Institute of Public Affairs   (714) 624-5212
c/o Claremont College
Box 10
Claremont, CA 91711   Lizanne Fleming, Editor
 Research institutes and public service centers at California colleges and universities. **Price:** $18.50, plus $1.50 shipping.

Accent On Living
P.O. Box 700   (309) 378-2961
Bloomington, IL 61702-0700   Betty Garee, Editor
 Magazine for people with physical disabilities, their families, and the professional and lay persons working with disabled people. Features motivational articles that emphasize success stories of handicapped people and contains new inventions and ideas for making daily living easier. **Frequency:** Quarterly. **Subscription:** $6.00.

Advanced Manufacturing Technology
Technical Insights, Inc.   (201) 568-4744
P.O. Box 1304
Fort Lee, NJ 07024-9967   Edward D. Flinn, Editor
 Designed to inform readers about "the cutting edge" of manufacturing technology, including products being currently developed. Deals with both traditional and electronics manufacturing technology, covering all phases of discrete products manufacturing: automation, robotics, laser, assembly, welding, soldering, finishing, material treatment, and new techniques for machining. Recurring features include news of research, book reviews, a calendar of events, and a monthly insert titled Plant Data Flow about the use of computers in manufacturing plants. **Frequency:** Semimonthly. **Price:** $470/yr., U.S. and Canada; $512 elsewhere. **Send Orders To:** P.O. Box 1304, Fort Lee, NJ 07024.

Advertising Age
Crain Communications, Inc.   (212) 210-0100
220 E. 42nd St.
New York, NY 10017   Fred Danzig, Editor
 Trade newspaper covering news, trends, and analysis related to the advertising community. **Frequency:** Weekly. **Subscription:** $64.00.

Adweek/East
A/S/M Communications, Inc.   (212) 529-5500
49 E. 21st St.
New York, NY 10010   Geoffrey Precourt, Editor
 Advertising news magazine. **Frequency:** Weekly. **Subscription:** $50.00.

AFHRL Newsletter
Air Force Human Resources Laboratory   (512) 536-3879
AFHRL/TSRE
Brooks AFB, TX 78235-5601   Leasley Besetsny, Editor
 Presents research and development techniques and findings related to manpower and personnel, education and training, simulation and aircrew training devices, and logistics and human factors that were developed by the AFHRL. Promotes the exchange of ideas between researchers and Air Force and other defense personnel. Recurring features include news of research and notices of publications available. **Frequency:** Quarterly. **Price:** Free.

Advertising Age and Ad Week: *there are no better sources than these weekly trade publications for the latest information on Fortune 500 new introductions and who's spending how much to promote and introduce what product.*

All types of business trade magazines and newsletters can be located in libraries that offer comprehensive business reference rooms. University technical libraries also usually have a wide range of these periodicals.

As valuable as these publications are, however, remember that there is no substitute for personal contacts. It's know how" combined with "know who" that makes the difference. I recommend a phone call to authors and editors as the best way to get information on R&D opportunities out of many publications.

Air Market News

General Publications, Inc. (301) 381-9295
P.O. Box 2368
Columbia, MD 21045 Jennifer Prill, Editor

Magazine on new products fot the aviation industry. **Frequency:** 6x/yr. **Subscription:** $20.00.

American Association of Small Research Companies—Members Directory

American Association of Small Research (215) 522-1500
 Companies
1200 Lincoln Ave., Suite 5
Prospect Park, PA 19076 Joanne Martin, Editor

About 400 small research and development companies covering most scientific disciplines. **Price:** $17.00.

American Industry

Publications for Industry (516) 487-0990
21 Russell Woods Rd.
Great Neck, NJ 11021 Jack S. Panes, Editor

Tabloid featuring new products of interest to managers of manufacturing plants with 100 or more employees. **Frequency:** Monthly. **Subscription:** $25.00.

Appliance New Product Digest

Dana Chase Publications, Inc. (312) 990-3484
1110 Jorie Blvd., CS 9019
Oak Brook, IL 60522-9019 James Stevens, Editor

New product magazine (tabloid) for production, engineering, and purchasing functions for companies producing consumer, commercial, and business appliances. **Frequency:** Quarterly. **Subscription:** $10.00; $20.00 foreign.

The Authority Report

Arkansas Science and Technology (501) 371-3554
 Authority
100 Main St., Suite 450
Little Rock, AR 72201 Kay Speed Kelly, Editor

Concentrates on issues related to the development of Arkansas' scientific and technological resources. Contains notices of publications available and related conferences as well as news of research, including research and grants funded by the Authority. Recurring features include interviews, reports of meetings, and news of educational opportunities. **Frequency:** Quarterly.

Behind Small Business

Dona M. Risdall (612) 881-5364
P.O. Box 37147
Minneapolis, MN 55431 Dona M. Risdall, Editor

Offers investment advice, marketing tips, public relations assistance, success stories, money-making ideas, and other advice for the start-up and operation of a small business. Gives small business owners the opportunity for free publicity on a national scale through press releases. Recurring features include editorials, news of research, book reviews, a calendar of events, and columns titled Small Business Publicity Handbook, Peace of Mind Investing, and Your Image. **Frequency:** Bimonthly. **Price:** $14/yr., U.S. and Canada; $18 elsewhere.

The Big Idea

Inventor's Association of St. Louis
P.O. Box 16544
St. Louis, MO 63105

Association newsletter.

Biochemical & Biophysical Research Communications

Academic Press, Inc. (619) 699-6825
1250 Sixth Ave.
San Diego, CA 92101 John N. Abelson, Editor

International scientific journal reporting on experimental results in the field of modern biology. Articles emphasize the innovative aspects of research reports. **Frequency:** Semimonthly. **Subscription:** $712.00 (U.S. and Canada); $879.00 (other countries).

Bioengineering News

Deborah J. Mysiewicz Publishers, Inc. (206) 928-3176
P.O. Box 1210
Port Angeles, WA 98362 Thomas G. Mysiewicz, Editor

Biotechnical newsletter. Covers developments in areas such as recombinant DNA, genetic engineering, hybridomas, monoclonal, antibodies, cell fusion, tissue culture and fermentation and purification systems. Carries items on law, patents, new companies, etc. **Frequency:** Weekly. **Subscription:** $495.00.

Bionics

Bionics, Box 1553 (517) 723-5298
Owosso, MI 48867 Jeff Campbell, Editor

Provides news of developments in the bionics, biosensor, and biomedical industries. Features recent inventions and technological advances. ** Quarterly. **Subscription:** $50.

Bioprocessing Technology

Technical Insights, Inc. (201) 568-4744
P.O. Box 1304
Fort Lee, NJ 07024-9967 Laurel A. Van Der Wende, Editor
Reports developments in the biological production of chemicals and energy. Lists new patent introductions. Recurring features include news of research, book reviews, notices of publications available, and a calendar of events. **Frequency:** Monthly. **Price:** $440/yr., U.S. and Canada; $482 elsewhere. **Send Orders To:** P.O. Box 1304, Fort Lee, NJ 07024.

BioScience

American Institute of Biological Sciences (202) 628-1500
730 11th St., N.W.
Washington, DC 20001-4584 Julie Ann Miller, Editor
Review journal for biologists, containing articles, book reviews, new product reviews, features, and announcements. **Frequency:** Monthly. **Subscription:** $42.00; $93.00 foreign; $8.75 single issue.

Biotechnology Law Report

Mary Ann Liebert, Inc. (215) 892-9580
416 N. Chester Rd.
Swarthmore, PA 19081-1108 Gerry Elman, Editor
Covers legal developments affecting the fields of biotechnology and genetic engineering. Discusses patent law, product liability, biomedical law, contract and licensing law, and international law. Describes pertinent legislation, regulatory actions, personnel and company changes, litigation resolution, and international developments. Publishes complete texts of significant court decisions, regulations, and legislation. Recurring features include book reviews, news of research, and a calendar of events. **Frequency:** Bimonthly. **Indexed:** Annually. **Price:** $375/yr., U.S. and Canada; $395 elsewhere.

BNA's Patent, Trademark & Copyright Journal

Bureau of National Affairs, Inc. (202) 452-4500
1231 25th St., N.W.
Washington, DC 20037 James D. Crowne, Editor
Monitors developments in the intellectual property field, including patents, trademarks, and copyrights. Covers proposed and enacted legislation, litigation, Patent and Trademark Office decisions, Copyright Office practices, activities of professional associations, government contracting, and international developments. **Frequency:** Weekly. **Price:** $784/yr.

Boardroom Reports—Breakthrough

Boardroom Reports, Inc. (212) 239-9000
330 W. 42nd St.
New York, NY 10036
Disseminates information on emerging technology in lay terms. Reports on the fields of electronics, biotechnology, medicine, business, and computers; identifies the breakthrough and often supplies data on its background, application, and developers. Recurring features include periodic ''special report'' supplements focusing on specific fields. **Frequency:** Semimonthly. **Price:** $69/yr., U.S. **Send Orders To:** RGK, 21 Bleeker St., Millburn, NJ 07041; (201) 379-4642.

Brand Names: Who Owns What

Facts on File, Inc. (212) 683-2244
460 Park Ave. South
New York, NY 10016
Over 750 firms and their 15,000 brand names. **Price:** $65.00.

Business & Acquisition Newsletter

Newsletters International, Inc. (713) 783-0100
2600 S. Gessner
Houston, TX 77063-3297 Len Fox, Editor
Contains highly confidential information about companies that want to buy or sell companies, divisions, subsidiaries, product lines, and patents. Includes information on sources of capital to finance purchases of such operations and suggestions on how to structure, negotiate, and complete such deals. **Frequency:** Monthly. **Price:** $300/yr., U.S.; $350 elsewhere.

Business Capital Sources

IWS, Inc. (516) 766-5850
24 Canterbury Road
Rockville Centre, NY 11570 Tyler G. Hicks, Editor
About 1,500 banks, insurance and mortgage companies, commercial finance, leasing, and venture capital firms that lend money for business investment. **Price:** $15.00.

Business Ideas & Shortcuts

I.B.I.S. (714) 552-8494
P.O. Box 4082
Irvine, CA 92716-4082 Peter Joseph, Editor
Contains special reports, practical ideas, tips, and guidelines on current business opportunities and shortcuts to profits. Covers specific topics, such as how to be a manufacturer without investing; how to get free national advertising; how to protect your business; and how to tap

I cannot put too great an emphasis on their value and the time you should dedicate to researching and reading. Many of the most valuable publications are free-of-charge, controlled-circulation publications for the business trade. Such periodicals make their money through the sale of advertising, not the sale of subscriptions. One of the best things about them is the abundance of trade advertisements, which provide great clues to the personalities of the companies and information on the products they promote. Contact the publishers to see if you qualify for a free subscription.

I love prospecting through publications. Often when I need creative stimulation, I will go to a nearby library and spend hours leafing through a variety of publications. If nothing else, this exercise tends to focus my direction.

It's a sure bet that even the most general circulation magazines will run a few stories each year on inventions and creativity and maybe even something related to your particular field of interest. The way I keep up with what is being covered is through the Reader's Guide to Periodical Literature (H.W. Wilson Co.) at my local library. The Reader's Guide tracks the stories appearing in hundreds of publications on a day-to-day basis.

overlooked sources of financing. Recurring features include columns titled Business Shortcut of the Month and Unique Ideas for Entrepreneurs. **Frequency:** Monthly. **Price:** $59/yr.

Business Ideas Newsletter

1051 Bloomfield Ave., 2-A   (201) 778-6677
Clifton, NJ 07012   Dan Newman, Editor
     Newsletter containing information on new products, promotion, and advertising of interest to businesspeople, advertising managers, and marketing professionals. **Frequency:** 10x/yr. **Subscription:** $40.00.

Business Organizations, Agencies, and Publications Directory

Gale Research Inc.   (313) 961-2242
835 Penobscot Bldg.
Detroit, MI 48226   Sandra A. MacRitchie, Editor
     More than 22,000 organizations and publications of all kinds which are helpful in business, including trade, business, commercial, and labor associations; government agencies; commodity and stock exchanges; United States and foreign diplomatic offices; regional planning and development agencies; convention, fair, and trade organizations; franchise companies; banks and savings and loans; hotel/motel systems; publishers; newspapers; information centers; computer information services; research centers; graduate schools of business; special libraries, periodicals, directories, indexes, etc. **Price:** $298.00.

Business Week—R&D Scoreboard Issue

McGraw-Hill, Inc.   (212) 512-2000
1221 Avenue of the Americas
New York, NY 10020
     Price: $2.00.

Canadian Research

Sentry Communications   (416) 490-0220
245 Fairview Mall Dr., Suite 500
Willowdale, ON, Canada M2J 4T1   Tom Gale, Editor
     Magazine covering research and development in the life and physical sciences in Canada and around the world. **Frequency:** Monthly. **Subscription:** $30.00.

Catalog of Government Inventions Available for Licensing to U.S. Businesses

Center for the Utilization of Federal   (703) 487-4838
     Technology
National Technical Information Service
Department of Commerce
5285 Port Royal Road
Springfield, VA 22161   Edward J. Lehman, Editor
     About 1,800 federal inventions, developed during the previous year, that are available for licensing. **Price:** $36.00, plus $3.00 shipping.

Category Reports

International Product Alert   (716) 374-6326
Marketing Intelligence Service
33 Academy St.
New York, NY 14512   Donna Maschiano, Editor
     New product reporting service (newsletter) in the following categories: foods, beverages, snacks, health and beauty aids, and household, pets, and miscellaneous. **Frequency:** Monthly. **Subscription:** $1,000.00; overseas postage extra.

CDA Reporter

Commercial Development Association   (202) 429-9440
1133 15th St., N.W.
Washington, DC 20005   Trecey St. Pierre, Editor
     Covers the activities of the Association, whose members are engaged in "identifying, evaluating, and establishing profitable new businesses." Provides limited coverage of related industry events. Recurring features include news of members, a calendar of events, and the column titled President's Message. **Frequency:** Quarterly. **Price:** Included in membership.

Chemical Marketing Reporter

Schnell Publishing Co., Inc.   (212) 732-9820
100 Church St.
New York, NY 10007-2694   Harry Van, Editor
     International tabloid newspaper for the chemical process industries. Includes new technology. **Frequency:** Weekly. **Subscription:** $65.00.

China Exchange Newsletter

Committee on Scholarly Communication   (202) 334-2718
  with the People's Republic of China
National Academy of Sciences
2101 Constitution Ave., N.W.
Washington, DC 20418   Kathlin Smith, Editor
    Covers recent developments in science, technology, and scholarship in the People's Republic of China. Discusses exchange activities with the Committee and worldwide exchange activities of China, often listing participants. Includes an extensive bibliography of recent publications on China. **Frequency:** Quarterly. **Price:** Free.

Circuit News

JIT Publishing, Inc.   (714) 887-8083
P.O. Box 3709
San Bernardino, CA 92404   Robert J. Blanset, Editor
    News and current events in electronics and production, covering engineering, design, production, new products. **Frequency:** Monthly. **Subscription:** $50.00. $100.00 foreign.

The Communications Industries Report

International Communications Industries   (703) 273-7200
  Association
3150 Spring St.
Fairfax, VA 22031-2399   R. Jane Gould, Editor
    Trade newsletter for audio, audio-visual, and microcomputer dealers, agents, manufacturers, producers, and distributors. Focus is on current issues and new products. **Frequency:** Monthly. **Subscription:** $15.00.

Companies and Their Brands

Gale Research Inc.   (313) 961-2242
835 Penobscot Bldg.
Detroit, MI 48226   Donna Wood, Editor
    Over 45,000 companies that manufacture, distribute, import, or otherwise market consumer-oriented products. **Price:** $330.00.

Computer Systems News

CMP Publications, Inc.   (516) 365-4600
600 Community Dr.
Manhasset, NY 11030   Mike Azzara, Editor
    Computer trade newspaper covering all aspects of the industry: news, trends, new products, prople, business events, jobs, mergers/acquisitions, hardware/software developments, and product/technology updates. **Frequency:** Weekly. **Subscription:** $15.00.

Contacto

National Eye Research Foundation   (312) 564-4652
910 Skokie Blvd.
Northbrook, IL 60062   Catherine Stahl, Editor
    Magazine containing general clinical information on eye and health care. New product and services column highlights companies with innovative items. **Frequency:** Quarterly. **Subscription:** $125.00.

Copyright Law Reports

Commerce Clearing House, Inc.   (312) 583-8500
4025 W. Peterson Ave.
Chicago, IL 60646   Allen E. Schechter, Editor
    Copyright law publication. **Frequency:** Monthly. **Subscription:** $395.00.

Copyright Management

Communications Publishing Group, Inc.
1505 Commonwealth Ave., No. 32
Boston, MA 02135   Jerry Cohen, Editor
    Discusses legal, tax, management, and business aspects of the copyright industries: publishing, entertainment, records and tapes, and software. Recurring features include editorials, news of research, letters to the editor, book reviews, and a calendar of events. **Frequency:** Monthly. Indexed: Annually. **Price:** $257/yr., U.S. and Canada; $307 elsewhere.

Corporate Finance Sourcebook

National Register Publishing Company,   (312) 441-2344
  Inc.
Macmillan, Inc.
3004 Glenview Road
Wilmette, IL 60091   Cathy Patruno, Publication Manager
    Securities research analysts; major private lenders; merger and acquisition specialists; investment banking personnel; commercial banks; United States-based foreign banks; commercial finance firms; leasing companies; foreign investment bankers in the United States; pension managers; banks that offer master trusts; cash managers; business insurance brokers; business real estate specialists; lists about 2,500 firms. **Price:** $212.00.

Commerce Business Daily. *In order to alert potential sources to emerging government R&D interests, agencies must, by law, publish advance notice of R&D opportunities in the* Commerce Business Daily *(CBD). Published Monday through Saturday (except on federal holidays), copies of CBD may be obtained at the Department of Commerce field offices and at most major urban and university libraries. If you want your own copies, contact the Superintendent of Documents, Government Printing Office, Washington, D.C. 20402; (202) 275-3054. The GPO takes Visa and Mastercard.*

Must Reads:
Business Week,
Forbes, Fortune,
Newsweek, Time,
The Wall Street
Journal. These
publications are the
best way to track the
ins and outs of senior
executives and the
ups and downs of
corporate earnings,
trends, and
competition.

Corporate Venturing News

Venture Economics, Inc.   (617) 449-2100
75 Second Ave., Suite 700
Needham, MA 02158   Paul A. Ferran, Editor

Concerned with corporate development strategies, particularly strategic alliances and business relationships such as joint ventures, research and development contracts, licensing agreements, and supplier/purchaser relationships. Lists recent deals by industry, profiles companies, and provides articles on strategic partnering. Includes coverage of internal venturing activities. Recurring features include interviews, news of research, and columns titled Deals in the News, Deal Log, and For Your Information. **Frequency:** 18/yr. Indexed: Semiannually. **Price:** $345/yr. for institutions, U.S. and Canada; $370 for institutions elsewhere.

Coup

Box 1553
Owosso, MI 48867

Venture capital newsletter. **Frequency:** Quarterly. **Subscription:** $100.00. $25.00 single issue.

CPIA Bulletin

Chemical Propulsion Information Agency   (301) 935-5000
Johns Hopkins Rd.
Laurel, MD 20707   Cathy McDermott, Editor

Updates and summarizes the progress of research-development-test-and-evaluation programs on chemical guns and rocket propulsion; CPIA products and services; rockets, missiles, and guns; ramjets, boosters, and sustainers; and space vehicles. Recurring features include notices of the Joint Army-Navy-NASA-Air Force (JANNAF) Interagency Propulsion Committee and columns titled People in Propulsion, Readers Write, and Recent CPIA Publications and Literature Searches. **Frequency:** Bimonthly. **Price:** Free.

CRS Venture Directory—Florida

CRS Publishing Division   (305) 591-9475
7270 N.W. 12th St. Suite 760
Miami, FL 33126   David M. Sowers, Managing Director

Organizations and individuals seeking venture capital funding in Florida; companies and government agencies offering services and information related to venture capital; financial institutions offering venture capital funding nationally. **Price:** $49.95.

C3I News

Washington Defense Reports, Inc.   (703) 941-6600
7043 Wimsatt Rd.
Springfield, VA 22151   Clay Wick, Editor

Focuses on the area of command, control, and communications intelligence (C3I). Provides information on government contracts, research and development programs, effectiveness of new systems, and related political and legislative issues. Recurring features include news of research and reports of meetings. **Frequency:** Monthly. **Price:** $120/yr., U.S. and Canada; $129 elsewhere. **Send Orders To:** P.O. Box 34312, Bethesda, MD 20817.

Current Controversy

Institute for Scientific Information   (215) 386-0100
3501 Market St.
Philadelphia, PA 19104   Thomas G. DiRenzo, Editor

"Designed to provide concise, balanced overviews of current and emerging issues in science and technology." Consists of abstracts with citations of articles and research in controversial areas of science and technology. **Frequency:** 12/yr. **Price:** $50/yr., U.S.; $55, Canada, Mexico, and Europe; $60 elsewhere.

Current Energy Patents

P.O. Box 62   (615) 576-1170
Oak Ridge, TN 37831   Lila Smith, Editor

Worldwide coverage of energy patents. **Frequency:** Monthly. **Subscription:** $85.00.

Defense Monitor

Center for Defense Information   (202) 862-0700
1500 Massachusetts Ave., N.W.
Washington, DC 20005   Rear Admiral Eugene Carroll, Co-Editor

Concerned with U.S. military issues such as nuclear and conventional weapons; research, development, and procurement; armed force levels; foreign commitments; arms control and SALT; annual military budget; and economic and political implications. Provides analysis and conclusions for a single subject in each issue, and recommends desirable changes and areas for further study. **Frequency:** 10/yr. Cumulative index available for May 1972 to February 1988. **Price:** $25/yr.

Defense R&D

Pasha Publications, Inc.   (703) 528-1244
1401 Wilson Blvd., Suite 900
Arlington, VA 22209   Harry Baisden, Editor

Tracks trends in defense research and development by scrutinizing Department of Defense programs and budgets. Highlights particularly promising technologies and dramatic technological innovations in each issue. Recurring features include a calendar of events and a column titled R&D Briefs. **Frequency:** Biweekly. **Price:** $167/yr., U.S. and Canada; $182 elsewhere.

Dental Lab Products

7400 Skokie Blvd.   (312) 674-0110
Skokie, IL 60077-3339   Jeanne K. Matson, Editor

Tabloid serving dental laboratory owners and managers. Includes new product introduction, new literature, technical training seminars and videotapes, major laboratory conferences, and technique features. **Frequency:** 6x/yr. **Subscription:** $18.00.

Design Engineering

McLean Hunter Ltd.   (416) 596-5819
777 Bay St.
Toronto, ON M5W 1A7   Steve Purwitsky, Editor

Magazine on product design engineering. **Frequency:** Monthly. **Subscription:** $26.00.

Designfax

International Thompson Industrial Press,   (216) 248-1125
    Inc.
6521 Davis Industrial Pkwy.
Solon, OH 44139   David Curry, Editor

Magazine covering design engineering. **Frequency:** Monthly. **Subscription:** $20.00.

Development Directory

Editorial Pkg., 100 Neck Road   (203) 245-9513
Madison, CT 06443

Organizations, foundations, academic institutions, government agencies, professional associations, and firms pursuing research and development interests worldwide. **Price:** $60.00.

Directory of American Research and Technology

Jaques Cattell Press   (212) 337-7050
R. R. Bowker Company
245 W. 17th Street
New York, NY 10011   Beverley McDonough, Senior Editor

11,270 publicly and privately owned industrial research facilities. **Price:** $185.00 (ISSN 0073-7623).

Directory of Financing Sources for Mergers, Buyouts & Acquisitions

Venture Economics in Massachusetts   (616) 431-8100
16 Laurel Ave.
Wellesley Hills, MA 02181   Keith Stephenson, Editor

Over 250 sources of acquisition financing, including banks, asset-based lenders, small business investment companies, insurance companies, and venture capital firms. **Price:** $85.00, plus $3.00 shipping.

Directory of Intellectual Property Lawyers and Patent Agents

Clark Boardman Company Ltd.   (212) 929-7500
435 Hudson St.
New York, NY 10014   Lynn M. LoPucki, Editor

More than 13,000 patent agents, lawyers, and law firms specializing in intellectual property law, including acquisitions and divestitures, biotechnology, consumer products, and tax aspects. **Price:** $145.00, cloth; $125.00, paper.

Directory of Operating Small Business Investment Companies

Small Business Administration   (202) 653-6672
1441 L St., N.W., Room 808
Washington, DC 20416   John R. Wilmeth, Editor

About 570 operating small business investment companies holding regular licenses and licenses under the section of the Small Business Investment Act covering minority enterprise SBICs. **Price:** Free.

Directory of Public High Technology Corporations

American Investor, Inc.   (215) 925-2761
311 Bainbridge St.
Philadelphia, PA 19147   Ronald P. Smolin, Editor

2,000 high-technology publicly held corporations in all aspects of computer technology, electronics, aerospace, telecommunications, medical devices and services, biotechnology, artificial intelligence, pharmaceuticals, optics and electro-optics, lasers, chemicals, materials, environmental control, robotics, scientific instruments, and technical services. **Price:** Base edition, $195.00 (1987 edition) (ISSN 0738-7369).

Directory of Scientific Resources in Georgia

Economic Development Laboratory   (404) 894-3863
Georgia Tech Research Institute
Atlanta, GA 30332

Over 660 research facilities in Georgia, including industrial laboratories, consulting engineering firms, government laboratories, and colleges and universities. **Price:** Free to Georgia residents; $10.00 to others.

D*esignfax is dedicated to product design engineering. It has a fine section on new materials, such as long-fiber composites, epoxy putty compounds, ceramic coatings, and optical release films. The magazine has high advertisement content and provides reader service cards for more information.*

Electronic Engineering Times. *Billing itself as the "Industry Newspaper for Engineers and Technical Management," Electronic Engineering Times is one of the better free publications in its field. It serves up the latest news on technologies, business and new products. Its reporters cover the major trade shows.*

Electronics. *This is a monthly magazine that reports on a wide range of subjects from corporate reorganizations and news analysis to new product and technology introductions. It includes executive profiles, reviews, and news of upcoming meetings.*

Directory of Venture Capital Clubs
International Venture Capital Institute, Inc.   (203) 323-3143
Baxter Associates, Inc.
Box 1333
Stamford, CT 06904
   Approximately 95 venture capital clubs comprised of entrepreneurs, inventors, and small businessmen. **Price:** $7.50 per issue; $35.00 per year.

East/West Technology Digest
Welt Publishing Company   (202) 371-0555
1413 K St., N.W., Suite 800
Washington, DC 20005   Jerry Orvedahl, Co-Editor
   Supplies information on new technology and license offers, with addresses for obtaining further information. Recurring features include news of research. **Frequency:** Monthly. **Price:** $99/yr.

Edison Entrepreneur
Thomas Edison Program, Ohio State Department of Development   (614) 466-3086
77 South High St., 26th Floor
Columbus, OH 43215
   Covers a variety of subjects dealing with research and development throughout the state of Ohio. Recurring features include interviews and news of research. **Frequency:** Quarterly. **Price:** Free.

EE: Electronic/Electrical Product News
Sutton Publishing Co., Inc.   (914) 949-8500
707 Westchester Ave.
White Plains, NY 10604   Ed Walter, Editor
   New product tabloid. **Frequency:** Monthly.

Electric Light & Power
PennWell Publishing Co.   (312) 382-2450
1250 S. Grove Ave., Suite 302
Barrington, IL 60010   Robert A. Lincicome, Editor
   Tabloid providing news of electrical utility industry developments and activities and coverage of new products and technology. **Frequency:** Monthly. **Subscription:** $38.00. $4.00 plus postage, single issue.

Electronic Engineering Times
CMP Publications Inc.   (516) 562-5882
600 Community Dr.
Manhasset, NY 11030   Steve Weitzner, Editor
   Tabloid newspaper reporting on electronic news, developments, and products. **Frequency:** Weekly. **Subscription:** Free (controlled).

Electronics
VNU Business Publications, Inc.   (201) 393-6000
Ten Holland Dr.
Hasbrouck Heights, NJ 07604   J. Robert Lineback, Editor
   Magazine providing managers of electronic O.E.M. companies with reports on new technology for engineers and business trends for managers in the growth markets of computers, communication systems, industrial equipment, military components, medical equipment, and testing and measurement instrumentation. **Frequency:** Monthly. **Subscription:** $60.00. $6.00 per issue.

Electronics and Technology Today
Moorshead Publications Ltd.   (416) 445-5600
Toronto, ON, Canada M3B 3M8   Halvor W. Moorshead, Editor
   "Canada's magazine for high-tech discovery." **Frequency:** Monthly. **Subscription:** $22.95.

Electronics of America
P.O. Box 848   (904) 788-8617
Holly Hill, FL 32017   G.E. Hopper, Editor
   Lists new electronic products, technical developments, techniques, and research programs initiated by American manufacturers and institutions. Includes technical description, price, and name and address of manufacturer. Recurring features include occasional book reviews and special reports on specific aspects of the electronics industry. **Frequency:** Weekly. **Price:** $205/yr.

Emerging High Tech Ventures
Technical Insights, Inc.   (201) 568-4744
32 N. Dean Street
Englewood, NJ 07631
   About 40 companies involved in high technology research and development for applications in genetics, light robotics, acoustics, computers, artificial intelligence, and medical fields. **Price:** $890.00. **Send Orders To:** Technical Insights, Inc. Box 1304, Fort Lee, NJ 07024.

Encyclopedia of Associations: National Organizations of the U.S.

Gale Research Inc.　　　　　　　　　(313) 961-2242
835 Penobscot Bldg.
Detroit, MI 48226　　　　　　　　　Susan Martin, Editor

22,000 nonprofit United States membership organizations of national scope divided into 18 classifications: trade, business, and commercial; agricultural organizations and commodity exchanges; legal, governmental, public administration, and military; scientific, engineering, and technical; educational; cultural; social welfare; health and medical; public affairs; fraternal, foreign interest, nationality, and ethnic; religious organizations; veterans, hereditary, and patriotic; hobby and avocational; athletic and sports; labor unions, associations, and federations; chambers of commerce and trade and tourism; Greek and non-Greek letter societies, associations and federations, fan clubs. **Price:** Volume 1, "National Organizations of the United States," $240.00; volume 2, "Geographic and Executive Index," $220.00; "New Association and Projects," (supplement), $220.00; interedition updating service, $195.00.

Engineering Journal

Delta Communications, Inc.　　　　　(312) 670-5424
400 N. Michigan Ave., Suite 1216
Chicago, IL 60611

Magazine featuring articles from its readers on new developments or techniques in steel design, research, the design and/or construction of new projects, steel fabrication methods, or new products or techniques of significance to the uses of steel in building and bridge construction. **Frequency:** Quarterly. **Subscription:** $11.00. $3.50 single issue.

Entrepreneur Magazine

Entrepreneur, Inc.　　　　　　　　(714) 261-2325
2392 Morse Ave.
Irvine, CA 92714　　　　　　　　　Rieva Lesonsky, Editor

Small business magazine. **Frequency:** Monthly. **Subscription:** $19.97. $2.95 single issue.

Entrepreneurial Manager's Newsletter

Center for Entrepreneurial Management,　(212) 633-0060
　Inc.
180 Varick St., Penthouse
New York, NY 10014　　　　　　　Lee Levin, Editor

"Designed to provide accurate and authoritative information relative to subjects of concern to entrepreneurial managers." Covers management, taxes, finance, marketing, information sources, and educational programs. Recurring features include news of seminars, book reviews, news of research and survey results, and columns titled Entrepreneurs's Hall of Fame, Mind Your Own Business, Resources, Accounting, Personal, Miscellaneous, and The Business Exchange. **Frequency:** Monthly. **Price:** Included in membership; $71/yr. for nonmembers.

Entrepreneurship: Theory and Practice

The John F. Buagh Center for　　　　(817) 755-2265
　Entrepreneurship
Baylor University Box 8011
Waco, TX 76798-8011　　　　　　　Dr. D. Ray Bagby, Editor

Academic journal on small business management, entrepreneurship, and family-owned businesses. Formerly known as American Journal of Small Business. **Frequency:** 4x/yr. **Subscription:** $35.00; $18.00 individuals; add $15.00 outside North America.

EUREKA!

Thunderbird Technical Group
Box 30062
Albuquerque, NM 87190

Association newsletter.

EUREKA!: The Canadian Inventors Newsletter

Canadian Industrial Innovation Centre/
　Waterloo
156 Columbus St. W.
Waterloo, ON, Canada N2L 3L3

Association newsletter.

Explosives & Pyrotechnics

Applied Physics Laboratory　　　　　(215) 448-1555
Franklin Research Center, 20th & Race
　Sts.
Philadelphia, PA 19103　　　　　　R.H. Thompson, Editor

Emphasizes educational and technology transfer aspects of basic and applied science in the field of explosives and pyrotechnics. Covers new developments and applications, training and safety techniques, and related meetings and seminars. Recurring features include reviews of new books and technical reports, and listings of U.S. manufacturers and their products. **Frequency:** Monthly. **Price:** $40/yr.

Trade Journals and Newsletters. Although the news is not as fresh as you'll find in the daily paper, industry trade magazines and newsletters are excellent sources for in-depth information. And they carefully track and report a very broad range of executive assignments, not just the upper-most echelon.

It was through a trade magazine, for example, that I obtained the name of the senior vice president of research and development who licensed our first product for manufacture.

Export/Exportador

Johnson International Publishing Corp.   (212) 689-0120
386 Park Ave., S.
New York, NY 10016   Robert Weingarten, Editor
    New product and merchandising magazine (printed in English and Spanish). **Frequency:** Bimonthly.

Extended Care Product News

Health Management Publications   (215) 337-4466
649 S. Henderson Road
King of Prussia, PA 19406   Laurie Gustafson, Editor
    Health care magazine (tabloid) focusing on new products and news in the field of ostomy, incontinence, nutrition, skin care, and wound care. Geared toward industry purchasing directors. **Frequency:** Quarterly. **Subscription:** Controlled.

Federal Research in Progress (FEDRIP)

Office of Product Management   (703) 487-4929
National Technical Information Service
Department of Commerce
5285 Port Royal Road
Springfield, VA 22161   Linda J. LaGarde, Product Manager

Federal Research Report

Business Publishers, Inc.   (301) 587-6300
951 Pershing Dr.
Silver Spring, MD 20910   Leonard A. Eiserer, Ph.D., Editor
    Provides information on research and development funds available from federal agencies and bureaus or associations that provide support money for research and development. Lists items in categories including environment/energy, transportation, medicine/health, education, and social sciences. **Frequency:** Weekly. **Price:** $160/yr.

The FLC News

Federal Laboratory Consortium for   (209) 251-3830
   Technology Transfer
1945 N. Fine Ave., Suite 109
Fresno, CA 93727   D.M. DelaBarre, Editor
    Summarizes current projects involving domestic technology transfer among FLC-member laboratories. Describes technological innovations developed in member laboratories and reports on related funding and grant awards. Recurring features include news of research and educational opportunities, reports of meetings, notices of publications available, and a calendar of events. **Frequency:** Bimonthly. **Price:** Free.

FOCUS

Women Inventors Project
22 King St. S.
Waterloo, ON, Canada N2J 1N8
    Association newsletter.

For Your Eyes Only

Tiger Publications   (806) 655-2009
P.O. Box 8759
Amarillo, TX 79114-8759   Stephen V. Cole, Editor
    Digests the specialty military press. Carries reports from published and unpublished sources on military events, developments, arms sales, technology, and research programs. Recurring features include book reviews and statistics. **Frequency:** Biweekly. **Price:** $55/yr., U.S. and Canada; $72 elsewhere.

Foreign Technology: An Abstract Newsletter

National Technical Information Service   (703) 487-4630
U.S. Department of Commerce
5285 Port Royal Road
Springfield, VA 22161   Albert Eggerton, Editor
    Carries abstracts of reports on the field of foreign technology. Covers biomedical technology; civil, construction, structural, and building engineering; communications; computer, electro, and optical technology; energy, manufacturing, and industrial engineering; and physical and materials sciences. Recurring features include notices of publications available and a form for ordering reports from NTIS. **Frequency:** Weekly. **Indexed:** Annually. **Price:** $135/yr., U.S., Canada, and Mexico; $185 elsewhere.

Foreign Trade Fairs New Products Newsletter

Printing Consultants, Publishers   (201) 686-2382
Box 636, Federal Sq.
Newark, NJ 07101   John E. Felbar, Editor
    Provides descriptions of and manufacturer's addresses for new foreign products. **Frequency:** Monthly. **Price:** $45/yr.

French Advances in Science & Technology

Science and Technology Office   (202) 944-6246
Embassy of France
4101 Reservoir Rd., N.W.
Washington, DC 20007-2176   Jane Alexander, Editor

Explores recent French basic and applied research and development in the fields of aerospace and transportation; telecommunications and computers; health, genetic engineering, and environmental protection; astronomy and nuclear science; and materials and manufacturing. Recurring features include editorials by French decision-makers in government, industry and education, in-depth special reports, news in brief, and sections titled News From France and International Cooperation. **Frequency:** Quarterly. **Price:** Free.

Get Rich News

Get Rich News Inc.   (407) 586-0978
P.O. Box 126
Lake Worth, FL 33460   Brian Hogan, Editor

Magazine (tabloid) containing mass market business news for lower to middle income entrepreneurs. **Frequency:** Monthly. **Subscription:** $9.95. $1.25 single issue.

Gorman's New Product News

Gorman Publishing Company   (312) 693-3200
8750 W. Bryn Mawr Ave.
Chicago, IL 60631   Martin J. Friedman, Editor

Reports on new consumer product introductions in the U.S. and abroad that will be sold in food and drug stores. Gives a brief description of each product and lists the manufacturer. **Frequency:** Monthly. **Price:** $295/yr.

Government Data Systems

Media Horizons, Inc.   (212) 645-1000
50 W. 23rd St.
New York, NY 10010   Mark Baven, Editor

Magazine examing computer and systems solutions within government and focusing on computer applications, new products, and product reviews. **Frequency:** 8x/yr. **Subscription:** $25.00. $3.00 per issue.

Government Inventions for Licensing: An Abstract Newsletter

National Technical Information Service   (703) 487-4630
U.S. Department of Commerce
5285 Port Royal Road
Springfield, VA 22161   Albert Eggerton, Editor

Abstracts reports on mechanical devices and equipment and other government inventions in chemistry, nuclear technology, biology and medicine, metallurgy, and electrotechnology, as well as optics and lasers, patent applications, and miscellaneous instruments. Recurring features include a form for ordering reports from NTIS. **Frequency:** Weekly. **Indexed:** Annually. **Price:** $235/yr., U.S., Canada, and Mexico; $345 elsewhere.

Government Research Directory

Gale Research Inc.   (313) 961-2242
835 Penobscot Bldg.
Detroit, MI 48226   Annette Piccirelli, Editor

About 4,000 research and development facilities operated by or partly or fully funded by the United States government, including research centers, bureaus, and institutes; research and development installations; testing and experiment stations; and major research-supporting service units; units are active in all areas of physical, social, and life sciences, technology, etc. **Price:** Base edition, $375.00; supplement, $210.00.

GRID

Gas Research Institute   (312) 399-8100
8600 W. Bryn Mawr
Chicago, IL 60631   C. Drugan, Editor

Reports on gas energy research and development sponsored by the Institute. Carries announcements of technical reports available. **Frequency:** Quarterly. **Price:** Free.

Healthcare Technology & Business Opportunities

Biomedical Business International, Inc.   (714) 755-5757
1524 Brookhollow Dr.
Santa Ana, CA 92706   Michael Gibb, Editor

Features patents, business and joint venture opportunities and technology, products for development or sale, as well as editorial material relevant to R & D management. Also contains international opportunities from primary, hard-to-find sources, and carries complete contact information for reader convenience. Recurring features include columns titled New Technology, Business Opportunities, Patents, and Information Resources. **Frequency:** Monthly. **Price:** $325/yr.

Hi-Tech Alert

Communication Research Associates, (301) 445-3230
Inc.
10606 Mantz Rd.
Silver Spring, MD 20903 Michael R. Naver, Editor

Provides "fresh, timely, usable news of hi-tech developments in plain English." Monitors the areas of electronic mail, office automation, desktop publishing, computer-aided research, online information services, and personal computer applications. Recurring features include notices of publications available and news of educational opportunities. **Frequency:** Monthly. **Price:** $98/yr. for individuals, $108 for institutions, U.S. and Canada; $108 for individuals, $118 for institutions elsewhere.

High Tech Ceramics News

Business Communications Company, Inc. (203) 853-4266
25 Van Zant St.
Norwalk, CT 06855-1781 Thomas Abraham, Editor

Analyzes products, patents and trends in the high-tech ceramics industry. **Frequency:** Monthly. **Subscription:** $295.

High-Tech Materials Alert

Technical Insights, Inc. (201) 568-4744
32 N. Dean St.
Englewood, NJ 07631 Alan Brown, Editor

Alerts research directors and business executives to advances in materials research, development, testing, manufacture, and application. Emphasizes technology transfer; covers a wide range of materials, including metals, glasses, ceramics, plastics, and electronics materials. Recurring features include news of research, listings of new patents, and a calendar of events. **Frequency:** Monthly. **Price:** $467/yr., U.S. and Canada; $514 elsewhere. **Send Orders To:** P.O. Box 1304, Fort Lee, NJ 07024.

High Tech Tomorrow

High Tech Information, Inc.
330 W. 42nd St.
New York, NY 10036 Laurie Meisler, Editor

Provides an overview of developments in the field of high technology for potential investors. Covers the areas of computer-aided technologies, mainframes/electronics, biotechnology, and computer-aided drafting and design. Rates and makes recommendations for buying technology stocks. **Frequency:** Monthly. **Price:** $95/yr., U.S. and Canada; $120 elsewhere.

Hispanic Business Magazine

360 S. Hope Ave., Suite 300C (805) 682-5843
P.O. Box 30794
Santa Barbara, CA 93130-0794 Joel Russell, Editor

Business magazine catering to Hispanic professionals, executives and entrepreneurs. **Frequency:** Monthly. **Subscription:** $12.00.

Home Business News

12221 Beaver Pike (614) 988-2331
Jackson, OH 45640 Ed Simpson, Editor

"The voice of America's homebased business owner." Magazine for small business entrepreneurs operating from their homes. **Frequency:** 6x/yr. **Subscription:** $18.00. $3.00 single issue.

HomeBased Entrepreneur Newsletter

JEB Publications (312) 324-5802
5520 S. Cornell
Chicago, IL 60637 Joanne Esters-Brown, Editor

Furnishes guidance and information for homebased business owners. Features items on marketing, advertising, financing, taxes, recordkeeping, insurance, starting a business, and other issues of interest to entrepreneurs. Recurring features include book reviews and columns titled Business Basics Q & A, Entrepreneur's Corner, Business Planning, Ideas, and News and Notes. **Frequency:** Monthly. **Price:** $18/yr., U.S. and Canada; $24 elsewhere. **Send Orders To:** P.O. Box 19036, Chicago, IL 60619.

I.C. Asia

Dataquest, Inc. (408) 971-0910
Dun & Bradstreet Corp.
1290 Ridder Park Dr.
San Jose, CA 95131-2398 Patricia S. Cox, Editor

Provides interpretation and analysis of information generally available to the public on Japanese and Asian high technology industries. Covers major industry events, the companies themselves, the products manufactured, materials and equipment, and new technology. Presents indicators of the integrated circuit industry in Asia. Recurring features include editorials and news of research. **Frequency:** Semimonthly. **Price:** $430/yr.

IANE Newsletter
Inventors Association of New England
P.O. Box 335
Ann Arbor, MI 48103
    Association newsletter.

ICSB Bulletin
International Council for Small Business   (213) 743-2098
Entrepreneur Program, BRI 6
Graduate School of Business
    Administration
University of Southern California
Los Angeles, CA 90089-1421   Alan L. Carsrud, Editor
    Deals with management assistance projects and programs for entrepreneurial ventures and
    small business. Covers innovations in entrepreneurship and small business management
    development and focuses on material of international impact. Recurring features include
    editorials, news of research, government policy, a calendar of events, and columns titled
    President's Message and News From Abroad. **Frequency:** Quarterly. **Price:** Free.

Idea Exchange
Alaska Inventors Association
P.O. Box 241801
Anchorage, AK 99524
    Association newsletter.

Idea: A Resource Publication for Inventors and New Product Designers
P.O. Box 268
Stillwater, MN 55082
    Newsletter.

IMPACT Compressors
IMPACT Publications   (517) 688-9654
P.O. Box 93
Somerset, MI 49281   E.L. Farrah, Editor
    Reviews new compressor products, recent patents, industry software, and literature in the field.
    Recurring features include notices of publications available, news of research, a calendar of
    events, reports of meetings, book reviews, and news of educational opportunities. **Frequency:**
    10/yr. Indexed: Annually. **Price:** $55/yr., U.S.; $65, Canada; $75 elsewhere.

IMPACT Pumps
IMPACT Publications   (517) 688-9654
P.O. Box 93
Somerset, MI 49281   E.L. Farrah, Editor
    Provides data and major claim of newly issued U.S. pump patents. Supplies designers with ideas
    to improve existing designs or develop ones for specific needs. **Frequency:** 10/yr. **Price:** $95/yr.

IMPACT Valves
IMPACT Publications   (517) 688-9654
P.O. Box 93
Somerset, MI 49281   E. L. Farrah, Editor
    Reviews data and the major claim of newly issued U.S. valve patents. Provides designers with
    new ideas to improve existing designs, and develop new applications for specific needs
    Frequency: 10/yr. **Price:** $125/yr.

In Business
JG Press, Inc.   (215) 967-4135
P.O. Box 323
Emmaus, PA 18049   Jerome Goldstein, Editor/Publisher
    Small business management magazine. **Frequency:** 6x/yr. **Subscription:** $21.00.

INCOM News
Inventor's Council of Michigan
c/o Metropolitan Center for High
    Technology
2727 2nd Ave.
Detroit, MI 48201
    Association newsletter.

Incubator Times
Office of Private Sector Initiatives   (202) 653-7880
U.S. Small Business Administration
1441 L St., N.W., Rm. 720A
Washington, DC 20416   Samantha Silva, Editor
    Relates information on incubator projects and initiatives for entrepreneurs and small
    businesses around the country. Discusses new and existing legislative programs, strategies for
    incubator operators, and economic development success stories. Recurring features include a
    calendar of events and columns titled Innovators in the Field, Legislative Update, A Helping
    Hand, A Closer Look, and Ask the Network. **Frequency:** 4/yr. **Price:** Free.

Indiana Inventor
Indiana Inventors Association, Inc.
P.O. Box 2388
Indianapolis, IN 46206
Association newsletter.

Industrial Equipment News
11 Penn Plaza, Suite 1005   (212) 868-5661
New York, NY 10001   Mark F. Devlin, Editor
Magazine containing new product information for manufacturing industries. **Frequency:** Monthly. **Subscription:** $35.00.

Industrial Product Bulletin
P.O. Box 1952   (201) 361-9060
Dover, NJ 07801-0952   Anita LaFord, Editor
Magazine reporting new product information on manufacturing equipment, maintenance supplies, and high technology innovations. **Frequency:** 12x/yr. **Subscription:** $45.00.

Industrial Product Ideas
Sentry Communications   (416) 490-0220
245 Fairfiew Mall Dr., Suite 500
Willowdale, ON, Canada M2J 4T1   Michael Shelley, Editor
New product tabloid. **Frequency:** 8x/yr.

Industrial Research and Development Magazine
Technical Publishing   (312) 381-1840
1301 S. Groove Street
Barrington, IL 60010

Ingenuity
1630 16th Lane
Lake Worth, FL 33463
Serves as a "nationwide communication network to and among inventors." Explores the inventing process as well as the researching, developing, financing, marketing, licensing, and distributing of new inventions. **Frequency:** Quarterly. **Price:** $10/yr., U.S.; $22 elsewhere.

InKnowVation Newsletter
Innovation Development Institute   (617) 595-2920
45 Beach Bluff Ave., Suite 300
Swampscott, MA 01907
Focuses on SBIR (Small Business Innovative Research) program opportunities for small businesses entering early-stage, high-risk research and development ventures. Discusses how SBIR programs can be used to obtain initial funding or to test promising ideas.

Innovation News
Aremco Products, Inc.   (914) 762-0685
P.O. Box 429
Ossining, NY 10562-0429   Brenda T. Lyons, Editor
Features technical research, updates on process equipment, and photographs of ceramic materials. Covers high temperature design, including process information and case histories. ** Biennially.

Inside R&D
Technical Insights, Inc.   (201) 568-4744
32 N. Dean St.
Englewood, NJ 07631   Richard Consolas, Editor
Describes research and development breakthroughs in industry, government, and academic labs, with an emphasis on technology transfer. Emphasizes how the results of the research covered can be applied practically by industry. Recurring features include news of research and columns titled Managing Innovation, Technology Transfer, and Future Tech. **Frequency:** Weekly. **Price:** $457/yr., U.S. and Canada; $529 elsewhere. **Send Orders To:** P.O. Box 1304, Fort Lee, NJ 07024.

Instruments & Computers: Applications in the Laboratory
154 E. Boston Post Road   (914) 698-6655
Mamaroneck, NY 10543-2826   M. Lovetta Francis, Editor
Journal featuring test reports on new laboratory software and systems, original papers on programming for lab research and experimentation, and new products related to lab software, instruments, and computers. **Frequency:** 12x/yr. **Subscription:** $75.00.

Integrated Manufacturing Technology
American Society of Mechanical
  Engineers
345 E. 47th St.
New York, NY 10017   Donald Porteous, Advertising Manager
Journal of engineering research and development for designers, researchers, developers, managers, and users of integrated manufacturing technology and related management and economic issues. **Frequency:** Quarterly.

Intelligence
Edward Rosenfeld   (212) 222-1123
P.O. Box 20008
New York, NY 10025   Edward Rosenfeld, Editor
> Covers technologies that affect the future of computing and offers viewpoints. Concentrates on business, research and government activities in artificial intelligence, neural networks, parallel processing, pattern recognition, expert systems, natural language interfaces, voice and speech technologies, art and graphics, optical storage, and industrial policy. Recurring features include editorials and news of research. **Frequency:** Monthly. **Price:** $295/yr., U.S. and Canada. **Telecommunication Access:** Alternate telephone (800)638-7257.

International Invention Register
Catalyst
Box 547
Fallbrook, CA 92028   Dudley Rosborough, Editor
> **Price:** $18.00 per year.

The International Lawyer
International Law & Practice Section,   (312) 988-5000
    American Bar Assn.
750 N. Lake Shore Dr.
Chicago, IL 60611   Marla Hillery, Editor
> Journal featuring practical papers on legal issues. Emphasizes international trade, licensing, direct investment, finance, taxation, litigation and dispute resolution. **Frequency:** Quarterly. **Subscription:** $23.00; $28.00 outside the U.S.

International New Product Newsletter
Transcommunications International, Inc.
Box 1146
Marblehead, MA 01945   Pamela H. Michaelson, Editor
> Provides "advance news of new products and processes, primarily from sources outside the U.S." Emphasizes new products which can cut costs and improve efficiency. Recurring features include the column Special Licensing Opportunities which lists new products and processes that are available for manufacture under license, or are for sale or import. **Frequency:** Monthly. **Price:** $150/yr., U.S.; $210 elsewhere.

International New Products Newsletter
U.S. International Marketing Co., Inc.   (213) 925-2918
17057 Bellflower Blvd., Suite 205
Bellflower, CA 90706   R. Mervyn Heaton, Editor
> Newsletter focusing on new products. **Frequency:** 6x/yr. **Subscription:** $48.00.

International Product Alert
Marketing Intelligence Service, Ltd.   (716) 374-6326
33 Academy St.
Naples, NY 14512   Sherie Meeker-Barton, Editor
> Provides concise reports on new products in 18 countries outside of the U.S. and Canada. Covers foods, beverages, non-prescription drugs, cosmetics, toiletries, pet products, and miscellaneous household items. Also lists products that are extensions of existing lines and lists packaging changes. Recurring features include occasional copies of advertising. **Frequency:** Semimonthly. **Price:** $600/yr.

International Venture Capital Institute—Directory of Venture Capital Clubs
International Venture Capital Institute   (203) 323-3143
Box 1333
Stamford, CT 06904   Carroll A. Greathouse, President
> Over 100 venture capital clubs; international coverage. **Price:** $9.95 per issue; $14.95 per year.

International Wealth Success
Tyler G. Hicks, 24 Canterbury Road   (516) 766-5850
Rockville Centre, NY 11570   Tyler G. Hicks, Editor
> Covers methods of making money in a successful business, including sources of business capital, real estate income methods, mail order, import/export, franchising, and licensing of products. Recurring features include news of export/import opportunities and a column on capital sources. **Frequency:** Monthly. **Price:** $24/yr.

Invent!
Mindsight Publishing   (805) 388-3097
3201 Corte Malpaso, Suite 304
Camarillo, CA 93010   David Alan Foster, Editor
> International magazine for inventors, innovators, designers, engineers, and entrepreneurs. **Frequency:** Bimonthly. **Subscription:** $35.00. $6.00 per single issue.

E_d _Zimmer of Ann Arbor, Michigan, publishes a chatty monthly newsletter, the Inventor Entrepreneur Network. With more than 4,000 subscribers, the newsletter has been described as "a kind of lonely-hearts newsletter where inventors, investors, small manufacturers and service providers can meet and sometimes find happiness together."_

"I can't judge what's dumb" says Zimmer. "If Dr. Fad (Washington D.C. entrepreneur Ken Hakuta) had walked in here with his Wacky WallWalker, and if I were telling him the truth, I would've told him to forget the damn thing. That slimy little piece of plastic that made $22 million for him."

To provide grist for his mill, Zimmer offers inventors and other contributors free phone service into Ann Arbor over an 800 number (1-800-468-8871) from 8 a.m. to 5 p.m., weekdays.

Inventing and Patenting Sourcebook

Gale Research Inc.   (313) 961-2242
835 Penobscot Bldg.
Detroit, MI 48226   Richard C. Levy
    Edited by Robert Huffman. Provides inventors, innovators, and marketers with a comprehensive and practical "how-to" guide to developing, patenting, licensing, and marketing their ideas and concepts. Contains directory to Project XL, list of 13,000 registered patent attorneys and agents, and the complete index to U.S. Patent Classifications. **Price:** $80.

Invention Development Society Newsletter

8230 S.W. 8th St.
Oklahoma, OK 73128
    Association newsletter.

Invention Manufacturing Opportunity

Inventors' Council
53 W. Jackson, Suite 1041
Chicago, IL 60604
    Association newsletter.

Inventor Entrepreneur Network Newsletter

Zimmer Foundation
6175 Jackson Rd.
Ann Arbor, MI 48103
    Eight-page newsletter features essays on such topics as raising money, selling the product, and the benefits of licensing. **Frequency:** Monthly. **Subscription:** Free.

Inventors' Digest

Affiliated Inventors Foundation, Inc.   (719) 635-1234
2132 E. Bijou St.
Colorado Springs, CO 80909-5950   Joanne H. Hayes, Editor
    Magazine for inventors and others interested in the invention process, development and marketing. **Frequency:** 6x/yr. **Subscription:** $15.00; $32.00 overseas.

Inventor's Gazette

Inventors Association of America   (714) 980-6446
P.O. Box 1531
Rancho Cucamonga, CA 91730
    Provides information on inventions, inventors, and patents. ** Monthly. **Subscription:** $24.

Inventors News

Inventors Clubs of America, Inc.   (404) 938-5089
P.O. Box 450261
Atlanta, GA 30345   Alexander T. Marinaccio, Editor
    Designed to keep inventors abreast with the latest information on patents and trademarks. Carries news of the Club and information on exhibits and inventions. Recurring features include editorials, news of research, letters to the editor, news of members, a calendar of events, and a column titled Patents for Sale. **Frequency:** Monthly. **Price:** $50/yr., U.S. and Canada; $60 elsewhere.

Investing Licensing & Trading Conditions Abroad

Business International Corp.   (212) 460-0630
215 Park Ave., S.
New York, NY 10003   Robert Harris, Editor
    International business magazine. Describes, country by country, the principal laws and regulations that govern establishing a direct investment abroad, licensing of foreign firms and exporting to foreign markets. Covers 53 countries. **Frequency:** Monthly. **Subscription:** $1,575.

IPS Industrial Products & Services

Clifford Elliot Publishing   (416) 842-2884
Royal Life Bldg.
277 Lakeshore Road., E. Suite 209
Oakville, ON, Canada L6J 6J3   Carol Radford, Editor
    Magazine covering new products and services with emphasis on the Canadian market. **Frequency:** 6x/yr. **Subscription:** $15.00; $30.00 U.S.

IVCI Venture Capital Digest

International Venture Capital Institute   (203) 323-3143
Box 1333
Stamford, CT 06904
    Price: $40.00.

Journal of Business Venturing

Elsevier Science Publishing Company, Inc.   (212) 370-5520
Journal Information Center
52 Vanderbilt Ave.
New York, NY 10017   Ian C. MacMillan, Editor
> Journal presenting empirically based research on entrepreneurship, either as independent start-ups or within existing corporations. **Frequency:** Quarterly. **Subscription:** $52.00; $96.00 institutions; add $16.00 outside the U.S.

Journal of the Copyright Society of the U.S.A.

Copyright Society of the U.S.A.   (212) 998-6194
c/o New York University Law School
40 Washington Sq. S.
New York, NY 10012
> Journal focusing on copyright law. **Frequency:** Quarterly. **Subscription:** $125.00.

Journal of Small Business Management

Bureau of Business Research   (304) 293-5837
West Virginia University
College of Business and Economics
P.O. Box 6025
Morgantown, WV 26506-6025   J.H. Thompson, Editor
> Magazine dedicated to the development of entrepreneurship and small business management through education, research, and the free exchange of ideas. **Frequency:** Quarterly. **Subscription:** $25.00; $30.00 institutions. $7.50 per issue.

Journal of Proprietary Rights

Law and Business, Inc.   (201) 894-8538
910 Sylvan Ave.
Englewood Cliffs, NJ 07632
> Covers trends involving patent, trade secret, trademark and intellectual property protection issues, including practical solutions. ** Monthly. **Subscription:** $275.

KAI Developments

Kansas Association of Inventors
2015 Lakin
Great Bend, KS 67530
> Association newsletter.

Laboratory Times

Sentry Communications   (416) 490-0220
245 Fairview Mall Dr., Suite 500
Willowdale, ON M2J 4T1   Doug Dingeldein, Editor
> Professional tabloid presenting new product reviews for the laboratory market. **Frequency:** 6x/yr. **Subscription:** Controlled circulation to qualified laboratory personnel.

Lasers/Electro-Optics Patents Newsletter

Communications Publishing Group, Inc.   (617) 651-9904
P.O. Box 767
Natick, MA 01760   Jeffrey L. Swartz, Editor
> Identifies newly published U.S. and international patent documents in laser and electro-optics. Provides insights into research and development results and industry marketing strategies. **Frequency:** Monthly. **Price:** $297/yr.

Les Nouvelles

Licensing Executives Society International   (216) 771-2600
71 E. Ave., Suite S
Norwalk, CT 06851   Jack Stuart Ott, Editor
> Concerned with licensing and related subjects. Covers technology, patents, trade marks, and licensing "know-how" world-wide. **Frequency:** Quarterly. **Price:** Included in membership. **Telecommunication Access:** Alternate telephone (203)852-7168.

Licensing Law and Business Report

Clark Boardman Company, Ltd.   (212) 929-7500
435 Hudson St.
New York, NY 10014
> Focuses on a specific area within the licensing field in each issue. Provides analysis of court cases and recent developments affecting the design of licensing agreements worldwide, including such topics as antitrust law, tax considerations, and technology management consulting. Recurring features include an annual table of cases. **Frequency:** 6/yr. **Price:** $125/yr.

The Licensing Letter

New Market Enterprises (602) 948-1527
P.O. Box 1665
Scottsdale, AZ 85252 Arnold R. Bolka, Editor

Concerned with all aspects of licensed merchandising, "the business of associating someone's name, likeness or creation with someone else's product or service, for a consideration." Recurring features include statistics, news of research, a calendar of events, mechanics, properties, and lists of licensors and licensees. **Frequency:** Monthly. **Price:** $125/yr., U.S. and Canada; $170 elsewhere.

Licensing Today

International Thomson Retail Press (212) 686-7744
345 Park Ave., S.
New York, NY 10010 James K. Willcox, Editor

Newsletter covering the licensing industry, with an emphasis on toy licensing. Supplement to Toy & Hobby World. **Frequency:** 6x/yr. **Subscription:** $60.00. Included in subscription to Toy & Hobby World.

Lookout Nonfoods

Marketing Intelligence Service, Ltd. (716) 374-6326
33 Academy St.
Naples, NY 14512 Tom Vierhile, Editor

Carries photographs and detailed descriptions of the most innovative products, package design, line extensions, and marketing background in consumer goods categories. Includes nonprescription drugs, cosmetics and toiletries, and miscellaneous household items. Also copies advertising support, including layouts and storyboards. Recurring features include product information: name of manufacturer, ingredients, nutritional information, directions for use, background data, and marketing strategies. **Frequency:** Semimonthly. Indexed: Annually. **Price:** $600/yr.

Manufacturing Technology: An Abstract Newsletter

National Technical Information Service (703) 487-4630
 (NTIS)
U.S. Department of Commerce
5285 Port Royal Road
Springfield, VA 22161

Reports on Computer Aided Design, Computer Aided Manufacturing, technology transfer, and other matters related to manufacturing technology. Also provided information on subjects such as planning, marketing and economics, and research program administration. **Frequency:** Weekly. Indexed: Annually. **Price:** $135/yr., U.S., Canada, and Mexico; $185 elsewhere.

Marketeer

1602 E. Glen Ave. (309) 688-8106
Peoria, IL 61614 V.B. Cook, Editor

Magazine focusing on new product merchandising for industries. **Frequency:** Monthly. **Subscription:** $5.00.

Martindale-Hubbell Law Directory

Martindale-Hubbell, Inc. (201) 464-6800
Box 1001
Summit, NJ 07901

Lawyers and law firms in the United States and its possessions, Canada, and abroad; includes a biographical section by firm, and a separate list of patent lawyers and attorneys in government service. **Price:** $195.00.

MCIC Current Awareness Bulletin

Metals and Ceramics Information Center (614) 424-5000
Battelle Columbus Division
505 King Ave.
Columbus, OH 43201-2693 Harold Hucek, Editor

Features articles and reference information on metals and ceramics of interest to the Department of Defense. Carries abstracts and critiques of reports primarily related to Defense Department funded materials development and research. Recurring features include new developments update, a state-of-the-art summary, and a calendar of meetings and symposia. **Frequency:** Monthly. **Price:** Free.

Medical Device Patents Letter

Washington Business Information, Inc. (703) 247-3424
117 N. 19th St., Suite 200
Arlington, VA 22209-1798 Jeffrey Yohn, Editor

Monitors patent status of new medical devices in the U.S. and worldwide. Reports new licenses granted and new product options. **Frequency:** Monthly. **Subscription:** $537.

Metalworking Production and Purchasing

Action Communications Inc. (416) 477-3222
135 Spy Ct.
Markham, ON L3R 5H6 Maurice Holtham, Editor
    Metalworking industry journal (tabloid) emphasizing new products. **Frequency:** 6x/yr. **Subscription:** $30.00.

MIC MEMO

Minnesota Inventors Congress
Box 71
Redwood Falls, MN 56283
    Association newsletter.

Military Fiber Optic News

Phillips Publishing, Inc. (202) 429-1888
1850 M St., N.W., Suite 810
Washington, DC 20036 Calvin Biesecker, Editor
    Covers specialized fiber optic applications in the defense industry and in the federal government. Tracks developments relating to military fiber optics projects, including the Strategic Defense Initiative (SDI). Also deals with networks and standards and includes updates of more than 100 projects currently underway in the military that include fiber optics. **Frequency:** Biweekly. **Price:** $397/yr., U.S. and Canada; $432 elsewhere. **Telecommunication Access:** Alternate telephone (800)558-8851.

Military Research Letter

Callahan Publications (703) 356-1925
P.O. Box 3751
Washington, DC 20007 Vincent F. Callahan, Editor
    Provides information on contracting opportunities for military research, development, testing, and evaluation. Includes news of installations, programs, legislation, and new developments. **Frequency:** Semimonthly. **Price:** $175/yr., U.S. and Canada.

Military Robotics

L&B, Ltd. (202) 723-1600
19 Rock Creek Church Road, N.W.
Washington, DC 20011-6005 Joseph A. Lovece, Co-Editor
    Contains "timely and accurate information in the area of government and defense applications of robotics." Covers remotely-piloted aircraft, unmanned submarines, teleoperated combat vehicles, cruise missiles, unmanned spacecraft, and teleoperated and autonomous weapons. Lists solicitations and contract awards. **Frequency:** Biweekly. **Price:** $325/yr., U.S. and Canada; $350 elsewhere.

Minorities and Women in Business

Venture X, Inc. (919) 722-3927
1701 Link Road
Winston-Salem, NC 27103 John D. Enoch, Editor
    Magazine networks with major corporations and small businesses owned and operated by minority and female entrepreneurs. **Frequency:** Bimonthly. **Subscription:** $12.00 yearly; $29.00 for three years.

Minority Business Entrepreneur

924 N. Market St. (213) 673-9398
Inglewood, CA 90302 Jeanie M. Barnett, Editor
    Business magazine aimed primarily at Black and Hispanic readership. **Frequency:** Bimonthly. **Subscription:** $12.00.

NASA Tech Briefs

Associated Business Publications Co. (212) 490-3999
41 E. 42nd St., Suite 921
New York, NY 10017-5391 Joseph Pramberger, Editor
    Publication transferring technology to American industry and government in the fields of electronics, computers, physical sciences, materials, mechanics, machinery, fabrication technology, math and information sciences, and the life sciences. **Frequency:** Monthly, except July/Aug. and Nov./Dec. **Subscription:** Free.

National Association of Investment Companies—Membership Directory

National Association of Investment (202) 347-8600
   Companies
915 15th St., N.W., Suite 700
Washington, DC 20005 Benita M. Gore, Publications Director
    About 150 venture capital firms for minority small businesses; licensed by the Small Business Administration. **Price:** $3.39, postpaid.

National Invention Center News

80 W. Bowery St., Suite 201
Akron, OH 44308
    Newsletter.

N*ASA Tech Briefs. The National Aeronautics and Space Administration sponsors the publication of* Tech Briefs, *a monthly high-gloss magazine dedicated to the transfer of technology from the space program to industry.*

If you have applied NASA *technology to your products and processes, you can receive free publicity in* NASA Spinoffs, *an annual publication designed to tell consumers how* NASA *technologies are being applied by industry. To find out if you qualify, contact Linda Watts at (301) 621-0241.*

National Venture Capital Association Membership Directory

National Venture Capital Association   (703) 528-4370
1655 N. Fort Myer Dr., Suite 700
Arlington, VA 22209   Molly M. Myers, Editor

    Nearly 225 venture capital firms, including subsidiaries of banks and insurance companies. **Price:** Free; send self-addressed, business envelope, stamped with $1.50 postage.

NBIA Review

National Business Incubation Association   (614) 593-4331
One President St.
Athens, OH 45701-2923   Dinah Adkins, Co-Editor

    Serves as an information exchange and network for individuals interested in business incubation, a concept based on shared resources and services among entrepreneurs and small businesses. Carries Association reports, interviews with executives and entrepreneurs, and notices of business opportunities. Recurring features include a calendar of events. **Frequency:** 4/yr. **Price:** Included in membership.

NCPLA Newsletter

National Council of Patent Law   (202) 659-2811
   Associations
1819 H St., N.W., Suite 1100
Washington, DC 20006   Charles P. Baker, Editor

    Provides legal information concerning the patent, trademark, and copyright fields and news of the member state and local associations. **Frequency:** Quarterly. **Price:** $50/yr.

NCST Quarterly Briefing

National Coalition for Science and   (202) 833-2322
   Technology
2000 P St., N.W., Suite 305
Washington, DC 20036   Deborarh A. Cohn, Editor

    Concentrates on the political activities of the National Coalition for Science & Technology regarding items such as animal rights, technology transfer, technology innovation, and other areas in which science policy affects society. Recurring features include book reviews, a calendar of events, and columns titled The Chairman's Column, The Director's Column, and Nest Action Alert. **Frequency:** Quarterly. **Price:** Included in membership; $30/yr. for nonmembers.

The Nevada Inventor

Nevada Inventor Association
c/o Institute for Business and Industry,
   Truckee Meadows Com. College
4001 S. Virginia St.
Reno, NV 89502

    Association newsletter.

New From Europe

Prestwick Publications, Inc.   (305) 427-2924
390 N. Federal Hwy., No. 401
Deerfield Beach, FL 33441   Roy H. Roecker, Editor

    Contains market forecasts, trends, and descriptions of new products and technologies from Europe. Descriptions include the developer's name and address and an explanation of why the new product is superior to existing products or processes. Provides an overview of the European economy, its research and new product emphasis, and governmental actions that will affect future market activity. Recurring features include news of research. **Frequency:** Monthly. **Price:** $275/yr.

New From Japan

Prestwick Publications, Inc.   (305) 427-2924
390 N. Federal Hwy., No. 401
Deerfield Beach, FL 33441   Roy H. Roecker, Editor

    Describes new Japanese products and technologies and explains why they are superior to existing products or processes. Covers consumer products, energy conserving processes and products, manufacturing methods, and electronic products. Recurring features include news of research. **Frequency:** Monthly. **Price:** $275/yr.

New From U.S.

Prestwick Publications, Inc.   (305) 427-2924
390 N. Federal Hwy., No. 401
Deerfield Beach, FL 33441   Roy H. Roecker, Editor

    Describes new products and technologies researched and developed in the U.S. Examines a single product and its use and applications in depth in each issue. **Frequency:** Monthly. **Price:** $275/yr.

New Mexico R&D Forum

New Mexico Research and Development   (505) 277-3661
  Institute
Research and Development
  Communications Office
University of New Mexico
457 Washington, S.E., Suite M
Albuquerque, NM 87108   Richard W. Cole, Editor
    Concerned with developments relating to technology in New Mexico. Reports on new projects
    being funded by the Institute, profiles technology firms and laboratories in New Mexico, and
    contains news briefs on pertinent legislation. Recurring features include information on
    workshops and useful publications, news of research, book reviews, and a calendar of events.
    Frequency: Monthly. **Price:** Free.

New Product Development

Point Publishing Company, Inc.   (201) 295-8258
P.O. Box 1309
Point Pleasant, NJ 08742   Jim Betts, Editor
    Concentrates on issues relating to new product research and development, idea generation,
    marketing, distribution, design, and other aspects of product development within national and
    international companies. Recurring features include interviews, news of research, reports of
    meetings, book reviews, and a calendar of events. **Frequency:** Monthly. **Price:** $75/yr., U.S. and
    Canada; $85 elsewhere.

New Technology Week

King Communications Group, Inc.   (202) 638-4260
627 National Press Bldg.
Washington, DC 20045   Richard McCormack, Editor
    Carries news on evolving technologies, especially those in defense-related fields. Follows
    legislation and government agency action affecting defense and high-tech industries. Lists
    recipients of foundation and research grants in the U.S. Recurring features include a calendar of
    events and news of employment opportunities. **Frequency:** Weekly. **Price:** $495/yr.

Newsletter for Independent Businessowners

Earl D. Brodie   (415) 986-4834
465 California St.
San Francisco, CA 94104   Earl D. Brodie, Editor
    Offers specific recommendations for dealing with the wide range of problems facing
    independent business owners. Discusses topics in the areas of finance, manufacturing,
    production, wholesaling and retailing, services, and personnel, as well as the overall economic
    scene. **Frequency:** Semimonthly. **Price:** $85/yr.

NTIAC Newsletter

Nondestructive Testing Information   (512) 684-5111
  Analysis Center
Defense Information Analysis Center
P.O. Box 28510
San Antonio, TX 78284   F.A. Iddings, Editor
    Features articles on methods, applications, and happenings in the field of nondestructive testing
    of defense systems. Carries items on new equipment and techniques, listings of contract awards
    and negotiations, literature surveys, and conference reports. Recurring features include news of
    research, calls for papers, and a calendar of events. **Frequency:** Quarterly. **Price:** Free.

Official Gazette of the United States Patent and Trademark Office: Patents

Patent and Trademark Office   (703) 557-3158
Washington, DC 20231
    Price: $18.00 per issue; $375.00 per year (S/N 003-004-80001-1; ISSN 0098-1133). **Send Orders To:**
    Government Printing Office, Washington, DC 20402.

Official Gazette of the United States Patent and Trademark Office: Trademarks

Patent and Trademark Office   (703) 557-3158
Washington, DC 20231
    Price: $7.00 per issue; $246.00 per year (S/N 003-004-80002-0). **Send Orders To:** Government
    Printing Office, Washington, DC 20402.

Oil, Gas & Petrochem Equipment

PennWell Publishing Company   (918) 835-3161
1412 S. Sheridan
Tulsa, OK 74112   J.B. Avants, Editor
    Tabloid of new products and services for the petroleum industry in the fields of drilling, refining,
    production, petrochemical manufacturing, natrual gas processing, pipeline, enhanced oil
    recovery, maintenance, safety, and instrumentation. **Frequency:** Monthly.

OTTO News

Ohio Technology Transfer Organization (614) 422-5485
Ohio State University
1712 Neil Ave.
Columbus, OH 43210 Barbara J. Ayres, Editor

Provides information about members and the Organization, which "serves the needs of Ohio's business and industry by brokering information from Ohio's two-year colleges and universities to Ohio's businesses, federal laboratories, and other sources. **Frequency:** Quarterly. **Price:** Free.

Partners

Partners of the Americas (202) 628-3300
1424 K St., N.W., No. 700
Washington, DC 20005 Doreen Buscemi Cubie, Editor

Carries news of the association, a non-profit organization which sponsors technical assistance projects and exchanges between the U.S. and Latin America. Reports on projects in agriculture, public health, education, and development. Recurring features include relevant clippings from local newspapers and notices of workshops and conferences. **Frequency:** 5-6/yr.

Patent Attorneys

American Business Directories, Inc. (402) 331-7169
American Business Lists, Inc.
5707 S. 86th Circle
Omaha, NE 68127

Price: $90.00, payment with order. Significant discounts offered for standing orders.

The Patent Trader

Tucker Communications, Inc. (914) 763-3700
P.O. Box 1000
Cross River, NY 10518 Carll Tucker, Editor

Community newspaper. **Frequency:** Weekly. **Subscription:** $45.00.

PharmIndex

Skyline Publishers, Inc. (503) 228-6568
P.O. Box 1029
Portland, OR 97207 Frank D. Portash, Editor

Compiles information on new, changed, and forthcoming pharmaceutical products, including description, adverse reactions, warnings, cautions, pharmacology, and related products. Divides information into areas such as Hormones, Ear, Nose and Throat, Geriatric Therapy, and Immunological Agents. Recurring features include reviews of continuing education programs and information on new and changed products, package sizes, drug prices, discontinued items, and investigational drugs. **Frequency:** Monthly. **Indexed:** Monthly; cumulated annually. **Price:** $79/yr.

Photocopy Authorization Report

Copyright Clearance Center (617) 744-3350
27 Congress St.
Salem, MA 01970

Publicizes Center services and activities and reports general news of the copyright community. Recurring features include news of research and columns titled Coverage Update, Items of Interest, and Readers Ask. **Frequency:** Quarterly. **Price:** Included in membership; $10/yr. for nonmembers, U.S.; $12 for nonmembers elsewhere.

Playthings

51 Madison Ave. (212) 689-4411
New York, NY 10010 Frank Reysen, Editor

Magazine focusing on toys, games, hobbycraft, and licensing. **Frequency:** Monthly. **Subscription:** $20.00.

Pratt's Guide to Venture Capital Sources

Venture Economics, Inc. (617) 431-8100
16 Laurel Ave.
Wellesley Hills, MA 02181

Over 700 venture capital firms, principally in the United States; small business investment corporations (SBICs); corporate venture groups; and selected consultants and "deal men." **Price:** $125.00. **Send Orders To:** Bernan Associates - UNIPUB, 4611-F Assembly Drive, Lanham, MD 20706 (800-233-0506).

Product Alert

Marketing Intelligence Service, Ltd. (716) 374-6326
33 Academy St.
Naples, NY 14512 Diane Beach, Editor

Reports on new consumer goods launched in American retailing, including foods and beverages, non-prescription drugs, cosmetics and toiletries, and miscellaneous household items. Lists products that are an extension of an existing product line, package changes, and marketing plans. Recurring features include pictures as well as descriptions of the products. **Frequency:** Weekly. **Indexed:** Annually. **Price:** $600/yr.

Product Design and Development

Chilton Company, Chilton Way
Radnor, PA 19089

(215) 964-4354
Robert Bierwirth, Editor

Magazine on design and development of durable goods. **Frequency:** Monthly.

Project Summaries

Division of Science Resources Studies
National Science Foundation
1800 G Street, N. W.
Washington, DC 20550

(202) 655-4000

Millicent Gough, Editor

About 70 projects in information collection and analysis sponsored by the National Science Foundation in the fiscal year. **Price:** Free.

Prospectus

Women Entrepreneurs
1275 Market St., No. 1300
San Francisco, CA 94103-1424

(415) 929-0129

Offers women business owners support, recognition, and access to vital information and resources. Monitors legislative developments affecting business and covers programs, workshops, and technical assistance educational seminars conducted by the organization. Recurring features include news of members. **Frequency:** Monthly. **Price:** Included in membership.

Public Information Contact Directory

American Association for the
   Advancement of Science
Office of Communications
1333 H. Street, N.W.
Washington, DC 20005

(202) 326-6400

Carol L. Rogers, Head of
   Communications

Public information contacts at more than 400 colleges and universities, foundations, government agencies and laboratories, museums, nonprofit and industrial research institutions, and scientific and related organizations in the United States, Canada, and Puerto Rico. **Price:** $10.00; payment must accompany order.

Pump News

IMPACT Publications
P.O. Box 93
Somerset, MI 80517

(517) 688-9654

E. L. Farrah, Editor

Reviews new pump products, software, and current literature in the field. Recurring features include news of research, a calendar of events, reports of meetings, news of educational opportunities, book reviews, notices of publications available, notices of upcoming seminars, and technical reports. **Frequency:** 10/yr. Indexed: Annually in December. **Price:** $30/yr., U.S.; $40, Canada; $50 elsewhere.

Quarterly Counselor

Vidas & Arrett, P.A.
2925 Multifoods Tower
33 S. 6th St.
Minneapolis, MN 55402

(612) 339-8801

Oliver F. Arrett, Editor

Focuses on intellectual property law. Supplies general information and helpful guidelines concerning patent, trademark, copyright, and trade secret law. **Frequency:** Quarterly. **Price:** Free.

Research and Development

Cahners Publishing
1350 E. Touhy Ave.
P.O. Box 5080
Des Plaines, IL 60017-5080

(312) 635-8800

Robert R. Jones, Editor

Magazine serving research scientists, engineers, and technical managers. Reports significant advances, problems, and trends that affect the performance, funding, and administration of research. **Frequency:** Monthly. **Subscription:** $45.00.

Research & Development Directory

Government Data Publications
1661 McDonald Ave.
Brooklyn, NY 11230

(718) 627-0819

Siegfried Lobel, Editor

Firms which received research and development contracts from the federal government during preceding fiscal year. **Price:** $15.00.

Research & Development Telephone Directory

Cahners Publishing Company
275 Washington St.
Newton, MA 02158

(617) 964-3030

About 4,000 manufacturers, distributors, and suppliers of products and equipment to industrial research facilities. **Price:** $15.00 (ISSN 0160-4074).

Research & Invention

Research Corp.   (602) 296-6400
6840 E. Broadway Blvd.
Tucson, AZ 85710-2815   W. Stevenson Bacon, Editor

Examines academic scientific and technological research and invention. Provides information on patenting and licensing inventions and on foundation developments, programs, and personnel. Recurring features include columns titled Patent Highlights, Patent Pitfalls, and Grants Update. **Frequency:** Quarterly. **Price:** Free. ISSN 0276-0401.

Research & Technology Management

The Industrial Research Institute, Inc.   (212) 683-7626
100 Park Ave., Suite 3600
New York, NY 10017   Michael F. Wolff, Editor

Magazine for research and development managers. **Frequency:** 6x/yr. **Subscription:** $33.00; $55.00 institutions, libraries and companies. $10.00 per issue.

Research Centers Directory

Gale Research Inc.   (313) 961-2242
835 Penobscot Bldg.
Detroit, MI 48226   Karen Ann Hill, Editor

About 10,000 university-related and other nonprofit research organizations which are established on a permanent basis and carry on continuing research programs in all areas of study; includes research institutes, laboratories, experiment stations, computing centers, and other facilities and activities; coverage includes Canada. **Price:** Base edition, $380.00; supplement service, $240.00.

Research Horizons

Research Communications Office   (404) 894-6987
Georgia Institute of Technology
223 Centennial Research Bldg.
Atlanta, GA 30332   Mark Hodges, Editor

Reports highlights of engineering research conducted by Georgia Tech Research Institute and related Georgia Tech academic departments. A recent issue included articles on a remote life detection device, a study of elite women runners, a new synthetic pigment of ultramarine blue, and research and alcohol and drug abuse in the workplace. **Frequency:** Quarterly. **Price:** Free.

Research Money

Evert Communications, Ltd.   (613) 728-4621
982 Wellington St.
Ottawa, ON, Canada K1Y 2X8   Vincent Wright, Editor

Supplies "reports and analyses of the forces driving science and technology investment in Canada," with special emphasis on government policies, granting programs, and other incentives for industry and universities. Tracks major expenditures on research and development and highlights areas where research monies are available. Recurring features include interviews, news of research, reports of meetings, news of educational opportunities, and a calendar of events. **Frequency:** 20/yr. **Price:** $275/yr.

Research Services Directory

Gale Research Inc.   (313) 961-2242
835 Penobscot Bldg.
Detroit, MI 48226   Robert J. Huffman, Editor

Over 3,000 laboratories, consultants, firms, data collection and analysis centers, individuals, and facilities in the private sector which conduct research in all areas of business, government, humanities, social science, and science and technology. **Price:** $290.00.

Research-Technology Management

Sheridan Press   (717) 632-3535
Fame Ave.
Hanover, PA 17331   Michael Wolff, Editor

Journal about management of research and development companies. **Frequency:** 6x/yr. **Subscription:** $55.00.

The Review of Scientific Instruments

American Institute of Physics   (212) 661-9404
335 45th St.
New York, NY 10017

Magazine focussing on instruments and methods. **Frequency:** Monthly. **Subscription:** $455.00.

Rights Alert

Knowledge Industry Publications, Inc. (914) 328-9157
701 Westchester Ave.
White Plains, NY 10604   Janet Bailey, Editor

Concerned with the protection of proprietary rights "affecting all major information and entertainment markets, including print, data, software, video, film and music." Analyzes current suits, court decisions, and legislative actions. **Frequency:** Monthly. **Price:** $175/yr., U.S. and Canada; $185 elsewhere.

RMIC Newsletter, The Rocky Mountain Inventor's and Entrepreneur's Congress Newsletter
3405 Penrose Place, Suite 104
Boulder, CO 80301
    Association newsletter.

Robotics Patents Newsletter
Communications Publishing Group, Inc.   (617) 651-9904
309 W. Central St., Suite 226
Natick, MA 01760   Jeffrey L. Swartz, Editor
    Focuses on recently published U.S. and international patent documents in the field of robotics. Supplies evaluations of research and development results; predicts industry marketing strategies. **Frequency:** Monthly. **Price:** $357/yr.

Rocky Mountain High Technology Directory
Leading Edge Communications, Inc.   (303) 752-2400
2620 South Parker Road, Suite 185
Aurora, CO 80014   Charles Koelsch, Managing Editor
    About 2,000 manufacturers and research and development firms in Arizon, Colorado, New Mexico, Montana, Nevada, Utah, and Wyoming engaged in high technology activities, including work with aerospace equipment and systems, biotechnology devices and materials, communications, computers, electronics, genetics, instruments, material handling systems, medical diagnostics, medical electronics, microelectronics, office automation, pharmaceuticals, robotics, video equipment, and other categories. Also lists about 60 venture capital firms, law, accounting, and merger and acquisition firms. **Price:** $129.00, plus $4.00 shipping.

Roster of Attorneys and Agents Registered to Practice before the United States Patent and Trademark Office
Patent and Trademark Office   (703) 557-3341
Washington, DC 20231
    Price: Free (S/N 003-004-00609-0). **Send Orders To:** Government Printing Office, Washington, DC 20402.

SBIC Directory and Handbook of Small Business Finance
International Wealth Success, Inc.   (516) 766-5850
24 Canterbury Road
Rockville Centre, NY 11570   Tyler G. Hicks, Editor in Chief
    Over 400 small business investment companies (SBIC's) which lend money for periods from 5 to 20 years to small businesses. **Price:** $15.00, payment with order.

Science Trends
Trends Publishing, Inc.   (202) 393-0031
National Press Bldg.
Washington, DC 20045   Arthur Kranish, Editor
    Reports on developments in general science, in education and throughout society. Covers research and development, current trends, information on scientific and technical publications, and the high-technology outlook. Recurring features include news of research, book reviews, items on publications available, calls for papers, and notices of conferences, seminars, and symposia. **Frequency:** Weekly; monthly in July and August. **Price:** $560/yr., U.S. and Canada.

Semiconductors/ICs Patents Newsletter
Communications Publishing Group, Inc.   (617) 651-9904
209 W. Central St., Suite 226
Natick, MA 01760   Steven Weissman, Editor
    Newsletter updating U.S. and international patent activity in the semiconductor field. **Frequency:** Monthly. **Subscription:** $357.00; $407.00 foreign.

Small Business Guide to Federal R&D Funding Opportunities
Office of Small Business Research and   (202) 357-7464
   Development
National Science Foundation
1800 G St.
Washington, DC 20550
    Federal agencies and their major components with significant research and development programs. **Price:** Free (S/N 038-000-00522-7). **Send Orders To:** Government Printing Office, Washington DC 20402.

Small Business: The Magazine for Canadian Entrepreneurs
McLean Hunter Ltd.   (416) 596-5914
777 Bay St.
McLean Hunter Bldg., 4th Fl., Suite 412
Toronto, ON, Canada M5W 1A7   Randall Litchfield, Editor
    Magazine addressing the needs and concerns of Canada's entrepreneurs. **Frequency:** 10x/yr. **Subscription:** $19.95; $12.50 students. $2.50 single copy. $3.50 special June issue.

Society of Mississippi Inventors Newsletter
P.O. Box 5111
Jackson, MS 39269
    Association newsletter.

Software Protection
Law & Technology Press   (213) 372-1678
P.O. Box 3280
Manhattan Beach, CA 90266
    Addresses current issues in the development and maintenance of computer software and database security for attorneys, government agencies, and software company executives. Recurring features include letters to the editor, book reviews, case summaries, and a calendar of events. **Frequency:** Monthly. **Price:** $197/yr., U.S. and Canada; $230 elsewhere.

Space R&D Alert
Aerospace Communications   (212) 927-8919
350 Cabrini Blvd.
New York, NY 10040   Jeffrey K. Manber, Editor
    Devoted to the transfer of technology from research centers to the commercial space market. Covers patents, product developments, conferences, and reports in the materials processing, satellite communications, and space technology industries. Recurring features include news of research and a calendar of events. **Price:** $195/yr.

Spacenews Capsules
QW Communications Company
P.O. Box 6591
Penacook, NH 03303-6591   Michael A. O'Bryant, Editor
    Contains short news pieces on the space industry, covering the space shuttle, space stations, space vehicles, remote sensing and satellites, launchers and rockets, space defense and the Strategic Defense Initiative (SDI), commercialization of space, and planetary exploration. Also contains information on people in the industry, a publications review, a list of information sources, and related abstracts. **Frequency:** Monthly. **Price:** $15/yr., U.S.; $18 elsewhere.

Stack Gas Control Patents
IMPACT Publications   (303) 586-5636
P.O. Box 1972
Estes Park, CO 80517   Arthur L. Anderson, Editor
    Reviews newly issued U.S. patents related to air pollution control, particularly in the areas of sulfur oxide control, electrostatic precipitators, filters, hydrogen sulfide control, nitrous oxide control, and scrubber systems. **Frequency:** Monthly. **Price:** $75/yr.

State & Regional Directory
Pennsylvania Chamber of Business and   (717) 255-3252
    Industry
222 N. Third St.   Susan E. Smith, Director of Chamber
Harrisburg, PA 17101   Services
    About 1,500 Pennsylvania organizations, including civic, cultural, educational, health and welfare, professional, research, taxpayer, trade associations, and venture capital sources. **Price:** $25.00, postpaid (ISSN 0098-5368).

Status Report of the Energy-Related Inventions Program
Office of Energy-Related Inventions   (301) 975-5500
National Bureau of Standards   A. J. Vitullo, Manager, Inventions
Gaithersburg, MD 20899   Program
    Inventors of items recommended for possible Department of Energy support. **Price:** Free; limited supply.

SUNY Research Foundation—Research Newsletter
State University of New York (SUNY)   (518) 434-7180
State University Plaza
Albany, NY 12246-0001   Sara Wiest, Editor
    Carries scientific and educational research news of the State University of New York. Recurring features include articles on campus research and columns titled Research Reflections, Perspective on Research, Campus Commentary, Technology Transfer Trends (SUNY patents and inventions), the SUNY Press, Research Newsline, and Research Foundation Update. **Frequency:** Bimonthly during the academic year. **Price:** Free.

Supergrowth Technology USA
21st Century Research   (201) 868-0881
8200 Blvd. E.
North Bergen, NJ 07047   Maria R. Hendrie, Editor
    Ranks the top 100 fastest growing high technology billion dollar markets, including data on current and previous month's rank, rapid growth market segment, and estimated annual compound growth in years ahead. Identifies the most promising new ventures financed by leading venture capital firms and monitors initial public offerings of venture-backed firms and institutional acquisition rates. **Frequency:** Monthly. **Price:** $750/yr., U.S. and Canada; $795 elsewhere.

Tech Notes
National Technical Information Service   (703) 487-4630
U.S. Department of Commerce
5285 Port Royal Rd.
Springfield, VA 22161   Edward J. Lehmann, Editor
Presents fact sheets on recently developed federal government technology, selected as having potential commercial or practical application, in the following fields: agriculture and food, computers, electrotechnology, energy, engineering, environmental science and technology, manufacturing, machinery and tools, materials, medicine and biology, natural resources technology and engineering, physical sciences, and transportation. Recurring features include news of research. **Frequency:** Monthly. **Price:** $175/yr., U.S., Canada, and Mexico; $350 elsewhere. **Telecommunication Access:** Alternate telephone (703)487-4630.

Technology Forecasts and Technology Surveys
PWG Publications   (213) 273-3486
205 S. Beverly Dr., Suite 208
Beverly Hills, CA 90212   Irwin Stambler, Editor
Covers new developments in advanced technology and predicts future trends in areas such as sales volumes, consumer demand, new technological advances, and developments in the methodology for forecasting future trends. Concerned with a range of technologies, including electronics, computers, medical technology, chemicals, pulp and paper, food, and materials. **Frequency:** Monthly. **Price:** $144/yr., U.S. and Mexico; $147 Canada; $158 elsewhere.

Technology Management News
Communications Publishing Group, Inc.
1505 Commonwealth Ave., No. 32
Boston, MA 02135   Jeffrey L. Swartz, Editor
Concerned with inventions of significance and the patenting and marketing processes. Discusses developments in law, taxes, patenting and licensing. Recurring features include announcements of noteworthy publications, a calendar of events, and a column titled Subscribers' Forum. **Frequency:** Semimonthly. **Indexed:** Annually. **Price:** $257/yr., U.S.; $307 elsewhere.

Technology Mart
Thomas Publishing Company
One Penn Plaza, 50 W. 34th St.
New York, NY 10001

Technology NY
Anderson Research & Communications   (518) 283-8109
P.O. Box 535
Troy, NY 12180   Olga K. Anderson, Editor
Provides a comprehensive analysis on all aspects of state-wide technology developments in companies and universities. Focuses on specific products, projects, and corporate ventures, including news briefs on capital availability and news concerning research and development issues. Recurring features include news of relevant legislative and regulatory activity, people and company news, reports on regional economic development, special reports, and a calendar of events. **Frequency:** Monthly. **Indexed:** Annually. **Price:** $87/yr.

Technology Transfer Society—Newsletter
Technology Transfer Society
611 N. Capitol Ave.
Indianapolis, IN 46204   Neil A. MacDonald, Editor
Serves the communication and information needs of technology transfer professionals and policymakers. Focuses on issues and methodologies critical to the effective transfer of technology. **Frequency:** 12/yr. **Price:** Included in membership.

Technology Update
Predicasts, Inc.   (216) 795-3000
11001 Cedar Ave.
Cleveland, OH 44106   Cynthia Lenox, Editor
Compiles "technology news abstracted from more than 1000 industry and trade journals, government reports, research studies and other documents." Covers areas such as technical management, agriculture, chemistry, health and medicine, energy, engineering, transportation, communications, environment, lifestyle and leisure, and education. **Frequency:** Weekly. **Price:** $200/yr., U.S.; $225 elsewhere. **Telecommunication Access:** Alternate telephone (800)321-6388.

TOWERS Club, U.S.A.—Newsletter
TOWERS Club, U.S.A.   (206) 574-3084
P.O. Box 2038
Vancouver, WA 98668-2038   Jerry Buchanan, Editor
Intended for freelance writers, publishers, entrepreneurs, and those engaged in marketing their own creative efforts. Provides an exchange of news, quotes, and clippings. Recurring features include columns titled News, Tips, and Sources, Readin' Jerry's Mail, and M/O Mini Clinic. **Frequency:** Monthly, except August and December. **Price:** $60/yr.

Trademark Design Register

Trademark Register (202) 662-1233
National Press Bldg.
Washington, DC 20045
> Owners of over 15,000 registered logos, symbols, and design trademarks. **Price:** $274.00; payment must accompany order.

Trademark Reporter

The U.S. Trademark Association (212) 986-5880
6 E. 45th St.
New York, NY 10017
> Legal journal focusing on trademarks. **Frequency:** 6x/yr. **Subscription:** $80.00 membership only.

TRADEMARKSCAN —FEDERAL

Thomson & Thomson (617) 479-1600
500 Victory Road
North Quincy, MA 02171
> More than 780,000 pending and active federal trademark registrations on file in the United States Patent and Trademark Office (USPTO).

TRADEMARKSCAN —STATE

Thomson & Thomson (617) 479-1600
500 Victory Road
North Quincy, MA 02171
> Provides information on trademarks registered with the Secretaries of State of all 50 U.S. states and in Puerto Rico.

20/20

Jobson Publishing Co. (212) 685-4848
352 Park Ave., S.
New York, NY 10010 Pat McMillan, Editor
> New products, marketing, and merchandising tabloid directed to retailers of eyewear, including opticians, optometrists, managers of optical chains, and dispensing ophthalmologists. **Frequency:** 12x/yr. **Subscription:** $50.00.

Two's News

Technocracy, Inc. (213) 428-4915
Section 2, Regional Division 11833
435 E. Market St.
Long Beach, CA 90805 Audrey C. Adams, Editor
> Reports on the activities of Section 11833-2 of Technocracy, Inc., as well as on trends and events in the field of technological change. **Frequency:** 8/yr. **Price:** $6/yr.

United States Patents Quarterly

Bureau of National Affairs, Inc. (202) 452-4200
1231 25th St., N.W.
Washington, DC 20037 Cynthia J. Bolbach, Editor
> Reports all important decisions dealing with patents, trademarks, copyrights, unfair competition, trade secrets, and computer chip protection. **Frequency:** Weekly. **Indexed:** Monthly; cumulated annually. **Price:** $928/yr.

U.S. Executive Report

18 Blooms Corners Road (914) 986-7755
Warwick, NY 10990 Francesca Lupton, Editor
> Magazine containing articles authored by leading U.S. CEO's on new technologies, financial trends, business management, and marketing innovations for their counterparts in Europe and the Pacific rim. **Frequency:** Every six weeks (except Dec.). **Subscription:** $200.00.

USSR Technology Update

Delphic Associates (703) 556-0278
c/o Mary Heslin
7700 Leesburg Pike, No. 250
Falls Church, VA 22043 Mary Heslin, Editor
> Provides current information on Soviet activities in trade and technology. Presents articles on topics "ranging from fifth generation computer research to industrial automation." Compiles information from Soviet scientific and technical journals in areas including lasers, energy technology, fiber optics, low temperature physics, and computer technology. Also lists U.S. patents granted to Soviet and East European countries. **Frequency:** Biweekly. **Indexed:** Annually. **Price:** $375/yr. for individuals, $200 for institutions, U.S. and Canada; $400 for individuals, $200 for institutions elsewhere.

Valve News

IMPACT Publications (517) 688-9654
P.O. Box 93
Somerset, MI 49281 E. L. Farrah, Editor
> Examines new valve products, software, and technical developments. Includes literature reviews and manufacturers' addresses. Recurring features include news of research, a calendar of events, reports of meetings, news of educational opportunities, book reviews,

notices of publications available, notices of upcoming seminars, and technical reports. **Frequency:** 10/yr. **Indexed:** Annually in December. **Price:** $30/yr., U.S.; $40, Canada; $50 elsewhere.

Venture

521 Fifth Ave.   (212) 682-7373
New York, NY 10175   Jim Jubak, Editor
   Magazine about business owners and investors, containing profiles of entrepreneurs, companies, and examinations of industries and areas of opportunity. **Frequency:** Monthly. **Subscription:** $18.00. $3.00 per issue.

Venture Capital

Venture Capital Sons of America   (212) 838-5577
509 Madison Ave., Suite 812
New York, NY 10022   B. Henry Campbell, Editor
   Reports on all aspects of corporate financing, including venture capital, mergers and acquisitions, leveraged buyouts, private placements, spin-offs and divestitures, SOPs, and IPOs. Recurring features include editorials, news of research, book reviews, and columns titled Interview and Funding Sources. **Frequency:** Monthly. **Price:** $125/yr.; $10/single copy.

Venture Capital Journal

Venture Economics, Inc.   (617) 449-2100
75 Second Ave., Suite 700
Needham, MA 02194-2813   Jane Koloski Morris, Editor
   Magizine on new business development and venture capital investment. **Frequency:** Monthly. **Subscription:** $595.00.

Venture Capital Resource Directory

Office of Urban Assistance   (217) 782-7500
Illinois Department of Commerce and
   Community Affairs
620 E. Adams Street
Springfield, IL 62701   Dennis R. Whetstone, Editor
   Over 65 venture capital firms, clubs, and networks in Illinois; branch offices of the Department of Commerce and Community Affairs. **Price:** Free.

Venture Capital: Where to Find It

Nat'l. Assoc. of Small Business   (202) 833-8230
   Investment Companies
1156 15th St., N. W., Suite 1101
Washington, DC 20005   Eileen E. Denne, Editor
   About 400 member firms licensed as small business investment companies (SBICs) under the Small Business Investment Act of 1958; associate and sustaining members who are non-SBIC investors in small businesses or suppliers of services are included. **Price:** $2.00, payment with order; send self-addressed, business-size envelope.

Venture/Product News

Technology Information Operation   (518) 377-8857
Genium Publishing Corp.
1145 Catalyn St.
Schenectady, NY 12303   Robert A. Roy, Editor
   Carries "concise descriptions of products and processes from research and development firms, universities, industries, and government sources that are available for license/acquisition. Provides information on technology transfer, acquisition, business opportunities, and new products." Recurring features include licensing tips, meeting announcements, and descriptions of firms with unique product-related skills. **Frequency:** Monthly. **Price:** $250/yr., U.S. and Canada; $280 elsewhere.

Washington D.C. Area R&D Firms Directory

WJB Company   (301) 320-5076
7704 Massena Road
Bethesda, MD 20817   Walter J. Bank, Editor
   Over 400 firms located in the Washington, D. C., area engaged in research and development in all fields. **Price:** $45.00.

Western Association of Venture Capitalists—Directory of Members

Western Association of Venture   (415) 854-1322
   Capitalists
3000 Sand Hill Road, Bldg. 2, Suite 215
Menlo Park, CA 94025
   About 120 venture capital firms; coverage limited to the western United States. **Price:** $25.00.

Who's Who in Technology

Gale Research Inc.   (313) 961-2242
835 Penobscot Bldg.
Detroit, MI 48226   Amy Unterburger, Editor
   36,500 engineers, scientists, inventors, and researchers; volume 1, electronics and computer science; volume 2, mechanical engineering and materials science; volume 3, chemistry and

The Wall Street Journal. The Journal provides the most current, fast-breaking information on industry. It was through a piece in this newspaper that I learned what was going on at Proctor & Gamble's Crest brand, information that led to my selling P&G what was to become a $12 million premium program: the Crest Flourider.

The Women Inventors Project in Ontario has published *The Book for Women Who Invent or Want to.* It discusses the creative process from the idea to product stage, including the patent process and marketing. The book also covers how to create a network of women inventors. Contact the Women Inventors Project, P.O. Box 689, Waterloo, Ontario, Canada N2J 4B8.

plastics; volume 4, civil engineering, energy, and earth science; volume 5, physics and optics; volume 6, biotechnology. **Price:** Seven volume set, $545.00; biographical volumes, $95.00 each; index volume, $150.00.

Who's Who in Venture Capital
John Wiley & Sons, Inc.　　　　　　　　　　(212) 850-6331
605 Third Ave.
New York, NY 10158
　　Companies employing about 650 individuals involved in investment and venture capital. **Price:** $29.95.

The Woman Entrepreneur Tax Letter
Richard S. Greenwood
P.O. Box 43204
Detroit, MI 48243　　　　　　　　　　Richard S. Greenwood, Editor
　　Frequency: Monthly. **Price:** $8/yr.

World Biolicensing & Patent Report TM
Deborah J. Mysiewicz Publisher, Inc.　　　(206) 928-3176
P.O. Box 1210
Port Angeles, WA 98362　　　　　　　Thomas Mysiewicz, Editor
　　Newsletter reporting on licensing opportunities available in biotechnology. Lists U.S. and European patent applications with summaries of important applications. **Frequency:** 10x/yr. **Subscription:** Free with subscription to BioEngineering News.

World Electronic Developments
Prestwick Publications, Inc.　　　　　　　(305) 427-2924
390 N. Federal Hwy., No. 401
Deerfield Beach, FL 33441
　　Contains forecasts, trends, and descriptions of new developments in electronics in western Europe, Japan, and the United States. Covers automation and robotics, communications and information processing, computer processing and computer-aided design, and circuits and electronics. Recurring features include statistics and news of research. **Frequency:** Monthly. **Price:** $195/yr., U.S. and Canada; $225 elsewhere.

World Patent Information
Pergamon Journals, Inc.　　　　　　　　(914) 592-7700
Maxwell House, Fairview Park
Elmsford, NY 10523　　　　　　　　　V.S. Dodd, Editor
　　Journal serving as worldwide forum for the exchange of information among professionals in the patent information and documentation field. **Frequency:** Quarterly. **Subscription:** $115.00.

World Technology/Patent Licensing Gazette
Techni Research Associates, Inc.　　　　　(215) 657-1753
Willow Grove Plaza
Willow Grove, PA 19090　　　　　　Louis F. Schiffman, Editorial Director
　　Leading firms, private and government research laboratories, universities, inventors, consultants, and others that have new products, new process developments, and new technologies available for license or acquisition; also lists related seminars and meetings. **Price:** $120.00 per year.

World Weapons Review
Forecast Associates, Inc.　　　　　　　　(203) 426-0800
22 Commerce Rd.
Newtown, CT 06470
　　Identifies applications and problems relating to all types of weapons and weapons systems, from small arms to Intercontinental Ballistic Missiles (ICBMs), on an international scale. Discusses the impact of new weapons development; funding; reported arms sales, transfers, and assistance deals worldwide and their impact on balance of power; weapons retrofit and modernization programs; and the outlook for procurement of new weapons or weapons systems. **Frequency:** 24/yr. **Price:** $385/yr., U.S. and Canada; $460 elsewhere.

Instant Info: Electronic Sources

APIPAT
American Petroleum Institute (API)   (212) 587-9660
Central Abstracting and Indexing Service
275 7th Ave.
New York, NY 10001
> Provides worldwide coverage of patents related to petroleum refining, the petro-chemical industry, and synthetic fuels. **Type of Database:** Bibliographic. **Record Items:** Author(s); corporate author(s); title of item; date; patent information. **Contact:** Monica Pronin, Manager, Administration, Central Abstracting and Indexing Service.

Automated Patent Searching
MicroPatent   (203) 786-5500
25 Science Park
New Haven, CT 06511
> Contains complete patent data for all patents issued by the U.S. Patent and Trademark Office from 1973 to current. **Contact:** Toll-free: 800-648-6787.

BioPatents
BIOSIS   (215) 587-4800
2100 Arch St.
Philadelphia, PA 19103-1399
> Contains references to recently granted U.S. patents in biotechnology, biomedicine, agriculture, and food technology. **Type of Database:** Bibliographic. **Record Items:** Patent number, patent title, U.S. Patent Classification number, inventor, inventor address, assignee, journal name, date granted, publication year. **Contact:** Toll-free 800-523-4806.

BNA Patent, Trademark, & Copyright Daily
The Bureau of National Affairs, Inc.   (202) 452-4132
  (BNA)
BNA ONLINE
1231 25th St., N.W.
Washington, DC 20037
> Provides details on legal, regulatory, and legislative developments affecting patent, copyright, trademark, and unfair competition laws. **Contact:** BNA ONLINE Help Desk. Toll-free: 800-862-4636.

BNA's Patent, Trademark & Copyright Journal (Online)
The Bureau of National Affairs, Inc.   (202) 452-4132
  (BNA)
BNA ONLINE
1231 25th St., N.W.
Washington, DC 20037
> Interprets and analyzes developments in intellectual property issues. Includes coverage of activities of American Intellectual Proprerty Law Association, U.S. Trademark Association, American Bar Association, and others; and the policies, proccedures, statements, announcements, and practices of relevant agencies (Copyright Office and Patent and Trademark Office) **Type of Database:** Full-text. **Contact:** BNA ONLINE Help Desk. Toll-free 800-862-4636.

BIZ. ORBIT. CompuServe. DIALOG. NEXIS. These words look and sound like ITT cable addresses, but represent just five of the major players in an industry encompassing more than 4,000 online databases. The abundance of available electronic information is such that choosing the appropriate database can be a complicated business. Hence, we have selected some 80 databases, both national and international, of particular interest to the inventor. You'll find databases on patenting, trademarks, copyrights, entrepreneurship, invention marketing, and other aspects of the inventing business. For those interested in European patents, a number of databases are described. Many of the publications cited in Hot Press may also be available online; check with the appropriate publisher for details. For more information on databases, see also the chapter entitled It All Begins with a Search.

Your access to online files may be as close as your own computer terminal or as far away as your nearest public library. The inventor is required to stay informed about myriad fields of research and development, many outside one's speciality. Before the advent of online databases this was a tough, time-consuming task. But with the present network of electronic tracking systems, the inventor is capable of monitoring fast-breaking events as they occur. All one requires is a personal computer connected to a phone line and a password to the particular system. And, of course, you can now maintain select databases on your PC utilizing CD-ROM technology.

BRANDY

Toyo Information Systems Co., Ltd.   (03) 575-4021
Shinbashi-Sanwa-Toyo Bldg.
1-11-7, Shinbashi, Minato-ku
Tokyo 105, Japan

    Contains trademarks and patents registered in Japan. **Type of Database:** Full-text. **Contact:** Mr. K. Miyazaki, Database Service Section, Marketing and Sales Department, Toyo Information Systems Co., Ltd. Tlx: 252 2817 TISTOK J. Fax: 03 575 4337.

BREV

Belgium Ministere des Affaires
   Economiques
Office de la Propriete Industrielle (OPRI)
24-26, rue J.A. De Mot
B-1040 Brussels, Belgium

    Contains references to all sectors of patentable activities in Belgium. **Type of Database:** Bibliographic. **Record Items:** International Patent Classification codes and English translation; descriptive text. **Contact:** Dr. Francois-Dominique Declerck, Data Base Manager. Tlx: 20627 COMHAN. Fax: 02 2310256.

Canadian Patent Reporter (CPR)

Canada Law Book Inc.   (416) 773-6472
240 Edward St.
Aurora, ON, Canada L4G 3S9

    Provides case law data concerning copyright, patent, trademark, design, and intellectual property decisions. **Type of Database:** Full-text. **Contact:** Lorna Luke, Manager, CAN/LAW Projects, Canada Law Book Inc. Toll-free (in Canada): 800-263-2037.

Canadian Trade Marks

STM Systems Corporation, Electronic   (613) 727-5445
   Publishing
955 Green Valley Crescent
Ottawa, ON, Canada K2C 3V4

    Provides information on registered and pending marks in Canada. includes the complete text of all registered trademarks and of trademark applications. **Contact:** Sean McCafferty, Marketing Manager. Tlx: 053 4393. Fax: 613-727-0715.

Chinese Patent Abstracts in English Data Base

International Patent Documentation
   Center (INPADOC)
Mollwaldplatz 4
A-1041 Vienna, Austria

    Contains bibliographic information and English-language abstracts of all patents published in the People's Republic of China since the opening of the Chinese Patent Office on April 1, 1985. **Type of Database:** Bibliographic. **Record Items:** The initial publication for a given invention in China (called a patent basic), plus any subsequent Chinese patent documents concerning the same invention; application country, date, or number; authors (inventors); International Patent Classification (IPC) Code; number of patents; patent assignee; patent country code; publication date or year; patent country code and number. **Contact:** Dipl.-Kfm Norbert Fux, Director, Sales Department, International Patent Documentation Center. Tlx: 136 337 INPA A. Fax: 0222 5053386.

CIBERPAT

Registro de la Propiedad Industrial (RPI)
Departamento de Informacion
   Tecnologica
Calle Panama, 1
28036 Madrid, Spain

    Provides bibliographic information and abstracts on patents and models registered in Spain. **Type of Database:** Bibliographic. **Record Items:** International classification, applicant, description of invention, significant date, and foreign country where the invention is registered. **Contact:** Rosina Vazquez de Parga, Chief, Publications and Services. Tlx: 47020 RPI E. Fax: 2592428.

CLAIMS/CITATION

IFI/Plenum Data Corporation   (703) 683-1085
302 Swann Ave.
Alexandria, VA 22301

    Provides more than 5,000,000 patent references cited during the patent examination process against each United States patent, and references to the patents in which it has subsequently been cited. **Type of Database:** Bibliographic. **Contact:** Harry M. Allcock, Vice President, IFI/Plenum Data Corporation. Toll-free 800-368-3093.

CLAIMS/CLASS

IFI/Plenum Data Corporation   (703) 683-1085
302 Swann Ave.
Alexandria, VA 22301

    Provides a classification code and title dictionary for all classes and subclasses of the U.S. Patent Classification System. **Type of Database:** Full-text. **Record Items:** Uniterm code/text, USC code, general or compound term code, CDB fragment code/text, molecular formula. **Contact:** Harry M. Allcock, Vice President, IFI/Plenum Data Corporation.

CLAIMS/Comprehensive Data Base

IFI/Plenum Data Corporation   (703) 683-1085
302 Swann Ave.
Alexandria, VA 22301

> Provides bibliographic and major claim information on chemical patents issued by the United State Patent and Trademark Office since 1950; general, electrical, and mechanical patents since 1963; and design patents since 1980; abstracts are included since 1971. **Type of Database:** Bibliographic. **Record Items:** Abstract, claim text, CAS Registry Numbers, application country/date/number, controlled term, document type, family member country/date/number, field availability, file segment, fragment code, IPC code, inventors, patent assignee/country/number, publication date, role indicator, title, uniterm code, UPC code. **Contact:** Harry M. Allcock, Vice President, IFI/Plenum Data Corporation.

CLAIMS/Reassignment & Reexamination

IFI/Plenum Data Corporation   (703) 683-1085
302 Swann Ave.
Alexandria, VA 22301

> Provides information on patents whose ownership has been reassigned from the original assignee to another company or individual since 1975, and patents reexamined since 1981 by the U.S. Patent and Trademark Office at the request of a second party who has raised substantial new questions regarding the patentability of the patent's claims. **Type of Database:** Bibliographic. **Record Items:** Reexamination request number/date, requestor and requestor location, reexamination certificate date/number/text; reassignment date/type, new patent assignee; expiration date. **Contact:** Harry M. Allcock, Vice President, IFI/Plenum Data Corporation.

CLAIMS/UNITERM

IFI/Plenum Data Corporation   (703) 683-1085
302 Swann Ave.
Alexandria, VA 22301

> Provides bibliographic and major claim information on chemical patents issued by the United States Patent and Trademark Office since 1950; general, electrical, and mechanical patents since 1963; and design patents since 1980. **Type of Database:** Bibliographic. **Record Items:** Title, abstract, claim text, country code, application country/date/number, inventor name/country, class code, CAS Registry Number, document type, International Patent Classification (IPC) code, patent assignee, patent assignee country/code, issue date, patent number, publication year. **Contact:** Harry M. Allcock, Vice President, IFI/Plenum Data Corporation.

CLAIMS/U.S. Patent Abstracts

IFI/Plenum Data Corporation   (703) 683-1085
302 Swann Ave.
Alexandria, VA 22301

> Provides bibliographic and major claim information on chemical patents issued by the United States Patent and Trademark Office since 1950; general, electrical, and mechanical patents since 1963; design patents since l980; abstracts are included since 1971. **Type of Database:** Bibliographic. **Record Items:** Title, abstract, claim text, country code, application country/date/number, inventor name/country, class code, CA reference number, document type, International Patent Classification (IPC) code, patent assignee, patent assignee country/code, issue date, patent number, publication year, CAS Registry Number. **Contact:** Harry M. Allcock, Vice President, IFI/Plenum Data Corporation.

CLINPAT

Registro de la Propiedad Industrial (RPI)
Departamento de Informacion
  Tecnologica
Calle Panama, 1
28036 Madrid, Spain

> Contains the full text of the International Patent Classification. **Type of Database:** Full-text. **Contact:** Rosina Vazquez de Parga, Chief, Publications and Services. Tlx: 47020 RPI E. Fax: 2592428.

Compu-Mark Rechtsstandlexicon

Telesystemes   (014) 582-6464
Questel
83-85, blvd. Vincent Auriol
F-75013 Paris, France

> Contains all trademarks currently in force in West Germany. **Type of Database:** Bibliographic. **Record Items:** Trademark name, registration number, international class, legal status. **Contact:** In the United States: Questel, Inc., 5201 Leesburg Pike, Suite 603, Falls Church, VA 22041.

Years ago, electronic databases were used and understood only by specially training information specialists. Not any more. The community of online information is a democratic one, with many databases requiring 20 minutes of practice and study before your first fruitful forays online.

Compu-Mark U.K. On-Line
Compu-Mark (UK) Ltd. (071) 278-4646
150 Southhampton Row, New Premier
  House, Suite 3
London WC1B 5AL, England
> Covers more than 250,000 registrations, applications, and pending applications filed with the British Patent Office. **Type of Database:** Full-text. **Record Items:** Trademark; registration of application number; Part B marks; class(es) of goods and/or services; status: pending unpublished applications, active published applications/registrations, inactive marks (abandoned, cancelled, or expired); WHO, INN's, or ISO pesticide names; year and page of publication in Trade Marks Journal; owner; codes indicating changes to any aspect of the trademark's status. **Contact:** David I. Sheppard, Manager, Compu-Mark (UK) Ltd. Tlx: 25105 COMPUK G. Fax: 01-278 5934.

Compu-Mark U.S. On-Line
Compu-Mark U.S. (301) 907-9600
7201 Wisconsin Ave.
Bethesda, MD 20814
> Covers all trademarks contained in the U.S. federal and state registers. **Type of Database:** Full-text. **Record Items:** Trademark; owner; registration of application numbers, with code indicating federal or state origins; class(es) of goods or services by U.S. and international classification; status: pending applications, active registrations, inactive (abandoned, cancelled, or expired), state registration, affidavit(s); publication date in official gazette; texts to provide information on changes of ownership, assignment, other. **Contact:** Edward Green, Customer Relations, Compu-Mark U.S. Telephone (202) 737-7900.

Computerized Administration of Patent Documents Reclassified According to the IPS (CAPRI)
International Patent Documentation
  Center (INPADOC)
Mollwaldplatz 4
A-1041 Vienna, Austria
> Contains references to worldwide patent documentation issued before 1973 which have been or are being reclassified according to the International Patent Classification. **Type of Database:** Bibliographic. **Record Items:** Country of publication; type of document; document number; IPC symbols; edition(s) in which the given IPC symbol is valid (IPC 1, 2, 3.). **Contact:** Dipl.-Kfm. Norbert Fux, Director, Sales Department, International Patent Documentation Center.

COMPUTERPAT
Pergamon ORBIT InfoLine, Inc. (703) 442-0900
8000 Westpark Dr.
McLean, VA 22102
> Contains data for all U.S. digital data processing patent documents as classified by the U.S. Patent and Trademark Office in subclasses 364/200 and 364/900, beginning the year the first patent for this technology was issued. **Type of Database:** Bibliographic. **Record Items:** Author(s); author's address; corporate author(s); title of item; date; cited references by source item: bibliographic description; country codes for priority data and cited references; abstract; patent information; patent claim. **Contact:** Toll free 800-456-7248.

Current Patents Fast-Alert
Current Patents Ltd. (071) 631-0341
34-42 Cleveland St.
London W1P 5SB, England
> Provides information on pharmaceutical patents and patent applications registered internationally and in the U.S. **Contact:** James Drake, Managing Director.

Database on Legal Precedents Regarding Intellectual Property Rights
Kinki University
Industrial & Law Information Institute
Kowakae 3-4-1, Higashiosaka-shi
Osaka 577, Japan
> Contains the complete text of legal precedents for laws concerning intellectual property rights.

Deutsche Patent Datenbank (PATDPA)
Deutsches Patentamt
Zweibrueckenstr. 12
D-8000 Munich 2, Federal Republic of
  Germany
> Contains bibliographic information, abstracts, and graphics from patent documents published by the Deutsches Patentamt. **Type of Database:** Bibliographic. **Record Items:** Title of invention; inventor; system number; application country, date, kind, number, and type; document type; entry date and week; family member country, publication date, kind, number, and publication type; International Patent Classification; language; patent country; publication date. **Contact:** A. Dollt, Library Information, Deutsches Patentamt.

DYNIS

Control Data Canada, Ltd. (613) 598-0200
Information Services Group
130 Albert St., Suite 1105
Ottawa, ON, Canada K1P 5G4
> Contains information on more than 300,000 registered and pending trademark applications in Canada. **Type of Database:** Full-text. **Contact:** Jean Millette, Sales Administration, Control Data Canada, Ltd. Telephone: (613) 598-0217.

ECLATX

Institut National de la Propriete (014) 293-2120
 Industrielle (INPI)
26 bis, rue de Leningrad
F-75800 Paris Cedex 8, France
> Contains the complete text of the 4th edition of the International Patent Classification. Includes all codes and terms that are included in the 86,700 groups and subgroups of the classification scheme. **Type of Database:** Full-text. **Contact:** Catherine Pagis, Marketing Manager, Institut National de la Propriete Industrielle.

EDOC

Institut National de la Propriete (014) 293-2120
 Industrielle (INPI)
26 bis, rue de Leningrad
F-75800 Paris Cedex 8, France
> Provides cross-referenced numbers to patents issued by different countries for the same invention. **Type of Database:** Bibliographic. **Contact:** Catherine Pagis, Marketing Manager, Institut National de la Propriete Industrielle.

EPAT

Institut National de la Propriete (014) 293-2120
 Industrielle (INPI)
26 bis, rue de Leningrad
F-75800 Paris Cedex 8, France
> Lists patents applied for and published in the European Patent Office's printed European Patent Bulletin **Type of Database:** Bibliographic. **Record Items:** Depositor; title; technical section; country of origin; place of publication; date of filing; date of publication; publication and registration numbers; applicant name; inventor name; representative priority rights; title; technical sections according to the International Patent Classification (IPC); opposition; designated states; and original titles in French, English, and German. **Contact:** Catherine Pagis, Marketing Manager, Institut National de la Propriete Industrielle. Tlx: 290368 F.

ESPACE CD-ROM

European Patent Office (EPO)
Erhardstr. 27
D-8000 Munich 2, Federal Republic of
 Germany
> Contains the complete text of European patent applications on CD-ROM. **Contact:** Fax: 089 23995143.

Federal Applied Technology Database (FATD)

U.S. National Technical Information (703) 487-4838
 Service (NTIS)
Center for the Utilization of Federal
 Technology (CUFT)
5285 Port Royal Rd., Room 304F
Springfield, VA 22161
> Contains the following: descriptions of federal laboratory resources available to technology-oriented professionals, including facilities and equipment for sharing, expertise, and special services; technology fact sheets covering expertise and technologies selected as having better than average potential; and U.S. government-owned inventions available for licensing by U.S. businesses. Also covers technologies that companies can use to develop new products and processes. **Type of Database:** Bibliographic. **Record Items:** Title, author, information type, year, journal announcement, report number, availability, subject classification codes, descriptors, abstract. **Contact:** Edward J. Lehmann, Center for the Utilization of Federal Technology. Telephone (703) 487-4805.

Financial Times Business Reports

Financial Times Business Information (071) 925-2323
 (FTBI)
Financial Times Electronic Publishing
126 Jermyn St.
London SW1Y 4UJ, England
> Covers specialist financial, business, technology, and media markets. **Type of Database:** Full-text; bibliographic. **Record Items:** Trade names, company and organizational names, personal names and affiliation, country, topic, article type, section codes, length of text. **Contact:** Tlx: FTCONF 27347 G.

FIRST CD-ROM

European Patent Office (EPO)
Erhardstr. 27
D-8000 Munich 2, Federal Republic of
  Germany

Provides the first pages of all European patent applications on CD-ROM. **Contact:** Fax: 089 23995143.

FPAT

Institut National de la Propriete   (014) 293-2120
  Industrielle (INPI)
26 bis, rue de Leningrad
F-75800 Paris Cedex 8, France

Lists French patents applied for and published in the printed Bulletin Officiel de la Propriete Industrielle. **Type of Database:** Bibliographic. **Record Items:** Depositor; title; technical section; country of origin; place of publication; date of filing; date of publication; publication and registration numbers; applicant name; inventor name; representative priority rights; title; technical sections according to the International Patent Classification (IPC). **Contact:** Catherine Pagis, Marketing Manager, Institut National de la Propriete Industrielle.

Government-Industry Data Exchange Program (GIDEP)

U.S. Navy   (714) 736-4677
Naval Fleet Analysis Center
GIDEP Operations Center
Corona, CA 91720

Facilitates the exchange of technical data between government and private industry on parts, components, materials, and processes. **Type of Database:** Numeric; full-text. **Contact:** Edwin T. Richards, Program Manager for Reliability, Government-Industry Data Exchange Program.

INPADOC Data Base (IDB)

International Patent Documentation
  Center (INPADOC)
Mollwaldplatz 4
A-1040 Vienna, Austria

Contains references to worldwide patent documentation for 55 countries which accounts for 96 percent of the world's currently published patent documents. **Type of Database:** Bibliographic. **Record Items:** Country of publication; type of document; document publication date; country of priority; number of application (serves as the basis of priority); priority date; International Patent Classification (IPC) symbol (if present). For certain countries, the following are also provided: inventor name; name of owner; applicant name; invention title; national classification symbol; and other legally related domestic application. **Contact:** Dipl.-Kfm. Norbert Fux, Director, Sales Department, International Patent Documentation Center.

INPAMAR

Registro de la Propiedad Industrial (RPI)
Departamento de Informacion
  Tecnologica
Calle Panama, 1
28036 Madrid, Spain

Contains current references, national marks, trade names, business signs, and international marks registered in Spain. **Type of Database:** Full-text. **Contact:** Rosina Vazquez de Parga, Chief, Publications and Services. Tlx: 47020 RPI E. Fax: 2592428.

INPI-MARQUES

Institut National de la Propriete   (014) 293-2120
  Industrielle (INPI)
26 bis, rue de Leningrad
F-75800 Paris Cedex 8, France

Covers all French trademarks currently in use. **Type of Database:** Full-text. **Record Items:** Trademark name; designated products; classification code; agent; applicant; filing number and date; publication number; renewal information; and design and color. **Contact:** Catherine Pagis, Marketing Manager, Institut National de la Propriete Industrielle.

ISC Data Bank

Invention Submission Corporation   (412) 288-1300
903 Liberty Ave.
Pittsburgh, PA 15222

Contains information on new product ideas and interested manufacturers. Enables inventors to identify potential corporate backers of their new products. **Contact:** Peter Geiringer, Sales Director.

Japan High Tech Review

Kyodo News International, Inc. (KNI)   (212) 397-3726
50 Rockefeller Plaza, Suite 803
New York, NY 10020

Contains news and analyses of Japan's high technology industries. **Type of Database:** Full-text. Emphasis is placed on electronics, telecommunications, and computer developments. **Contact:** Mr. Nakazato, Director, Online Services, Kyodo News International, Inc.

Japio

Japan Patent Information Organization   (035) 690-5555
Sato Dia Bldg.
1-7, Toyo 4-Chome, Koto-ku
Tokyo 135, Japan

> Contains abstracts and drawings of Japanese patent and utility model documents and covers design and trademark information as well. **Type of Database:** Bibliographic; graphic. **Contact:** Yasushi Furukawa, Manager, International Affairs Section, Japan Patent Information Organization. Tlx: 222 4152 JAPATIJ. Fax: 03 56905566.

JURINPI

Institut National de la Propriete
    Industrielle (INPI)
Bureau de Documentation Juridique et
    Technique
26 bis, rue de Leningrad
F-75800 Paris Cedex 8, France

> Contains references and abstracts of French legal decisions concerning patents issued since 1823 and trademarks registered since 1904. **Type of Database:** Bibliographic. **Record Items:** Jurisdiction, date of decision, parties involved, patent number and title, abstract, precedent cases, and bibliographic reference. **Contact:** Catherine Pagis, Marketing Manager. Telephone: 01 42945260. Tlx: 290368 F. Fax: 01 42935930.

LATIPAT

Registro de la Propiedad Industrial (RPI)
Departamento de Informacion
    Tecnologica
Calle Panama, 1
28036 Madrid, Spain

> Provides bibliographic data of granted Latin American inventions; currently covers Argentina, Mexico, and Colombia. **Type of Database:** Bibliographic. **Contact:** Rosina Vazques de Parga, Chief, Publications and Services. Tlx: 47020 RPI E. Fax: 2592428.

LEXIS Federal Patent, Trademark, & Copyright Library (PATCOP)

Mead Data Central, Inc./LEXIS   (513) 865-6800
9393 Springboro Pike, P.O. Box 933
Dayton, OH 45401

> Provides the complete text of court decisions and publications issues relating to patent, trademark, and copyright laws. **Type of Database:** Full-text. **Contact:** LEXIS Customer Service; toll-free 800-543-6862; 800-553-3685 in Canada.

LEXPAT

Mead Data Central, Inc. (MDC)   (513) 865-6800
9393 Springboro Pike
P.O. Box 933
Dayton, OH 45401

> Provides the complete text of utility patents issued by the U.S. Patent and Trademark Office (PTO) since January 1975, and plant and design patents since December 1976. **Type of Database:** Full-text. **Contact:** Toll-free 800-227-4908.

LitAlert

Research Publications, Inc.   (203) 397-2600
Rapid Patent Service
12 Lunar Dr., Drawer AB
Woodbridge, CT 06525

> Provides information on unpublished, unresolved, and current U.S. patent and trademark litigation. **Type of Database:** Bibliographic. **Record Items:** Patent or trademark number, patent or trademark title, classification title, patent publication date, patent or trademark document type, inventor, patent assignee, other patent or trademark numbers, trademark status date, trademark, court location, defendant, plaintiff, docket number, filing date, action date, description of the action, note, and names of law firms or attorneys. **Contact:** Eleanor Roberts, Rapid Patent Service. Toll-free: 800-336-5010.

Master Search TM

Tri Star Publishing   (215) 641-6000
475 Virginia Dr.
Fort Washington, PA 19034

> Contains the complete text of some 600,000 active U.S. trademark applications and registrations. **Type of Database:** Full-text. **Record Items:** Trademark text and image; status; status date; registration number and date; serial number; date filed and published; international class; U.S. class; use; foreign registration information; owner address information; disclaimers, translations, color lining, and related statements; and other items. **Contact:** R. Spencer Nickel, Product Manager, phone (215) 641-6203; toll-free 800-872-2828.

DROLS. You may gain online access to the Defense Technical Information Center information on planned, ongoing, and completed research activities via the Defense RDT&E Online System (DROLS). DROLS can communicate with any terminal (CRT or typewriter) which employs the standard AASCII asynchronous protocol. Terminal communications speeds are 300, 1200, or 2400 baud in even parity. Access is also available through the TYMNET data communications network.

The charge is $30 per connect hour (or proportionate share). Subscribers to the service must have a deposit account with the National Technical Information Service. Users will not be charged for time during which they input technical data into the DTIC databases.

DROLS availability is limited to unclassified access only. Users requiring classified access will be required to use specialized UNIVAC terminals with dedicated telephone lines. If you are interested in the service, contact: **Defense Technical Information Center ATTN: Online Support Office (DTIC-BLD) Bldg. 5, Cameron Station Alexandria, VA 22304-6145 (202) 274-7709.**

McGraw-Hill Publications Online

McGraw-Hill, Inc.   (609) 426-5523
Princeton-Hightstown Rd.
48th Floor
Hightstown, NJ 08520
    Contains the full text of 32 industry-specific magazines and newsletters which serve as a source of background information on companies and industry, the economy and international markets, labor and management, government, and technology. Publications include *Business Week*, *Aerospace Daily*, *Electrical Marketing*, and *Biotechnology Newswatch*. **Type of Database:** Full-text. **Record Items:** Title; journal name; publication date and year; page number; journal code; ISSN; section heading; byline; text; special feature (accompanying captions in graphs, tables, illustrations, and photographs). **Contact:** Andrea D. Broadbent, Manager, McGraw-Hill Publications Online; fax: (609) 426-5178.

New Technology Week

King Communications Group, Inc.   (202) 638-4260
627 National Press Bldg., N.W.
Washington, DC 20045
    Provides news and information on worldwide technological research and development in all areas.

New Trade Names in the Rubber and Plastics Industry

Rapra Technology Ltd.   (093) 925-0383
Information Centre
Shrewsbury, Shrops. SY4 4NR, England
    Contains citations and abstracts of worldwide literature on tradenames of particular interest to companies producing rubber and plastics materials and products, as well as to suppliers to the industry and users of its products. **Contact:** Paul Cantrill, Controller, Information Services; tlx: 35134; fax 0939 251118.

NUANS

Control Data Canada, Ltd.   (613) 598-0200
Information Services Group
130 Albert St., 11th Floor
Ottawa, ON, Canada K1P 5G4
    Contains corporate names and trademarks in Canada currently in use. **Type of Database:** Full-text. **Record Items:** Name, evaluation of corporate name based on phonetics, letter content, root word, coined words, synonyms, distinctive versus descriptive terms, line of business, geographical proximity. **Contact:** Mr. Jean Millette, NUANS Sales Administrator; phone (613) 598-0217; fax (613) 563-1716.

PAPERCHEM

Institute of Paper Science and   (404) 853-9500
  Technology
575 14th St., N.W.
Atlanta, GA 30318
    Covers worldwide literature dealing with pulp and paper technology, including patents, equipment and products, and processes. **Type of Database:** Bibliographic. **Record Items:** Author; editor; corporate author; article title; journal title; publication date; volume and issue numbers; page numbers; publisher; abstract; language; price. **Contact:** D. Gail Stahl, Database Manager. Telephone: (404) 853-9528. Toll-free: 800-558-6611. Fax: (404) 853-9510.

PATDATA

BRS Information Technologies   (703) 442-0900
8000 Westpark Dr.
McLean, VA 22102
    Contains detailed information and abstracts of U.S. utility patents issued since 1971 and all reissue patents which have been issued since 1975. Provides information on cited patents back to 1836. **Type of Database:** Bibliographic. **Record Items:** Author(s); author's address; title of item; date; abstract; patent information; patent assignee, application data, reissued patent, foreign priority, U.S. Classification Code, Cross-Ref-U.S. Classification Code, International Patent Classification Code, U.S. & foreign patents cited. **Contact:** BRS Customer Services; toll-free 800-345-4BRS; fax: (703) 889-4632.

PATOS European Legal Status and Alterations

Wila-Verlag KG   (089) 579-5220
Landsberger Str. 191a
Munich, FG D-8000
    Contains citations to the legal status of European patents, published patent applications, and EURO-PCT applications published by the European Patent Office. **Contact:** Frau Rapp. Fax: 089 5706693.

PATOS European Patent Applications

Wila-Verlag KG   (089) 579-5220
Landsberger Str. 191a
Munich, FG D-8000
    Contains citations to European patent applications submitted by the German Patent Office. **Contact:** Frau Rapp. Fax: 089 5706693.

PATOS European Patents

Wila-Verlag KG   (089) 579-5220
Landsberger Str. 191a
D-8000 Munich 21, Federal Republic of
   Germany
     Contains citations to European patents granted by the European Patent Office. **Contact:** Frau
     Rapp. Fax: 089 5706693.

PATOS German Patent Applications and Utility Models

Wila-Verlag KG   (089) 579-5220
Landsberger Str. 191a
D-8000 Munich 21, Federal Republic of
   Germany
     Contains citations to first patent applications and utility models published by the German Patent
     Office. Includes the complete main patent claim. **Contact:** Frau Rapp. Fax: 089 5706693.

PATOS German Patents

Wila-Verlag KG   (089) 579-5220
Landsberger Str. 191a
D-8000 Munich 21, Federal Republic of
   Germany
     Contains citations to German patents granted by the German Patent Office. **Contact:** Frau Rapp.
     Fax: 089 5706693.

PATOS PCT Applications

Wila-Verlag KG   (089) 579-5220
Landsberger Str. 191a
D-8000 Munich 21, Federal Republic of
   Germany
     Contains citations to international patent applications submitted to the Patent Cooperation
     Treaty (PCT). **Contact:** Frau Rapp. Fax: 089 5706693.

PD-basen

Danmarks Patentdirektoratet   (027) 717-171
Helgeshoj Alle 81
DK-2630 Taastrup, Denmark
     Contains references to patents, trademarks, and designs recognized in Denmark as well as to
     new applications for each. **Contact:** Tlx: 16046 DPO DK. Fax: 02 717170.

PHARM

Institut National de la Propriete   (014) 294-5263
   Industrielle
26 bis, rue de Leningrad
F-75800 Paris Cedex 8, France
     Indexes and abstracts published European and French patent applications and U.S. patents in
     the field of pharmaceutical chemistry and biology.

PharmPat

American Chemical Society (ACS)   (614) 447-3600
Chemical Abstracts Service (CAS)
2540 Olentangy River Rd.
P.O. Box 3012
Columbus, OH 43210
     Provides detailed abstracts for patent documents relating to drugs and other agents for the
     treatment or diagnosis of disease and indexed in the printed Chemical Abstracts and its online
     counterpart, CA File. **Type of Database:** Bibliographic. **Record Items:** Title of document; inventor;
     patent assignee and location; patent application country, number, date, and pagination; CODEN;
     patent information; designated states; patent application information; patent priority application
     information; language of patent; graphic image (structure diagram which appears with abstract
     in printed Chemical Abstracts); abstract, including overview, independent claims, additional
     claims, uses and advantages, an illustrative example, and specifics including testing
     procedures, process conditions, and other information; International Patent Classification (main
     and secondary); Chemical Abstracts section code and title; Chemical Abstracts section cross
     reference and code; index entries (substance and qualifiers); CAS Registry Number; formulation
     components and use. **Contact:** Ruth Ann Geiger, Customer Services Manager; toll-free 800-848-
     6538; in Ohio and Canada, 800-848-6533.

PTS New Product Announcements/Plus (NPA/Plus)

Predicasts   (216) 795-3000
11001 Cedar Ave.
Cleveland, OH 44106
     Contains the complete text of news releases that announce new product introductions and
     modifications, new and applied technologies, and facilities expansions as issued by the
     developing company or its marketing agent. **Type of Database:** Full-text. **Record Items:** Dateline,
     release date, word count, company name, telephone number, address, additional access
     numbers, contact name and phone, date, title, text, company name, product name and code, use
     name and code, tradename, geographic name and code, special features. **Contact:** Customer
     Service Department, Predicasts. Toll-free 800-321-6388.

Electronic databases are not the end-all by any means. The personal touch is often required to iron out the finer points and make sure that some original material has not been omitted. The data are, after all, input by human hands. Databases are a valuable resource, however, as they provide more information faster, easier, and more cost-effectively than manual searches.

PTS PROMT

Predicasts   (216) 795-3000
11001 Cedar Ave.
Cleveland, OH 44106

Provides citations and abstracts of journal articles, newspaper articles, and other sources of worldwide market and technology information relating to more than 120,000 companies and organizations. **Type of Database:** Bibliographic. **Record Items:** Article title; source journal title; publication date; page numbers; abstract; country code and name; product code and name; event code and description; company name; D-U-N-S number; ticker symbol; CUSIP number. **Contact:** Customer Service Department, Predicasts. Toll-free 800-321-6388.

SITADEX

Registro de la Propiedad Industrial (RPI)
Departamento de Informacion
  Tecnologica
Calle Panama, 1
28036 Madrid, Spain

Includes information on the legal status of all property rights registered in Spain. **Type of Database:** Full-text. **Contact:** Rosina Vazquez de Parga, Chief, Publications and Services. Tlx: 47020 RPI E. Fax: 2592428.

TECHNO-SEARCH

Nihon Data Base Development Company
No. 1 Moritoku Bldg. 3 F
7-7-27, Nishi-shinjuku, Shinjuku-ku
Tokyo 360, Japan

Contains the titles and abstracts of articles appearing in five major Japanese industrial and engineering newspapers. **Type of Database:** Bibliographic. **Record Items:** Source newspaper code; publication date; subject(s); keywords(s); technical field code; organization name; organizational and regional codes; length of article; title; abstract.

Technology Assessment and Forecast Reports Data Base

U.S. Patent and Trademark Office   (703) 557-5652
Technology Assessment and Forecast
  Program (TAF)/CM2-304
Washington, DC 20231

Contains statistical and other data relating to patents granted by the U.S. Patent and Trademark Office (PTO). **Type of Database:** Statistical. **Record Items:** Patent number; assignee for specific corporation or government; inventor's residence as state or country; date of patent application; independent inventors (street address, city, state and ZIP Code); abstract. **Contact:** Jane S. Myers, Manager; fax (703) 557-0668.

Technology Information Exchange-Innovation Network (TIE-IN)

Ohio State Department of Development   (614) 466-2115
Division of Technological Innovation
77 E. High St.
P.O. Box 1001
Columbus, OH 43272-4309

An inventory of research and development activity in Ohio. **Type of Database:** Bibliographic; directory. **Record Items:** Venture Opportunities: type of funding desired, amount of funding, and technology of interest. Ohio Patents: patent title, date assigned, company name, and investor's name. Corporate Research and Development: research category, location, company name, type of technology, and training interests. Faculty Research Interests: institution name, location, department, type of technology, and training interest. Sponsored University Grants: title, keywords, investigator name, awarding agency. Technical Publications of Ohio Authors: author, title, institutions, or corporate source. **Contact:** Keith Ewald, General Manager.

Technology Transfer Databank (TECTRA)

California State University   (916) 929-8454
School of Business and Public
  Administration
6000 J St.
Sacramento, CA 95819

Contains descriptions of cases exemplifying the successful transfer of technology from government-supported research laboratories to the public and/or private sectors. **Type of Database:** Directory. **Record Items:** Specific technology; name, address and telephone number of the person who generated the technology and that of the person using it; keyword and/or classification, such as year, laboratory, type of technology, and type of user. **Contact:** Dr. James A. Jolly, Director, Technology Transfer Databank.

Thomas New Industrial Products Database

Thomas Publishing Company   (212) 290-7291
One Penn Plaza
250 W. 34th St.
New York, NY 10119

Provides technical information on new industrial products and systems introduced by American and foreign manufacturers and sellers. **Type of Database:** Full-text. **Record Items:** Product name, synonymous names for the product, product features, attributes, performance specifications, press release publication date, trade names, model numbers, manufacturer name and address. **Contact:** Gary Craig, Manager, Thomas Online; tlx: 126266 NYK; fax (212) 290-7362.

Thomas Register Online

Thomas Publishing Company   (212) 290-7291
One Penn Plaza
250 W. 34th St.
New York, NY 10119

> Provides information on products made in the United States, the companies that make them, and where they are made. **Type of Database:** Directory. **Record Items:** Company name; address; telephone number and other communications information; assets; number of employees; parent company name and officers; products; trade names. **Contact:** Gary Craig, Manager, Thomas Online. Tlx: 126266 NYK; fax (212) 290-7362.

TMA Trademark Report

Tobacco Merchants Association of the   (609) 275-4900
   United States (TMA)
231 Clarkville Rd.
P.O. Box 8019
Princeton, NJ 08543-8019

> Contains a listing of tobacco-related trademark activity as reported in the U.S. Patent Office Official Gazette. **Type of Database:** Full-text. **Contact:** Dr. Thomas C. Slane, Director of Research. Fax: (609) 275-8379.

TMINT

Institut National de la Propriete   (014) 293-2120
   Industrielle (INPI)
26 bis, rue de Leningrad
F-75800 Paris Cedex 8, France

> Covers all 300,000 international trademarks in force, filed, and renewed with the World Intellectual Property Organization under the Madrid Agreement. **Type of Database:** Full-text. **Record Items:** Trademark name; designated products; classification code; registered owner; registration number, date, and duration of the mark; country of origin of registration; countries for which protection is claimed or refused; colors claimed and classification of figurative elements of device marks. **Contact:** Catherine Pagis, Marketing Manager, telephone: 01 45945260; tlx 290368 F.; fax 01 42935930.

Trade Marks (TMRK)

Canada Systems Group (CSG)   (613) 727-5445
Federal Systems Division
Electronic Publishing Division
Product Sales Directorate
Ottawa, ON, Canada K2C 3V4

> Provides information on more than 300,000 registered and pending marks in Canada. **Type of Database:** Directory. **Record Items:** Registration number and date, application number and date, priority date, registered owner, agent or representative, trademark name, disclaimer, products or services for which the trademark is registered, basis of claim, associated marks, and footnotes. Formatted fields include: wordmark, design, certification mark, distinguishing guise marks, date the trademark was first used, and filing, priority, and registration dates. **Contact:** David Macdonald, Marketing Manager, Electronic Publishing Division.

TRADEMARKSCAN-FEDERAL

Thomson & Thomson   (617) 479-1600
500 Victory Rd.
North Quincy, MA 02171-1545

> Provides information on all active registered and pending trademarks on file in the U.S. Patent and Trademark Office and all inactive trademarks since October 1983. **Type of Database:** Full-text. **Record Items:** Trademark text and enhancements; serial number; U.S. class number; status date; date of first use of mark; goods/services description; mark type; owner name; registration number; series code; status text and code; permuted trademark text and enhancements. **Contact:** Anthea Gotto, Manager, Online Marketing; toll-free 800-338-1867.

TRADEMARKSCAN-STATE

Thomson & Thomson   (617) 479-1600
500 Victory Rd.
North Quincy, MA 02171-1545

> Provides information on trademarks registered with the Secretaries of State of all 50 U.S. states and in Puerto Rico. **Type of Database:** Full-text. **Record Items:** Trademark text and enhancements; rotated trademark; design type; state of registration; U.S. class number; international class; goods/service description; mark type; registration number; status; date of registration; date of renewal; date of cancellation; date of first use; owner name. Corporate name records are not included. Trade names, assumed names, and fictitious names are also not generally included, but may be identified for some states. **Contact:** Anthea Gotto, Manager, Online Marketing; toll-free 800-338-1867.

Your local librarian might just be an online ace. If in doubt about which database to access for an inquiry or on search strategies, check out the nearest library and see what they can tell you.

TRANSIN

Transinove International, INPI   (014) 294-5250
26 bis, rue de Leningrad
F-75800 Paris Cedex 8, France
> Indexes offers and requests for patented technologies, new products and inventions, and innovative ideas in need of development from the private and public sectors worldwide. **Type of Database:** Bibliographic. **Record Items:** International Patent Classification (IPC) code; type of opportunity code; development stage; descriptive title; technology level and descriptors; description; technical skill descriptors; agency or company source; contact person name and address; publication reference number; file input date. **Contact:** Olivier Arondel, Assistant Manager, Transinove International.

UK Trade Marks (UKTM)

Great Britain Patent Office   (071) 829-6474
State House, 66-71 High Holborn
London WC1R 4TP, England
> Contains information on nearly 500,000 British trademarks that are active, pending, or lapsed since January 1976. **Type of Database:** Directory. **Record Items:** Name and address of owner, class of goods, goods specification. **Contact:** Mr. W. Preacher. Tlx: 266546.

U.S. Patent Classification System

U.S. Patent and Trademark Office   (703) 557-0400
Office of Documentation Planning and
  Support
Patent Documentation Organizations
Washington, DC 20231
> Constitutes 120,000 patent document classifications to assist patent examiners and others in the field in assessing the novelty, intrinsic merit, and utility of inventions deposited at the U.S. Patent and Trademark Office. **Contact:** Edward J. Earls, Director, Office of Documentation. Tlx: 710 955 0671. Fax: (703) 557-0668.

U.S. Patents Files

Derwent, Inc.   (703) 790-0400
1313 Dolley Madison Blvd.
McLean, VA 22101
> Contains patent information for inventions patented in the United States. **Type of Database:** Full-text. **Record Items:** Title of item; date; total number; abstract; patent information; inventor and assignee name, patent number, application number, attorney name, examiner name. **Contact:** Jeffrey L. Forman, Vice President, Marketing, Derwent, Inc. Toll-free 800-451-3451.

USCLASS

Derwent, Inc.   (703) 790-0400
1313 Dolley Madison Blvd.
McLean, VA 22101
> Contains classification information for nearly 5 million patents issued by the U S Patent and Trademark Office (PTO). **Type of Database:** Bibliographic. **Contact:** Jeffrey L. Forman, Vice President, Marketing, Derwent, Inc. Toll-free: 800-451-3451.

WESTLAW Intellectual Property Library

West Publishing Company   (612) 228-2500
50 W. Kellogg Blvd.
P.O. Box 64526
St. Paul, MN 55164-0526
> Contains the complete text of federal court decisions, statutes and regulations, specialized files, and texts and periodicals dealing with copyright, patent, and trademark law. **Type of Database:** Full-text. **Contact:** WESTLAW Customer Service. Toll-free: 800-WESTLAW.

World Bank of Technology

Dr. Dvorkovitz and Associates (DDA)   (904) 677-7033
P.O. Box 1748
Ormond Beach, FL 32074
> Covers licensable technology around the world. **Type of Database:** Directory. **Record Items:** Title, description, subject category, individual. **Contact:** Barbara Witkowski, Data Processing Manager.

World Patents Index (WPI)

Derwent Publications Ltd.
Rochdale House
128 Theobalds Rd.
London WC1X 8RP, England
> Supplies titles and other details of general, mechanical, electrical, and chemical patents, covering the patent literature of leading industrial countries and using IPC codes for areas such as human necessities, performing operations, transporting, chemistry, textiles, building, construction, mechanics, lighting, heating, instruments, nuclear science, and electricity. **Type of Database:** Bibliographic. **Record Items:** Author(s); corporate author(s); title of item; date; publisher; patent number; abstract; patent information; equivalents which are patent numbers in other countries and priorities (filing date and serial number for first filing). **Contact:** Martin Rees, Manager Online Marketing; tlx 267487 DERPUB G. Fax: 071-405 3630.

INDEX
TO DIRECTORY MATERIAL

O

P

Y

TABLE OF FIGURES

Patents

Trademarks

Copyrights

Miscellaneous

United States Patent [19]

DeLay, Jr.

[11] **Patent Number:** **4,557,395**

[45] **Date of Patent:** **Dec. 10, 1985**

[54] **PORTABLE CONTAINER WITH INTERLOCKING FUNNEL**

[75] Inventor: Victor A. DeLay, Jr., Largo, Fla.

[73] Assignee: E-Z Out Container Corp., Clearwater, Fla.

[21] Appl. No.: 717,439

[22] Filed: Mar. 28, 1985

[51] Int. Cl.⁴ B65D 3/04

[52] U.S. Cl. .. .. 220/86 R; 220/85 F; 220/1 C; 141/98

[58] Field of Search .... 220/86 R, 85 F, 1 C; 141/98, 331; 220/360

[56] **References Cited**

U.S. PATENT DOCUMENTS

1,554,589	9/1925	Long	220/1 C X
3,410,438	11/1968	Bartz	220/1 C
4,010,863	3/1977	Ebel	220/1 C
4,149,575	4/1979	Fisher	220/85 F X
4,162,020	7/1979	Kirkland	220/1 C X
4,296,838	10/1981	Cohen	220/1 C X
4,301,841	11/1981	Sandow	220/1 C X

Primary Examiner—Steven M. Pollard
Attorney, Agent, or Firm—Stanley M. Miller

[57] **ABSTRACT**

A portable, vented container for dirty oil, of the type having a small fill spout and having increased utility when used in conjunction with a funnel. A vent closure member and a funnel securing latch are integral with the funnel so that when the funnel is inverted and positioned in surmounting relation to the container, the vent closure member closes the vent and the securing latch is engaged by a fill spout cap which engagement secures the funnel against movement and hence maintains the vent closure as well. Removal of the fill spout cap releases the funnel, and positioning the funnel into its operative position relative to an automotive oil drain plug separates the vent closure portion of the funnel from the vent. An elongate extension member having a flexible medial portion is further provided.

20 Claims, 12 Drawing Figures

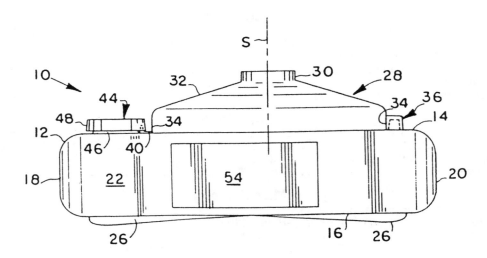

Fig. 1

U.S. Patent   Dec. 10, 1985   4,557,395

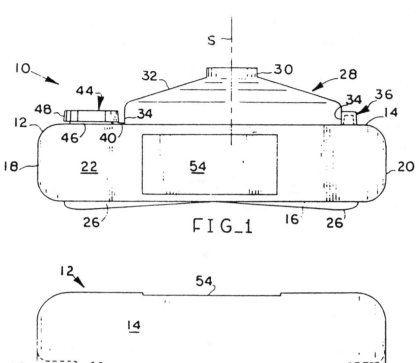

FIG_1

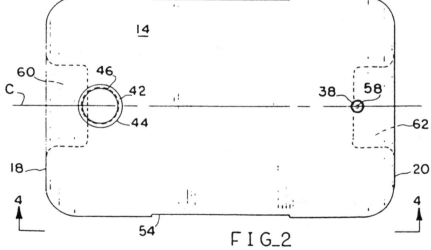

FIG_2

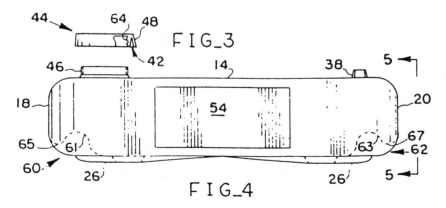

FIG_3

FIG_4

Fig. 1 (continued)

U.S. Patent   Dec. 10, 1985   4,557,395

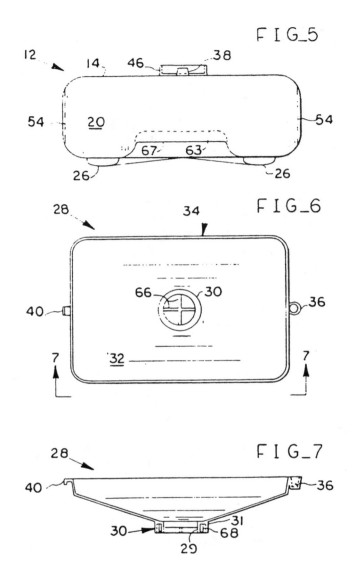

FIG_5

FIG_6

FIG_7

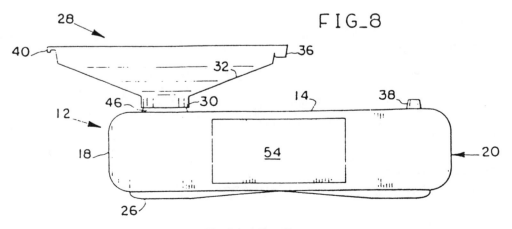

FIG_8

Fig. 1 (continued)

U.S. Patent   Dec. 10, 1985   4,557,395

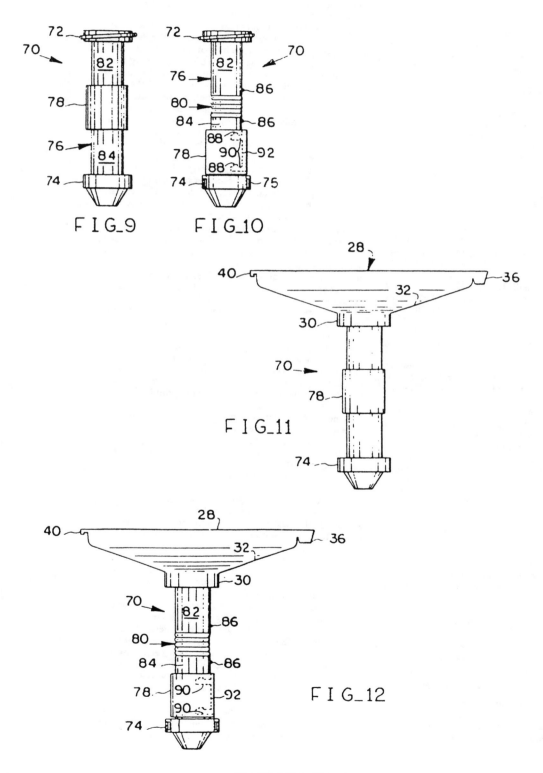

F I G _ 9   F I G _ 10

F I G _ 11

F I G _ 12

Fig. 1 (continued)

4,557,395

1

PORTABLE CONTAINER WITH INTERLOCKING FUNNEL

BACKGROUND OF THE INVENTION

1. Field of the Invention

This invention relates generally to containers having small fill spouts, and more particularly this invention relates to a vented container the vent of which is closed when the funnel is stored in latching engagement with the container body.

2. Description of the Prior Art

A thorough description of the prior art in the the field to which this invention pertains may be found in my co-pending application having a filing date of Sept. 14, 1983, Ser. No. 06/531,948. Moreover, the most pertinent prior art is believed to be the container for dirty oil disclosed in said application.

Other patents of interest are: U.S. Pat. Nos. 4,403,692 to Pollacco (1983); 822,854 to Cosgrave (1906); 2,576,154 to Trautvetter (1951); 4,098,393 to Meyers (1978); 4,217,940 to Wheeler and others (1980); and 4,301,841 to Sandow (1981).

Of the known containers, only the container provided by the present inventor and disclosed in the above-identified patent application contains a means whereby the funnel of the container can be conveniently stored when not in use.

Containers having small fill spouts are normally vented to allow the air inside the container to escape as liquid fluids are charged thereinto. Typically, the vent is provided in the form of an upstanding coupling which is provided with a closure member in the form of a cap which may or may not be attached to the coupling itself. Where the cap is attached to the coupling, its loss is safeguarded against but still the user of the container must remember to open and close the vent as needed. Vent caps that are not attached to their couplings are usually lost.

There is a need, therefore, for a vent cap that is safeguarded against loss, and which also opens and closes the vent as needed without requiring the user thereof to remember to open and close such vent.

Another common problem with small-mouthed containers is that the funnels which must be used therewith are often lost. Pollacco solves this problem by permanently securing his funnel to his container. This storage expedient is unsatisfactory because it is important to maintain funnels of the type used to fill automotive crankcases in a substantially clean condition as the introduction of dirt into a crankcase can damage engine parts.

Therefore, there is a need for a funnel storage apparatus capable of storing a funnel in an inverted position when it is not in use. The storage apparatus that is needed would also safeguard against the loss of the funnel.

The art has heretofore developed elongate funnel extension members of the type disclosed by Cosgrave, Trautvetter, and the present inventor, but the same are inflexible and thus inadequate and lacking in utility in certain specific environments.

SUMMARY OF THE INVENTION

The longstanding but heretofore unfulfilled need for a portable container for dirty oil having the desireable features of a self-opening and self-closing vent, a funnel that is storable in an inverted position and which is also

2

secured against loss, is now fulfilled by the invention disclosed hereinafter and summarized as follows.

The container is of parallelepiped form and has finger-receiving recesses formed in its opposite ends, on the underside thereof, which recesses are grasped by an individual when transporting the container.

The top of the container includes a large, imperforate medial portion against which the rim of the funnel is seated when the funnel is in its storage position.

A fill spout of small diameter projects upwardly from the top of the container, and is disposed near the periphery of the container so the medial portion of the container can receive the stored funnel, as aforesaid.

A sleeve member which defines a vent opening projects upwardly from the top of the container as well, but is disposed in longitudinally spaced relation to the fill spout so that it is near the periphery of the container opposite from the fill spout.

The longitudinal axis of symmetry of the container bisects the finger-receiving recesses or handles, the fill spout, the vent-defining sleeve, and the funnel when the latter is in its stored position. In this manner, the container is stable when transported.

The funnel has an integral vent closure member that projects outwardly from the rim of the funnel, in radial relation to the funnel's axis of symmetry. A latch member used to secure the stored funnel against movement is also formed integral to the funnel, extends radially with respect to said axis from the rim thereof, and is positioned in opposition to the vent closure member.

The funnel's size and the amount of space between the fill spout and the vent opening are selected so that when the funnel is inverted and placed in the center of the medial portion of the top wall of the container, and properly rotated about its axis of symmetry, the vent closure member will align with and seal the vent opening and the latch which is opposed to the vent closure member will be positioned in close proximity to the fill spout.

A novel fill spout closure member in the form of a double-walled cap, when brought into screw threaded engagement with the fill spout, will seal the spout and simultaneously overlie the funnel latch to secure the funnel against displacement.

The novel cap's first wall is internally threaded and thus adapted for screw threaded engagement with the externally threaded fill spout. It outer wall defines an annular recess having an open bottom, which recess surrounds the first wall and which receives the funnel latch therewithin. The annular configuration of the recess eliminates any need for aligning the cap with respect to the latch.

In this manner, the act of inverting the funnel and placing it in its storage position on the top wall of the funnel will close the vent if the proper alignment is made. Once the vent has been closed, no further alignment is required as the sealing of the fill spout by the novel cap will also secure the funnel as desired.

Thus, when the funnel is deployed into its operative configuration, the user of the invention need only remove the fill spout cap, as such will release the funnel from its stored position. The act of placing the funnel's spout into the container's fill spout then serves to open the vent.

A funnel extension member having a flexible medial portion is also disclosed hereinafter. A slideably mounted rigid sleeve member serves to delete the flexi-

Fig. 1 (continued)

4,557,395

3

bility function of the extension member when desired when such sleeve member is positioned in registration with the flexible portion of the member. However, the flexibility of the member is restored upon slidingly displacement of the sleeve away from the flexible medial portion.

An important object of this invention, therefore, is to provide a container for dirty oil that includes a funnel as an attachment to the container so that the funnel is not easily misplaced.

Another object is to provide an attachment means that protects the sloping inside walls of the funnel contamination when the funnel is stored.

Another object of this invention is to provide a means whereby the vent of a container can be automatically opened and closed at the time the container's funnel is placed into its operative position and its storage position, respectively.

Other objects will become apparent as this description proceeds.

The invention accordingly comprises the features of construction, combination of elements and arrangement of parts that will be exemplified in the construction hereinafter set forth, and the scope of the invention will be indicated in the claims.

BRIEF DESCRIPTION OF THE DRAWINGS

For a fuller understanding of the nature and objects of the invention, reference should be made to the following detailed description, taken in connection with the accompanying drawings, in which:

FIG. 1 is a side elevational view of the container with the funnel stored in its inverted position thereatop;

FIG. 2 is a top plan view of the container body member;

FIG. 3 is a partially cut away side elevational view of the novel fill spout closure means;

FIG. 4 is a side elevational view taken along line 4—4 of FIG. 2;

FIG. 5 is an end view taken along line 5—5 of FIG. 4;

FIG. 6 is a top plan view of the novel funnel member;

FIG. 7 is a side elevational view of the funnel member taken along line 7—7 of FIG. 6;

FIG. 8 is a side elevational view, like that of FIG. 4, which shows the funnel member engaging the fill spout of the container body;

FIG. 9 is a side elevational view of the novel funnel downspout extension member with the rigid sleeve in its locked position;

FIG. 10 is a side elevational view of the funnel downspout extension member with the rigid sleeve in its unlocked position;

FIG. 11 is a side elevational view showing the extension member operatively coupled to the funnel member with the sleeve in its locked position; and

FIG. 12 is a side elevational view showing the extension member operatively coupled to the funnel member with the sleeve in its unlocked position.

Similar reference numerals refer to similar parts throughout the several views of the drawings.

DETAILED DESCRIPTION OF THE PREFERRED EMBODIMENT

Referring now to FIG. 1, it will there be seen that an illustrative embodiment of the invention is designated by the reference numeral 10 as a whole. The container body 12 has a parallelepiped construction when seen in

4

perspective. Visible in FIG. 1 are the container's top wall 14, bottom wall 16, its left and right end walls 18, 20, a side wall 22, and support members collectively designated 26.

The novel funnel is indicated generally by the numeral 28. Funnel 28 includes downspout 30, sloping or converging walls 32, and an annular rim 34.

A vent closure member 36 is integrally formed with the rim 34 and extends therefrom as shown. The closure member 36 overlies a vent shroud 38 which is shown in phantom lines in FIG. 1.

A latch 40 is also integrally formed with the funnel rim 34 and is on the opposite side thereof relative to the vent closure member 36. The latch 40 has an "L" shape as shown. The horizontal leg of the latch abuts the top wall 14 of the container 12 and extends radially with respect to the axis of symmetry S of the funnel 28. It terminates in an upstanding leg (shown in phantom lines in FIG. 1) that extends into a cavity 42, which cavity 42 is an annular recess as shown in FIG. 2.

Referring again to FIG. 1, fill spout cap 44 is internally threaded to mate with the external threads of the fill spout 46. The annular latch-receiving recess 42 is formed by the provision of annular wall 48 that surrounds the spout 46, said annular wall depending to the periphery of the top wall of cap 44. The diameter of the top wall of cap 44 is greater than the diameter of the fill spout 46 by an amount substantially equal to the width of the latch-receiving recess 42.

The placement of the upstanding portion of latch 40 in the annular cavity 42 maintains the funnel 28 in its inverted, stored position until the cap 44 is removed.

The space designated 54 in FIG. 1 is a display space and accommodates a label which may have imprinted thereon the trademark of the device and other information.

Returning now to FIG. 2, it will there be seen that the longitudinal axis of symmetry of the device 10 is indicated by the centerline C. It bisects the vent 58 which is formed in the top wall 14 of the container 12 and which is surrounded by vent shroud 38, the fill spout 46, and the longitudinally spaced handles 60, 62 of the invention. The width of the handles 60, 62 is sufficient to accommodate four fingers of a human hand. Both of the label-accommodating recesses 54, 54 mentioned in connection with the description of FIG. 1 are shown in FIG. 2 as well.

The vent closure member 36 slideably and snugly engages the outer walls of the shroud 38, thereby closing the vent opening 58, when funnel 28 is in the inverted storaage position, as aforesaid.

FIG. 3 shows the internal threads 64 on the cap 44 and the annular wall 48 that depends to the periphery of the cap top wall to define the annular cavity 42 into which the upstanding portion of latch 40 extends.

The externally threaded fill spout 46 is shown in FIG. 4, which FIG. shows the container 12 with funnel 28 and cap 44 separated therefrom.

The handles 60, 62 include concave surfaces 61, 63, respectively, and convex surfaces 65, 67, the former of which are abutted by fingertips when the container is carried and the latter of which provide a comfortable rounded weight bearing surface.

An end view of the container 12 is provided in FIG. 5.

A top view of the novel funnel 28 appears in FIG. 6. A strainer 66 formed by a pair of cross bars is formed where the downwardly sloping walls 32 of the funnel 28

Fig. 1 (continued)

4,557,395

5

merge with the funnel's downspout. The generally rectangular planform of the funnel 28 conforms to the planform of the container body 12 as shown in FIG. 2, but the corresponding dimensions of the funnel are smaller.

The downspout 30 of funnel 28 is internally threaded as indicated by the reference numeral 68 appearing in FIG. 7, and is thus adapted for screw threaded engagement with the externally threaded fill spout 46. Accordingly, the downspout 30 of the funnel 28 is coupled to fill spout 46 when it is desired to charge the container with dirty oil. This operative positioning of the funnel 28 and fill spout 46 is depicted in FIG. 8. A comparison of FIGS. 1 and 8 indicates that the removal of cap 44 from spout 46 releases latch 40 so that funnel 28 can be separated from its engagement with top wall 14 of container 12, restored to its upright configuration, and coupled with the spot 46. The separation of the funnel 28 and the container body top wall 14 also separates the vent closure member 36 from vent shroud 38, which separation exposes vent 58 (FIG. 2) to ambient. The internal threads 68 of downspout 30 are formed in outer wall 31 thereof. An inner wall 29 is spaced radially inwardly of outer wall 31, and is concentric therewith. Accordingly, dirty oil contacts inner wall 29 only.

The truncate downspout 30 of funnel 28 is provided because some vehicle are built close to the ground. However, other vehicles are built higher from the ground and the use of a downspout extension member becomes advisable.

An improved downspout extension member is shown in FIGS. 9–12, and is designated 70 as a whole. It includes an externally threaded adapter 72 which is coupled to the internally threaded downspout 30 of funnel 28 when in use, as shown in FIGS. 11 and 12. Another adapter 74 at the lower end of the extension member 70 is internally threaded as at 75 (FIG. 10) to mate with the external threads of the fill spout 46. An elongate medial portion 76 interconnects the upper and lower adapters 72 and 74.

A slideably mounted rigid sleeve member 78 is shown mid-length of the medial portion 76 in FIG. 9. When the sleeve member 78 is locked into this position by means disclosed hereinafter, the novel extension member 70 can be used in the same manner as conventional downspout extension members, which use is depicted in FIG. 11.

However, when the sleeve 78 is unlocked and slideably displaced to its lowermost position, which position is depicted in FIG. 10, such displacement frees a flexible member 80 from confinement so that it is free to bend. More specifically, upper portion 82 of the downspout extension member medial portion 76 and lower portion 84 thereof may be displaced from their axial alignment with each other, i.e., their respective axes of longitudinal symmetry may be made oblique to one another. As shown in FIG. 12, when the flexible member 80 is free, funnel 28 can be moved in any direction relative to lower coupling 74, or vice versa.

FIGS. 10 and 12 both show the means employed to lock and unlock sleeve 78 as desired. A pair of vertically spaced beads, collectively designated 86, are formed on upper and lower portions 82, 84 of the extension member medial portion 76. A pair of vertically spaced bead-receiving cavities, collectively designated 88, are formed internally of sleeve member 78, so that the sleeve 78 is locked into overlying relation to the flexible member 80 when beads 86 are disposed therein.

6

To unlock the sleeve 78, the user of the inventive apparatus grasps sleeve 78 and slides it upwardly by a distance equal to the depth of the bead-receiving cavities 88. Each bead 86 will then be positioned in channels 90 which are also formed internally of sleeve 78. The user of the device then rotates the sleeve 78 until the beads 86 have traveled the length of the arcuate channels 90, which length could be a quarter of an inch, for example. This rotation of sleeve 78 will bring the beads 86 into registration with a vertically extending channel 92 so that the sleeve 78 can be moved to the position shown in FIGS. 10 and 12.

It will thus be seen that the objects set forth above, and those made apparent from the foregoing description, are effectively attained and since certain changes may be made in the above construction without departing from the scope of the invention, it is intended that all matters contained in the foregoing description or shown in the accompanying drawings shall be interpreted as illustrative and not in a limiting sense.

It is also to be understood that the following claims are intended to cover all of the generic and specific features of the invention herein described, and all statements of the scope of the invention which, as a matter of language, might be said to fall therebetween.

Now that the invention has been described,

What is claimed is:

1. A container of the type having a small fill spout and having increased utility when used in conjunction with a funnel, comprising:

a container body member of generally parallelepiped configuration,

a fill spout formed in a top wall of said container body member and projecting upwardly therefrom,

a vent means in the form of an aperture formed in said top wall,

a funnel member having a rim, converging sidewalls, and a downspout,

said fill spout and funnel downspout adapted for releasable engagement with one another,

a vent closure member secured to said funnel rim and projecting outwardly therefrom,

said vent closure member closing said vent when brought into registration therewith.

2. The container of claim 1, further comprising,

a fill spout closure means in the form of a cap member,

a latch member secured to and projecting outwardly from said funnel rim,

said cap member adapted to releasably engage said latch member when said funnel member is inverted and disposed atop said container top wall and when said cap member is releasably engaged to said fill spout.

3. The container of claim 2, wherein said vent closure member and said latch member are secured to said rim in opposed relation to each other.

4. The container of claim 3, further comprising,

a sleeve-shaped shroud member disposed in surrounding relation to said aperture and projecting upwardly from said container top wall,

said vent closure member adapted to engage said shroud member when said funnel is inverted and said vent closure member is brought into releasable engagement with said shroud member.

5. The container of claim 4, further comprising,

a first handle means formed in said container body member at a first end thereof,

Fig. 1 (continued)

4,557,395

7

a second handle means formed in said container body member at a second end thereof which is longitudinally spaced from said first end,

each of said first and second handle means defined by a concavity formed in the bottom wall of said container body member and by a convexity contiguous thereto and continuous therewith, said convexity merging with an end wall of said container body member.

6. The container of claim 5, wherein the depth of the concavity forming a handle means is greater than the height of the convexity contiguous thereto.

7. The container of claim 5, wherein said first and second handle means are disposed transverse to and are bisected by the longitudinal axis of symmetry of said container body member.

8. The container of claim 3, wherein said cap member has a top wall having a diameter greater than the outer diameter of said fill spout, wherein an annular wall depends to the periphery of said cap top wall, wherein an annular cavity is defined between said fill spout and said depending wall, and wherein said latch member is specifically configured to enter into said annular cavity when brought into registration therewith.

9. The container of claim 8, wherein said latch member has a generally L-shaped configuration.

10. The container of claim 3, wherein said fill spout, said vent and said funnel member, latch member and vent closer member are collectively aligned with the longitudinal axis of symmetry of said container body member when said funnel member is inverted, when said vent closure member is disposed in engaging relation to said vent, and when said latch member is disposed in engaging relation to said fill spout cap.

11. The container of claim 1, wherein a strainer means is positioned within said funnel member at the juncture of said converging sidewalls and said downspout.

12. The container of claim 1, wherein said funnel member has a generally rectangular configuration when seen in plan view, and wherein said latch member and vent closure member are disposed mid-length of the opposite truncate sidewalls of said funnel member.

13. The container of claim 1, wherein said fill spout is externally threaded and wherein said funnel member downspout is internally threaded.

8

14. The container of claim 1, further comprising, an elongate funnel downspout extension member having a first end adapted to releasably engage said funnel downspout and a second end adapted to releasably engage said fill spout,

and said downspout extension member having a flexible medial portion.

15. The container of claim 14, further comprising, a rigid sleeve-shaped locking member, having a length greater than the length of said flexible medial portion and having an inside diameter slightly greater than the outside diameter of said downspout extension member, disposed in ensleeving relation to said flexible medial portion and restricting said downspout extension member from flexing at said medial portion.

16. The container of claim 15, further comprising, means for selectively locking and unlocking said sleeve member into and out of its restricting engagement with said medial portion, respectively.

17. The container of claim 16, wherein said means for selectively locking and unlocking said sleeve member includes a pair of vertically spaced bead members formed on said downspout extension member, one of which is positioned above said flexible medial portion and one of which is positioned below said flexible medial portion, and wherein said sleeve member has a pair of cooperatively spaced bead-receiving cavities formed therein, which cavities are interconnected by a vertical slot and which cavities are formed at the end of associated channels orthogonal to said vertical slot.

18. The container of claim 13, wherein said funnel member downspout further comprises a cylindrical outer wall within which said internal threads are formed, and a cylindrical inner wall spaced radially inwardly of said outer wall so that dirty oil contacts only said inner wall when the container is used.

19. The container of claim 18, wherein said downspout inner wall is concentric with said downspout outer wall.

20. The container of claim 19, wherein the spacing between said downspout outer and inner walls is sufficient to receive therebetween said externally threaded fill spout.

* * * * *

Fig. 1 (continued)

UNITED STATES PATENT OFFICE

HENRY F. BOSENBERG, OF NEW BRUNSWICK, NEW JERSEY, ASSIGNOR TO LOUIS C. SCHUBERT, OF NEW BRUNSWICK, NEW JERSEY

CLIMBING OR TRAILING ROSE

Application filed August 6, 1930. Serial No. 473,410.

My invention relates to improvements in roses of the type known as climbing or trailing roses in which the central or main stalks acquire considerable length and when given 5 moderate support "climb" and branch out in various directions.

In roses it is very desirable to have a long period of blooming. This has been acquired in non-climbing roses of the type ordinarily 10 called monthly roses or everblooming roses. My invention now gives the true everblooming character to climbing roses.

The following description and accompanying illustrations apply to my improvements 15 upon the well known variety Dr. Van Fleet, with which my new plant is identical as respects color and form of flower, general climbing qualities, foliage and hardiness, but from which it differs radically in flowering habits 20 —but the same everblooming habits may be attained by breeding this new quality into other varieties of climbing roses.

Figure I shows (1) a flower that is just dropping its petals, (2) a bud about to open, 25 (3) a terminal bud just forming on a large side shoot, and (4) a new shoot which has not yet finished its growth and formed buds at its terminus. This shoot would not appear on the branch illustrated until several weeks 30 later than the stage of development shown, when it would grow out ordinarily from the axil of the first or second leaf below the bloomed-off flower. (5) shows a second way in which new flowering shoots form, by 35 branching off on a short stem immediately or closely adjacent to the blossom that has just finished blooming. Figure II shows a further method of branching and bud formation in cases where the bloom has been 40 cut off, but the formation of new flowering shoots is not dependent upon pruning off the old blossoms. It is evident that this succession of blooms continuously or intermittently supplied by new shoots branching 45 out throughout the summer and fall gives the true everblooming character. When grown in the latitude of New Brunswick, New Jersey, my new climbing rose named "The New Dawn" and illustrated herewith in 50 exact drawings from photographs, provides a succession of blossoms on a single plant from about the end of May to the middle of November, or until stopped by frost.

No claim is made as to novelty in color or other physical characteristics of the individ- 55 ual blossoms, nor as to the foliage or growing habits of this rose other than as described above.

I claim:

A climbing rose as herein shown and de- 60 scribed, characterized by its everblooming habit.

In testimony whereof I affix my signature hereunto.

HENRY F. BOSENBERG. 65

Fig. 2

Aug. 18, 1931. H. F. BOSENBERG Plant Pat. 1

CLIMBING OR TRAILING ROSE

Filed Aug. 6, 1930

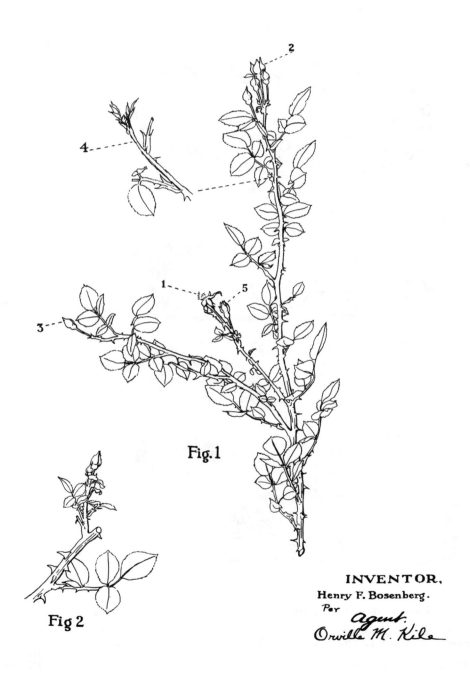

Fig. 1

Fig 2

INVENTOR,

Henry F. Bosenberg.

Per *agent.*

Orville M. Kile

Fig. 2 (Continued)

United States Patent

Des. 237,427

Patented Nov. 4, 1975

237,427

SHOE

William H. Thornberry, Newtown, Conn., assignor to
Uniroyal, Inc.

Filed July 26, 1974, Ser. No. 492,307

Term of patent 14 years

Int. Cl. D2—*04*

U.S. Cl. D2—310

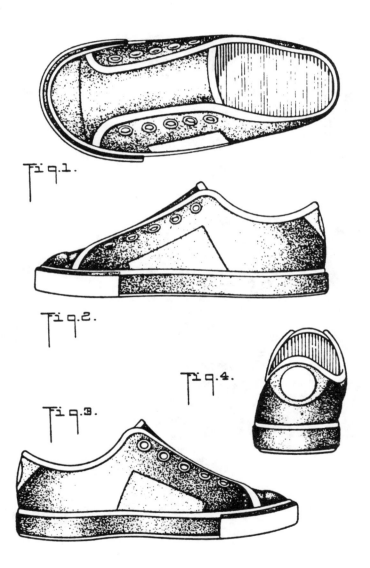

Fig.1.

Fig.2.

Fig.4.

Fig.3.

FIG. 1 is a plan view of a shoe embodying my new
design;

FIG. 2 is a side elevational view of the FIG. 1 article;

FIG. 3 is a side elevational view of the FIG. 1 article;
and

FIG. 4 is an end elevational view of the FIG. 1 article.

I claim:

The ornamental design for a shoe, substantially as
shown and described.

References Cited

UNITED STATES PATENTS

D. 118,131	12/1939	Pick	D2—313
D. 173,699	12/1954	Hosker	D2—310
D. 226,461	3/1973	Nelson	D2—309.

LOIS S. LANIER, Primary Examiner

Fig. 3

Patent Ownership Graphs (Source: PTO)

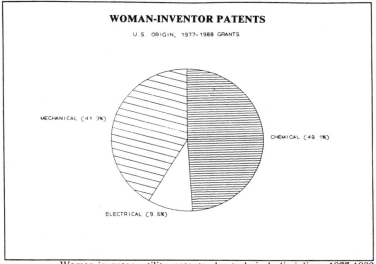

Woman-inventor utility patents: by technical discipline, 1977-1988 grants

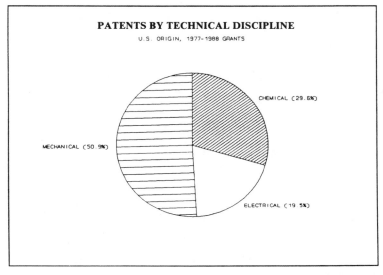

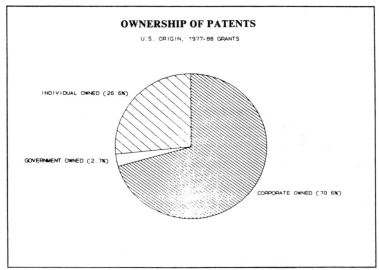

Fig. 4A

TOP PATENT—RECEIVING CORPORATIONS

SOURCE: PTO

Corporation / Metric	PRE 76	1976	1977	1978	1979	1980	1981	1982	1983	1984	1985	1986	1987	1988	1989	TOTAL
HITACHI, LTD. PATENTS GRANTED															1053	1053
PATENTED APPLICATIONS							3	5	6	10	32	195	481	310	11	1053
TOSHIBA CORPORATION PATENTS GRANTED															961	961
PATENTED APPLICATIONS							1	2	4	5	24	123	441	336	25	961
CANON KABUSHIKI KAISHA PATENTS GRANTED															949	949
PATENTED APPLICATIONS						1		2	2	11	36	195	347	295	60	949
FUJI PHOTO FILM CO., LTD PATENTS GRANTED															884	884
PATENTED APPLICATIONS									1	13	31	104	423	296	16	884
GENERAL ELECTRIC COMPANY PATENTS GRANTED															818	818
PATENTED APPLICATIONS								3	10	11	23	90	345	312	24	818
MITSUBISHI DENKI KABUSHIKI KAISHA PATENTS GRANTED															767	767
PATENTED APPLICATIONS								6	3	8	11	58	315	349	17	767
U.S. PHILIPS CORPORATION PATENTS GRANTED															745	745
PATENTED APPLICATIONS					2		2	6	4	12	17	75	317	290	20	745
SIEMENS AKTIENGESELLSCHAFT PATENTS GRANTED															656	656
PATENTED APPLICATIONS		1		1			1	2	5	7	12	56	291	266	14	656
INTERNATIONAL BUSINESS MACHINES CORPORATION PATENTS GRANTED															623	623
PATENTED APPLICATIONS							1	1	2	8	25	99	277	196	14	623
EASTMAN KODAK COMPANY PATENTS GRANTED															589	589
PATENTED APPLICATIONS									2	2	4	30	196	323	32	589
NEC CORPORATION PATENTS GRANTED															480	480
PATENTED APPLICATIONS									1	5	22	57	220	167	8	480
BAYER AKTIENGESELLSCHAFT PATENTS GRANTED															468	468
PATENTED APPLICATIONS							1	2	1	8	10	60	191	188	7	468
WESTINGHOUSE ELECTRIC CORP. PATENTS GRANTED															452	452
PATENTED APPLICATIONS					1	1	1	5	2	14	7	25	176	209	11	452
E. I. DU PONT DE NEMOURS AND COMPANY PATENTS GRANTED															443	443
PATENTED APPLICATIONS								1	5	7	11	44	171	195	9	443
DOW CHEMICAL COMPANY PATENTS GRANTED															431	431
PATENTED APPLICATIONS								2	7	7	27	45	161	175	7	431
GENERAL MOTORS CORPORATION PATENTS GRANTED															412	412
PATENTED APPLICATIONS			1					3	2	1	4	10	98	277	16	412
TEXAS INSTRUMENTS, INCORPORATED PATENTS GRANTED															400	400
PATENTED APPLICATIONS						1	2	2	6	1	12	55	163	145	13	400
MOTOROLA INC. PATENTS GRANTED															384	384
PATENTED APPLICATIONS				1						3	7	31	141	188	13	384
AMERICAN TELEPHONE AND TELEGRAPH COMPANY PATENTS GRANTED															381	381
PATENTED APPLICATIONS							2	1	4	6	16	63	149	130	10	381
MINNESOTA MINING AND MANUFACTURING COMPANY PATENTS GRANTED															381	381
PATENTED APPLICATIONS								2	4	4	5	31	165	163	7	381
MOBIL OIL CORP. PATENTS GRANTED															376	376
PATENTED APPLICATIONS								1	4	8	11	16	177	151	8	376
SHELL OIL COMPANY PATENTS GRANTED															370	370
PATENTED APPLICATIONS								1	1	2	3	23	149	174	17	370
MATSUSHITA ELECTRIC INDUSTRIAL CO., LTD. PATENTS GRANTED															364	364
PATENTED APPLICATIONS									1	3	5	65	171	114	5	364
BASF AKTIENGESELLSCHAFT PATENTS GRANTED															354	354
PATENTED APPLICATIONS									2	4	10	27	161	145	5	354
HONDA GIKEN KOGYO KABUSHIKI KAISHA (HONDA MOTOR CO., LTD.) PATENTS GRANTED															352	352
PATENTED APPLICATIONS						1			1	3	5	28	166	144	4	352
MINOLTA CAMERA CO., LTD. PATENTS GRANTED															350	350
PATENTED APPLICATIONS											4	17	150	167	12	350
CIBA-GEIGY CORPORATION PATENTS GRANTED															346	346
PATENTED APPLICATIONS									1	3	4	40	171	122	5	346
SHARP KABUSHIKI KAISHA (SHARP CORPORATION) PATENTS GRANTED															339	339
PATENTED APPLICATIONS							1	4	6	3	14	57	131	112	11	339

Fig. 4B

	PRE 76	1976	1977	1978	1979	1980	1981	1982	1983	1984	1985	1986	1987	1988	1989	TOTAL
SONY CORPORATION																
PATENTS GRANTED															320	320
PATENTED APPLICATIONS							1	2	2	1	17	47	129	114	7	320
HUGHES AIRCRAFT COMPANY																
PATENTS GRANTED															298	298
PATENTED APPLICATIONS	1								1	4	18	52	116	103	3	298
XEROX CORPORATION																
PATENTS GRANTED															283	283
PATENTED APPLICATIONS								4	5	11	8	8	113	130	4	283
FUJITSU LIMITED																
PATENTS GRANTED															279	279
PATENTED APPLICATIONS					1		1		1	2	8	37	124	96	9	279
ALLIED-SIGNAL INC.																
PATENTS GRANTED															274	274
PATENTED APPLICATIONS								4	6	6	7	24	110	111	6	274
NISSAN MOTOR COMPANY, LIMITED																
PATENTS GRANTED															274	274
PATENTED APPLICATIONS								2	6	3	9	38	110	99	7	274
RICOH CO., LTD.																
PATENTS GRANTED															271	271
PATENTED APPLICATIONS										1	4	25	111	125	5	271
TOYOTA JIDOSHA K.K.																
PATENTS GRANTED															270	270
PATENTED APPLICATIONS								1		2	8	42	113	103	1	270
HOECHST AKTIENGESELLSCHAFT																
PATENTS GRANTED															257	257
PATENTED APPLICATIONS									1	1	7	39	116	89	4	257
AMP INCORPORATED																
PATENTS GRANTED															243	243
PATENTED APPLICATIONS								1	2	3	4	11	72	139	11	243
HEWLETT-PACKARD COMPANY																
PATENTS GRANTED															242	242
PATENTED APPLICATIONS										1	6	28	117	83	7	242
UNITED TECHNOLOGIES CORPORATION																
PATENTS GRANTED															231	231
PATENTED APPLICATIONS							1		1	2	3	23	107	91	3	231
UNITED STATES OF AMERICA, DEPARTMENT OF ENERGY																
PATENTS GRANTED															215	215
PATENTED APPLICATIONS						1		4	2	3	7	21	82	91	4	215
BOEING COMPANY																
PATENTS GRANTED															213	213
PATENTED APPLICATIONS							1	2	4	5	19	20	97	65		213
HONEYWELL INC.																
PATENTS GRANTED															213	213
PATENTED APPLICATIONS								2	3	5	6	23	71	95	8	213
ROBERT BOSCH GMBH																
PATENTS GRANTED															200	200
PATENTED APPLICATIONS								2	3	5	4	14	88	80	4	200
AMOCO CORPORATION																
PATENTS GRANTED															198	198
PATENTED APPLICATIONS									3	5	4	6	90	88	2	198
OLYMPUS OPTICAL CO., LTD.																
PATENTS GRANTED															195	195
PATENTED APPLICATIONS											5	16	76	96	2	195
MERCK + CO., INC.																
PATENTS GRANTED															173	173
PATENTED APPLICATIONS									2	4	3	24	79	57	4	173
GTE PRODUCTS CORPORATION																
PATENTS GRANTED															168	168
PATENTED APPLICATIONS								1	3	5	7	7	51	82	12	168
PIONEER ELECTRONIC CORPORATION																
PATENTS GRANTED															164	164
PATENTED APPLICATIONS								1	1			12	82	62	6	164
FORD MOTOR COMPANY																
PATENTS GRANTED															163	163
PATENTED APPLICATIONS							1	1		1	4	6	83	61	6	163
UNITED STATES OF AMERICA, ARMY																
PATENTS GRANTED															163	163
PATENTED APPLICATIONS							2	2	2	3	2	11	34	88	19	163
UOP INC.																
PATENTS GRANTED															161	161
PATENTED APPLICATIONS										3	3	15	71	68	1	161
AISIN SEIKI KABUSHIKI KAISHA																
PATENTS GRANTED															158	158
PATENTED APPLICATIONS						1				1	2		7	79	64	158
IMPERIAL CHEMICAL INDUSTRIES PLC																
PATENTS GRANTED															157	157
PATENTED APPLICATIONS					1			1	1	3	6	21	72	50	2	157
BROTHER KOGYO KABUSHIKI KAISHA																
PATENTS GRANTED															156	156
PATENTED APPLICATIONS												14	77	61	4	156
SUMITOMO CHEMICAL COMPANY, LIMITED																
PATENTS GRANTED															146	146
PATENTED APPLICATIONS										1	3	14	70	53	5	146

Fig. 4B (Continued)

PTO FEE SCHEDULE

FEE CODE DESCRIPTION   **FEE**

PATENT FEES

Group 1 – Patent Filing Fees
101 BASIC FILING FEE – UTILITY	370.00
102 INDEPENDENT CLAIMS IN EXCESS OF THREE	36.00
103 CLAIMS IN EXCESS OF TWENTY	12.00
104 MULTIPLE DEPENDENT CLAIM	120.00
105 SURCHARGE – LATE FILING FEE OR OATH/DECL.	120.00
106 DESIGN FILING FEE	150.00
107 PLANT FILING FEE	250.00
108 REISSUE FILING FEE	370.00
109 REISSUE INDEPENDENT CLAIMS OVER PATENT	36.00
110 REISSUE CLAIMS IN EXCESS OF TWENTY & PATENT	12.00
139 NON-ENGLISH SPECIFICATION	30.00

Group 2 – Small Entity Patent Filing Fees
201 BASIC FILING FEE – UTILITY	185.00
202 INDEPENDENT CLAIMS IN EXCESS OF THREE	18.00
203 CLAIMS IN EXCESS OF TWENTY	6.00
204 MULTIPLE DEPENDENT CLAIM	60.00
205 SURCHARGE – LATE FILING FEE OR OATH	60.00
206 DESIGN FILING FEE	75.00
207 PLANT FILING FEE	125.00
208 REISSUE FILING FEE	185.00
209 REISSUE INDEPENDENT CLAIMS OVER PATENT	18.00
210 REISSUE CLAIMS IN EXCESS OF TWENTY & PATENT	6.00

Group 3 – Patent Extension Fees
115 EXTENSION – ONE MONTH	62.00
116 EXTENSION – TWO MONTHS	180.00
117 EXTENSION – THREE MONTHS	430.00
118 EXTENSION – FOUR MONTHS	680.00

Group 4 – Small Entity Patent Extension Fees
215 EXTENSION – ONE MONTH	31.00
216 EXTENSION – TWO MONTHS	90.00
217 EXTENSION – THREE MONTHS	215.00
218 EXTENSION – FOUR MONTHS	340.00

Group 5 – Patent Appeals/Interference Fees
119 NOTICE OF APPEAL	140.00
120 FILING A BRIEF	140.00
121 REQUEST FOR ORAL HEARING	120.00

Group 6 – Small Entity Patent Appeals/Interference Fees
219 NOTICE OF APPEAL	70.00
220 FILING A BRIEF	70.00
221 REQUEST FOR ORAL HEARING	60.00

Group 7 – Patent Petition Fee
PETITIONS TO THE COMMISSIONER:
122 –NOT ALL INVENTORS; NOT THE INVENTOR	120.00

Fig. 5

```
123 -CORRECTION OF INVENTORSHIP IN APPL. .......................................  120.00
124 -NOT PROVIDED FOR QUESTIONS ...........................................  120.00
125 -SUSPEND RULES.............................................................  120.00
160 -EXPEDITED LICENSE........................................................  120.00
161 -CHANGE SCOPE OF LICENSE ...............................................  120.00
162 -RETROACTIVE LICENSE.....................................................  120.00
163 -REFUSING MAINTENANCE FEE..............................................  120.00
164 -REINSTATEMENT OF EXPIRED PATENT......................................  120.00
165 -INTERFERENCE.............................................................  120.00
166 -RECONSIDER INTERFERENCE PET. DECISION .............................  120.00
167 -LATE FILING OF INTERFERENCE SETTLEMENT.............................  120.00
168 -REFUSAL TO PUBLISH SIR .................................................  120.00
127 -ACCESS TO ASSIGNMENT RECORD ........................................  120.00
128 -ACCESS TO APPLICATION...................................................  120.00
129 -LATE PRIORITY PAPERS ....................................................  120.00
131 -SUSPEND ACTION ..........................................................  120.00
132 -DIVISIONAL REISSUES.......................................................  120.00
133 -ACCESS TO INTERFERENCE AGREEMENT....................................  120.00
134 -AMENDMENT AFTER ISSUE FEE PAID .....................................  120.00
135 -WITHDRAWAL FROM ISSUE..................................................  120.00
136 -DEFER ISSUE................................................................  120.00
137 -ISSUE TO LATE RECORDED ASSIGNEE .....................................  120.00
130 PET. TO COMM. TO MAKE APPL. SPECIAL.................................   80.00
138 PET. TO COMM. - PUBLIC USE PROCEEDING ..............................  1200.00
140 PET. - REVIVE ABAND. APPL. - UNAVOIDABLE ...........................   62.00
141 PET. - REVIVE ABAND. APPL. - UNINTENTIONAL .........................  620.00
146 PET. - CORRECTION OF INVENTORSHIP IN PATENT........................  120.00
```

Group 8 - Small Entity Patent Petition Fees
```
240 PET. - REVIVE ABAND. APPL. - UNAVOIDABLE .....................   31.00
241 PET. - REVIVE ABAND. APPL. - UNINTENTIONAL ...............  310.00
```

Group 9 - Patent Issue Fees
```
142 UTILITY ISSUE FEE .........................................................  620.00
143 DESIGN ISSUE FEE ..........................................................  220.00
144 PLANT ISSUE FEE............................................................  310.00
148 STATUTORY DISCLAIMER.....................................................   62.00
```

Group 10 - Small Entity Patent Issue Fees
```
242 UTILITY ISSUE FEE .........................................................  310.00
243 DESIGN ISSUE FEE ..........................................................  110.00
244 PLANT ISSUE FEE............................................................  155.00
248 STATUTORY DISCLAIMER.....................................................   31.00
```

Group 11 - Patent Post-Allowance Fees
```
112 SIR - PRIOR TO EXAMINER'S ACTION .....................................  400.00
113 SIR - AFTER EXAMINER'S ACTION ........................................  800.00
145 CERTIFICATE OF CORRECTION ............................................   60.00
147 RE-EXAMINATION ...........................................................  2000.00
111 EXTENSION OF THE TERM OF PATENT......................................  600.00
```

Group 12 - Patent Maintenance Fees - Applications
Filed December 12, 1980 - August 27, 1982
```
170 DUE AT 3.5 YEARS...........................................................  245.00
171 DUE AT 7.5 YEARS...........................................................  495.00
172 DUE AT 11.5 YEARS..........................................................  740.00
176 SURCHARGE - LATE PAYMENT WITHIN SIX MONTHS .......................  120.00
```

Group 13 - Patent Maintenance Fees - Applications
Filed On Or After August 26, 1982
```
173 DUE AT 3.5 YEARS...........................................................  490.00
174 DUE AT 7.5 YEARS...........................................................  990.00
```

Fig. 5

175 DUE AT 11.5 YEARS.. 1480.00
177 SURCHARGE – LATE PAYMENT WITHIN SIX MONTHS 120.00
178 SURCHARGE AFTER EXPIRATION 550.00

**Group 14 – Small Entity Patent Maintenance Fees –
Applications Filed On Or After August 27, 1982**
273 DUE AT 3.5 YEARS.. 245.00
274 DUE AT 7.5 YEARS.. 495.00
275 DUE AT 11.5 YEARS.. 740.00
277 SURCHARGE – LATE PAYMENT WITHIN SIX MONTHS 60.00

Group 15 – Patent Service Fees
501 COPY OF PATENT 1.50
503 COPY OF PLANT PATENT 10.00
506 COPY OF OFFICE RECORDS, (each thirty pages).......................... 10.00
500 COPY OF UTILITY PATENT IN COLOR 20.00
535 PATENT COPY – EXPEDITED SERVICE 3.00
536 PATENT COPY EXPEDITED SERVICE VIA EOS 25.00
504 COPY OF APPLICATION AS FILED, CERTIFIED 10.00
505 COPY OF FILE WRAPPER, CERTIFIED 170.00
533 COPY OF PATENT ASSIGNMENT, CERTIFIED 5.00
537 CERTIFIED COPY OF PATENT APPLICATION EXPEDITED................... 20.00
508 CERTIFYING OFFICE RECORDS 3.00
509 SEARCH OF RECORDS.. 15.00
513 PATENT DEPOSITORY LIBRARY 50.00
514 LIST OF PATENTS IN SUBCLASS 2.00
528 UNCERTIFIED STATEMENT.. 5.00
532 COPY OF NON-U.S. DOCUMENT 10.00
510 COMPARING COPIES PER DOCUMENT... 10.00
534 DUPLICATE OR CORRECTED FILING RECEIPT 15.00
516 FILING A DISCLOSURE DOCUMENT 6.00
522 BOX RENTAL.. 50.00
526 INTERNATIONAL TYPE SEARCH REPORT 30.00
517 SEARCHING, INVENTOR RECORDS, TEN YEARS 10.00
524 COPISHARE CARD PER PAGE .... .15
518 RECORDING PATENT ASSIGNMENT.. 8.00
520 PUBLICATION IN OFFICIAL GAZETTE 20.00
521 DUPLICATE USER PASS.. 10.00
523 LOCKER RENTALS .... .25
525 UNSPECIFIED OTHER SERVICES AT COST
529 RETAINING ABANDONED APPLICATION 120.00
530 HANDLING FEE – OMITTED SPEC./DRAWING................................ 15.00
531 HANDLING FEE FOR WITHDRAWAL OF SIR................................... 120.00

Group 16 – Patent Enrollment Service Fees
609 ADMISSION TO EXAMINATION... 270.00
610 REGISTRATION TO PRACTICE.. 90.00
611 REINSTATEMENT TO PRACTICE.. 10.00
612 COPY OF CERTIFICATE OF GOOD STANDING 10.00
613 CERTIFICATE OF GOOD STANDING – FRAMING............................ 100.00
615 REVIEW OF DECISION OF DIRECTOR, OED 100.00
616 REGRADING OF EXAMINATION 100.00

Group 17 – Patent PCT Fees – International Stage
150 TRANSMITTAL FEE 170.00
151 PCT SEARCH FEE – NO U.S. APPLICATION 550.00
153 PCT SEARCH – PRIOR U.S. APPLICATION 380.00
152 SUPPLEMENTAL SEARCH PER ADDITIONAL INVENTION 150.00
190 PRELIMINARY EXAM. FEE – ISA WAS U.S. 400.00
191 PRELIMINARY EXAM. FEE – ISA NOT U.S. 600.00
192 ADDITIONAL INVENTION – ISA WAS U.S..................................... 130.00
193 ADDITIONAL INVENTION – ISA NOT U.S. 200.00

Fig. 5

Group 18 – Patent PCT Fees – National Stage
956 IPEA – U.S. 330.00
958 INTERNATIONAL SEARCHING AUTHORITY – U.S. 370.00
960 PTO NOT ISA OR IPEA 500.00
962 CLAIMS MEET ART. 33(1)–(4) – IPEA – U.S. 50.00
964 CLAIMS – EXTRA INDEPENDENT (over three) 36.00
966 CLAIMS – EXTRA TOTAL (over twenty) 12.00
968 CLAIMS – MULTIPLE DEPENDENT 120.00
154 SURCHARGE – LATE FILING FEE OR OATH/DEC. 120.00
156 ENGLISH TRANSL. – AFTER TWENTY MONTHS 30.00

Group 19 – Small Entity Patent PCT Fees –
National Stage
957 IPEA – U.S. 165.00
959 INTERNATIONAL SEARCHING AUTHORITY – U.S. 185.00
961 PTO NOT ISA OR IPEA 250.00
963 CLAIMS MEET ART. 33(1)–(4) – ipea – U.S. 25.00
965 CLAIMS – EXTRA INDEPENDENT (over three) 18.00
967 CLAIMS – EXTRA TOTAL (over twenty) 6.00
969 CLAIMS – MULTIPLE DEPENDENT 60.00
254 SURCHARGE – LATE FILING FEE OR OATH/DEC. 60.00

Group 20 – Patent PCT Fees to WIPO
800 BASIC FEE (first thirty pages) 485.00
801 BASIC SUPPLEMENTAL FEE
        (for each page over thirty) 10.00
803 HANDLING FEE 150.00
DESIGNATION FEE PER COUNTRY 120.00

Group 21 – Patent PCT Fees to EPO
802 INTERNATIONAL SEARCH 1160.00

TRADEMARK FEES

Group 22 – Trademark Filing Fees
301 APPLICATION FOR REGISTRATION, PER CLASS 175.00

Group 23 – Trademark Post Registration Fees
302 APPLICATION FOR RENEWAL, PER CLASS 300.00
303 SPECIAL HANDLING FOR LATE RENEWAL 100.00
304 PUBLICATION OF MARK UNDER 12c, PER CLASS 100.00
309 FILING 8 AFFIDAVIT, PER CLASS 100.00
310 FILING 15 AFFIDAVIT, PER CLASS 100.00
311 FILING COMBINED 8 & 15 AFFIDAVIT, PER CLASS 200.00

Group 24 – Trademark Amended Registration Fees
305 ISSUING NEW CERTIFICATE OF REGISTRATION 100.00
306 CERT. OF CORRECTION, REGISTRANT'S ERROR 100.00
307 FILING DISCLAIMER TO REGISTRATION 100.00
308 FILING AMENDMENT TO REGISTRATION 100.00

Group 25 – Trademark Petition Fee
312 PETITION TO THE COMMISSIONER, PER CLASS 100.00

Group 26 – Trademark Trial and Appeal Board Fees
313 PETITION FOR CANCELLATION, PER CLASS 200.00
314 NOTICE OF OPPOSITION, PER CLASS 200.00
315 EX PARTE APPEAL, PER CLASS 100.00

Group 27 – Trademark Service Fees
401 PRINTED COPY OF EACH REGISTERED MARK 1.50
403 CERTIFY TM RECORDS, PER CERTIFICATE 3.50

Fig. 5

404 PHOTOCOPIES OF TM RECORDS, PER PAGE .... .30
405 RECORDING ASSIGNMENT, PER MARK, PER DOCUMENT 8.00
407 ABSTRACTS OF TITLE, PER REGISTRATION 12.00
408 COPY OF REG. MARK WITH TITLE OR STATUS 6.50
410 MAKE CERTIFICATION SPECIAL 25.00
409 UNSPECIFIED OTHER SERVICES AT COST
424 COPISHARE CARD, PER PAGE .... .15

Group 28 – [Reserved]

GENERAL FEES

Group 29 – Finance Service Fees
607 ESTABLISH DEPOSIT ACCOUNT 10.00
608 SERVICE CHARGE FOR BELOW MIN. BALANCE 20.00
617 PROCESSING RETURNED CHECKS 50.00

Group 30 – Computer Service Fees
618 COMPUTER RECORDS AT COST

Fig. 5

PATENT APPLICATION TRANSMITTAL LETTER	ATTORNEY'S DOCKET NO.

TO THE COMMISSIONER OF PATENTS AND TRADEMARKS:

Transmitted herewith for filing is the patent application of _____

_____

for _____

_____

Enclosed are:

☐ _____ sheets of drawing.

☐ an assignment of the invention to _____

_____

☐ a certified copy of a _____ application.

☐ associate power of attorney.

☐ verified statement to establish small entity status under 37 CFR 1.9 and 1.27. _____

CLAIMS AS FILED

FOR:	NO. FILED	NO. EXTRA		SMALL ENTITY RATE	FEE		OTHER THAN A SMALL ENTITY RATE	FEE
BASIC FEE					$	OR		$
TOTAL CLAIMS	-20-	•		x $6=	$	OR	x $12=	$
INDEP. CLAIMS	-3-	••		x $17=	$	OR	x $34=	$
MULTIPLE DEPENDENT CLAIM PRESENT				+$55=	$	OR	+$110=	$
				TOTAL	$	OR	TOTAL	$

• If the difference in col. 1 is less than zero, enter "0" in col. 2

☐ Please charge my Deposit Account No. _____ in the amount of $ _____
☐ A duplicate copy of this sheet is enclosed.

☐ A check in the amount of $ _____ to cover the filing fee is enclosed.

☐ The Commissioner is hereby authorized to charge payment of the following fees associated with this communication or credit any overpayment to Deposit Account No. _____. A Duplicate copy of this sheet is enclosed.

    ☐ Any additional filing fees required under 37 CFR 1.16.

    ☐ Any patent application processing fees under 37 CFR 1.17

☐ The Commissioner is hereby authorized to charge payment of the following fees during the pendency of this application or credit any overpayment to Deposit Account No. _____. A duplicate copy of this sheet is enclosed.

    ☐ Any filing fees under 37 CFR 1.16 for presentation of extra claims.

    ☐ Any patent application processing fees under 37 CFR 1.17.

    ☐ The issue fee set in 37 CFR 1.18 at or before mailing of the Notice of Allowance, pursuant to 37 CFR 1.311(b).

_____   _____
date   signature

Patent and Trademark Office - U.S. DEPARTMENT of COMMERCE

UNITED STATES DEPARTMENT OF COMMERCE
Patent and Trademark Office
Address: COMMISSIONER OF PATENTS AND TRADEMARKS
Washington, D.C. 20231

SERIAL NUMBER	FILING DATE	FIRST NAMED APPLICANT	ATTY. DOCKET NO.

DATE MAILED:

NOTICE OF INFORMAL APPLICATION
(Attachment to Office Action)

This application does not conform with the rules governing applications for the reason(s) checked below. The period within which to correct these requirements and avoid abandonment is set in the accompanying Office action.

A. A new oath or declaration, identifying this application by the serial number and filing date is required. The oath or declaration does not comply with 37 CFR 1.63 in that it:

1. ☐ was not executed in accordance with either 37 CFR 1.66 or 1.68.

2. ☐ does not identify the city and state or foreign country of residence of each inventor.

3. ☐ does not identify the citizenship of each inventor.

4. ☐ does not state whether the inventor is a sole or joint inventor.

5. ☐ does not state that the person making the oath or declaration:

    a. ☐ has reviewed and understands the contents of the specification, including the claims, as amended by any amendment specifically referred to in the oath or declaration.

    b. ☐ believes the named inventor or inventors to be the original and first inventor or inventors of the subject matter which is claimed and for which a patent is sought.

    c. ☐ acknowledges the duty to disclose information which is material to the examination of the application in accordance with 37 CFR 1.56(a).

6. ☐ does not identify the foreign application for patent or inventor's certificate on which priority is claimed pursuant to 37 CFR 1.55, and any foreign application having a filing date before that of the application on which priority is claimed, by specifying the application serial number, country, day, month, and year of its filing.

7. ☐ does not state that the person making the oath or declaration acknowledges the duty to disclose material information as defined in 37 CFR 1.56(a) which occurred between the filing date of the prior application and filing date of the continuation-in-part application which discloses and claims subject matter in addition to that disclosed in the prior application (37 CFR 1.63(d)).

8. ☐ does not include the date of execution.

9. ☐ does not use permanent ink, or its equivalent in quality, as required under 37 CFR 1.52(a) for the: ☐ signature ☐ oath/declaration.

10. ☐ contains non-initialed alterations (See 37 CFR 1.52(c) and 1.56).

11. ☐ does not contain the clause regarding "willful false statements..." as required by 37 CFR 1.68.

12. ☐ Other:

B. Applicant is required to provide:

1. ☐ A statement signed by applicant giving his or her complete name. A full name must include at least one given name without abbreviation as required by 37 CFR 1.41(a).

2. ☐ Proof of authority of the legal representative under 37 CFR 1.44.

3. ☐ An abstract in compliance with 37 CFR 1.72(b).

4. ☐ A statement signed by applicant giving his or her complete post office address (37 CFR 1.33(a)).

5. ☐ A copy of the specification written, typed, or printed in permanent ink, or its equivalent in quality as required by 37 CFR 1.52(a).

6. ☐ Other:

Fig. 7

UNITED STATES DEPARTMENT OF COMMERCE
Patent and Trademark Office
Address: COMMISSIONER OF PATENTS AND TRADEMARKS
Washington, D.C. 20231

SERIAL NUMBER	FILING DATE	FIRST NAMED APPLICANT	ATTY. DOCKET NO.

DATE MAILED:

Notice of Incomplete Application

A filing date has NOT been assigned to the above identified application papers for the reason(s) shown below.

1. ☐ The specification (description and claims):
   - a. ☐ is missing
   - b. ☐ has pages _____ missing.
   - c. ☐ does not include a written description of the invention.
   - d. ☐ does not include at least one claim in compliance with 35 U.S.C. 112.

A complete specification in compliance with 35 U.S.C. 112 is required.

2. ☐ A drawing of Figure(s) _____ described in the specification is required in compliance with 35 U.S.C. 111.

3. ☐ A drawing of applicant's invention is required since it is necessary for the understanding of the subject matter of the invention in compliance with 35 U.S.C. 113.

4. ☐ The inventor's name(s) is missing. The full names of all inventors are required in compliance with 37 CFR 1.41.

5. ☐ Other items missing but not required for a filing date:

All of the above-noted omissions, unless otherwise indicated, must be submitted within TWO MONTHS of the date of this notice or the application will be returned or otherwise disposed of. Any fee which has been submitted will be refunded less a $50.00 handling fee. See 37 CFR 1.53(c).

The filing date will be the date of receipt of all the items required above, unless otherwise indicated. Any assertions that the items required above were submitted, or are not necessary for a filing date, must be by way of a petition directed to the attention of the Office of the Assistant Commissioner for Patents accompanied by the $140.00 petition fee (37 CFR 1.17(h)). If the petition alleges that no defect exists, a request for refund of the petition fee may be included in the petition.

Direct the response to, and questions about, this notice to the undersigned, Attention: Application Branch, and include the above Serial Number and Receipt Date.

Enclosed:

   ☐ "General Information Concerning Patents". See page _____.
   ☐ Copy of a patent to assist applicant in making corrections.
   ☐ "Notice to File Missing Parts of Application", Form PTO-1532.
   ☐ Other: _____

For: **Manager, Application Branch**
(703) 557-_____

FORM PTO-1123 (REV. 4-87)

Fig. 8

UNITED STATES DEPARTMENT OF COMMERCE
Patent and Trademark Office
Address: COMMISSIONER OF PATENTS AND TRADEMARKS
Washington, D.C. 20231

SERIAL NUMBER	FILING DATE	FIRST NAMED APPLICANT	ATTY. DOCKET NO.

DATE MAILED:

NOTICE TO FILE MISSING PARTS OF APPLICATION—
NO FILING DATE
(Attachment to Form PTO-1123)

In order to avoid payment by applicant of the surcharge required if items 1 and 3-6 are filed after the filing date the following items are also brought to applicant's attention at this time.

If all missing parts of this form and on the "Notice of Incomplete Application" are filed together, the total amount owed by applicant as a ☐ large entity ☐ small entity (verified statement filed) is $_____.

1. ☐ The statutory basic filing fee is: ☐ missing ☐ insufficient. Applicant as a ☐ large entity ☐ small entity must submit $_____ to complete the basic filing fee and MUST ALSO SUBMIT THE SURCHARGE, IF REQUIRED, AS INDICATED BELOW.

2. ☐ Additional claim fees of $_____ as a ☐ large entity, ☐ small entity, including any required multiple dependent claim fee, are required. Applicant must submit the additional claim fees or cancel the additional claims for which fees are due. NO SURCHARGE IS REQUIRED FOR THIS ITEM.

3. ☐ The oath or declaration:
   ☐ is missing.
   ☐ does not cover items required on the "Notice of Incomplete Application".
   An oath or declaration in compliance with 37 CFR 1.63, referring to the above Serial Number and Receipt Date is required. A SURCHARGE, IF REQUIRED, MUST ALSO BE SUBMITTED AS INDICATED BELOW.

4. ☐ The oath or declaration does not identify the application to which it applies. An oath or declaration in compliance with 37 CFR 1.63, identifying the application by the above Serial Number and Receipt Date is required. A SURCHARGE, IF REQUIRED, MUST ALSO BE SUBMITTED AS INDICATED BELOW.

5. ☐ The signature to the oath or declaration is: ☐ missing; ☐ a reproduction; ☐ by a person other than the inventor or a person qualified under 37 CFR 1.42, 1.43, or 1.47. A properly signed oath or declaration in compliance with 37 CFR 1.63, referring to the above Serial Number and Recipt Date is required. A SURCHARGE, MUST ALSO BE SUBMITTED AS INDICATED BELOW.

6. ☐ The signature of the following joint inventor(s) is missing from the oath or declaration:
   _____. Applicant(s) should provide, if possible, an oath or declaration signed by the omitted inventor(s), identifying this application by the above Serial Number and Receipt Date. A SURCHARGE, IF REQUIRED, MUST ALSO BE SUBMITTED AS INDICATED BELOW.

7. ☐ A $20.00 processing fee is required for returned checks. (37 CFR 1.21(m)).

8. ☐ Other:

Required items 1-7 above SHOULD be filed, if possible, with any items required on the "Notice of Incomplete Application" enclosed with this form. If concurrent filing of all required items is not possible, items 1-7 above must be filed no later than two months from the filing date of this application. The filing date will be the date of receipt of the items required on the "Notice of Incomplete Application." If items 1 and 3-6 above are submitted after the filing date, THE PAYMENT OF A SURCHARGE OF $110.00 for large entities, or $55.00 for small entities who have filed a verified statement claiming such status, is required. (37 CFR 1.16(e)).

Applicant must file all the required items 1-7 indicated above within two months from any filing date granted to avoid abandonment. Extensions of time may be obtained by filing a petition accompanied by the extension fee under the provisions of 37 CFR 1.136(a).

Direct the response to, and any questions about, this notice to the undersigned, Attention: Application Branch.

A copy of this notice __MUST__ be returned with response.

For: Manager, Application Branch
(703) 557-3254

FORM PTO-1532 (REV 7-87)

For Office Use Only	
☐ 102	☐ 202
☐ 103	☐ 203
☐ 104	☐ 204
☐ 105	☐ 205

OFFICE COPY

Fig. 9

UNITED STATES DEPARTMENT OF COMMERCE
Patent and Trademark Office
Address: COMMISSIONER OF PATENTS AND TRADEMARKS
Washington, D.C. 20231

SERIAL NUMBER	FILING DATE	FIRST NAMED APPLICANT	ATTY DOCKET NO

DATE MAILED:

NOTICE TO FILE MISSING PARTS OF APPLICATION— FILING DATE GRANTED

A filing date has been granted to this application. However, the following parts are missing.

If all missing parts are filed within the period set below, the total amount owed by applicant as a ☐ large entity, ☐ small entity (verified statement filed), is $ _____.

1. ☐ The statutory basic filing fee is: ☐ missing. ☐ insufficient. Applicant as a ☐ large entity, ☐ small entity, must submit $ _____ to complete the basic filing fee and MUST ALSO SUBMIT THE SURCHARGE AS INDICATED BELOW.

2. ☐ Additional claim fees of $ _____ as a ☐ large entity, ☐ small entity, including any required multiple dependent claim fee, are required. Applicant must submit the additional claim fees or cancel the additional claims for which fees are due. NO SURCHARGE IS REQUIRED FOR THIS ITEM.

3. ☐ The oath or declaration:
    ☐ is missing.
    ☐ does not cover items omitted at the time of execution.

   An oath or declaration in compliance with 37 CFR 1.63, identifying the application by the above Serial Number and Filing Date is required. A SURCHARGE MUST ALSO BE SUBMITTED AS INDICATED BELOW.

4. ☐ The oath or declaration does not identify the application to which it applies. An oath or declaration in compliance with 37 CFR 1.63 identifying the application by the above Serial Number and Filing Date is required. A SURCHARGE MUST ALSO BE SUBMITTED AS INDICATED BELOW.

5. ☐ The signature to the oath or declaration is: ☐ missing; ☐ a reproduction; ☐ by a person other than the inventor or a person qualified under 37 CFR 1.42, 1.43, or 1.47. A properly signed oath or declaration in compliance with 37 CFR 1.63, identifying the application by the above Serial Number and Filing Date is required. A SURCHARGE MUST ALSO BE SUBMITTED AS INDICATED BELOW.

6. ☐ The signature of the following joint inventor(s) is missing from the oath or declaration: _____. Applicant(s) should provide, if possible an oath or declaration signed by the omitted inventor(s), identifying this application by the above Serial Number and Filing Date. A SURCHARGE MUST ALSO BE SUBMITTED AS INDICATED BELOW.

7. ☐ The application was filed in a language other than English. Applicant must file a verified English translation of the application and a fee of $26.00 under 37 CFR 1.17(k), unless this fee has already been paid NO SURCHARGE UNDER 37 CFR 1.16(e) IS REQUIRED FOR THIS ITEM.

8. ☐ A $20.00 processing fee is required for returned checks. (37 CFR 1.21(m)).

9. ☐ Your filing receipt was mailed in error because check was returned.

10. ☐ Other:

A Serial Number and Filing Date have been assigned to this application. However, to avoid abandonment under 37 CFR 1.53(d), the missing parts and fees identified above in items 1 and 3-6 must be timely provided ALONG WITH THE PAYMENT OF A SURCHARGE OF $110.00 for large entities or $55.00 for small entities who have filed a verified statement claiming such status. The surcharge is set forth in 37 CFR 1.16(e). Applicant is given ONE MONTH FROM THE DATE OF THIS LETTER, OR TWO MONTHS FROM THE FILING DATE of this application, WHICHEVER IS LATER, within which to file all missing parts and pay any fees. Extensions of time may be obtained by filing a petition accompanied by the extension fee under the provisions of 37 CFR 1.136(a).

Direct the response to, and any questions about, this notice to the undersigned, Attention: Application Branch.

A copy of this notice __MUST__ be returned with response.

For: Manager, Application Branch
(703) 557-3254

FORM PTO-1533 (REV. 7-87)

OFFICE COPY

For Office Use Only	
☐ 102	☐ 202
☐ 103	☐ 203
☐ 104	☐ 204
☐ 105	☐ 205

Fig. 10

DECLARATION FOR PATENT APPLICATION

Docket No. _____

As a below named inventor, I hereby declare that:

My residence, post office address and citizenship are as stated below next to my name.

I believe I am the original, first and sole inventor (if only one name is listed below) or an original, first and joint inventor (if plural names are listed below) of the subject matter which is claimed and for which a patent is sought on the invention entitled
_____, the specification of which

(check one) ☐ is attached hereto.
      ☐ was filed on _____ as
      Application Serial No. _____
      and was amended on _____ (if applicable).

I hereby state that I have reviewed and understand the contents of the above identified specification, including the claims, as amended by any amendment referred to above.

I acknowledge the duty to disclose information which is material to the examination of this application in accordance with Title 37, Code of Federal Regulations, §1.56(a).

I hereby claim foreign priority benefits under Title 35, United States Code, §119 of any foreign application(s) for patent or inventor's certificate listed below and have also identified below any foreign application for patent or inventor's certificate having a filing date before that of the application on which priority is claimed:

Prior Foreign Application(s)

Priority Claimed

(Number)	(Country)	(Day/Month/Year Filed)	Yes	No
(Number)	(Country)	(Day/Month/Year Filed)	Yes	No
(Number)	(Country)	(Day/Month/Year Filed)	Yes	No

I hereby claim the benefit under Title 35, United States Code, §120 of any United States application(s) listed below and, insofar as the subject matter of each of the claims of this application is not disclosed in the prior United States application in the manner provided by the first paragraph of Title 35, United States Code, §112, I acknowledge the duty to disclose material information as defined in Title 37, Code of Federal Regulations, §1.56(a) which occurred between the filing date of the prior application and the national or PCT international filing date of this application:

(Application Serial No.)	(Filing Date)	(Status—patented, pending, abandoned)
(Application Serial No.)	(Filing Date)	(Status—patented, pending, abandoned)

I hereby appoint the following attorney(s) and/or agent(s) to prosecute this application and to transact all business in the Patent and Trademark Office connected therewith:
_____.

Address all telephone calls to _____ at telephone no. _____.
Address all correspondence to _____
_____
_____
_____

I hereby declare that all statements made herein of my own knowledge are true and that all statements made on information and belief are believed to be true; and further that these statements were made with the knowledge that willful false statements and the like so made are punishable by fine or imprisonment, or both, under Section 1001 of Title 18 of the United States Code and that such willful false statements may jeopardize the validity of the application or any patent issued thereon.

Full name of sole or first inventor _____
Inventor's signature _____ Date _____
Residence _____ Citizenship _____
Post Office Address _____

Full name of second joint inventor, if any _____
Second Inventor's signature _____ Date _____
Residence _____ Citizenship _____
Post Office Address _____

(Supply similar information and signature for third and subsequent joint inventors.)

Form PTO-FB-A110 (8-83)   Fig. 11

POWER OF ATTORNEY OR AUTHORIZATION OF AGENT, NOT ACCOMPANYING APPLICATION

TO THE COMMISSIONER OF PATENTS AND TRADEMARKS:

The undersigned having, on or about the .... day of ...., 19 .... , made application for letters patent for an improvement in ...., Serial Number..................................., hereby appoints .... of .... , State of .... , Registration No.................................... , his attorney (or agent), to prosecute said application, and to transact all business in the Patent and Trademark Office connected therewith.

....
(Signature)

REVOCATION OF POWER OF ATTORNEY OR AUTHORIZATION OF AGENT

TO THE COMMISSIONER OF PATENTS AND TRADEMARKS:

The undersigned having, on or about the .... day of ...., 19...................... , appointed.. , of.. , State of .... , his attorney (or agent) to prosecute an application for letters patent which application was filed on or about the... day of.. , 19 ..., for an improvement in .... , Serial Number .... , hereby revokes the power of attorney (or authorization of agent) then given.

....
(Signature)

Fig. 12 (A, B)

Applicant or Patentee:_____ Attorney's
Serial or Patent No.:_____ Docket No.:_____
Filed or Issued:_____
For:_____

VERIFIED STATEMENT (DECLARATION) CLAIMING SMALL ENTITY STATUS
(37 CFR 1.9(f) & 1.27(c)) - NONPROFIT ORGANIZATION

I hereby declare that I am an official empowered to act on behalf of the nonprofit organization identified below:

NAME OF ORGANIZATION _____

ADDRESS OF ORGANIZATION _____

_____

TYPE OF ORGANIZATION

[ ] UNIVERSITY OR OTHER INSTITUTION OF HIGHER EDUCATION

[ ] TAX EXEMPT UNDER INTERNAL REVENUE SERVICE CODE (26 U.S.C. 501(a) and 501(c)(3)

[ ] NONPROFIT SCIENTIFIC OR EDUCATIONAL UNDER STATUTE OF STATE OF THE UNITED STATES OF AMERICA

(NAME OF STATE _____)

(CITATION OF STATUTE _____)

[ ] WOULD QUALIFY AS TAX EXEMPT UNDER INTERNAL REVENUE SERVICE CODE (26 U.S.C. 501(a) and 501(c) IF LOCATED IN THE UNITED STATES OF AMERICA

[ ] WOULD QUALIFY AS NONPROFIT SCIENTIFIC OR EDUCATIONAL UNDER STATUTE OF STATE OF THE UNITED STATES OF AMERICA IF LOCATED IN THE UNITED STATES OF AMERICA

(NAME OF STATE _____)

(CITATION OF STATUTE _____)

I hereby declare that the nonprofit organization identified above qualifies as a small business concern as defined in 37 CFR 1.9(e) for purposes of paying reduced fees under section 41(a) and (b) of Title 35, United States Code with regard to the invention entitled _____ by inventor(s)_____

described in

[ ] the specification filed herewith

[ ] application serial no._____ , filed _____

[ ] patent no._____ , issued _____

I hereby declare that rights under contract or law have conveyed to and remain with the nonprofit organization with regard to the above identified invention.

If the rights held by nonprofit organization are not exclusive, each individual, concern or organization having rights to the invention is listed below* and no rights to the invention are held by any person, other than the inventor, who would not qualify as a small business concern under 37 CFR 1.9(d) or by any concern which would not qualify as a small business concern under 37 CFR 1.9(d) or a nonprofit organization under 37 CFR 1.9(e).

*NOTE: Separate verified statements are required from each named person, concern or organization having rights to the invention averring to their status as small entities. (37 CFR 1.27)

NAME _____

ADDRESS _____

[ ] INDIVIDUAL   [ ] SMALL BUSINESS CONCERN   [ ] NONPROFIT ORGANIZATION

NAME _____

ADDRESS _____

[ ] INDIVIDUAL   [ ] SMALL BUSINESS CONCERN   [ ] NONPROFIT ORGANIZATION

I acknowledge the duty To file, in this application or patent, notification of any change in status resulting in loss of entitlement to small entity status prior to paying, or at the time of paying, the earliest of the issue fee or any maintenance fee due after the date an which status as a small entity is no longer appropriate. (37 CFR 1.28(b))

I hereby declare that all statements made herein of my own knowledge are true and that all statements made on information and belief are believed to be true; and further that these statements were wade with the knowledge that willful false statements and the like so made are punishable by fine or imprisonment, or both, under section 1001 of Title 18 of the United States Code, and that such willful false statements may jeopardize the validity of the application, any patent issuing thereon, or any patent to which this verified statement is directed.

NAME OF PERSON SIGNING _____

TITLE IN ORGANIZATION _____

ADDRESS OF PERSON SIGNING _____

_____

SIGNATURE _____ DATE _____

Fig. 13

Rev. 11, Apr. 1989

VERIFIED STATEMENT (DECLARATION) BY A NON-INVENTOR
SUPPORTING A CLAIM BY ANOTHER FOR SMALL ENTITY STATUS

I hereby declare that I am making this verified statement to support a claim by _____ for small entity status for purposes of paying reduced fees under section 41(a) and (b) of Title 35, United States Code, with regard to the invention entitled _____ by inventor(s) _____

[ ] the specification filed herewith
[ ] application serial number _____, filed _____
[ ] patent number _____, issued _____ .

I hereby declare that I would qualify as an independent inventor as defined in 37 CFR 1.9(c) for purposes of paying fees under section 41(a) and (b) of Title 35, United States Code, if I had made the above identified invention.

I have not assigned, granted, conveyed or licensed and am under no obligation under contract or law to assign, grant, convey or license, any rights to the invention to any person who could not be classified as an independent inventor under 37 CFR 1.9(c) if that person had made the invention, or to any concern which would not qualify as a small business concern under 37 CFR 1.9(d) or a nonprofit organization under 37 CFR 1.9(a).

Each person, concern or organization to which I have assigned, granted, conveyed, or licensed or am under an obligation under contract or law to assign, grant, convey, or license any rights in the invention is listed below:

[ ] No such person, concern, or organization
[ ] Persons, concerns or organizations listed below*

* Note: Separate verified statements are required from each named person, concern or organization having rights to the invention averring to their status as small entities. (37 CFR 1.27)

NAME _____
ADDRESS _____
[ ] INDIVIDUAL   [ ] SMALL BUSINESS CONCERN   [ ] NONPROFIT ORGANIZATION

NAME _____
ADDRESS _____
[ ] INDIVIDUAL   [ ] SMALL BUSINESS CONCERN   [ ] NONPROFIT ORGANIZATION

NAME _____
ADDRESS _____
[ ] INDIVIDUAL   [ ] SMALL BUSINESS CONCERN   [ ] NONPROFIT ORGANIZATION

I acknowledge the duty To file, in this application or patent, notification of any change in status resulting in loss of entitlement to small entity status prior to paying, or at the time of paying, the earliest of the issue fee or any maintenance fee due after the date an which status as a small entity is no longer appropriate. (37 CFR 1.28(b))

I hereby declare that all statements mad
e herein of my own knowledge are true and that all statements made on information and belief are believed to be true; and further that these statements were wade with the knowledge that willful false statements and the like so made are punishable by fine or imprisonment, or both, under section 1001 of Title 18 of the United States Code, and that such willful false statements may jeopardize the validity of the application, any patent issuing thereon, or any patent to which this verified statement is directed.

NAME OF PERSON SIGNING _____
ADDRESS OF PERSON SIGNING _____
SIGNATURE _____ DATE _____

Applicant or Patentee:_____ Attorney's
Serial or Patent No.:_____ Docket No.:_____ Filed or Is-
sued:_____
For:_____

VERIFIED STATEMENT (DECLARATION) CLAIMING SMALL ENTITY STATUS
(37 CFR 1.9(f) & 1.27(c)) - SMALL BUSINESS CONCERN

I hereby declare that I am

[ ] the owner of the small business concern identified below:
[ ] an official of the small business concern empowered to act on behalf of the concern identified below:
NAME OF CONCERN _____
ADDRESS OF CONCERN _____
_____

    I hereby declare that the above identified small business concern qualifies as a small business concern as defined in 13 CFR 121.12, and reproduced in 37 CFR 1.9(d), for purposes of paying reduced fees under section 41(a) and (b) of Title 35, United States Code, in that the number of employees of the concern, including those of its affiliates, does not exceed 500 persons. For purposes of this statement, (1) the number of employees of the business concern is the average over the previous fiscal year of the concern of the persons employed on a full-time, part-time or temporary basis during each of the pay periods of the fiscal year, and (2) concerns are affiliates of each other when either, directly or indirectly, one concern controls or has the power to control the other, or a third party or parties controls or has the power to control both.

    I hereby declare that rights under contract or law have been conveyed to and remain with the small business concern identified above with regard to the invention, entitled _____
_____by inventor(s)

described in

[ ] the specification filed herewith
[ ] application serial no._____ , filed _____
[ ] patent no._____ , issued _____
    If the rights held by the above identified small business concern are not exclusive, each individual, concern or organization having rights to the invention is listed below and no rights to the invention are held by any person, other than the inventor, who would not qualify as a small business concern under 37 CFR 1.9(d) or by any concern which would not qualify as a small business concern under 37 CFR 1.9(d) or a nonprofit organization under 37 CFR 1.9(e). NOTE: Separate verified statements are required from each named person, concern or organization having rights to the invention averring to their status as small entities. (37 CFR 1.27)

NAME _____
ADDRESS _____
  [ ] INDIVIDUAL   [ ] SMALL BUSINESS CONCERN   [ ] NONPROFIT ORGANIZATION

NAME _____
ADDRESS _____
  [ ] INDIVIDUAL   [ ] SMALL BUSINESS CONCERN   [ ] NONPROFIT ORGANIZATION

    I acknowledge the duty To file, in this application or patent, notification of any change in status resulting in loss of entitlement to small entity status prior to paying, or at the time of paying, the earliest of the issue fee or any maintenance fee due after the date an which status as a small entity is no longer appropriate. (37 CFR 1.28(b))

    I hereby declare that all statements made herein of my own knowledge are true and that all statements made on information and belief are believed to be true; and further that these statements were wade with the knowledge that willful false statements and the like so made are punishable by fine or imprisonment, or both, under section 1001 of Title 18 of the United States Code, and that such willful false statements may jeopardize the validity of the application, any patent issuing thereon, or any patent to which this verified statement is directed.

NAME OF PERSON SIGNING _____
TITLE OF PERSON OTHER THAN OWNER _____
ADDRESS OF PERSON SIGNING _____
_____

SIGNATURE _____ DATE _____

Applicant or Patentee:_____ Attorney's

Serial or Patent No.:_____ _____ Docket No.:_____

Filed or Issued:_____

For:_____

VERIFIED STATEMENT (DECLARATION) CLAIMING SMALL ENTITY STATUS
(37 CFR 1.9(f) & 1.27(c)) - INDEPENDENT INVENTOR

As a below named inventor, I hereby declare that I qualify as an independent inventor as defined in 37 CFR 1.9(c) for purposes of paying reduced fees under section 41(a) and (b) of Title 35, United States Code, to the Patent and Trademark Office with regard to the invention entitled _____

described in

    [ ] the specification filed herewith

    [ ] application serial number _____, filed _____

    [ ] patent number _____, issued _____ .

I have not assigned, granted, conveyed or licensed and am under no obligation under contract or law to assign, grant, convey or license, any rights to the invention to any person who could not be classified as an independent inventor under 37 CFR 1.9(c) if that person made the invention, or to any concern which would not qualify as a small business concern under 37 CFR 1.9(d) or a nonprofit organization under 37 CFR 1.9(a).

Each person, concern or organization to which I have assigned , granted, conveyed, or licensed or am under an obligation under contract or law to assign, grant, convey, or license any rights in the invention is listed below:

    [ ] No such person, concern, or organization

    [ ] Persons, concerns or organizations listed below*

    * Note: Separate verified statements are required from each named person, concern or organization having rights to the invention averring to their status as small entities. (37 CFR 1.27)

NAME _____

ADDRESS _____

    [ ] INDIVIDUAL   [ ] SMALL BUSINESS CONCERN   [ ] NONPROFIT ORGANIZATION

NAME _____

ADDRESS _____

    [ ] INDIVIDUAL   [ ] SMALL BUSINESS CONCERN   [ ] NONPROFIT ORGANIZATION

NAME _____

ADDRESS _____

    [ ] INDIVIDUAL   [ ] SMALL BUSINESS CONCERN   [ ] NONPROFIT ORGANIZATION

    I acknowledge the duty To file, in this application or patent, notification of any change in status resulting in loss of entitlement to small entity status prior to paying, or at the time of paying, the earliest of the issue fee or any maintenance fee due after the date an which status as a small entity is no longer appropriate. (37 CFR 1.28(b))

    I hereby declare that all statements made herein of my own knowledge are true and that all statements made on information and belief are believed to be true; and further that these statements were wade with the knowledge that willful false statements and the like so made are punishable by fine or imprisonment, or both, under section 1001 of Title 18 of the United States Code, and that such willful false statements may jeopardize the validity of the application, any patent issuing thereon, or any patent to which this verified statement is directed.

_____   _____   _____

NAME OF INVENTOR   NAME OF INVENTOR   NAME OF INVENTOR

_____   _____   _____

Signature of inventor   Signature of inventor   Signature of inventor

_____   _____   _____

Date   Date   Date

Fig. 16

Rev. 11, Apr. 1989

INFORMATION DISCLOSURE CITATION

(Use several sheets if necessary)

ATTY. DOCKET NO.	SERIAL NO.
APPLICANT	
FILING DATE	GROUP

U.S. PATENT DOCUMENTS

*EXAMINER INITIAL		DOCUMENT NUMBER	DATE	NAME	CLASS	SUBCLASS	FILING DATE IF APPROPRIATE

FOREIGN PATENT DOCUMENTS

	DOCUMENT NUMBER	DATE	COUNTRY	CLASS	SUBCLASS	TRANSLATION YES	NO

OTHER DOCUMENTS *(Including Author, Title, Date, Pertinent Pages, Etc.)*

EXAMINER	DATE CONSIDERED

*EXAMINER: Initial if reference considered, whether or not citation is in conformance with MPEP 609; Draw line through citation if not in conformance and not considered. Include copy of this form with next communication to applicant.

Form PTO-FB-A820
(also form PTO-1449)

Patent and Trademark Office - U.S. DEPARTMENT of COMMERCE

Fig. 17

FORM PTO-892 (REV. 3-78)	U.S. DEPARTMENT OF COMMERCE PATENT AND TRADEMARK OFFICE	SERIAL NO.	GROUP ART UNIT	ATTACHMENT TO PAPER NUMBER	
NOTICE OF REFERENCES CITED		APPLICANT(S)			

U.S. PATENT DOCUMENTS

*		DOCUMENT NO.	DATE	NAME	CLASS	SUB-CLASS	FILING DATE IF APPROPRIATE
	A						
	B						
	C						
	D						
	E						
	F						
	G						
	H						
	I						
	J						
	K						

FOREIGN PATENT DOCUMENTS

*		DOCUMENT NO.	DATE	COUNTRY	NAME	CLASS	SUB-CLASS	PERTINENT SHTS. DWG	PP. SPEC.
	L								
	M								
	N								
	O								
	P								
	Q								

OTHER REFERENCES (Including Author, Title, Date, Pertinent Pages, Etc.)

	R	
	S	
	T	
	U	

EXAMINER	DATE	

* A copy of this reference is not being furnished with this office action.
(See Manual of Patent Examining Procedure, section 707.05 (a).)

Fig. 18

☐ This application has been examined   ☐ Responsive to communication filed on _____   ☐ This action is made final.

A shortened statutory period for response to this action is set to expire_____ month(s), _____ days from the date of this letter.
Failure to respond within the period for response will cause the application to become abandoned.   35 U.S.C. 133

Part I   THE FOLLOWING ATTACHMENT(S) ARE PART OF THIS ACTION:

1. ☐ Notice of References Cited by Examiner, PTO-892.   2. ☐ Notice re Patent Drawing, PTO-948.
3. ☐ Notice of Art Cited by Applicant, PTO-1449.   4. ☐ Notice of Informal Patent Application, Form PTO-152.
5. ☐ Information on How to Effect Drawing Changes, PTO-1474.   6. ☐ _____

Part II   SUMMARY OF ACTION

1. ☐ Claims _____ are pending in the application.

     Of the above, claims _____ are withdrawn from consideration.

2. ☐ Claims _____ have been cancelled.

3. ☐ Claims _____ are allowed.

4. ☐ Claims _____ are rejected.

5. ☐ Claims _____ are objected to.

6. ☐ Claims _____ are subject to restriction or election requirement.

7. ☐ This application has been filed with informal drawings under 37 C.F.R. 1.85 which are acceptable for examination purposes.

8. ☐ Formal drawings are required in response to this Office action.

9. ☐ The corrected or substitute drawings have been received on _____ . Under 37 C.F.R. 1.84 these drawings
     are ☐ acceptable. ☐ not acceptable (see explanation or Notice re Patent Drawing, PTO-948).

10. ☐ The proposed additional or substitute sheet(s) of drawings, filed on _____ has (have) been ☐ approved by the
     examiner. ☐ disapproved by the examiner (see explanation).

11. ☐ The proposed drawing correction, filed on _____, has been ☐ approved. ☐ disapproved (see explanation).

12. ☐ Acknowledgment is made of the claim for priority under U.S.C. 119. The certified copy has ☐ been received ☐ not been received
     ☐ been filed in parent application, serial no. _____ ; filed on _____ .

13. ☐ Since this application appears to be in condition for allowance except for formal matters, prosecution as to the merits is closed in
     accordance with the practice under Ex parte Quayle, 1935 C.D. 11; 453 O.G. 213.

14. ☐ Other

EXAMINER'S ACTION

Fig. 19

AMENDMENT TRANSMITTAL LETTER			ATTORNEY'S DOCKET NO.
SERIAL NO.	FILING DATE	EXAMINER	GROUP ART UNIT
INVENTION			

TO THE COMMISSIONER OF PATENTS AND TRADEMARKS:

Transmitted herewith is an amendment in the above-identified application.

Small entity status of this application under 37 CFR 1.27 has been established by a verified statement previously submitted.

A verified statement to establish small entity status under 37 CFR 1.9 and 1.27 is enclosed.

No additional fee is required.

The fee has been calculated as shown below:

	(1) CLAIMS REMAINING AFTER AMENDMENT		(2) HIGHEST NO. PREVIOUSLY PAID FOR	(3) PRESENT EXTRA	SMALL ENTITY			OR	OTHER THAN A SMALL ENTITY	
					RATE	ADDIT FEE			RATE	ADDIT FEE
TOTAL	*	MINUS	**	-	x $6 =	$			x $12 =	$
INDEP	*	MINUS	***	-	x $17 =	$			x $34 =	$
FIRST PRESENTATION OF MULTIPLE DEP. CLAIM					+$55 =	$			+$110 =	$
					TOTAL ADDIT. FEE	$	OR		TOTAL	$

  * If the entry in Col 1 is less than the entry in Col. 2, write "0" in Col. 3.

 ** If the "Highest No Previously Paid For" IN THIS SPACE is less than 20, enter "20".

*** If the "Highest No Previously Paid For" IN THIS SPACE is less than 3, enter "3".

The "Highest No Previously Paid For" (Total or Indep.) is the highest number found in the appropriate box in Col 1

Please charge my Deposit Account No. _____ in the amount of $ _____ .
A duplicate copy of this sheet is enclosed.

A check in the amount of $ _____ to cover the filing fee is enclosed.

The Commissioner is hereby authorized to charge payment of the following fees associated with this communication or credit any overpayment to Deposit Account No. _____ . A Duplicate copy of this sheet is enclosed.

Any additional filing fees required under 37 CFR 1.16.

Any patent application processing fees under 37 CFR 1.17

_____   _____
(date)   (signature)

Form PTO-FB-A520 (10-85)
(also form PTO-1083)   Fig. 20

Patent and Trademark Office - U.S. DEPARTMENT of COMMERCE

Channels of Ex Parte Review

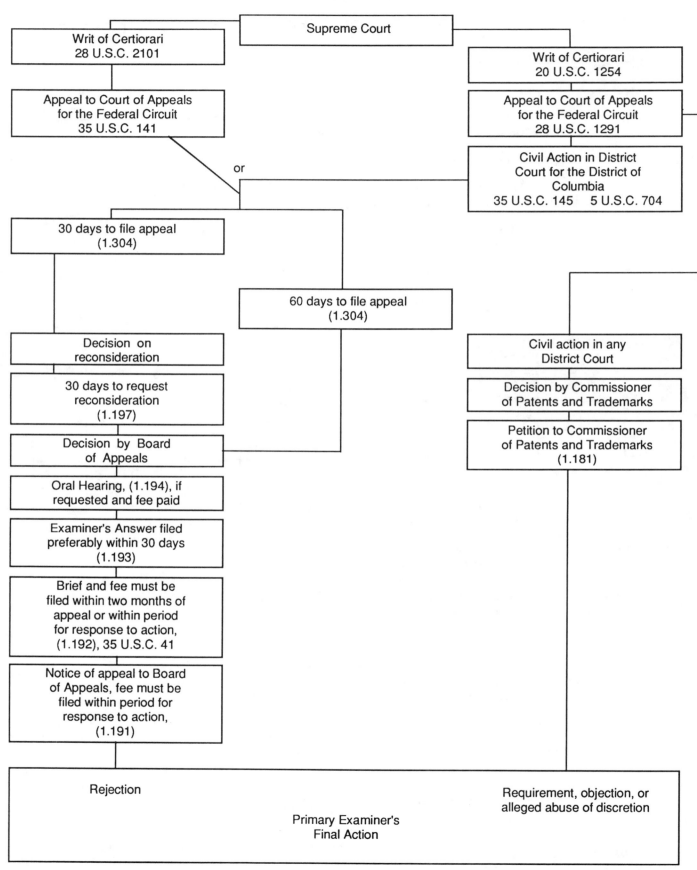

Fig. 21

ASSIGNMENT OF PATENT

(No special form is prescribed for assignments, which may contain various provisions depending upon the agreement of the parties. The following two formsare specimens of assignments which have been used in some cases.)

WHEREAS, I,................................, of.............................. , did obtain Letters Patent of the United States for an improvement in .... No................................, dated .... ; and whereas, I am now the sole owner of said patent; and,

WHEREAS,................................ , of.............................. , whose post-office address is .... , City of.............................., and State of .... , is desirous of acquiring the entire interest in the same; NOW, THEREFORE, in consideration of the sum of........................ dollars ($........................), the receipt of which is hereby acknowledged, and other good and valuable considerations, I, ..., by these presents do sell, assign, and transfer unto the said .... the entire right, title, and interest in and to the said Letters Patent aforesaid; the same to be held and enjoyed by the said ...., for his own used and hehoof, and for his legal representa-

tives and assigns, to the full end of the term for which said Letters Patent are granted, as fully and entirely as the same would have been held by me had this assignment and sale not been made.

EXecuted, this .... day of ...., 19 ...., at .... STATE... County of .... ss:

Before me personally appeared said... and acknowledged the foregoing instruments to be his free act and deed this............................. day of.............................,

(NOTARY PUBLIC)

[SEAL]

ASSIGNMENT OF APPLICATION

Whereas, I,................................ , of.............................. have invented certain new anduseful Improvements in ...., for which an application for United States Letters Patent was filed on..., Serial No.................., [if the application has been prepared but not yet filed, state "for which an application for United States Letters Patent was executed on ....," instead] and

Wheareas,................................ , of.............................. , whose post-office address is .... , is desirous of acquiring the entire right, title and interest in the same;

Now, THEREFORE, in consideration of the sum of........................ dollars ($........................), the receipt whereof is hereby acknowledged, and other good and valuable consideration, I, the said ...., by these presents do sell, assign nd transfer unto said ...., the full and exclusive right to the said invention in the United States and the entire right, title, and interest in and to any and all Letters Patent which may be granted, therefore in the United States.

I hereby authorize and request the Commissioner of Patents and Trademarks to issue said Letters Patent to said ...., as the assignee of the entire right, title, and interest in and to the same, for his sole use and behoof; and for the use and behoof of his legal representatives, to the full end of the term for which said Letters Patent may be granted, as fully and entirely as the same would have been held by me had this assignment and sale not be made.

Excuted this .... day of........................ , 19 ...., at .... State of .... County of ....

ss:

Before me personally appeared said......;... and acknowledged the foregoing instrument to be his free act an deed this........................ day of .... , 19.....

....
(Notary Public)

[SEAL]

Fig. 22 (A,B)

Patent Worksheet

TRADE MARK (WORKING): _____

TAG LINE:_____

DESCRIPTION:_____

_____

_____

_____

PATENT NOTES:_____

_____

_____LODE #_____

WHO/DATE CONCEIVED:_____

WITNESSED:_____

SKETCH/PHOTO:

NOTES:_____

_____

_____

LICENSES/TIE-INS/SPIN-OFFS/ACCESSORIES/LINE CONCEPTS:____

_____

_____

_____

POTENTIAL MANUFACTURERS:_____

_____

_____

SEEN BY/DATE: _____   _____

_____   _____   _____

_____   _____   _____

Fig. 23

TRADEMARK EXAMINATION ACTIVITIES*

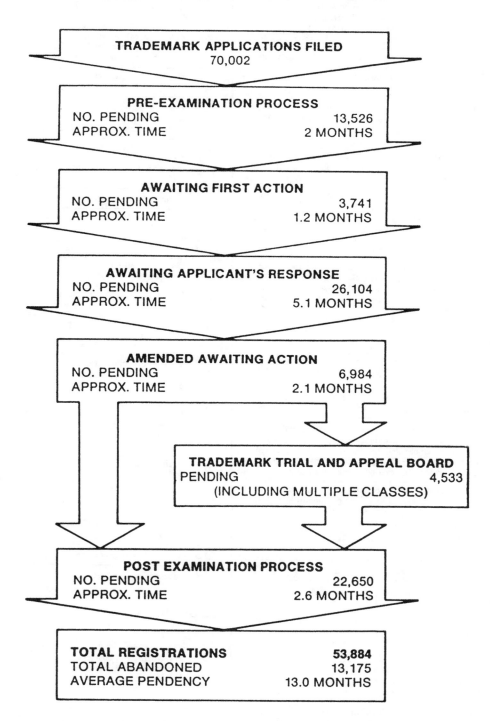

TRADEMARK APPLICATIONS FILED
70,002

PRE-EXAMINATION PROCESS
NO. PENDING 13,526
APPROX. TIME 2 MONTHS

AWAITING FIRST ACTION
NO. PENDING 3,741
APPROX. TIME 1.2 MONTHS

AWAITING APPLICANT'S RESPONSE
NO. PENDING 26,104
APPROX. TIME 5.1 MONTHS

AMENDED AWAITING ACTION
NO. PENDING 6,984
APPROX. TIME 2.1 MONTHS

TRADEMARK TRIAL AND APPEAL BOARD
PENDING 4,533
(INCLUDING MULTIPLE CLASSES)

POST EXAMINATION PROCESS
NO. PENDING 22,650
APPROX. TIME 2.6 MONTHS

TOTAL REGISTRATIONS **53,884**
TOTAL ABANDONED 13,175
AVERAGE PENDENCY 13.0 MONTHS

*This figure represents a simplified work-flow diagram with statistics on cases at various stages of processing at the end of FY 1987.

Fig. 24

International schedule of classes of goods and services

Goods

1 — Chemicals used in industry, science, photography, as well as in agriculture, horticulture, and forestry; unprocessed artificial resins; unprocessed plastics; manures; fire extinguishing compositions; tempering and soldering preparations; chemical substances for preserving foodstuffs; tanning substances; adhesives used in industry.

2 — Paints, varnishes, lacquers; preservatives against rust and against deterioration of wood; colourants; mordants; raw natural resins; metals in foil and powder form for painters, decorators, printers and artists.

3 — Bleaching preparations and other substances for laundry use; cleaning, polishing, scouring and abrasive preparations; soaps; perfumery, essential oils, cosmetics, hair lotions; dentifrices.

4 — Industrial oils and greases; lubricants; dust absorbing, wetting and binding compositions; fuels (including motor spirit) and illuminants; candles, wicks.

5 — Pharmaceutical, veterinary, and sanitary preparations; dietetic substances adapted for medical use, food for babies; plasters, materials for dressings material for stopping teeth, dental wax, disinfectants; preparations for destroying vermin; fungicides, herbicides.

6 — Common metals and their alloys; metal building materials; transportable buildings of metal; materials of metal for railway tracks; non-electric cables and wires of common metal; ironmongery, small items of metal hardware; pipes and tubes of metal; safes; goods of common metal not included in other classes; ores

7 — Machines and machine tools; motors (except for land vehicles); machine coupling and belting (except for land vehicles); agricultural implements; incubators for eggs.

8 — Hand tools and implements (hand operated); cutlery; side arms; razors.

9 — Scientific, nautical, surveying, electric, photographic, cinematographic, optical, weighing, measuring, signalling, checking (supervision), life-saving and teaching apparatus and instruments; apparatus for recording transmission or reproduction of sound or images; magnetic data carriers, recording discs; automatic vending machines and mechanisms for coin-operated apparatus; cash registers, calculating machines, data processing equipment and computers; fire-extinguishing apparatus.

10 — Surgical, medical, dental, and veterinary apparatus and instruments, artificial limbs, eyes and teeth; orthopedic articles; suture materials.

11 — Apparatus for lighting, heating, steam generating, cooking, refrigerating, drying, ventilating, water supply, and sanitary purposes.

12 — Vehicles; apparatus for locomotion by land, air or water.

13 — Firearms; ammunition and projectiles; explosives; fireworks.

14 — Precious metals and their alloys and goods in precious metals or coated therewith, not included in other classes; jewelry, precious stones; horological and other chronometric instruments.

15 — Musical instruments.

16 — Paper and cardboard and goods made from these materials, not included in other classes; printed matter; bookbinding material; photographs; stationery; adhesives for stationery or household purposes; artists' materials; paint brushes; typewriters and office requisites (except furniture); instructional and teaching material (except apparatus); plastic materials for packaging (not included on other classes); playing cards; printers' type; printing blocks.

17 — Rubber, gutta-percha, gum, asbestos, mica and goods made from these materials and not included in other classes; plastics in extruded form for use in manufacture; packing, stopping and insulating materials; flexible pipes, not of metal.

18 — Leather and imitations of leather, and goods made from these materials and not included in other classes; animal skins, hides; trunks and travelling bags; umbrellas, parasols and walking sticks; whips, harness and saddlery.

19 — Building materials (non-metallic); non-metallic rigid pipes for building; asphalt, pitch and bitumen; non-metallic transportable buildings; monuments, not of metal.

20 — Furniture, mirrors, picture frames; goods (not included in other classes) of wood, cork, reed, cane, wicker, horn, bone, ivory, whalebone, shell, amber, mother-of-pearl, meerschaum and substitutes for all these materials, or of plastics.

21 — Household or kitchen utensils and containers (not of precious metal or coated therewith); combs and sponges; brushes (except paint brushes); brush-making materials; articles for cleaning purposes; steel wool; unworked or semi-worked glass (except glass used in building); glassware, porcelain and earthenware, not included in other classes.

22 — Ropes, string, nets, tents, awnings, tarpaulins, sails, sacks; and bags (not included other classes); padding and stuffing materials (except of rubber or plastics); raw fibrous textile materials.

23 — Yarns and threads, for textile use.

24 — Textile and textile goods, not included in other classes; bed and table covers.

25 — Clothing, footwear, headgear.

26 — Lace and embroidery, ribbons and braid; buttons, hooks and eyes, pins and needles; artificial flowers.

27 — Carpets, rugs, mats and matting; linoleum and other materials for covering existing floors; wall hangings (non-textile).

28 — Games and playthings; gymnastic and sporting articles not included in other classes; decorations for Christmas trees.

29 — Meats, fish, poultry and game; meat extracts; preserved, dried and cooked fruits and vegetables; jellies, jams; eggs, milk and milk products; edible oils and fats; salad dressings; preserves.

30 — Coffee, tea, cocoa, sugar, rice, tapioca, sago, artificial coffee; flour, and preparations made from cereals, bread, pastry and confectionery, ices; honey, treacle; yeast, baking-powder; salt, mustard, vinegar, sauces, (except salad dressings) spices; ice.

31 — Agricultural, horticultural and forestry products and grains not included in other classes; living animals; fresh fruits and vegetables; seeds, natural plants and flowers; foodstuffs for animals, malt.

32 — Beers; mineral and aerated waters and other non-alcoholic drinks; fruit drinks and fruit juices; syrups and other preparations for making beverages.

33 — Alcholic beverages (except beers).

34 — Tobacco; smokers' articles; matches.

Services

35 — Advertising and business.

36 — Insurance and financial.

37 — Construction and repair.

38 — Communication.

39 — Transportation and storage.

40 — Material treatment.

41 — Education and entertainment.

42 — Miscellaneous.

Fig. 25

TRADEMARK/SERVICE MARK APPLICATION, PRINCIPAL REGISTER, WITH DECLARATION	MARK (Identify the mark)
	CLASS NO. (If known)

TO THE ASSISTANT SECRETARY AND COMMISSIONER OF PATENTS AND TRADEMARKS:

APPLICANT NAME:

APPLICANT BUSINESS ADDRESS:

APPLICANT ENTITY: (Check one and supply requested information)

☐ Individual - Citizenship: (Country) _____

☐ Partnership - Partnership Domicile: (State and Country) _____
Names and Citizenship (Country) of General Partners: _____
_____

☐ Corporation - State (Country, if appropriate) of Incorporation: _____

☐ Other: (Specify Nature of Entity and Domicile) _____
_____

GOODS AND/OR SERVICES:

Applicant requests registration of the above-identified trademark/service mark shown in the accompanying drawing in the United States Patent and Trademark Office on the Principal Register established by the Act of July 5, 1946 (15 U.S.C. 1051 et. seq., as amended.) for the following goods/services: _____
_____
_____

BASIS FOR APPLICATION: (Check one or more, but NOT both the first AND second boxes, and supply requested information)

☐ Applicant is using the mark in commerce on or in connection with the above identified goods/services. (15 U.S.C. 1051(a), as amended.) Three specimens showing the mark as used in commerce are submitted with this application.
- Date of first use of the mark anywhere: _____
- Date of first use of the mark in commerce which the U.S. Congress may regulate: _____
- Specify the type of commerce: _____
  (e.g., interstate, between the U.S. and a specified foreign country)
- Specify manner or mode of use of mark on or in connection with the goods/services: _____
_____
  (e.g., trademark is applied to labels, service mark is used in advertisements)

☐ Applicant has a bona fide intention to use the mark in commerce on or in connection with the above identified goods/services. (15 U.S.C. 1051(b), as amended.)
- Specify intended manner or mode of use of mark on or in connection with the goods/services: _____
_____
  (e.g., trademark will be applied to labels, service mark will be used in advertisements)

☐ Applicant has a bona fide intention to use the mark in commerce on or in connection with the above identified goods/services, and asserts a claim of priority based upon a foreign application in accordance with 15 U.S.C. 1126(d), as amended.
- Country of foreign filing: _____ - Date of foreign filing: _____

☐ Applicant has a bona fide intention to use the mark in commerce on or in connection with the above identified goods/services and, accompanying this application, submits a certification or certified copy of a foreign registration in accordance with 15 U.S.C. 1126(e), as amended.
- Country of registration: _____ - Registration number: _____

Note: Declaration, on Reverse Side, MUST be Signed

<table>
<tr>
<td>

AMENDMENT TO ALLEGE USE UNDER 37 CFR 2.76, WITH DECLARATION

</td>
<td>

MARK (Identify the mark)

SERIAL NO.

</td>
</tr>
</table>

TO THE ASSISTANT SECRETARY AND COMMISSIONER OF PATENTS AND TRADEMARKS:

APPLICANT NAME:

Applicant requests registration of the above-identified trademark/service mark in the United States Patent and Trademark Office on the Principal Register established by the Act of July 5, 1946 (15 U.S.C. 1051 et. seq., as amended). Three specimens showing the mark as used in commerce are submitted with this amendment.

☐   Check here if Request to Divide under 37 CFR 2.87 is being submitted with this amendment.

Applicant is using the mark in commerce on or in connection with the following goods/services:

_____

_____

(NOTE: Goods/services listed above may not be broader than the goods/services identified in the application as filed)

Date of first use of mark anywhere: _____

Date of first use of mark in commerce
which the U.S. Congress may regulate: _____

Specify type of commerce: (e.g., interstate, between the U.S. and a specified foreign country) _____

_____

Specify manner or mode of use of mark on or in connection with the goods/services: (e.g., trademark is applied to labels, service mark is used in advertisements) _____

_____

The undersigned being hereby warned that willful false statements and the like so made are punishable by fine or imprisonment, or both, under 18 U.S.C. 1001, and that such willful false statements may jeopardize the validity of the application or any resulting registration, declares that he/she is properly authorized to execute this Amendment to Allege Use on behalf of the applicant; he/she believes the applicant to be the owner of the trademark/service mark sought to be registered; the trademark/ service mark is now in use in commerce; and all statements made of his/her own knowledge are true and all statements made on information and belief are believed to be true.

_____
Date

_____
Telephone Number

_____
Signature

_____
Print or Type Name and Position

| STATEMENT OF USE UNDER 37 CFR 2.88, WITH DECLARATION | MARK (Identify the mark) |
| | SERIAL NO. |

TO THE ASSISTANT SECRETARY AND COMMISSIONER OF PATENTS AND TRADEMARKS:

APPLICANT NAME:

NOTICE OF ALLOWANCE ISSUE DATE:

Applicant requests registration of the above-identified trademark/service mark in the United States Patent and Trademark Office on the Principal Register established by the Act of July 5, 1946 (15 U.S.C. 1051 et. seq., as amended). Three (3) specimens showing the mark as used in commerce are submitted with this statement.

☐   Check here only if a Request to Divide under 37 CFR 2.87 is being submitted with this Statement.

Applicant is using the mark in commerce on or in connection with the following goods/services: (Check One)

☐   Those goods/services identified in the Notice of Allowance in this application.

☐   Those goods/services identified in the Notice of Allowance in this application except: (Identify goods/services to be deleted from application) _____

_____

_____

Date of first use of mark anywhere: _____

Date of first use of mark in commerce
which the U.S. Congress may regulate: _____

Specify type of commerce: (e.g., interstate, between the U.S. and a specified foreign country) _____

_____

Specify manner or mode of use of mark on or in connection with the goods/services: (e.g., trademark is applied to labels, service mark is used in advertisements) _____

_____

The undersigned being hereby warned that willful false statements and the like so made are punishable by fine or imprisonment, or both, under 18 U.S.C. 1001, and that such willful false statements may jeopardize the validity of the application or any resulting registration, declares that he/she is properly authorized to execute this Statement of Use on behalf of the applicant; he/she believes the applicant to be the owner of the trademark/service mark sought to be registered; the trademark/service mark is now in use in commerce; and all statements made of his/her own knowledge are true and all statements made on information and belief are believed to be true.

_____   _____
Date   Signature

_____   _____
Telephone Number   Print or Type Name and Position

<table>
<tr><td>REQUEST FOR EXTENSION OF TIME
UNDER 37 CFR 2.89 TO FILE A STATEMENT
OF USE, WITH DECLARATION</td><td>MARK (Identify the mark)

SERIAL NO.</td></tr>
</table>

TO THE ASSISTANT SECRETARY AND COMMISSIONER OF PATENTS AND TRADEMARKS:

APPLICANT NAME:

NOTICE OF ALLOWANCE MAILING DATE:

Applicant requests a six-month extension of time to file the Statement of Use under 37 CFR 2.88 in this application.

☐   Check here if a Request to Divide under 37 CFR 2.87 is being submitted with this request.

Applicant has a continued bona fide intention to use the mark in commerce in connection with the following goods/services: (Check one below)

☐   Those goods/services identified in the Notice of Allowance in this application.

☐   Those goods/services identified in the Notice of Allowance in this application except: (Identify goods/services to be **deleted** from application) _____

_____

This is the _____ request for an Extension of Time following mailing of the Notice of Allowance.
(Specify first - fifth)

If this is not the first request for an Extension of Time, check one box below. If the first box is checked, explain the circumstance(s) of the non-use in the space provided:

☐   Applicant has not used the mark in commerce yet on all goods/services specified in the Notice of Allowance; however, applicant has made the following ongoing efforts to use the mark in commerce on or in connection with each of the goods/services specified above:

_____

_____

If additional space is needed, please attach a separate sheet to this form

☐   Applicant believes that it has made valid use of the mark in commerce, as evidenced by the Statement of Use submitted with this request; however, if the Statement of Use is found by the Patent and Trademark Office to be fatally defective, applicant will need additional time in which to file a new statement.

The undersigned being hereby warned that willful false statements and the like so made are punishable by fine or imprisonment, or both, under 18 U.S.C. 1001, and that such willful false statements may jeopardize the validity of the application or any resulting registration, declares that he/she is properly authorized to execute this Request for Extension of Time to File a Statement of Use on behalf of the applicant; he/she believes the applicant to be the owner of the trademark/service mark sought to be registered; and all statements made of his/her own knowledge are true and all statements made on information and belief are believed to be true.

_____   _____
Date   Signature

_____   _____
Telephone Number   Print or Type Name and Position

PTO Form 1581 (REV. 9/89)
OMB No. 0651-0023

Fig. 29

DEPARTMENT OF COMMERCE/Patent and Trademark Office

Filling Out Application Form TX

Detach and read these instructions before completing this form. Make sure all applicable spaces have been filled in before you return this form.

BASIC INFORMATION

When to Use This Form: Use Form TX for registration of published or unpublished non-dramatic literary works, excluding periodicals or serial issues. This class includes a wide variety of works: fiction, non-fiction, poetry, textbooks, reference works, directories, catalogs, advertising copy, compilations of information, and computer programs. For periodicals and serials, use Form SE.

Deposit to Accompany Application: An application for copyright registration must be accompanied by a deposit consisting of copies or phonorecords representing the entire work for which registration is to be made. The following are the general deposit requirements as set forth in the statute:

Unpublished Work: Deposit one complete copy (or phonorecord).

Published Work: Deposit two complete copies (or phonorecords) of the best edition.

Work First Published Outside the United States: Deposit one complete copy (or phonorecord) of the first foreign edition.

Contribution to a Collective Work: Deposit one complete copy (or phonorecord) of the best edition of the collective work.

The Copyright Notice: For published works, the law provides that a copyright notice in a specified form "shall be placed on all publicly distributed copies from which the work can be visually perceived." Use of the copyright notice is the responsibility of the copyright owner and does not require advance permission from the Copyright Office. The required form of the notice for copies generally consists of three elements: (1) the symbol "©", or the word "Copyright," or the abbreviation "Copr."; (2) the year of first publication; and (3) the name of the owner of copyright. For example: "© 1981 Constance Porter." The notice is to be affixed to the copies "in such manner and location as to give reasonable notice of the claim of copyright."

For further information about copyright registration, notice, or special questions relating to copyright problems, write:

Information and Publications Section, LM-455
Copyright Office
Library of Congress
Washington, D.C. 20559

LINE-BY-LINE INSTRUCTIONS

1 SPACE 1: Title

Title of This Work: Every work submitted for copyright registration must be given a title to identify that particular work. If the copies or phonorecords of the work bear a title (or an identifying phrase that could serve as a title), transcribe that wording *completely* and *exactly* on the application. Indexing of the registration and future identification of the work will depend on the information you give here.

Previous or Alternative Titles: Complete this space if there are any additional titles for the work under which someone searching for the registration might be likely to look, or under which a document pertaining to the work might be recorded.

Publication as a Contribution: If the work being registered is a contribution to a periodical, serial, or collection, give the title of the contribution in the "Title of this Work" space. Then, in the line headed "Publication as a Contribution," give information about the collective work in which the contribution appeared.

2 SPACE 2: Author(s)

General Instructions: After reading these instructions, decide who are the "authors" of this work for copyright purposes. Then, unless the work is a "collective work," give the requested information about every "author" who contributed any appreciable amount of copyrightable matter to this version of the work. If you need further space, request additional Continuation sheets. In the case of a collective work, such as an anthology, collection of essays, or encyclopedia, give information about the author of the collective work as a whole.

Name of Author: The fullest form of the author's name should be given. Unless the work was "made for hire," the individual who actually created the work is its "author." In the case of a work made for hire, the statute provides that "the employer or other person for whom the work was prepared is considered the author."

What is a "Work Made for Hire"? A "work made for hire" is defined as: (1) "a work prepared by an employee within the scope of his or her employment"; or (2) "a work specially ordered or commissioned for use as a contribution to a collective work, as a part of a motion picture or other audiovisual work, as a translation, as a supplementary work, as a compilation, as an instructional text, as a test, as answer material for a test, or as an atlas, if the parties expressly agree in a written instrument signed by them that the work shall be considered a work made for hire." If you have checked "Yes" to indicate that the work was "made for hire," you must give the full legal name of the employer (or other person for whom the work was prepared). You may also include the name of the employee along with the name of the employer (for example: "Elster Publishing Co., employer for hire of John Ferguson").

"Anonymous" or "Pseudonymous" Work: An author's contribution to a work is "anonymous" if that author is not identified on the copies or phonorecords of the work. An author's contribution to a work is "pseudonymous" if that author is identified on the copies or phonorecords under a fictitious name. If the work is "anonymous" you may: (1) leave the line blank; or (2) state "anonymous" on the line; or (3) reveal the author's identity. If the work is "pseudonymous" you may: (1) leave the line blank; or (2) give the pseudonym and identify it as such (for example: "Huntley Haverstock, pseudonym"); or (3) reveal the author's name, making clear which is the real name and which is the pseudonym (for example: "Judith Barton, whose pseudonym is Madeline Elster"). However, the citizenship or domicile of the author **must** be given in all cases.

Dates of Birth and Death: If the author is dead, the statute requires that the year of death be included in the application unless the work is anonymous or pseudonymous. The author's birth date is optional, but is useful as a form of identification. Leave this space blank if the author's contribution was a "work made for hire."

Author's Nationality or Domicile: Give the country of which the author is a citizen, or the country in which the author is domiciled. Nationality or domicile **must** be given in all cases.

Nature of Authorship: After the words "Nature of Authorship" give a brief general statement of the nature of this particular author's contribution to the work. Examples: "Entire text"; "Coauthor of entire text"; "Chapters 11-14"; "Editorial revisions"; "Compilation and English translation"; "New text."

Fig. 30

3 SPACE 3: Creation and Publication

General Instructions: Do not confuse "creation" with "publication." Every application for copyright registration must state "the year in which creation of the work was completed." Give the date and nation of first publication only if the work has been published.

Creation: Under the statute, a work is "created" when it is fixed in a copy or phonorecord for the first time. Where a work has been prepared over a period of time, the part of the work existing in fixed form on a particular date constitutes the created work on that date. The date you give here should be the year in which the author completed the particular version for which registration is now being sought, even if other versions exist or if further changes or additions are planned.

Publication: The statute defines "publication" as "the distribution of copies or phonorecords of a work to the public by sale or other transfer of ownership, or by rental, lease, or lending"; a work is also "published" if there has been an "offering to distribute copies or phonorecords to a group of persons for purposes of further distribution, public performance, or public display." Give the full date (month, day, year) when, and the country where, publication first occurred. If first publication took place simultaneously in the United States and other countries, it is sufficient to state "U.S.A."

4 SPACE 4: Claimant(s)

Name(s) and Address(es) of Copyright Claimant(s): Give the name(s) and address(es) of the copyright claimant(s) in this work even if the claimant is the same as the author. Copyright in a work belongs initially to the author of the work (including, in the case of a work made for hire, the employer or other person for whom the work was prepared). The copyright claimant is either the author of the work or a person or organization to whom the copyright initially belonging to the author has been transferred.

Transfer: The statute provides that, if the copyright claimant is not the author, the application for registration must contain "a brief statement of how the claimant obtained ownership of the copyright." If any copyright claimant named in space 4 is not an author named in space 2, give a brief, general statement summarizing the means by which that claimant obtained ownership of the copyright. Examples: "By written contract"; "Transfer of all rights by author"; "Assignment"; "By will." Do not attach transfer documents or other attachments or riders.

5 SPACE 5: Previous Registration

General Instructions: The questions in space 5 are intended to find out whether an earlier registration has been made for this work and, if so, whether there is any basis for a new registration. As a general rule, only one basic copyright registration can be made for the same version of a particular work.

Same Version: If this version is substantially the same as the work covered by a previous registration, a second registration is not generally possible unless: (1) the work has been registered in unpublished form and a second registration is now being sought to cover this first published edition; or (2) someone other than the author is identified as copyright claimant in the earlier registration, and the author is now seeking registration in his or her own name. If either of these two exceptions apply, check the appropriate box and give the earlier registration number and date. Otherwise, do not submit Form TX; instead, write the Copyright Office for information about supplementary registration or recordation of transfers of copyright ownership.

Changed Version: If the work has been changed, and you are now seeking registration to cover the additions or revisions, check the last box in space 5, give the earlier registration number and date, and complete both parts of space 6 in accordance with the instructions below.

Previous Registration Number and Date: If more than one previous registration has been made for the work, give the number and date of the latest registration.

6 SPACE 6: Derivative Work or Compilation

General Instructions: Complete space 6 if this work is a "changed version," "compilation," or "derivative work," and if it incorporates one or more earlier works that have already been published or registered for copyright, or that have fallen into the public domain. A "compilation" is defined as "a work formed by the collection and assembling of preexisting materials or of data that are selected, coordinated, or arranged in such a way that the resulting work as

a whole constitutes an original work of authorship." A "derivative work" is "a work based on one or more preexisting works." Examples of derivative works include translations, fictionalizations, abridgments, condensations, or "any other form in which a work may be recast, transformed, or adapted." Derivative works also include works "consisting of editorial revisions, annotations, or other modifications" if these changes, as a whole, represent an original work of authorship.

Preexisting Material (space 6a): For derivative works, complete this space **and** space 6b. In space 6a identify the preexisting work that has been recast, transformed, or adapted. An example of preexisting material might be: "Russian version of Goncharov's 'Oblomov'." Do not complete space 6a for compilations.

Material Added to This Work (space 6b): Give a brief, general statement of the new material covered by the copyright claim for which registration is sought. **Derivative work** examples include: "Foreword, editing, critical annotations"; "Translation"; "Chapters 11-17." If the work is a **compilation**, describe both the compilation itself and the material that has been compiled. Example: "Compilation of certain 1917 Speeches by Woodrow Wilson." A work may be both a derivative work and compilation, in which case a sample statement might be: "Compilation and additional new material."

7 SPACE 7: Manufacturing Provisions

Due to the expiration of the Manufacturing Clause of the copyright law on June 30, 1986, this space has been deleted.

8 SPACE 8: Reproduction for Use of Blind or Physically Handicapped Individuals

General Instructions: One of the major programs of the Library of Congress is to provide Braille editions and special recordings of works for the exclusive use of the blind and physically handicapped. In an effort to simplify and speed up the copyright licensing procedures that are a necessary part of this program, section 710 of the copyright statute provides for the establishment of a voluntary licensing system to be tied in with copyright registration. Copyright Office regulations provide that you may grant a license for such reproduction and distribution solely for the use of persons who are certified by competent authority as unable to read normal printed material as a result of physical limitations. The license is entirely voluntary, nonexclusive, and may be terminated upon 90 days notice.

How to Grant the License: If you wish to grant it, check one of the three boxes in space 8. Your check in one of these boxes, together with your signature in space 10, will mean that the Library of Congress can proceed to reproduce and distribute under the license without further paperwork. For further information, write for Circular R63.

9,10,11 SPACE 9, 10, 11: Fee, Correspondence, Certification, Return Address

Deposit Account: If you maintain a Deposit Account in the Copyright Office, identify it in space 9. Otherwise leave the space blank and send the fee of $10 with your application and deposit.

Correspondence (space 9): This space should contain the name, address, area code, and telephone number of the person to be consulted if correspondence about this application becomes necessary.

Certification (space 10): The application can not be accepted unless it bears the date and the **handwritten signature** of the author or other copyright claimant, or of the owner of exclusive right(s), or of the duly authorized agent of author, claimant, or owner of exclusive right(s).

Address for Return of Certificate (space 11): The address box must be completed legibly since the certificate will be returned in a window envelope.

Fig. 30 (continued)

FORM TX
UNITED STATES COPYRIGHT OFFICE

REGISTRATION NUMBER

TX   TXU

EFFECTIVE DATE OF REGISTRATION

Month	Day	Year

DO NOT WRITE ABOVE THIS LINE. IF YOU NEED MORE SPACE, USE A SEPARATE CONTINUATION SHEET.

1

TITLE OF THIS WORK ▼

PREVIOUS OR ALTERNATIVE TITLES ▼

PUBLICATION AS A CONTRIBUTION If this work was published as a contribution to a periodical, serial, or collection, give information about the collective work in which the contribution appeared. **Title of Collective Work ▼**

If published in a periodical or serial give: **Volume ▼**   **Number ▼**   **Issue Date ▼**   **On Pages ▼**

2

NOTE

Under the law, the "author" of a "work made for hire" is generally the employer, not the employee (see instructions). For any part of this work that was "made for hire" check "Yes" in the space provided, give the employer (or other person for whom the work was prepared) as "Author" of that part, and leave the space for dates of birth and death blank.

a

NAME OF AUTHOR ▼

DATES OF BIRTH AND DEATH
Year Born ▼   Year Died ▼

Was this contribution to the work a "work made for hire"?
☐ Yes
☐ No

AUTHOR'S NATIONALITY OR DOMICILE
Name of Country
OR { Citizen of ▶_____
Domiciled in ▶_____

WAS THIS AUTHOR'S CONTRIBUTION TO THE WORK
Anonymous? ☐ Yes ☐ No
Pseudonymous? ☐ Yes ☐ No
If the answer to either of these questions is "Yes," see detailed instructions.

NATURE OF AUTHORSHIP Briefly describe nature of the material created by this author in which copyright is claimed. ▼

b

NAME OF AUTHOR ▼

DATES OF BIRTH AND DEATH
Year Born ▼   Year Died ▼

Was this contribution to the work a "work made for hire"?
☐ Yes
☐ No

AUTHOR'S NATIONALITY OR DOMICILE
Name of country
OR { Citizen of ▶_____
Domiciled in ▶_____

WAS THIS AUTHOR'S CONTRIBUTION TO THE WORK
Anonymous? ☐ Yes ☐ No
Pseudonymous? ☐ Yes ☐ No
If the answer to either of these questions is "Yes," see detailed instructions.

NATURE OF AUTHORSHIP Briefly describe nature of the material created by this author in which copyright is claimed. ▼

c

NAME OF AUTHOR ▼

DATES OF BIRTH AND DEATH
Year Born ▼   Year Died ▼

Was this contribution to the work a "work made for hire"?
☐ Yes
☐ No

AUTHOR'S NATIONALITY OR DOMICILE
Name of Country
OR { Citizen of ▶_____
Domiciled in ▶_____

WAS THIS AUTHOR'S CONTRIBUTION TO THE WORK
Anonymous? ☐ Yes ☐ No
Pseudonymous? ☐ Yes ☐ No
If the answer to either of these questions is "Yes," see detailed instructions.

NATURE OF AUTHORSHIP Briefly describe nature of the material created by this author in which copyright is claimed. ▼

3

YEAR IN WHICH CREATION OF THIS WORK WAS COMPLETED This information must be given in all cases. ◀ Year

DATE AND NATION OF FIRST PUBLICATION OF THIS PARTICULAR WORK Complete this information ONLY if this work has been published. Month ▶_____ Day ▶_____ Year ▶_____ ◀ Nation

4

COPYRIGHT CLAIMANT(S) Name and address must be given even if the claimant is the same as the author given in space 2.▼

TRANSFER If the claimant(s) named here in space 4 are different from the author(s) named in space 2, give a brief statement of how the claimant(s) obtained ownership of the copyright.▼

See instructions before completing this space.

APPLICATION RECEIVED

ONE DEPOSIT RECEIVED

TWO DEPOSITS RECEIVED

REMITTANCE NUMBER AND DATE

DO NOT WRITE HERE OFFICE USE ONLY

MORE ON BACK ▶
- Complete all applicable spaces (numbers 5-11) on the reverse side of this page.
- See detailed instructions.
- Sign the form at line 10.

DO NOT WRITE HERE

Page 1 of_____pages

Fig. 30 (continued)

EXAMINED BY _____

CHECKED BY _____

☐ CORRESPONDENCE
Yes

☐ DEPOSIT ACCOUNT
FUNDS USED

FOR
COPYRIGHT
OFFICE
USE
ONLY

DO NOT WRITE ABOVE THIS LINE. IF YOU NEED MORE SPACE, USE A SEPARATE CONTINUATION SHEET.

PREVIOUS REGISTRATION Has registration for this work, or for an earlier version of this work, already been made in the Copyright Office?

☐ Yes ☐ No If your answer is "Yes," why is another registration being sought? (Check appropriate box) ▼

☐ This is the first published edition of a work previously registered in unpublished form.

☐ This is the first application submitted by this author as copyright claimant.

☐ This is a changed version of the work, as shown by space 6 on this application.

If your answer is "Yes," give: **Previous Registration Number** ▼   **Year of Registration** ▼

5

DERIVATIVE WORK OR COMPILATION Complete both space 6a & 6b for a derivative work; complete only 6b for a compilation.
a. Preexisting Material Identify any preexisting work or works that this work is based on or incorporates. ▼

b. Material Added to This Work Give a brief, general statement of the material that has been added to this work and in which copyright is claimed. ▼

6

See instructions
before completing
this space.

—space deleted—

7

REPRODUCTION FOR USE OF BLIND OR PHYSICALLY HANDICAPPED INDIVIDUALS   A signature on this form at space 10, and a check in one of the boxes here in space 8, constitutes a non-exclusive grant of permission to the Library of Congress to reproduce and distribute solely for the blind and physically handicapped and under the conditions and limitations prescribed by the regulations of the Copyright Office: (1) copies of the work identified in space 1 of this application in Braille (or similar tactile symbols); or (2) phonorecords embodying a fixation of a reading of that work; or (3) both.

a ☐ Copies and Phonorecords   b ☐ Copies Only   c ☐ Phonorecords Only

8

See instructions.

DEPOSIT ACCOUNT If the registration fee is to be charged to a Deposit Account established in the Copyright Office, give name and number of Account.
Name ▼   **Account Number** ▼

9

CORRESPONDENCE Give name and address to which correspondence about this application should be sent. Name/Address/Apt/City/State/Zip ▼

Area Code & Telephone Number ▶

Be sure to
give your
daytime phone
◀ number.

CERTIFICATION* I, the undersigned, hereby certify that I am the

Check one ▶
☐ author
☐ other copyright claimant
☐ owner of exclusive right(s)
☐ authorized agent of _____
   Name of author or other copyright claimant, or owner of exclusive right(s) ▲

of the work identified in this application and that the statements made by me in this application are correct to the best of my knowledge.

10

Typed or printed name and date ▼ If this is a published work, this date must be the same as or later than the date of publication given in space 3.

_____   date ▶ _____

👉 Handwritten signature (X) ▼

MAIL CERTIFI-CATE TO

Name ▼

Number/Street/Apartment Number ▼

City/State/ZIP ▼

Certificate will be mailed in window envelope

YOU MUST:
• Complete all necessary spaces
• Sign your application in space 10
SEND ALL 3 ELEMENTS IN THE SAME PACKAGE:
1. Application form
2. Non-refundable $10 filing fee in check or money order payable to *Register of Copyrights*
3. Deposit material
MAIL TO:
Register of Copyrights
Library of Congress
Washington, D.C. 20559

11

Filling Out Application Form VA

Detach and read these instructions before completing this form. Make sure all applicable spaces have been filled in before you return this form.

BASIC INFORMATION

When to Use This Form: Use Form VA for copyright registration of published or unpublished works of the visual arts. This category consists of "pictorial, graphic, or sculptural works," including two-dimensional and three-dimensional works of fine, graphic, and applied art, photographs, prints and art reproductions, maps, globes, charts, technical drawings, diagrams, and models.

What Does Copyright Protect? Copyright in a work of the visual arts protects those pictorial, graphic, or sculptural elements that, either alone or in combination, represent an "original work of authorship." The statute declares: "In no case does copyright protection for an original work of authorship extend to any idea, procedure, process, system, method of operation, concept, principle, or discovery, regardless of the form in which it is described, explained, illustrated, or embodied in such work."

Works of Artistic Craftsmanship and Designs: "Works of artistic craftsmanship" are registrable on Form VA, but the statute makes clear that protection extends to "their form" and not to "their mechanical or utilitarian aspects." The "design of a useful article" is considered copyrightable "only if, and only to the extent that, such design incorporates pictorial, graphic, or sculptural features that can be identified separately from, and are capable of existing independently of, the utilitarian aspects of the article."

Labels and Advertisements: Works prepared for use in connection with the sale or advertisement of goods and services are registrable if they contain "original work of authorship." Use Form VA if the copyrightable material in the work you are registering is mainly pictorial or graphic; use Form TX if it consists mainly of text. **NOTE:** Words and short phrases such as names, titles, and slogans cannot be protected by copyright, and the same is true of standard symbols, emblems, and other commonly used graphic designs that are in the public domain. When used commercially, material of that sort can sometimes be protected under state laws of unfair competition or under the Federal trademark laws. For information about trademark registration, write to the Commissioner of Patents and Trademarks, Washington, D.C. 20231.

Deposit to Accompany Application: An application for copyright registration must be accompanied by a deposit consisting of copies representing the entire work for which registration is to be made.

> **Unpublished Work:** Deposit one complete copy.

> **Published Work:** Deposit two complete copies of the best edition.

> **Work First Published Outside the United States:** Deposit one complete copy of the first foreign edition.

> **Contribution to a Collective Work:** Deposit one complete copy of the best edition of the collective work.

The Copyright Notice: For published works, the law provides that a copyright notice in a specified form "shall be placed on all publicly distributed copies from which the work can be visually perceived." Use of the copyright notice is the responsibility of the copyright owner and does not require advance permission from the Copyright Office. The required form of the notice for copies generally consists of three elements: (1) the symbol "©", or the word "Copyright," or the abbreviation "Copr."; (2) the year of first publication; and (3) the name of the owner of copyright. For example: "© 1981 Constance Porter." The notice is to be affixed to the copies "in such manner and location as to give reasonable notice of the claim of copyright."

For further information about copyright registration, notice, or special questions relating to copyright problems, write:

Information and Publications Section, LM-455
Copyright Office, Library of Congress, Washington, D.C. 20559

PRIVACY ACT ADVISORY STATEMENT Required by the Privacy Act of 1974 (P.L. 93-579)
The authority for requesting this information is title 17, U.S.C., secs. 409 and 410. Furnishing the requested information is voluntary. But if the information is not furnished, it may be necessary to delay or refuse registration and you may not be entitled to certain relief, remedies, and benefits provided in chapters 4 and 5 of title 17, U.S.C.
The principal uses of the requested information are the establishment and maintenance of a public record and the examination of the application for compliance with legal requirements.
Other routine uses include public inspection and copying, preparation of public indexes, preparation of public catalogs of copyright registrations, and preparation of search reports upon request.
NOTE: No other advisory statement will be given in connection with this application. Please keep this statement and refer to it if we communicate with you regarding this application.

LINE-BY-LINE INSTRUCTIONS

1 SPACE 1: Title

Title of This Work: Every work submitted for copyright registration must be given a title to identify that particular work. If the copies of the work bear a title (or an identifying phrase that could serve as a title), transcribe that wording *completely* and *exactly* on the application. Indexing of the registration and future identification of the work will depend on the information you give here.

Previous or Alternative Titles: Complete this space if there are any additional titles for the work under which someone searching for the registration might be likely to look, or under which a document pertaining to the work might be recorded.

Publication as a Contribution: If the work being registered is a contribution to a periodical, serial, or collection, give the title of the contribution in the "Title of This Work" space. Then, in the line headed "Publication as a Contribution," give information about the collective work in which the contribution appeared.

Nature of This Work: Briefly describe the general nature or character of the pictorial, graphic, or sculptural work being registered for copyright. Examples: "Oil Painting"; "Charcoal Drawing"; "Etching"; "Sculpture"; "Map"; "Photograph"; "Scale Model"; "Lithographic Print"; "Jewelry Design"; "Fabric Design."

2 SPACE 2: Author(s)

General Instructions: After reading these instructions, decide who are the "authors" of this work for copyright purposes. Then, unless the work is a "collective work," give the requested information about every "author" who contributed any appreciable amount of copyrightable matter to this version of the work. If you need further space, request additional Continuation Sheets. In the case of a collective work, such as a catalog of paintings or collection of cartoons by various authors, give information about the author of the collective work as a whole.

Name of Author: The fullest form of the author's name should be given. Unless the work was "made for hire," the individual who actually created the work is its "author." In the case of a work made for hire, the statute provides that "the employer or other person for whom the work was prepared is considered the author."

What is a "Work Made for Hire"? A "work made for hire" is defined as: (1) "a work prepared by an employee within the scope of his or her employment"; or (2) "a work specially ordered or commissioned for use as a contribution to a collective work, as a part of a motion picture or other audiovisual work, as a translation, as a supplementary work, as a compilation, as an instructional text, as a test, as answer material for a test, or as an atlas, if the parties expressly agree in a written instrument signed by them that the work shall be considered a work made for hire." If you have checked "Yes" to indicate that the work was "made for hire," you must give the full legal name of the employer (or other person for whom the work was prepared). You may also include the name of the employee along with the name of the employer (for example: "Elster Publishing Co., employer for hire of John Ferguson").

"Anonymous" or "Pseudonymous" Work: An author's contribution to a work is "anonymous" if that author is not identified on the copies or phonorecords of the work. An author's contribution to a work is "pseudonymous" if that author is identified on the copies or phonorecords under a fictitious name. If the work is "anonymous" you may: (1) leave the line blank; or (2) state "anonymous" on the line; or (3) reveal the author's identity. If the work is "pseudonymous" you may: (1) leave the line blank; or (2) give the pseudonym and identify it as such (for example: "Huntley Haverstock, pseudonym"); or (3) reveal the author's name, making clear which is the real name and which is the pseudonym (for example: "Henry Leek, whose pseudonym is Priam Farrel"). However, the citizenship or domicile of the author **must** be given in all cases.

Dates of Birth and Death: If the author is dead, the statute requires that the year of death be included in the application unless the work is anonymous or pseudonymous. The author's birth date is optional, but is useful as a form of identification. Leave this space blank if the author's contribution was a "work made for hire."

Fig. 31

Author's Nationality or Domicile: Give the country of which the author is a citizen, or the country in which the author is domiciled. Nationality or domicile **must** be given in all cases.

Nature of Authorship: Give a brief general statement of the nature of this particular author's contribution to the work. Examples: "Painting"; "Photograph"; "Silk Screen Reproduction"; "Co-author of Cartographic Material"; "Technical Drawing"; "Text and Artwork."

3 SPACE 3: Creation and Publication

General Instructions: Do not confuse "creation" with "publication." Every application for copyright registration must state "the year in which creation of the work was completed." Give the date and nation of first publication only if the work has been published.

Creation: Under the statute, a work is "created" when it is fixed in a copy or phonorecord for the first time. Where a work has been prepared over a period of time, the part of the work existing in fixed form on a particular date constitutes the created work on that date. The date you give here should be the year in which the author completed the particular version for which registration is now being sought, even if other versions exist or if further changes or additions are planned.

Publication: The statute defines "publication" as "the distribution of copies or phonorecords of a work to the public by sale or other transfer of ownership, or by rental, lease, or lending"; a work is also "published" if there has been an "offering to distribute copies or phonorecords to a group of persons for purposes of further distribution, public performance, or public display." Give the full date (month, day, year) when, and the country where, publication first occurred. If first publication took place simultaneously in the United States and other countries, it is sufficient to state "U.S.A."

4 SPACE 4: Claimant(s)

Name(s) and Address(es) of Copyright Claimant(s): Give the name(s) and address(es) of the copyright claimant(s) in this work even if the claimant is the same as the author. Copyright in a work belongs initially to the author of the work (including, in the case of a work made for hire, the employer or other person for whom the work was prepared). The copyright claimant is either the author of the work or a person or organization to whom the copyright initially belonging to the author has been transferred.

Transfer: The statute provides that, if the copyright claimant is not the author, the application for registration must contain "a brief statement of how the claimant obtained ownership of the copyright." If any copyright claimant named in space 4 is not an author named in space 2, give a brief, general statement summarizing the means by which that claimant obtained ownership of the copyright. Examples: "By written contract"; "Transfer of all rights by author"; "Assignment"; "By will." Do not attach transfer documents or other attachments or riders.

5 SPACE 5: Previous Registration

General Instructions: The questions in space 5 are intended to find out whether an earlier registration has been made for this work and, if so, whether there is any basis for a new registration. As a rule, only one basic copyright registration can be made for the same version of a particular work.

Same Version: If this version is substantially the same as the work covered by a previous registration, a second registration is not generally possible unless: (1) the work has been registered in unpublished form and a second registration is now being sought to cover this first published edition; or (2) some-

one other than the author is identified as copyright claimant in the earlier registration, and the author is now seeking registration in his or her own name. If either of these two exceptions apply, check the appropriate box and give the earlier registration number and date. Otherwise, do not submit Form VA; instead, write the Copyright Office for information about supplementary registration or recordation of transfers of copyright ownership.

Changed Version: If the work has been changed, and you are now seeking registration to cover the additions or revisions, check the last box in space 5, give the earlier registration number and date, and complete both parts of space 6 in accordance with the instructions below.

Previous Registration Number and Date: If more than one previous registration has been made for the work, give the number and date of the latest registration.

6 SPACE 6: Derivative Work or Compilation

General Instructions: Complete space 6 if this work is a "changed version," "compilation," or "derivative work," and if it incorporates one or more earlier works that have already been published or registered for copyright, or that have fallen into the public domain. A "compilation" is defined as "a work formed by the collection and assembling of preexisting materials or of data that are selected, coordinated, or arranged in such a way that the resulting work as a whole constitutes an original work of authorship." A "derivative work" is "a work based on one or more preexisting works." Examples of derivative works include reproductions of works of art, sculptures based on drawings, lithographs based on paintings, maps based on previously published sources, or "any other form in which a work may be recast, transformed, or adapted." Derivative works also include works "consisting of editorial revisions, annotations, or other modifications" if these changes, as a whole, represent an original work of authorship.

Preexisting Material (space 6a): Complete this space **and** space 6b for derivative works. In this space identify the preexisting work that has been recast, transformed, or adapted. Examples of preexisting material might be "Grunewald Altarpiece"; or "19th century quilt design." Do not complete this space for compilations.

Material Added to This Work (space 6b): Give a brief, general statement of the **additional** new material covered by the copyright claim for which registration is sought. In the case of a derivative work, identify this new material. Examples: "Adaptation of design and additional artistic work"; "Reproduction of painting by photolithography"; "Additional cartographic material"; "Compilation of photographs." If the work is a compilation, give a brief, general statement describing both the material that has been compiled **and** the compilation itself. Example: "Compilation of 19th Century Political Cartoons."

7,8,9 SPACE 7, 8, 9: Fee, Correspondence, Certification, Return Address

Deposit Account: If you maintain a Deposit Account in the Copyright Office, identify it in space 7. Otherwise leave the space blank and send the fee of $10 with your application and deposit.

Correspondence (space 7): This space should contain the name, address, area code, and telephone number of the person to be consulted if correspondence about this application becomes necessary.

Certification (space 8): The application cannot be accepted unless it bears the date and the **handwritten signature** of the author or other copyright claimant, or of the owner of exclusive right(s), or of the duly authorized agent of the author, claimant, or owner of exclusive right(s).

Address for Return of Certificate (space 9): The address box must be completed legibly since the certificate will be returned in a window envelope.

MORE INFORMATION

Form of Deposit for Works of the Visual Arts

Exceptions to General Deposit Requirements: As explained on the reverse side of this page, the statutory deposit requirements (generally one copy for unpublished works and two copies for published works) will vary for particular kinds of works of the visual arts. The copyright law authorizes the Register of Copyrights to issue regulations specifying "the administrative classes into which works are to be placed for purposes of deposit and registration, and the nature of the copies or phonorecords to be deposited in the various classes specified." For particular classes, the regulations may require or permit "the deposit of identifying material instead of copies or phonorecords," or "the deposit of only one copy or phonorecord where two would normally be required."

What Should You Deposit? The detailed requirements with respect to the kind of deposit to accompany an application on Form VA are contained in the Copyright

Office Regulations. The following does not cover all of the deposit requirements, but is intended to give you some general guidance.

For an Unpublished Work, the material deposited should represent the entire copyrightable content of the work for which registration is being sought.

For a Published Work, the material deposited should generally consist of two complete copies of the best edition. Exceptions: (1) For certain types of works, one complete copy may be deposited instead of two. These include greeting cards, postcards, stationery, labels, advertisements, scientific drawings, and globes; (2) For most three-dimensional sculptural works, and for certain two-dimensional works, the Copyright Office Regulations require deposit of identifying material (photographs or drawings in a specified form) rather than copies; and (3) Under certain circumstances, for works published in five copies or less or in limited, numbered editions, the deposit may consist of one copy or of identifying reproductions.

Fig. 31 (continued)

FORM VA
UNITED STATES COPYRIGHT OFFICE

REGISTRATION NUMBER

_____
VA   VAU

EFFECTIVE DATE OF REGISTRATION

_____
Month   Day   Year

DO NOT WRITE ABOVE THIS LINE. IF YOU NEED MORE SPACE, USE A SEPARATE CONTINUATION SHEET.

1

TITLE OF THIS WORK ▼   **NATURE OF THIS WORK ▼** See instructions

PREVIOUS OR ALTERNATIVE TITLES ▼

PUBLICATION AS A CONTRIBUTION If this work was published as a contribution to a periodical, serial, or collection, give information about the collective work in which the contribution appeared.   **Title of Collective Work ▼**

If published in a periodical or serial give: **Volume ▼**   **Number ▼**   **Issue Date ▼**   **On Pages ▼**

2

a
NAME OF AUTHOR ▼   **DATES OF BIRTH AND DEATH**
Year Born ▼   Year Died ▼

Was this contribution to the work a "work made for hire"?
☐ Yes
☐ No

AUTHOR'S NATIONALITY OR DOMICILE
Name of Country
OR { Citizen of ▶_____
{ Domiciled in ▶_____

WAS THIS AUTHOR'S CONTRIBUTION TO THE WORK
Anonymous? ☐ Yes ☐ No
Pseudonymous? ☐ Yes ☐ No
If the answer to either of these questions is "Yes," see detailed instructions.

NATURE OF AUTHORSHIP Briefly describe nature of the material created by this author in which copyright is claimed. ▼

NOTE

Under the law, the "author" of a "work made for hire" is generally the employer, not the employee (see instructions). For any part of this work that was "made for hire" check "Yes" in the space provided, give the employer (or other person for whom the work was prepared) as "Author" of that part, and leave the space for dates of birth and death blank.

b
NAME OF AUTHOR ▼   **DATES OF BIRTH AND DEATH**
Year Born ▼   Year Died ▼

Was this contribution to the work a "work made for hire"?
☐ Yes
☐ No

AUTHOR'S NATIONALITY OR DOMICILE
Name of country
OR { Citizen of ▶_____
{ Domiciled in ▶_____

WAS THIS AUTHOR'S CONTRIBUTION TO THE WORK
Anonymous? ☐ Yes ☐ No
Pseudonymous? ☐ Yes ☐ No
If the answer to either of these questions is "Yes," see detailed instructions.

NATURE OF AUTHORSHIP Briefly describe nature of the material created by this author in which copyright is claimed. ▼

c
NAME OF AUTHOR ▼   **DATES OF BIRTH AND DEATH**
Year Born ▼   Year Died ▼

Was this contribution to the work a "work made for hire"?
☐ Yes
☐ No

AUTHOR'S NATIONALITY OR DOMICILE
Name of Country
OR { Citizen of ▶_____
{ Domiciled in ▶_____

WAS THIS AUTHOR'S CONTRIBUTION TO THE WORK
Anonymous? ☐ Yes ☐ No
Pseudonymous? ☐ Yes ☐ No
If the answer to either of these questions is "Yes," see detailed instructions.

NATURE OF AUTHORSHIP Briefly describe nature of the material created by this author in which copyright is claimed. ▼

3

YEAR IN WHICH CREATION OF THIS WORK WAS COMPLETED This information must be given in all cases. ◀ Year

DATE AND NATION OF FIRST PUBLICATION OF THIS PARTICULAR WORK
Complete this information ONLY if this work has been published.
Month ▶_____ Day ▶_____ Year ▶_____ ◀ Nation

4

COPYRIGHT CLAIMANT(S) Name and address must be given even if the claimant is the same as the author given in space 2.▼

TRANSFER If the claimant(s) named here in space 4 are different from the author(s) named in space 2, give a brief statement of how the claimant(s) obtained ownership of the copyright.▼

DO NOT WRITE HERE
OFFICE USE ONLY

APPLICATION RECEIVED

ONE DEPOSIT RECEIVED

TWO DEPOSITS RECEIVED

REMITTANCE NUMBER AND DATE

MORE ON BACK ▶ • Complete all applicable spaces (numbers 5-9) on the reverse side of this page.
• See detailed instructions.   • Sign the form at line 8.

DO NOT WRITE HERE
Page 1 of_____pages

Fig. 31 (continued)

EXAMINED BY _____ **FORM VA**

CHECKED BY _____

☐ CORRESPONDENCE
    Yes

☐ DEPOSIT ACCOUNT
    FUNDS USED

FOR
COPYRIGHT
OFFICE
USE
ONLY

DO NOT WRITE ABOVE THIS LINE. IF YOU NEED MORE SPACE, USE A SEPARATE CONTINUATION SHEET.

PREVIOUS REGISTRATION   Has registration for this work, or for an earlier version of this work, already been made in the Copyright Office?

☐ **Yes** ☐ **No**   If your answer is "Yes," why is another registration being sought? (Check appropriate box) ▼

☐ This is the first published edition of a work previously registered in unpublished form.

☐ This is the first application submitted by this author as copyright claimant.

☐ This is a changed version of the work, as shown by space 6 on this application.

If your answer is "Yes," give: **Previous Registration Number** ▼   **Year of Registration** ▼

5

DERIVATIVE WORK OR COMPILATION   Complete both space 6a & 6b for a derivative work; complete only 6b for a compilation.

a.   Preexisting Material   Identify any preexisting work or works that this work is based on or incorporates. ▼

_____

_____

b.   Material Added to This Work   Give a brief, general statement of the material that has been added to this work and in which copyright is claimed. ▼

_____

_____

6

See instructions
before completing
this space.

DEPOSIT ACCOUNT   If the registration fee is to be charged to a Deposit Account established in the Copyright Office, give name and number of Account.

Name ▼   **Account Number** ▼

7

CORRESPONDENCE   Give name and address to which correspondence about this application should be sent.   Name/Address/Apt/City/State/Zip ▼

_____

_____

_____

Area Code & Telephone Number ▶

Be sure to
give your
daytime phone
◀ number.

CERTIFICATION*   I, the undersigned, hereby certify that I am the

Check only one ▼

☐ author

☐ other copyright claimant

☐ owner of exclusive right(s)

☐ authorized agent of_____
                        Name of author or other copyright claimant, or owner of exclusive right(s) ▲

8

of the work identified in this application and that the statements made
by me in this application are correct to the best of my knowledge.

Typed or printed name and date ▼ If this is a published work, this date must be the same as or later than the date of publication given in space 3.

_____ date ▶ _____

☞ Handwritten signature (X) ▼

**MAIL
CERTIFI-
CATE TO**

Name ▼

Number/Street/Apartment Number ▼

City/State/ZIP ▼

**Certificate
will be
mailed in
window
envelope**

Have you:
● Completed all necessary
  spaces?
● Signed your application in space
  8?
● Enclosed check or money order
  for $10 payable to *Register of
  Copyrights?*
● Enclosed your deposit material
  with the application and fee?

MAIL TO: Register of Copyrights,
Library of Congress, Washington,
D.C. 20559.

9

APPLICATION FOR
Renewal Registration

HOW TO REGISTER A RENEWAL CLAIM:

- **First:** Study the information on this page and make sure you know the answers to two questions:

  (1) What are the renewal time limits in your case?

  (2) Who can claim the renewal?

- **Second:** Turn this page over and read through the specific instructions for filling out Form RE. Make sure, before starting to complete the form, that the copyright is now eligible for renewal, that you are authorized to file a renewal claim, and that you have all of the information about the copyright you will need.

- **Third:** Complete all applicable spaces on Form RE, following the line-by-line instructions on the back of this page. Use typewriter, or print the information in dark ink.

- **Fourth:** Detach this sheet and send your completed Form RE to: Register of Copyrights, Library of Congress, Washington, D.C. 20559. Unless you have a Deposit Account in the Copyright Office, your application must be accompanied by a check or money order for $6, payable to: *Register of Copyrights*. Do not send copies, phonorecords, or supporting documents with your renewal application.

WHAT IS RENEWAL OF COPYRIGHT? For works originally copyrighted between January 1, 1950 and December 31, 1977, the statute now in effect provides for a first term of copyright protection lasting for 28 years, with the possibility of renewal for a second term of 47 years. If a valid renewal registration is made for a work, its total copyright term is 75 years (a first term of 28 years, plus a renewal term of 47 years). Example: For a work copyrighted in 1960, the first term will expire in 1988, but if renewed at the proper time the copyright will last through the end of 2035.

SOME BASIC POINTS ABOUT RENEWAL:

(1) There are strict time limits and deadlines for renewing a copyright.

(2) Only certain persons who fall into specific categories named in the law can claim renewal.

(3) The new copyright law does away with renewal requirements for works first copyrighted after 1977. However, copyrights that were already in their first copyright term on January 1, 1978 (that is, works originally copyrighted between January 1, 1950 and December 31, 1977) **still have to be renewed** in order to be protected for a second term.

TIME LIMITS FOR RENEWAL REGISTRATION: The new copyright statute provides that, in order to renew a copyright, the renewal application and fee must be received in the Copyright Office "within one year prior to the expiration of the copyright." It also provides that all terms of copyright will run through the end of the year in which they would otherwise expire. Since all copyright terms will expire on December 31st of their last year, all periods for renewal registration will run from December 31st of the 27th year of the copyright, and will end on December 31st of the following year.

To determine the time limits for renewal in your case:

(1) First, find out the date of original copyright for the work. (In the case of works originally registered in unpublished form, the date of copyright is the date of registration; for published works, copyright begins on the date of first publication.)

(2) Then add 28 years to the year the work was originally copyrighted.

Your answer will be the calendar year during which the copyright will be eligible for renewal, and December 31st of that year will be the renewal deadline. Example: a work originally copyrighted on April 19, 1957, will be eligible for renewal between December 31, 1984, and December 31, 1985.

WHO MAY CLAIM RENEWAL: Renewal copyright may be claimed only by those persons specified in the law. Except in the case of four specific types of works, the law gives the right to claim renewal to the individual author of the work, regardless of who owned the copyright during the original term. If the author is dead, the statute gives the right to claim renewal to certain of the author's beneficiaries (widow and children, executors, or next of kin, depending on the circumstances). The present owner (proprietor) of the copyright is entitled to claim renewal only in four specified cases, as explained in more detail on the reverse of this page.

CAUTION: Renewal registration is possible only if an acceptable application and fee are **received** in the Copyright Office during the renewal period and before the renewal deadline. If an acceptable application and fee are not received before the renewal deadline, the work falls into the public domain and the copyright cannot be renewed. The Copyright Office has no discretion to extend the renewal time limits.

PRIVACY ACT ADVISORY STATEMENT
Required by the Privacy Act of 1974 (Public Law 93-579)

AUTHORITY FOR REQUESTING THIS INFORMATION:
- Title 17, U.S.C., Sec. 304

FURNISHING THE REQUESTED INFORMATION IS:
- Voluntary

BUT IF THE INFORMATION IS NOT FURNISHED:
- It may be necessary to delay or refuse renewal registration

- If renewal registration is not made, the copyright will expire at the end of its 28th year

PRINCIPAL USES OF REQUESTED INFORMATION:
- Establishment and maintenance of a public record
- Examination of the application for compliance with legal requirements

OTHER ROUTINE USES:
- Public inspection and copying

- Preparation of public indexes
- Preparation of public catalogs of copyright registrations
- Preparation of search reports upon request

NOTE:
- No other advisory statement will be given you in connection with this application
- Please retain this statement and refer to it if we communicate with you regarding this application

Fig. 32

INSTRUCTIONS FOR COMPLETING FORM RE

SPACE 1: RENEWAL CLAIM(S)

• **General Instructions:** In order for this application to result in a valid renewal, space 1 must identify one or more of the persons who are entitled to renew the copyright under the statute. Give the full name and address of each claimant, with a statement of the basis of each claim, using the wording given in these instructions.

• **Persons Entitled to Renew:**

A. The following persons may claim renewal in all types of works except those enumerated in Paragraph B, below:

1. The author, if living. State the claim as: *the author.*

2. The widow, widower, and/or children of the author, if the author is not living. State the claim as: *the widow (widower) of the author* · · · · · · · · · · · · · · ·
(Name of author)

and/or *the child (children) of the deceased author* · · · · · · · · · · · · ·
(Name of author)

3. The author's executor(s), if the author left a will and if there is no surviving widow, widower, or child. State the claim as: *the executor(s) of the author* · · · · · · · · · · · ·
(Name of author)

4. The next of kin of the author, if the author left no will and if there is no surviving widow, widower, or child. State the claim as: *the next of kin of the deceased author* · · · · · · · · · · · · *there being no will.*
(Name of author)

B. In the case of the following four types of works, the proprietor (owner of the copyright at the time of renewal registration) may claim renewal:

1. Posthumous work (a work as to which no copyright assignment or other contract for exploitation has occurred during the author's lifetime). State the claim as: *proprietor of copyright in a posthumous work.*

2. Periodical, cyclopedic, or other composite work. State the claim as: *proprietor of copyright in a composite work.*

3. "Work copyrighted by a corporate body otherwise than as assignee or licensee of the individual author." State the claim as: *proprietor of copyright in a work copyrighted by a corporate body otherwise than as assignee or licensee of the individual author.* (This type of claim is considered appropriate in relatively few cases.)

4. Work copyrighted by an employer for whom such work was made for hire. State the claim as: *proprietor of copyright in a work made for hire.*

SPACE 2: WORK RENEWED

• **General Instructions:** This space is to identify the particular work being renewed. The information given here should agree with that appearing in the certificate of original registration.

• **Title:** Give the full title of the work, together with any subtitles or descriptive wording included with the title in the original registration. In the case of a musical composition, give the specific instrumentation of the work.

• **Renewable Matter:** Copyright in a new version of a previous work (such as an arrangement, translation, dramatization, compilation, or work republished with new matter) covers only the additions, changes, or other new material appearing for the first time in that version. If this work was a new version, state in general the new matter upon which copyright was claimed.

• **Contribution to Periodical, Serial, or other Composite Work:** Separate renewal registration is possible for a work published as a contribution to a periodical, serial, or other composite work, whether the contribution was copyrighted independently or as part of the larger work in which it appeared. Each contribution published in a separate issue ordinarily requires a separate renewal registration. However, the new law provides an alternative, permitting groups of periodical contributions by the same individual author to be combined under a single renewal application and fee in certain cases.

If this renewal application covers a single contribution, give all of the requested information in space 2. If you are seeking to renew a group of contributions, include a reference such as "See space 5" in space 2 and give the requested information about all of the contributions in space 5.

SPACE 3: AUTHOR(S)

• **General Instructions:** The copyright secured in a new version of a work is independent of any copyright protection in material published earlier. The only "authors" of a new version are those who contributed copyrightable matter to it. Thus, for renewal purposes, the person who wrote the original version on which the new work is based cannot be regarded as an "author" of the new version, unless that person also contributed to the new matter.

• **Authors of Renewable Matter:** Give the full names of all authors who contributed copyrightable matter to this particular version of the work.

SPACE 4: FACTS OF ORIGINAL REGISTRATION

• **General Instructions:** Each item in space 4 should agree with the information appearing in the original registration for the work. If the work being renewed is a single contribution to a periodical or composite work that was not separately registered, give information about the particular issue in which the contribution appeared. You may leave this space blank if you are completing space 5.

• **Original Registration Number:** Give the full registration number, which is a series of numerical digits, preceded by one or more letters. The registration number appears in the upper right hand corner of the certificate of registration.

• **Original Copyright Claimant:** Give the name in which ownership of the copyright was claimed in the original registration.

• **Date of Publication or Registration:** Give only one date. If the original registration gave a publication date, it should be transcribed here; otherwise the registration was for an unpublished work, and the date of registration should be given.

SPACE 5: GROUP RENEWALS

• **General Instructions:** A single renewal registration can be made for a group of works if **all** of the following statutory conditions are met: (1) all of the works were written by the same author, who is named in space 3 and who is or was an individual (not an employer for hire); (2) all of the works were first published as contributions to periodicals (including newspapers) and were copyrighted on their first publication; (3) the renewal claimant or claimants, and the basis of claim or claims, as stated in space 1, is the same for all of the works; (4) the renewal application and fee are "received not more than 28 or less than 27 years after the 31st day of December of the calendar year in which all of the works were first published"; and (5) the renewal application identifies each work separately, including the periodical containing it and the date of first publication.

Time Limits for Group Renewals: To be renewed as a group, all of the contributions must have been first published during the same calendar year. For example, suppose six contributions by the same author were published on April 1, 1960, July 1, 1960, November 1, 1960, February 1, 1961, July 1, 1961, and March 1, 1962. The three 1960 copyrights can be combined and renewed at any time during 1988, and the two 1961 copyrights can be renewed as a group during 1989, but the 1962 copyright must be renewed by itself, in 1990.

Identification of Each Work: Give all of the requested information for each contribution. The registration number should be that for the contribution itself if it was separately registered, and the registration number for the periodical issue if it was not.

SPACES 6, 7 AND 8: FEE, MAILING INSTRUCTIONS, AND CERTIFICATION

• **Deposit Account and Mailing Instructions (Space 6):** If you maintain a Deposit Account in the Copyright Office, identify it in space 6. Otherwise, you will need to send the renewal registration fee of $6 with your form. The space headed "Correspondence" should contain the name and address of the person to be consulted if correspondence about the form becomes necessary.

• **Certification (Space 7):** The renewal application is not acceptable unless it bears the handwritten signature of the renewal claimant or the duly authorized agent of the renewal claimant.

• **Address for Return of Certificate (Space 8):** The address box must be completed legibly, since the certificate will be returned in a window envelope.

Fig. 32 (continued)

FORM RE

UNITED STATES COPYRIGHT OFFICE

REGISTRATION NUMBER

EFFECTIVE DATE OF RENEWAL REGISTRATION

....
(Month)   (Day)   (Year)

DO NOT WRITE ABOVE THIS LINE.   FOR COPYRIGHT OFFICE USE ONLY

(1) Renewal Claimant(s)	**RENEWAL CLAIMANT(S), ADDRESS(ES), AND STATEMENT OF CLAIM:** (See Instructions)
	1 Name .... Address .... Claiming as .... (Use appropriate statement from instructions)
	2 Name .... Address .... Claiming as .... (Use appropriate statement from instructions)
	3 Name .... Address .... Claiming as .... (Use appropriate statement from instructions)

(2) Work Renewed	**TITLE OF WORK IN WHICH RENEWAL IS CLAIMED:**
	RENEWABLE MATTER:
	CONTRIBUTION TO PERIODICAL OR COMPOSITE WORK: Title of periodical or composite work: .... If a periodical or other serial, give: Vol. .... No. .... Issue Date ....

(3) Author(s)	**AUTHOR(S) OF RENEWABLE MATTER:**

(4) Facts of Original Registration	**ORIGINAL REGISTRATION NUMBER:**	**ORIGINAL COPYRIGHT CLAIMANT:**

ORIGINAL DATE OF COPYRIGHT:

• If the original registration for this work was made in published form, give:

DATE OF PUBLICATION: ....
(Month)   (Day)   (Year)

{ OR }

• If the original registration for this work was made in unpublished form, give:

DATE OF REGISTRATION: ....
(Month)   (Day)   (Year)

Fig. 32 (continued)

	EXAMINED BY:	RENEWAL APPLICATION RECEIVED:	
	CHECKED BY:		FOR COPYRIGHT OFFICE USE ONLY
	CORRESPONDENCE ☐ Yes	REMITTANCE NUMBER AND DATE:	
	DEPOSIT ACCOUNT FUNDS USED: ☐		

DO NOT WRITE ABOVE THIS LINE. FOR COPYRIGHT OFFICE USE ONLY

RENEWAL FOR GROUP OF WORKS BY SAME AUTHOR: To make a single registration for a group of works by the same individual author published as contributions to periodicals (see instructions). give full information about each contribution. If more space is needed, request continuation sheet (Form RE/CON).

⑤
Renewal for Group of Works

1
Title of Contribution: . . . .
Title of Periodical: . . . . Vol. . . . . No. . . . . . . Issue Date . . . .
Date of Publication: . . . . Registration Number: . . . .
(Month) (Day) (Year)

2
Title of Contribution: . . . .
Title of Periodical: . . . . Vol. . . . . No . . . . . . Issue Date . . . .
Date of Publication: . . . . Registration Number: . . . .
(Month) (Day) (Year)

3
Title of Contribution: . . . .
Title of Periodical: . . . . Vol. . . . . No. . . . . . . Issue Date . . . .
Date of Publication: . . . . Registration Number: . . . .
(Month) (Day) (Year)

4
Title of Contribution: . . . .
Title of Periodical: . . . . Vol. . . . . No. . . . . . . Issue Date . . . .
Date of Publication: . . . . Registration Number: . . . .
(Month) (Day) (Year)

5
Title of Contribution: . . . .
Title of Periodical: . . . . Vol . . . . No. . . . . . . Issue Date . . . .
Date of Publication: . . . . Registration Number: . . . .
(Month) (Day) (Year)

6
Title of Contribution: . . . .
Title of Periodical: . . . . Vol. . . . . No. . . . . . . Issue Date . . . .
Date of Publication: . . . . Registration Number: . . . .
(Month) (Day) (Year)

7
Title of Contribution: . . . .
Title of Periodical: . . . . Vol. . . . . No. . . . . . . Issue Date . . . .
Date of Publication: . . . . Registration Number: . . . .
(Month) (Day) (Year)

DEPOSIT ACCOUNT: (If the registration fee is to be charged to a Deposit Account established in the Copyright Office, give name and number of Account.)

Name: . . . .
Account Number: . . . .

CORRESPONDENCE: (Give name and address to which correspondence about this application should be sent.)

Name: . . . .
Address: . . . .
(Apt.)
(City) (State) (ZIP)

⑥
Fee and Correspondence

CERTIFICATION: I. the undersigned. hereby certify that I am the: (Check one)
☐ renewal claimant   ☐ duly authorized agent of: . . . .
(Name of renewal claimant)
of the work identified in this application. and that the statements made by me in this application are correct to the best of my knowledge.

☛ Handwritten signature: (X) . . . .
Typed or printed name: . . . .
Date: . . . .

⑦
Certification (Application must be signed)

MAIL CERTIFICATE TO

. . . .
(Name)
. . . .
(Number, Street and Apartment Number)
. . . .
(City) (State) (ZIP code)

(Certificate will be mailed in window envelope)

⑧
Address for Return of Certificate

Fig. 32 (continued)

FORM CA

UNITED STATES COPYRIGHT OFFICE
LIBRARY OF CONGRESS
WASHINGTON, D.C. 20559

Application for Supplementary Copyright Registration

To Correct or Amplify Information Given in the Copyright Office Record of an Earlier Registration

What is "Supplementary Copyright Registration"? Supplementary registration is a special type of copyright registration provided for in section 408(d) of the copyright law.

Purpose of Supplementary Registration. As a rule, only one basic copyright registration can be made for the same work. To take care of cases where information in the basic registration turns out to be incorrect or incomplete, the law provides for "the filing of an application for supplementary registration, to correct an error in a copyright registration or to amplify the information given in a registration."

Earlier Registration Necessary. Supplementary registration can be made only if a basic copyright registration for the same work has already been completed.

Who May File. Once basic registration has been made for a work, any author or other copyright claimant, or owner of any exclusive right in the work, who wishes to correct or amplify the information given in the basic registration, may submit Form CA.

Please Note:

- Do not use Form CA to correct errors in statements on the copies or phonorecords of the work in question, or to reflect changes in the content of the work. If the work has been changed substantially, you should consider making an entirely new registration for the revised version to cover the additions or revisions.

- Do not use Form CA as a substitute for renewal registration. For works originally copyrighted between January 1, 1950 and December 31, 1977, registration of a renewal claim within strict time limits is necessary to extend the first 28-year copyright term to the full term of 75 years. This cannot be done by filing Form CA.

- Do not use Form CA as a substitute for recording a transfer of copyright or other document pertaining to rights under a copyright. Recording a document under section 205 of the statute gives all persons constructive notice of the facts stated in the document and may have other important consequences in cases of infringement or conflicting transfers. Supplementary registration does not have that legal effect.

How to Apply for Supplementary Registration:

First: Study the information on this page to make sure that filing an application on Form CA is the best procedure to follow in your case.

Second: Turn this page over and read through the specific instructions for filling out Form CA. Make sure, before starting to complete the form, that you have all of the detailed information about the basic registration you will need.

Third: Complete all applicable spaces on this form, following the line-by-line instructions on the back of this page. Use typewriter, or print the information in dark ink.

Fourth: Detach this sheet and send your completed Form CA to: Register of Copyrights, Library of Congress, Washington, D.C. 20559. Unless you have a Deposit Account in the Copyright Office, your application must be accompanied by a non-refundable filing fee in the form of a check or money order for $10 payable to: *Register of Copyrights.* Do not send copies, phonorecords, or supporting documents with your application, since they cannot be made part of the record of a supplementary registration.

What Happens When a Supplementary Registration is Made? When a supplementary registration is completed, the Copyright Office will assign it a new registration number in the appropriate registration category, and issue a certificate of supplementary registration under that number. The basic registration will not be expunged or cancelled, and the two registrations will both stand in the Copyright Office records. The supplementary registration will have the effect of calling the public's attention to a possible error or omission in the basic registration, and of placing the correct facts or the additional information on official record. Moreover, if the person on whose behalf Form CA is submitted is the same as the person identified as copyright claimant in the basic registration, the Copyright Office will place a note referring to the supplementary registration in its records of the basic registration.

Fig. 33

PLEASE READ DETAILED INSTRUCTIONS ON REVERSE

Please read the following line-by-line instructions carefully and refer to them while completing Form CA.

INSTRUCTIONS
For Completing FORM CA (Supplementary Registration)

PART A: BASIC INSTRUCTIONS

• *General Instructions:* The information in this part identifies the basic registration to be corrected or amplified. Each item must agree exactly with the information as it already appears in the basic registration (even if the purpose of filing Form CA is to change one of these items).

• *Title of Work:* Give the title as it appears in the basic registration, including previous or alternative titles if they appear.

• *Registration Number:* This is a series of numerical digits, preceded by one or more letters. The registration number appears in the upper right hand corner of the certificate of registration.

• *Registration Date:* Give the year when the basic registration was completed.

• *Name(s) of Author(s) and Name(s) of Copyright Claimant(s):* Give all of the names as they appear in the basic registration.

PART B: CORRECTION

• *General Instructions:* Complete this part **only** if information in the basic registration was incorrect at the time that basic registration was made. Leave this part blank and complete Part C, instead, if your purpose is to add, update, or clarify information rather than to rectify an actual error.

• *Location and Nature of Incorrect Information:* Give the line number and the heading or description of the space in the basic registration where the error occurs (for example: "Line number 3 . . . Citizenship of author").

• *Incorrect Information as it Appears in Basic Registration:* Transcribe the erroneous statement exactly as it appears in the basic registration.

• *Corrected Information:* Give the statement as it should have appeared.

• *Explanation of Correction (Optional):* If you wish, you may add an explanation of the error or its correction.

PART C: AMPLIFICATION

• *General Instructions:* Complete this part if you want to provide any of the following: (1) additional information that could have been given but was omitted at the time of basic registration; (2) changes in facts, such as changes of title or address of claimant, that have occurred since the basic registration; or (3) explanations clarifying information in the basic registration.

• *Location and Nature of Information to be Amplified:* Give the line number and the heading or description of the space in the basic registration where the information to be amplified appears.

• *Amplified Information:* Give a statement of the added, updated, or explanatory information as clearly and succinctly as possible.

• *Explanation of Amplification (Optional):* If you wish, you may add an explanation of the amplification.

PARTS D, E, F, G: CONTINUATION, FEE, MAILING INSTRUCTIONS AND CERTIFICATION

• *Continuation (Part D):* Use this space if you do not have enough room in Parts B or C.

• *Deposit Account and Mailing Instructions (Part E):* If you maintain a Deposit Account in the Copyright Office, identify it in Part E. Otherwise, you will need to send the non-refundable filing fee of $10 with your form. The space headed "Correspondence" should contain the name and address of the person to be consulted if correspondence about the form becomes necessary.

• *Certification (Part F):* The application is not acceptable unless it bears the handwritten signature of the author, or other copyright claimant, or of the owner of exclusive right(s), or of the duly authorized agent of such author, claimant, or owner.

• *Address for Return of Certificate (Part G):* The address box must be completed legibly, since the certificate will be returned in a window envelope.

PRIVACY ACT ADVISORY STATEMENT
Required by the Privacy Act of 1974 (Public Law 93-579)

AUTHORITY FOR REQUESTING THIS INFORMATION:
• Title 17, U.S.C., Sec. 408 (d)

FURNISHING THE REQUESTED INFORMATION IS:
• Voluntary

BUT IF THE INFORMATION IS NOT PROVIDED:
• It may be necessary to delay or refuse supplementary registration

PRINCIPAL USES OF REQUESTED INFORMATION:
• Establishment and maintenance of a public record
• Evaluation for compliance with legal requirements

OTHER ROUTINE USES:
• Public inspection and copying
• Preparation of public indexes

• Preparation of public catalogs of copyright registrations
• Preparation of search reports upon request

NOTE:
• No other Advisory Statement will be given you in connection with the application
• Please retain this statement and refer to it if we communicate with you regarding this application

Fig. 33 (continued)

FORM CA
UNITED STATES COPYRIGHT OFFICE

REGISTRATION NUMBER

TX	TXU	PA	PAU	VA	VAU	SR	SRU	RE

Effective Date of Supplementary Registration

. . . .
MONTH   DAY   YEAR

DO NOT WRITE ABOVE THIS LINE. FOR COPYRIGHT OFFICE USE ONLY

Ⓐ
Basic Instructions

TITLE OF WORK:

REGISTRATION NUMBER OF BASIC REGISTRATION:

YEAR OF BASIC REGISTRATION:

NAME(S) OF AUTHOR(S):

NAME(S) OF COPYRIGHT CLAIMANT(S):

Ⓑ
Correction

LOCATION AND NATURE OF INCORRECT INFORMATION IN BASIC REGISTRATION:

Line Number . . . . Line Heading or Description . . . .

INCORRECT INFORMATION AS IT APPEARS IN BASIC REGISTRATION:

CORRECTED INFORMATION:

EXPLANATION OF CORRECTION: (Optional)

Ⓒ
Amplification

LOCATION AND NATURE OF INFORMATION IN BASIC REGISTRATION TO BE AMPLIFIED:

Line Number . . . . Line Heading or Description . . . .

AMPLIFIED INFORMATION:

EXPLANATION OF AMPLIFIED INFORMATION: (Optional)

Fig. 33 (continued)

	EXAMINED BY:	FORM CA RECEIVED:	FOR COPYRIGHT OFFICE USE ONLY
	CHECKED BY:		
	CORRESPONDENCE: ☐ YES	REMITTANCE NUMBER AND DATE:	
	REFERENCE TO THIS REGISTRATION ADDED TO BASIC REGISTRATION: ☐ YES ☐ NO	DEPOSIT ACCOUNT FUNDS USED: ☐	

DO NOT WRITE ABOVE THIS LINE. FOR COPYRIGHT OFFICE USE ONLY

CONTINUATION OF: (Check which) ☐ PART B OR ☐ PART C

(D) Continuation

DEPOSIT ACCOUNT: If the registration fee is to be charged to a Deposit Account established in the Copyright Office, give name and number of Account.

Name . . . . Account Number . . . .

(E) Deposit Account and Mailing Instructions

CORRESPONDENCE: Give name and address to which correspondence should be sent:

Name . . . . Apt. No. . . . .

Address . . . .
　　　　　(Number and Street)　　　　　　　(City)　　　　(State)　　　　(ZIP Code)

CERTIFICATION ✱ I, the undersigned, hereby certify that I am the: (Check one)

☐ author ☐ other copyright claimant ☐ owner of exclusive right(s) ☐ authorized agent of: . . . .
　　　　　　　　　　　　　　　　　(Name of author or other copyright claimant, or owner of exclusive right(s))

of the work identified in this application and that the statements made by me in this application are correct to the best of my knowledge.

👉 Handwritten signature: (X) . . . .

Typed or printed name . . . .

Date: . . . .

✱ 17 USC §506(e) FALSE REPRESENTATION – Any person who knowingly makes a false representation of a material fact in the application for copyright registration provided for by section 409, or in any written statement filed in connection with the application, shall be fined not more than $2,500.

(F) Certification (Application must be signed)

. . . .
　　　　　(Name)

. . . .
(Number, Street and Apartment Number)

. . . .
(City)　　　(State)　　　(ZIP code)

MAIL CERTIFICATE TO

(Certificate will be mailed in window envelope)

(G) Address for Return of Certificate

U.S.DEPARTMENT OF COMMERCE
National Institute of Standards and Technology

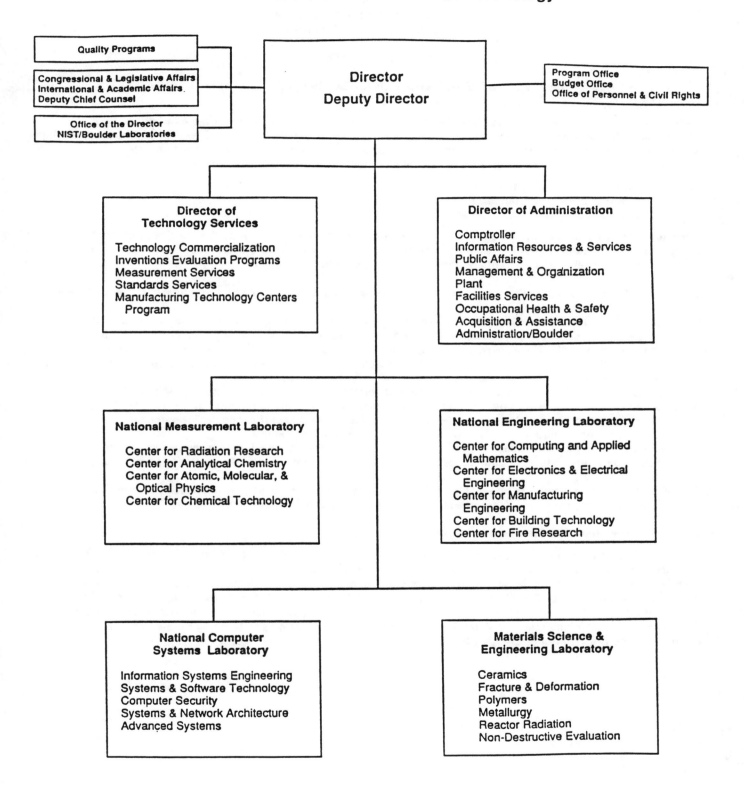

Quality Programs

Congressional & Legislative Affairs
International & Academic Affairs
Deputy Chief Counsel

Office of the Director
NIST/Boulder Laboratories

Director
Deputy Director

Program Office
Budget Office
Office of Personnel & Civil Rights

Director of
Technology Services

Technology Commercialization
Inventions Evaluation Programs
Measurement Services
Standards Services
Manufacturing Technology Centers
 Program

Director of Administration

Comptroller
Information Resources & Services
Public Affairs
Management & Organization
Plant
Facilities Services
Occupational Health & Safety
Acquisition & Assistance
Administration/Boulder

National Measurement Laboratory

Center for Radiation Research
Center for Analytical Chemistry
Center for Atomic, Molecular, &
 Optical Physics
Center for Chemical Technology

National Engineering Laboratory

Center for Computing and Applied
 Mathematics
Center for Electronics & Electrical
 Engineering
Center for Manufacturing
 Engineering
Center for Building Technology
Center for Fire Research

National Computer
Systems Laboratory

Information Systems Engineering
Systems & Software Technology
Computer Security
Systems & Network Architecture
Advanced Systems

Materials Science &
Engineering Laboratory

Ceramics
Fracture & Deformation
Polymers
Metallurgy
Reactor Radiation
Non-Destructive Evaluation

Fig. 34

U.S. Department of Commerce
National Bureau of Standards

OFFICE OF ENERGY-RELATED INVENTIONS

REQUEST FOR EVALUATION OF AN ENERGY-RELATED INVENTION

Instructions for Submission of Invention Disclosures and Substantiating Material for Evaluation.
After reading this page and the following page, complete page 3 of both OERI and Submitter copies. Check appropriate box on page 4 and sign, date and complete the Memorandum of Understanding. Retain the Submitter's Copy for your records. Detach the OERI copy (pages 3 and 4) and send with your invention disclosure to:

> Office of Energy-Related Inventions
> National Bureau of Standards
> Gaithersburg, Maryland 20899

A written disclosure of your invention, *in the English language,* must be attached to the OERI copy of this form. This disclosure should include an outline, a complete description of the invention and information to substantiate any claims for performance. Drawings or patents, where appropriate, should be included.

The quality of the evaluation will depend upon the quality of your submission. It should include or cover the following:

(1) **Purpose** of the invention. Discuss, if appropriate, where it can be used to best advantage: By industry? By individuals? By the Government? Emphasize the energy conservation or energy production potential.
(2) **The existing method(s)**, if any, of performing the function of the invention. Disadvantages of the existing method(s).
(3) **The new method**, using your invention. Details of the operation of the invention, identifying specific features which are new. If the invention is conceptual in nature, discuss typical applications.
(4) **Construction** of the invention, showing changes, deletions, improvement over the old method(s).
(5) **Data and calculations.** If tests have been conducted, detail the test conditions, controls, and results. Energy savings or efficiency estimates should be documented by calculations and data if available. Theoretical analyses should include the pertinent equations, definitions of terminology, and references.
(6) **Status of development:** Include information on stage of research, development, preproduction or production. Discuss proprietary nature, circumstances of public disclosure, instances of disclosure to government agencies, etc.
(7) **Difficulties** encountered or to be expected in exploiting your invention. Reasons why it has not been patented, manufactured, used, or accepted. **What needs to be done to bring the invention closer to use?**

Program Description and Statement of Policy

The Federal Nonnuclear Energy Research and Development Act of 1974 (Pub. L. 93-577) recognized the importance of encouraging invention and innovation in a national energy program. Section 14 of the Act directs the National Bureau of Standards (NBS) to give particular attention to the evaluation of promising energy-related inventions, particularly those from individual inventors and small companies. The Office of Energy-Related Inventions (OERI) was established at NBS to carry out the provisions of Section 14. Its duties include conducting analyses of submitted inventions to determine their technical and commercial feasibility for saving or producing energy, and bringing noteworthy concepts to the attention of the Department of Energy (DOE).

Fig. 35

The principal objective of the OERI effort is to assist DOE in identifying inventions that are ready to be moved into the private sector but may require business management assistance, or inventions that require further research and development (R&D), prototype fabrication, or laboratory tests in order to bring them to the point where they can compete with other DOE projects for program R&D funds. The evaluation of inventions submitted will, therefore, be performed principally as a service to DOE. Thus, the outcome of an evaluation will be either a recommendation for action by DOE in connection with the invention, or notification to the inventor that his invention is not being so recommended. It should be noted that a recommendation by OERI is no guarantee that DOE will provide assistance in developing a given invention.

A decision not to recommend action by DOE does not necessarily mean that the invention is considered scientifically unsound or without practical value. Also, a favorable evaluation by OERI should not be construed as being a ruling as to the patentability of any feature of an invention. The inventor should apply for a patent whenever such action is thought to be appropriate. OERI will provide no assistance in filing or prosecuting patent applications. Inventors interested in patent protection should discuss the matter with a registered patent attorney or agent.

To safeguard such proprietary rights as may exist in a submission, OERI will restrict access to invention disclosures to those persons having a need for purposes of administration or evaluation. Accordingly, in accepting invention disclosures for evaluation, an explicit statement is required (see page 4) that the information does or does not come within one of the exemptions of the Freedom of Information Act. If, for example, the disclosure contains information that is (a) a trade secret or (b) commercial or financial information that is privileged or confidential, such information falls within the exemption that is set out in 5 U.S.C. 552(b) (4). Thus, if the disclosure is protectable, the following or a similar statement should be applied to the title page or first page of the disclosure: "The disclosure contains information which is (a) a trade secret or (b) commercial or financial information that is privileged or confidential."

The Privacy Act of 1974 (Public Law 93-579), 5 U.S.C. 552a, requires that you be provided with certain information in connection with this form. You should know that:

a. The authority for collecting this data is the Federal Nonnuclear Energy Research and Development Act of 1974 (Public Law 93-577).
b. The furnishing of the information is entirely voluntary on your part.
c. The principal purpose for which the data will be used is to conduct an evaluation of your invention to determine its technical validity and potential for saving or producing energy.
d. The routine uses which may be made of the information submitted in this form are as follows:

1) Disclosure to those employees of the Office of Energy-Related Inventions or other Federal agencies having need for the information, either to perform evaluations or administer the evaluation program.
2) Disclosure to a contractor of the National Bureau of Standards having need for the information in the performance of a contract to perform evaluations of inventions and having agreed to hold the information in confidence.
3) Disclosure to a Member of Congress submitting a request involving your invention, when you have requested his assistance.
4) Disclosure to any persons with your written authorization.

Fig. 35 (Continued)

NIST-1019
(REV. 4-89)
OMB APPROVAL NUMBER 0693-0002
APPROVAL EXPIRES FEBRUARY 28, 1990

U.S. DEPARTMENT OF COMMERCE
NATIONAL INSTITUTE OF STANDARDS AND TECHNOLOGY

ENERGY-RELATED INVENTION EVALUATION REQUEST

FOLLOW ATTACHED INSTRUCTIONS AND SUBMIT THE OERI COPY OF THIS FORM AND OTHER DESCRIPTIVE MATERIAL OF INVENTION TO:

OFFICE OF ENERGY-RELATED INVENTIONS
NATIONAL INSTITUTE OF STANDARDS AND TECHNOLOGY
GAITHERSBURG, MARYLAND 20899

PLEASE PRINT OR TYPE ALL INFORMATION

NAME AND ADDRESS OF INVENTOR

THIS BOX IS FOR OFFICE USE ONLY

| DATE | ER NUMBER |

CLASSIFICATION

TELEPHONE NUMBER

TECHNICAL CATEGORY

NAME AND ADDRESS OF OWNER, IF DIFFERENT FROM ABOVE

| ANALYST | DATE |

HOW DID YOU LEARN OF THIS PROGRAM?

TELEPHONE NUMBER

REQUEST IS BEING SUBMITTED BY (check which)
☐ INVENTOR ☐ OWNER ☐ OTHER

NAME AND ADDRESS OF SUBMITTER, IF NOT INVENTOR OR OWNER

Inventor's Desktop Companion

OTHER (identify)

SIZE OF COMPANY INVOLVED
(Write number of employees; $ gross last year; N/N if none)
NUMBER OF EMPLOYEES | $ GROSS LAST YEAR

NAME OR TITLE OF THIS INVENTION

STATUS OF INVENTION DEVELOPMENT (check to indicate both the steps completed and the current status; highest number checked will indicate current status)
0 ☐ Concept Definition 1 ☐ Concept Development 2 ☐ Laboratory Test 3 ☐ Engineering Design 4 ☐ Working Model
5 ☐ Prototype Development 6 ☐ Prototype Test 7 ☐ Production Engineering 8 ☐ Limited Production/ Marketing 9 ☐ Production and Marketing

PATENT STATUS (check one)
0 ☐ Not patentable 1 ☐ Not applied for 2 ☐ Disclosure Document Program 3 ☐ Patent applied for 4 ☐ Patent granted
Patent Numbers

CHECK THE ITEM BELOW THAT MOST NEARLY DESCRIBES WHY YOU ARE REQUESTING EVALUATION

☐ 1. I wish the U.S. Government to provide funds to support development of the invention or new concept. Support is first needed for (write in):
_____

☐ 2. Development is complete. I need assistance to bring my invention or product into full utilization. Assistance is needed in (check whichever applies):
☐ General Marketing ☐ Selling to the Government ☐ Business Management ☐ Other _____

☐ 3. I only desire an opinion that the disclosure describes a technically valid invention. This information is for:
☐ Use in obtaining private development support ☐ Other (specify in disclosure)

☐ 4. The Small Business Administration suggested I request evaluation from NIST in connection with a loan application.

☐ 5. Other (specify) _____

☐ YES ☐ NO Has the invention been described to other agencies of the Government? (If yes, discuss in disclosure.)
☐ YES ☐ NO Has the invention been disclosed to any private companies, patent attorneys, etc.? (If yes, discuss in disclosure.)

OERI COPY (PLACE CARBON PAPER BEHIND THIS SHEET TO MAKE SUBMITTER'S COPY OF PAGE 3)
ELECTRONIC FORM

Fig. 35 (Continued)

MEMORANDUM OF UNDERSTANDING

I have read the Program Description and Statement of Policy on pages 1 and 2 of this form. As the owner, or with the authority from the owner who is listed on page 3, I have attached (or previously submitted) a disclosure of the identified invention for the purpose of evaluation by the National Institute of Standards and Technology pursuant to Section 14 of Public Law 93-577.

I understand that to help protect property rights in an unpatented invention, an appropriate statement or notation should be applied to the title page or first page of the invention description, and that if the description is so marked, the Government will consider all information that is in fact (a) trade secret or (b) commercial or financial information that is privileged or confidential, as coming within the exemption set out in 5 U.S.C. 552(b) (4). Accordingly, I have checked directly below, the box which is applicable to this invention.

	YES	NO
An appropriate statement has been applied to the information I have submitted.	☐	☐
Please apply an appropriate statement to all material I have submitted describing the invention to which this request pertains. (Example: This material contains commercial or financial information which is confidential.)	☐	☐
No statement is required because the information submitted is not confidential.	☐	

I also understand that NIST will evaluate the invention described in the invention disclosure on the following conditions:

1. The Government will, in the evaluation process, restrict access to the description to those persons, within or without the Government, who have a need for purposes of administration or evaluation and will restrict their use of this information to such purposes.

2. The information submitted will not be returned and may be retained as a Government record.

3. The Government may make additional copies of the material submitted if required to facilitate the review process.

4. The acceptance of the information for evaluation does not, in itself, imply a promise to pay, a recognition of novelty or originality, or a contractual relationship such as would render the Government liable to pay for use of the information submitted.

5. The provisions of this Memorandum of Understanding shall also apply to additions to the disclosure made by me incidental to the evaluation of the invention.

_____
Date

_____
Signature

_____
Status
(Owner, Business or Company Representative, Patent Attorney, Interested Party, etc.)

_____
Printed or Typed Name

Fig. 35 (Continued)

MEMORANDUM OF UNDERSTANDING

I have read the Program Description and Statement of Policy on pages 1 and 2 of this form. As the owner, or with the authority from the owner who is listed on Page 3, I have attached (or previously submitted) a disclosure of the identified invention for the purpose of evaluation by the National Bureau of Standards (NBS) pursuant to Section 14 of Public Law 93-577.

I understand that to help protect property rights in an unpatented invention an appropriate statement or notation should be applied to the title page or first page of the invention description, and that if the description is so marked, the Government will consider all information that is in fact (a) trade secret or (b) commercial or financial information that is privileged or confidential, as coming within the exemption set out in 5 U.S.C. 552(b) (4). Accordingly, I have checked directly below, the box which is applicable to this invention.

Yes   No

☐   ☐   An appropriate statement has been applied to the information I have submitted.

☐   ☐   Please apply an appropriate statement to all material I have submitted describing the invention to which this request pertains. (Example: This material contains commercial or financial information which is confidential.)

       ☐   No statement is required because the information submitted is not confidential.

I also understand that NBS will evaluate the invention described in the invention disclosure on the following conditions:

(a) The Government will, in the evaluation process, restrict access to the description to those persons, within or without the Government, who have a need for purposes of administration or evaluation and will restrict their use of this information to such purposes.

(b) The information submitted will not be returned and may be retained as a Government record.

(c) The Government may make additional copies of the material submitted if required to facilitate the review process.

(d) The acceptance of the information for evaluation does not, in itself, imply a promise to pay, a recognition of novelty or originality, or a contractual relationship such as would render the Government liable to pay for use of the information submitted.

(e) The provisions of this Memorandum of Understanding shall also apply to additions to the disclosure made by me incidental to the evaluation of the invention.

_____   _____
           Date   Signature

_____   _____
          Status   Printed or Typed Name
(Owner, Business or Company Representative,
Patent Attorney, Interested Party, etc.)

Fig. 35 (continued)

DEFENSE SMALL BUSINESS INNOVATION RESEARCH (SBIR) PROGRAM
PHASE I—FY 19__
COST PROPOSAL

Background:

The following items, as appropriate, should be included in proposals responsive to the DOD Solicitation Brochure.

Cost Breakdown Items (in this order, as appropriate):

1. Name of offeror
2. Home office address
3. Location where work will be performed
4. Title of proposed effort
5. Topic number and topic title from DOD Solicitation Brochure
6. Total Dollar amount of the proposal (dollars)
7. Direct material costs
   a. Purchased parts (dollars)
   b. Subcontracted items (dollars)
   c. Other
      (1) Raw material (dollars)
      (2) Your standard commercial items (dollars)
      (3) Interdivisional transfers (at other than cost) (dollars)
   d. Total direct material (dollars)
8. Material overhead (rate _____ %) × total direct material = dollars
9. Direct labor (specify)
   a. Type of labor, estimated hours, rate per hour and dollar cost for each type.
   b. Total estimated direct labor (dollars)
10. Labor overhead
    a. Identify overhead rate, the hour base and dollar cost.
    b. Total estimated labor overhead (dollars)
11. Special testing (include field work at Government installations)
    a. Provide dollar cost for each item of special testing
    b. Estimated total special testing (dollars)
12. Special equipment
    a. If direct charge, specify each item and cost of each
    b. Estimated total special equipment (dollars)
13. Travel (if direct charge)
    a. Transportation (detailed breakdown and dollars)
    b. Per Diem or subsistence (details and dollars)
    c. Estimated total travel (dollars)
14. Consultants
    a. Identify each, with purpose, and dollar rates
    b. Total estimated consultants costs (dollars)
15. Other direct costs (specify)
    a. Total estimated direct cost and overhead (dollars)
16. General and administrative expense
    a. Percentage rate applied
    b. Total estimated cost of G&A expense (dollars)
17. Royalties (specify)
    a. Estimated cost (dollars)
18. Fee or profit (dollars)
19. Total estimate cost and fee or profit (dollars)
20. The cost breakdown portion of a proposal must be signed by a responsible official, and the person signing must have typed name and title and date of signature must be indicated.
21. On the following items offeror must provide a yes or no answer to each question.
    a. Has any executive agency of the United States Government performed any review of your accounts or records in connection with any other government prime contract or subcontract within the past twelve months? If yes, provide the name and address of the reviewing office, name of the individual and telephone/extension.
    b. Will you require the use of any government property in the performance of this proposal? If yes, identify.
    c. Do you require government contract financing to perform this proposed contract? If yes, then specify type as advanced payments or progress payments.
22. Type of contract proposed, either cost-plus-fixed-free or firm-fixed price.

Fig. 36

U.S. DEPARTMENT OF DEFENSE
SMALL BUSINESS INNOVATION RESEARCH (SBIR) PROGRAM
PROPOSAL COVER SHEET
Failure to fill in all appropriate spaces may cause your proposal to be disqualified.

TOPIC NUMBER: _____

PROPOSAL TITLE: _____

_____

FIRM NAME: _____

MAIL ADDRESS: _____

_____

CITY: _____ STATE: _____ ZIP: _____

PROPOSED COST: _____ PHASE I OR II: _____ PROPOSED DURATION: _____
PROPOSAL   IN MONTHS

BUSINESS CERTIFICATION: YES   NO
▶ Are you a small business as described in paragraph 2.2?   ☐   ☐

▶ Are you a minority or small disadvantaged business as defined in paragraph 2.3?   ☐   ☐

▶ Are you a woman-owned small business as described in paragraph 2.4?   ☐   ☐

▶ Will you permit the government to disclose the information on Appendix B, if your proposal does not result   ☐   ☐
in an award, to any party that may be interested in contacting you for further information or possible
investment?

▶ Has this proposal been submitted to other US government agency/agencies; or DoD components, or other   ☐   ☐
SBIR Activity? If yes, list the name(s) of the agency, DoD component or other SBIR office in the spaces to
the left below. If it has been submitted to another SBIR activity list the Topic Numbers in the spaces to the
right below:

_____   _____

_____   _____

▶ Number of employees including all affiliates (average for preceding 12 months)   _____

PROJECT MANAGER/PRINCIPAL INVESTIGATOR   CORPORATE OFFICIAL (BUSINESS)

NAME: _____   NAME: _____

TITLE: _____   TITLE: _____

TELEPHONE: _____   TELEPHONE: _____

For any purpose other than to evaluate the proposal, this data except Appendix A and B shall not be disclosed outside the Government
and shall not be duplicated, used or disclosed in whole or in part, provided that if a contract is awarded to this proposer as a result of or in
connection with the submission of this data, the Government shall have the right to duplicate, use or disclose the data to the extent
provided in the funding agreement. This restriction does not limit the Government's right to use information contained in the data if it is
obtained from another source without restriction. The data subject to this restriction is contained on the pages of the proposal listed on the
line below.

PROPRIETARY INFORMATION: _____

DISCLOSURE PERMISSION STATEMENTS: All data on Appendix A are releasable. All data on Appendix B, of an awarded contract, are
also releasable.

_____ _____ _____ _____
SIGNATURE OF PRINCIPAL INVESTIGATOR   DATE   SIGNATURE OF CORPORATE BUSINESS OFFICIAL   DATE

Nothing on this page is classified or proprietary information/data.
Fig. 37

SMALL BUSINESS INNOVATION RESEARCH (SBIR) PROGRAM
PROJECT SUMMARY

TOPIC NUMBER: _____

PROPOSAL TITLE: _____

_____

FIRM NAME: _____

PHASE I or II PROPOSAL: _____

Technical Abstract (Limit your abstract to 200 words with no classified or proprietary information/data.)

Anticipated Benefits/Potential Commercial Applications of the Research or Development

List a maximum of 8 Key Words that describe the Project.

_____   _____

_____   _____

_____   _____

_____   _____

Fig. 38

AGREEMENT TO HOLD SECRET AND CONFIDENTIAL

The below described invention, idea or concept (hereinafter referred to as INVENTION is being submitted to _____ of _____(hereinafter referred to as COMPANY) by _____ _____ of _____ _____on_____,_____, 19_ (hereinafter referred to as INVENTOR) who is the inventor of record. The undersigned, in consideration of examining said INVENTION, with a purpose to opening negotiations to obtain a license to manufacture and sell said INVENTION, hereby agrees on behalf of himself/herself and said COMPANY that he/she represents, that:

1) He/she (during or after the termination of employment with said COMPANY) and said COMPANY, will keep said INVENTION and any information pertaining to it, in confidence.

2) He/she will not disclose said INVENTION or data related thereto to anyone save for employees of said COMPANY, sufficient information about said INVENTION to enable said COMPANY to continue with negotiations for said license, and that anyone in said COMPANY to whom said INVENTION is revealed, shall be informed of the confidential nature of the disclosure and shall agree to hold confidential the information, and be bound by the terms hereof, to the same extent as if they had signed this Agreement.

3) Neither he/she nor said COMPANY shall use any of the information provided to produce said INVENTION until agreement is reached with INVENTOR.

4) He/she has the authority to make this Agreement on behalf of said COMPANY.

It is understood, nevertheless, that the undersigned and said COMPANY shall not be prevented by the Agreement from selling any product heretofore sold by said COMPANY, or any product in the development or planning stage, as of the date first above written, or any product disclosed in any heretofore issued U.S. Letters Patent or otherwise known to the general public.

The terms of the preceding section releasing, under certain conditions, the obligation to hold the disclosure in confidence does not however, constitute a waiver of any patent, copyright or other rights which said Inventor or any licensee thereof may have against the undersigned or said COMPANY.

Fig. 39

IDEA SUBMISSION AGREEMENT I

While_____ wishes to take every opportunity to improve its products and add profitable ones to its line, it has found certain precautions necessary in accepting disclosures from persons not in its employ. For an idea to be considered, this form must be completed in full, signed and returned with any disclosure of an idea or invention.

(Date)_____,19_____

To

I am submitting to you, for your evaluation and permanent record, copies of certain ideas, suggestions or other materials having to do with

The information I am submitting to you consists of the following:

_____ *1) Description*

_____ *2) Drawing or sketches*

_____ *3) Samples*

_____ *4) Copy of a patent application(s)*

_____ *5) Other*

Please check the appropriate blank(s).

In doing so, I agree to the conditions printed on the back of this sheet and further agree that such conditions shall apply to any additional disclosures made incidental to the original material submitted.

(Signature)_____

(Name Printed)_____

(Address)_____

CONDITIONS OF SUBMISSION

1) All submissions or disclosures of ideas are voluntary on the part of the submitter. No confidential relationship is established by submission or implied

Fig. 40

from receipt or consideration of the submitted material.

2) Patented ideas and ideas covered by pending applications for patent are considered only with the understanding that the submitter agrees to rely for his/her protection solely on such rights as he/she may have under the patent laws of the United States.

3) Ideas which have not been covered by a patent or a pending application for patent are considered only with the understanding that the use to be made of such ideas and the compensation, if any, to be paid for them are matters resting solely in the discretion of the Company.

4) If the subject matter offered the Company is a proposed trademark, advertising slogan, or merchandising plan, susceptible to trademark or copyright protection, the Company will examine it only under the terms set forth in this Agreement. The submitter shall rely for his/her protection solely on such rights as he/she may have under the Copyright and Trademark Laws of the United States.

5) The foregoing conditions may not be modified or waived.

Fig. 40 (continued)

IDEA SUBMISSION AGREEMENT II

AGREEMENT made and entered into this _____ day of _____, 19_____, by and between _____
("Disclosure's" full name and address)
and _____ , a corporation organized under the laws of the State of _____, with offices located at _____, _____.

WHEREAS Discloser is a developer of, or has licensing rights to, concepts for _____ , and

WHEREAS, Discloser represents that he/she has developed a certain concept, device or other proprietary subject matter more specifically described at the end of this Agreement and on the attachment hereto (hereinafter referred to as the "Item"), and

WHEREAS, _____ desires to evaluate the commercial utility of the item, and

WHEREAS, in order to make this evaluation possible, it will be necessary for Discloser to disclose confidential information concerning the Item to _____.

NOW, THEREFORE, in consideration of the mutual promises hereinafter contained, and for other good and valuable consideration, the parties agree as follows:

1) Discloser shall make full disclosure with respect to the Item to employees of _____ or one of its affiliates (collectively, _____) and shall submit to _____ all relevant data in connection therewith. The disclosure by Discloser to _____ is solely to enable _____ to evaluate the Item in order to determine its commercial utility. _____ is under no obligation to market or produce the Item, unless and until a formal written agreement is entered into, and the obligations of _____ shall be only those which are set forth in any such agreement.

2) Discloser hereby represents to _____ that the Item is his own individual creation and wholly and solely the property of Discloser and that Discloser has not assigned, sold, licensed, mortgaged, pledged, or otherwise transferred or encumbered the Item or entered into any agreement to do any of the foregoing with respect to the Item. The execution and performance of this Agreement by Discloser does not violate any contract, agreement or other restriction to which Discloser is a party or by which it is bound or any rights of any third party.

3) The disclosure of the Item and all information incidental thereto is confidential and shall be received by _____ _____ in confidence. _____ shall not disclose such confidential information to others and shall take reasonable steps to prevent such disclosure. _____ agrees to use the same degree of care in protecting and safeguarding the confidentiality of the concepts and information disclosed hereunder as it uses for its own information of like

Fig. 41

importance. _____ shall not be liable for inadvertent disclosure or use of the Item by persons who are or have been in its employ, unless _____ fails to exercise the degree of care set forth above.

4) It is understood that _____s willingness to evaluate the Item is not to be construed as an admission of the Item's novelty, priority, or originality. Discloser understands that _____may have rights to the Item or particular elements thereof, due to prior access to information similar to the Item or elements thereof, including, by way of illustration and not limitation, prior patents, prior publication, prior submissions to _____by others, prior development by _____s personnel or representatives, prior use by _____ , prior knowledge, or prior sale. Accordingly, consideration of the Item by _____ shall not deprive _____of its existing rights, if any, with respect to the Item or any element thereof.

5) Without limiting the generality of the provisions of paragraph 4 hereof, the obligations of _____ hereunder are not applicable to such information which:

a) prior to disclosure by Discloser, was already know to _____ as evidenced by records kept in the ordinary course of business of _____or by proof of actual use by _____

b) was known to the public or generally available to the public prior to the date of disclosure.

c) becomes known to the public or is generally available to the public subsequent to the date of said disclosure through no act of _____ contrary to the obligations imposed by this Agreement.

d) is disclosed by Discloser to an unrelated third party without restriction.

e) is approved for public release by Discloser.

f) is rightfully received from a third party without similar restriction and without breach of this Agreement.

g) is independently developed by _____without breach of this Agreement.

h) is required to be disclosed by judicial or government action.

i) is disclosed in a judicial or governmental proceeding subject to a protective order.

_____ shall be free of any obligations restricting disclosure and use of the information provided by Discloser hereunder, subject to Discloser's patent rights, if any of the provisions of a) through i) of this paragraph 5 are applicable to the information disclosed.

6) Upon submission of the Item to _____ , _____ shall consider the Item and as promptly as practicable advise Discloser of_____s interest or lack of interest therein, all subject to the terms, conditions and provisions of this Agreement.

Fig. 41 (continued)

7) _____ shall not be obligated to take any action with regard to the Item other than pursuant to paragraphs 3 and 6 hereof.

8) _____ will, upon request, return any letters, drawings, descriptions, specifications or other materials submitted to it in connection with the Item.

9) The provisions of this Agreement shall apply to any additional or supplemental information pertaining to the Item provided by Discloser to _____.

10) This writing reflects the entire agreement between the parties concerning the Item, and no modification, amendment, waiver or cancellation of this Agreement or any provision hereof shall have any validity or effect whatsoever unless in writing and signed by both parties hereto.   Without limiting the generality of the foregoing, no agreement relating to the purchase or use of the Item by _____ or any of its affiliates, or relating to the terms of or consideration of such purchase or use, or relating to any compensation to, or reimbursement or any expenses of, Discloser, shall be binding upon either party hereto unless in writing and signed by both parties hereto.

11) This Agreement shall be governed by, construed and enforced in accordance with the internal laws of the State of _____, without reference to principles or conflict of laws.

12) This Agreement shall be binding upon, and inure to the benefit of, the Discloser and _____ (and the affiliates of _____) and their respective heirs, executors, administrators, successors, and assigns.

IN WITNESS WHEREOF, the parties have signed this Agreement on the respective dates hereinafter written.

The Item is generally described as follows:

SEE ATTACHED

_____
(company)

By: _____

_____
(Discloser)

Date:_____

Fig. 41 (continued)

NIST Agreement of Nondisclosure

I agree to handle Invention Disclosures received by me from the Office of Energy-Related Inventions pursuant to Section 14 of the Federal Nonnuclear Energy Research and Development Act of 1974.

I further agree that I shall hold in confidence for NIST any such Invention Disclosures provided to me by the Office of Energy-Related Inventions, and shall not disclose any Invention Disclosure or any portion thereof to anyone without the written authorization of the Contracting Officer.

My obligations under the Agreement of Nondisclosure shall not extend to any information or technical data

> a) which is now available or which later becomes available to the general public, other than by any breach of this agreement;

> b) which is obtained from any source other than the Officer of Energy-Related Inventions by proper means and without notice of any obligation to hold such information or technical data in confidence; or

> c) which is developed without the use of any Invention Disclosure provided to me by the Office of Energy-Related Inventions.

I further agree not to make, have made, or permit to be made, any copies of any Invention Disclosure or portions thereof, except with the written permission of Mr. George P. Lewett, Chief, Office of Energy-Related Inventions. Upon completion of the task called for in the letter, I shall return the Invention Disclosure and any copies thereof to Mr. Lewett.

If, upon examination of an Invention Disclosure, I feel that I have any financial interest or any relation with a third party which might be deemed likely to affect the integrity and impartiality of the performance of the task specified in the letter, I shall provide Mr. Lewett with a complete written disclosure of such interest or relationship prior to undertaking the task and shall not proceed with the task without the written authorization of Mr. Lewett.

If any invention or discovery is conceived or first actually reduced to practice by me in the course of or under this task, I shall promptly furnish Mr. Lewett with complete information thereon; and NIST shall have the sole power to determine whether or not and where a patent application shall be filed, and to determine the disposition of the title and rights in and to any invention or discovery and any patent application or patent that may result. The judgment of NIST on these matters shall be accepted as final and I agree to execute all documents and do all things necessary or proper to carry out the judgment of NIST.

_____
(signature)

_____
(typed name)

_____
(date)

Fig. 42